ALGERIA:

THE

TOPOGRAPHY AND HISTORY, POLITICAL, SOCIAL, AND NATURAL,

OF

FRENCH AFRICA.

BY

JOHN REYNELL MORELL.

LONDON
DARF PUBLISHERS LIMITED

1984

First published in 1854 by Nathaniel Cooke

This reprint published by
Darf Publishers Limited

First edition 1854
New impression 1984

Printed and bound in Great Britain by
A. Wheaton & Co., Ltd, Exeter

ISBN 1 85077 017 4

RANSOM OF CHRISTIAN PRISONERS.

p. 382.

PREFACE.

AFRICA, the land of mysterious memories and monstrous realities, the progenitor of pyramids, baobab-trees, negroes and boas, lies now between two fires. The rattle of Minié rifles is beginning to be heard at the Cape, and its echo resounds from the Atlas. Kabyles and Kaffirs are measuring their strength with France and England, and the issue cannot be doubtful. Having once tasted the sweets of conquest, neither of the two great Western Powers will be disposed to resign them in a hurry. Rather may we look to their grasping at their neighbour's goods, till some fine day finds French sentinels fraternising with the Cape corps on the Niger, and the Mountains of the Moon surveyed by *badauds de Paris* and honest cockneys.

As to the advantages derivable from European colonies in Africa, South or North, they are yet a matter of expectation. Hitherto the moderns have certainly suffered more and done less than the Romans in African campaigns. Algeria, the granary of Rome, has been the grave of the French soldier; and yet a nursery for a goodly crop of iron men of the Changarnier stamp, who have done brave service in the streets of Paris. The French Regency may be looked upon as an issue to relieve the apoplectic symptoms of the mother country, and a drain for her floating capital; but as to any positive returns derived by France for her outlay in that quarter, we confess ourselves unable to discover them, except in the shape of cotton and the above African chiefs, who have sharpened their wits and whetted their swords, as well as their appetite for slaughter, in Algerian razzias. That the future will show better things, is our firm belief. Algeria and Morocco, under an enlightened sway, and pacified, might in all probability yield glorious crops, and afford a noble field for commercial speculation. Nor is the day probably very distant when Cape Madeira and other Cape liquids, as well as solids, will find their way in great abundance into the English market. That we have not exag-

a

gerated the corn and olive crops of Barbary, and the cotton and warrior crops of Algeria, will appear from facts in the sequel of this work.

In this age of wonders, the greatest wonder is, that the multitude still follow the broad road of doubt, that the word 'impossible' is not offensive to all ears polite.

Thomas Grey was branded a madman and died a beggar because he was a fast man, and his thoughts were too locomotive for his generation; and yet we deny social progress and doubt Utopias. The Crystal Palace has extinguished Aladdin's lamp, and the dreams of the Arabian Nights are eclipsed by the daylight of science. The earth is girt with telegraphs; and yet we cannot conceive that the hour is at hand when humanity will be electrified by the spirit of liberty. We live in a golden age, but we cannot place faith in a coming commonwealth; serenaded by sirens, and rocked to sleep by the Muses, we yet laugh at the idea of a future harmony. We cannot get up the steam of faith in a dawning Millennium, and *our* clairvoyance is dazzled by the excessive light of the coming day.

The regeneration of man is daguerreotyped in characters of light, but we are blind to God's photographic art; the age of reason and the reign of love is rapped out by unearthly hands on our parlour tables, but we are deaf to the summons of the seventh heaven.

To tunnel the Atlantic and electrify China were thought sober prose, and shares in a gas-company at Jeddo would be at a premium to-morrow; but to irrigate the desert, set free the Poles, and make Europe Christian, is too much for our faith.

With faith in our hearts, science in our heads, and ready hands, we can exalt the valleys and make low the mountains. If France is true to herself, with Algeria at her doors, she will better herself and bless the nations. The wilderness will blossom as the rose, springs will gush forth in the desert, and flower-beds will cover the marshes; and we may anticipate the day, without any stretch of fancy, when ostrich expresses will furrow the Sahara, and teams of zebras or quaggas run daily from Algiers to the Cape *viâ* the Niger. We starve amidst plenty; with our lips to the brim, we die of thirst: beggars are we, though Midas's wand is in our hands.

A wise combination and economy, a perennial exodus, and, above all, construction substituted for our destructive habits, would make the world roll in riches and revel in luxuries.

Instead of sitting down by the stagnant waters of Conservatism and weep-ing, we should be up and doing, and putting our hands manfully to the wheel, sorrow and sighing would flee away, we should wipe away all tears, the lion and the lamb would lie down together, and a little child would lead them.

The French have much and can teach much of the wonders of science to their Arab brothers. Yet lack they one thing, which I ween they might learn better in the tabernacle of the wilderness than in the Madeleine—the power of faith.

If they unite these two levers, they will not only remove mountains, but raise the earth. True science ends where the Arab begins, in a child-like belief in the infinite power of God and the inexhaustible resources of his creation. Finality is destruction to science and death to religion.

We have in some measure outgrown the age of speculation; we are beginning to drop theories, and to be alive at length to the all-sufficiency of facts. It will soon be too late to write down, talk down, and preach down discoveries; nor will the magistrate or the priest be able to fulminate excommunication against new truths. Science, facts, and machinery are beginning to explode the conservative prejudices of the fore world, and to free the mind from the thraldom of custom and circumstances. And though the world is thus on the move towards the broad daylight of truth, we need not fear that it will extinguish the poetry of the past or the mysteries of nature. God's facts are ever full of poetry, and man's highest wisdom is at best such foolishness before high heaven, that he need never fear the danger of exhausting the secrets of the universe.

Imbued with the spirit of the times, the author has endeavoured impartially to collect, compare, and condense as many useful *facts* as possible in this volume. His object has been to make his countrymen familiar with an important and interesting region and people hitherto little known to us. The present critical position of the Ottoman empire adds additional interest to all territories verging on its frontier, and all tribes having an affinity with its population; and Algeria being the only French colony of note, and nearly equal to France in size, and having been once the granary and glory of Rome, has appeared to him well worthy of careful study on many grounds. Amongst the Arab tents, moreover, the reader will find many traces and footmarks of holy men and apostolic times. A classical soil and the cradle of Hannibal, the sunny shores of Tunis and Hippo are also dyed with the blood of a noble army of Christian martyrs; and thus this historical land possesses all the attributes calculated to secure the interest of the student and the traveller.

In short, it has been the author's endeavour to make the book as practically useful as possible, whether it falls into the hands of the many or lies on the desk of the few. He might easily have expanded his matter to an inconvenient bulk; but his limits and convenience restricted him to more moderate dimensions,—a circumstance which will probably be far from exciting regret in the reader.

In consulting the best and latest authorities on the subject, he has found almost all his materials in French works. The principal English books that have appeared on Algeria within the last twenty years are chiefly confined to temporary and local observations.

Those who wish to obtain the amplest details respecting all branches connected with the colony are referred to the volumes of the *Exploration Scientifique,* and the *Tableau de la Situation* for 1850, from which the author has gathered his most important facts, having likewise, in most cases, conformed to their spelling of Arabic names.

In giving the angle of the slopes of mountains and rivers, &c., the decimetres, centimetres, &c. have been merely reduced into inches and decimals of inches. The reader will find it convenient to bear in mind that a metre is rather more than a yard, or about 39 inches; and the author has considerately translated throughout the French measures into corresponding English measures, in order to prepare the tender British intellect for the grievous transition to the decimal system.

Hampstead, 1st January, 1854.

CONTENTS.

———◆———

PART I.—TOPOGRAPHY.

CHAPTER I.

THE HISTORICAL EXODUS.

PAGE

Problematical countries—The march of discovery—African character-
istics and mysteries—Caffres and Kabyles—General survey of North-
Western Africa—Its topography generalised—Herodotus—The desert
—The successive tides and strata of humanity—France in Africa—
The fall of Carthage a warning for the ages 19

CHAPTER II.

PHYSICAL GEOGRAPHY.

The Zones—Tell and Sahara—Orology—The Atlas chains—The Aouress—
Potamology—Primary Basins—The Shellif—Lake Melrir—El H'od'na
and the Ouad Mzab—Secondary basins—Natural hydraulics—The
lakes—Tertiary basins—General organic laws 29

CHAPTER III.

POLITICAL GEOGRAPHY.

Political divisions of North-West Africa—Political divisions of Algeria—
Latitudes and longitudes—Arab mensuration—Turkish divisions and
subdivisions—Scientific French division—Six districts—Distinction of
Tell and Sahara—Esoteric analysis—Exoteric delimitation—Surface—
Arab appellations—Zones and departments 43

c

A–A*

CHAPTER IV.

ALGIERS.

PAGE

Principal features of the province of Algiers—The Shellif—The Haratch, the Massafran, and the Isser—The Mitidja—The Sahel—Sidi-Ferruch —Cape Matifou—Algiers—The old Port—The new Port—Streets— Houses—Bazaars—The Casbah—The Faubourgs 57

CHAPTER V.

STATISTICS OF ALGIERS.

Religious edifices—Baths—Fountains—Drains—New civil edifices—Historical statistics of Algiers—The poetry of Eastern life—Antagonism of the social states of Europe and Africa—New military edifices and defences 84

CHAPTER VI.

PRECINCTS OF ALGIERS.

Precincts of Algiers—The two Mustaphas—Jardin d'Essai—Buffarick— Model farm—Maison Carrée—The Café of Hammah—The Consulate of Sweden—Ayoun Beni Menad—Pointe Pescade 97

CHAPTER VII.

INTERIOR OF THE PROVINCE.

Characteristics of Algerian scenery—Interior of the province—Blidah— The Chiffa—Medeah—Milianah—The River of Silver—Teniet-el-Had —Boghar—The Koubber Romeah—Scherschell—Tenes—The Darha— Orleansville—Aumale—The Oases of the Beni-Mzab—The Bedouin Tribes 116

CHAPTER VIII.

EXCURSIONS.

PAGE

Excursions—The orange-groves of Blidah—Coleah, its delightful neighbourhood and Moorish population—the Col de Mouzaia—M. Lamping's
expedition to the South—The Atlas—The Arabs—The Little Desert—
Sergeant Blandan — Mère Gaspard — Milianah — Expedition to the
Ouarscnis under Changarnier—The march—The bivouac—The blockade—Teniet-el-Had 132

CHAPTER IX.

PROVINCE OF ORAN. THE COAST.

Outline of the coast—Mostaganem—Arzeu—Oran—Nemours—Oran—
Mers-el-Kebir—The Gulf of Arzeu—Antiquities—St. Marie—Origin of
Mostaganem 149

CHAPTER X.

PROVINCE OF ORAN. INTERIOR.

Outline—Tlemsen—Mascara—Tagadempt—Mazounah—A tour through
the province—St. Denis—Mascara—Sidi Bel Abbess—Tlemsen—Nemours—The far South—Tiaret 168

CHAPTER XI.

PROVINCE OF CONSTANTINA. COAST.

The coast—Djidjelli—Collo—Philippeville—Bona—The port—The town
—The buildings—The population—Sanitary condition—Mount Edough
—Trip to La Calle—An Arab tribe—La Calle—Bastion de France . 194

CHAPTER XII.

PROVINCE OF CONSTANTINA. INTERIOR.

PAGE

Interior of the province—Broad outline—Analysis—Baron Baude—Natural features of Numidia—St. Marie—Constantina—Madame Prus—Borrer — Guelma — Gerard the Lion-king — Constantina — Betna—Aoures—El-Gantra—Biskra—The Oases   222

CHAPTER XIII.

GREAT KABYLIA.

Authorities—Broad outline—The different Kabylias—Great Kabylia—Etymology—History—Analysis of its topography—Bugia — Its roadstead—Its tribes—Expedition of Marshal Bugeaud—The Zaouias of Sidi-Ben-Ali-Cherif—Kuelaa—Dellys   250

PART II.

STATISTICS AND HISTORY,

Political, Social, and Natural.

CHAPTER XIV.

THE KABYLES.

Native population of Algeria—Characteristics of the Kabyles contrasted with the Arabs—Superstitions—Industry—Manufactures—Manners—Weddings—Women—Administration—Laws—Authorities—The Marabouts—The Zaouias—The Anaya—Illustrations of Scriptural and Classical antiquity   269

CHAPTER XV.

THE ARABS.

PAGE

Agriculturists and Bedouins—Tents—Furniture—Women—Distinctions of Arab life—Patriarchalism—Feudalism—Douars—Horses—Falconry —Illustrations—Markets—Legends—Scriptural Customs—The Arabs of Constantina—Administration of the Tribes—Bedouin Officials— Statistics—Bureaux Arabes 300

CHAPTER XVI.

MOORS, TURKS, KOULOUGLIS, JEWS, ETC.

Etymology—Moorish women—Toilette—Weddings—Divorces—Turks— Their government—Their costume—Yousouf—The Koulouglis—Their characteristics and laziness—The Jews—Their servility and persecution—The corporations 324

CHAPTER XVII.

THE NEGROES.

Utility of Slavery—Mahometan and Christian slavery—Degraded state of the Niger basin—The Slave-trade in Africa—The Blacks in Morocco— Unfortunate results of the attempt to stop the Slave-trade—The Djelep —Native Arts and Sciences 339

CHAPTER XVIII.

EUROPEAN POPULATION AND GENERAL STATISTICS.

European settlers—The French colonists—General character of European settlers—Latest tables—The component nations—Spaniards—Maltese —Italians—Native population . - 348

CHAPTER XIX.

COLONISATION.

PAGE

General survey of colonisation—Government decrees on rural property—
Concessions in land—Decree of the President, 1851—State of general
colonisation in the colony—Province of Algiers: Civil territory,
Military territory—Province of Oran: Civil territory, Military terri-
tory—Province of Constantina: Civil territory, Military territory—
New projects—Penitential colony at Lambessa—Agricultural colonies
—St. Denis and Robertville, &c. 354

CHAPTER XX.

CIVIL AND RELIGIOUS GOVERNMENT.

The Ool-ama—Three classes of them—Sheikhs—Khatebs and Imams—The
Mufti—The Santons, and other orders—The Dey's Ministers—The
Kaids—The Kadis—French civil administration—French tribunals—
Mussulman tribunals and schools 377

CHAPTER XXI.

THE FRENCH ARMY.

Roman razzias—Strength—Native troops—Zouaves—Spahis—French—
Chasseurs d'Afrique—Sanitary statistics, &c.—The African chiefs—
Changarnier—Cavaignac—Canrobert 388

CHAPTER XXII.

THE HISTORY OF ALGERIA AND BARBARY.

The reign of Mythos—The Semitic and Indo-Germanic conflict—The
Phœnicians—The spirit of Carthage—The first Punic war—The mercenaries—The second Punic war—Hannibal—Cannæ—Scipio—Zama

—The fall of Carthage—Jugurtha—Metellus—Marius—Juba—Christian Africa—Donatists—Circumcellions—Tertullian—Cyprian—St. Augustine—The Vandals—Belisarius—The Arabs—Their dynasties—The two Barbarossas—Charles V.—Piracy—Lord Exmouth—The French invasion—Rovigo—Trezel—Abd-el-Kader—The cave of Khartani—Capture of Abd-el-Kader—His liberation—Zaatcha—Laghouat . 404

APPENDIX.

SECTION I.

Antiquities of Algeria 447

SECTION II.

Language 461

SECTION III.

Commerce and Agriculture 464

SECTION IV.

Natural History, Geology, &c. 480

PART I.

TOPOGRAPHY.

FRENCH AFRICA.

CHAPTER I.

The Historical Exodus.

"Trascorser poi le piagge ove i Numidi
Menar già vita pastorale, erranti.
Trovar Bugia ed Algieri, infami nidi
Di corsari ; ed Oran trovar piu avanti ;
E costeggiar di Tingitana i lidi,
Nutrice di leoni e d' elefanti ;
C' or di Marocco è il regno, e quel di Fessa ;
E varcar la Granata, incontro ad essa."
          Il Gerusalemme del Tasso, l. 51. c. 21.*

PROBLEMATIQAL COUNTRIES—THE MARCH OF DISCOVERY—AFRICAN CHARACTER-
ISTICS AND MYSTERIES—CAFFRES AND KABYLES—GENERAL SURVEY OF NORTH-
WESTERN AFRICA — ITS TOPOGRAPHY GENERALISED — HERODOTUS — THE
DESERT—THE SUCCESSIVE TIDES AND STRATA OF HUMANITY—FRANCE IN
AFRICA—THE FALL OF CARTHAGE A WARNING FOR THE AGES.

WE read in the venerable pages that record the creation of the world and
of humanity, how God spake unto the latter, and said, " Be fruitful
and multiply, and replenish the earth, and subdue it." Ever since the early
days when these memorable words were spoken by the Almighty, we find,

----

* "They view where once the rude Numidian swain
    Pursued a wandering life from plain to plain.
  Algiers and Bugia then they reach, the seat
    Of impious corsairs ; next Oran they greet ;
  And now by Mauritani's strand proceed,
    Where elephants and hungry lions breed.
  Morocco here and Fez their cities rear ;
    To these opposed Granada's lands appear."
          Hoole's Tasso, l. 149-155, b. 15, p. 302.

from the twilight of tradition to the daylight of history, that as ages have rolled onwards, the children of men have, as a steady current, inundated the length and breadth of the habitable globe. Asia, the cradle and nursery of the human race, was probably first peopled ; and as adventure, or curiosity, or war, or the want of space, urged the bolder spirits to move on, the tide of human beings swept into the neighbouring hemisphere, and ultimately reached the remotest coral islands of Polynesia. Though this chronic exodus and perpetual emigration was undoubtedly checked for a season by one or more diluvial catastrophes, yet whenever the generations succeeding those that had been submerged or sufferers had recovered from the injuries thus received, they invariably rushed onwards once more in this progressive movement, till at length every spot of land that could offer a home or sustenance to man had been subdued and visited. Some of these early colonists appear to have always, or generally, maintained a friendly or hostile intercourse with the parent races and regions ; whilst other, more forward wanderers, have deviated so widely from the beaten track of nations, as to have lost all connexion with, or memory of, their early home.

Severed from the mother-country by pathless wastes or icy fields, they gradually lost most traces of affinity with the parent stock, and the ebony skin and uncouth utterance gave but few signs of relationship with pale faces and the musical Sanskrit. A mystery came at length to shroud these strange progenies in fabulous forms, till they and their country became an enchanted sphere.

Though modern science and discovery have done much to clear up the mystery, and restore the severed links of nationality, yet the salt wastes of Mongolia, and the icy horrors of the pole, still bid defiance to the heroism of blue-jackets and the scientific fanaticism of Asiatic Societies.

But among all the problems and vetos for the exploring mania of modern times, no portion of the globe has offered so fatal and fabulous a field as Africa. So deadly is its very air to the Indo-European races, except at its extremities, that the Caucasian man, treading its wastes and jungled forests, is inevitably doomed. It is true that many Semitic tribes seem to have assimilated better with the climate, but it is at the expense of their intellectual life ; and save in the Moorish monarchies of Northern Africa, experience proves that the Arab conquerors of this burning hemisphere have speedily been scorched almost to the grovelling level of the Negro.

As regards European and scientific travellers, it may justly be pronounced that its shores are their *ne plus ultra*. Niger expeditions, the sickly Congo, and the statistics of Sierra Leone, shew in clear figures the uncongeniality of the African climate to European constitutions. Hence all exploring expeditions into the heart of this *terra incognita* have been more or less failures. Holocausts of brave spirits have been the martyrs to scientific fanaticism on its fatal plains. Nor is it climate alone that offers serious impediments to the adventurous traveller. The comparative deficiency of large rivers and extensive mountain-chains, which operates directly on the climate in aggravating its heat and dryness, acts indirectly as a serious obstacle to commercial and all other intercourse and transit. But among the obstructions that have hitherto checked the course of European adventure and travel in Africa, we must especially place the exceptional and conservative character of its populations. If we turn to the British possessions at the Cape, we form an acquaintance with the bloodthirsty Zoola, the treacherous Caffre, and the Boschman, who, with the inhabitants of the Andaman Islands and of Tierra del Fuego, appears to represent the lowest degradation to which human nature can descend. Higher up, on the east coast, we meet with the atrocious populations of Arkeeko,* who seem to blend in perfection all the vices of savage and civilised life. Penetrating into the interior, we encounter the superstitious and jealous Abyssinians, who in their bloody banquets and forays acquire and strengthen that ferocity which naturally appals and deters the helpless traveller. Their neighbours the Shangallas and the Gallas, with their poisoned arrows and licentious customs, would shake the firmness of all wanderers save such a spirit as Bruce. If we except Egypt, the whole of Northern Africa forms no exception, but, on the contrary, powerfully corroborates our view of the character of its population. The Kabyles of the Atlas exceed most races in cruelty and charity ;† and the Bedouins of the Sahara are notorious for hospitality, perfidy, and bigotry. The shipwrecked crews who have tasted of Arab clemency on the coasts of the Desert, and the fate of French prisoners in Algeria, can best attest the sympathies and warmth of the Arab heart for suffering humanity. The reader will shortly be presented with some striking proofs of the accuracy of these remarks.

The western part of Central Africa has long been eminently repulsive in a moral point of view, from the bloodthirsty tyranny of its chiefs and people, and the atrocious practice of kidnapping and selling neighbours and countrymen into bondage. The ferocity of the king and people of Dahomey and Ashantee contributed for many years to deter the approach of the adventurous traveller even more powerfully than the deadly sun of Guinea ; and though recent events have made some alterations in this respect, exploring expeditions into Central Africa are still attended with imminent personal risk, as well from the unfriendly elements as from the

---

* Major Head's Life of Bruce, p. 209.

† For an explanation of this apparent enigma, the reader is referred to the chapter on the Kabyles. It will be sufficient here to observe, that this singular people, though ferocious and blood-thirsty in battle, have numerous institutions analogous to the monastic systems and freemasonry of Europe, fostering learning and dispensing brotherly love throughout the land.

inhospitable character of its people. Thus a barrier seems to have been placed by the hands of the Almighty to break off all intercourse between the poison of European civilisation and the conservative barbarism of this mysterious hemisphere. Yet there is much to attract the interest of the philosopher and fix the attention of the naturalist in this strange land of prodigies. Nature appears there in a new, a larger, and a more exuberant character, and deals largely in anomalies and monsters. From the days of Herodotus downwards, Africa has been the chosen home of the marvellous ; and though much that has been related and received concerning its prodigies must be attributed to the credulity of an unscientific age, enough remains to justify us in pronouncing it the parent of paradoxes. Thus in the human race, anomalous in its psychical and physical developments, it presents us with the Negro shading into the Boschman and the Hottentot. Passing to the inferior mammalia, we have the camelopard, the quagga, and the multitude of strange beasts that Gordon Cumming has found teeming and roaming through the wilds and wastes of the Cape district. Again, among birds, we observe the anomalous struthious species, which though extinct in New Zealand, yet multiplies and flourishes in the plains of Africa. Nor is the vegetable kingdom deficient in anomalies, presenting us with the gigantic baobab; and in the geological aspect of the continent the eye is astonished at the endless oceans of sand, and startled by the almost unparalleled variety of stratification found in Algeria.

Much more might be added to prove how deserving this vast continent is of the study of scientific men. The ancient Abyssinian church, with its theocratic hierarchy and oriental traditions; the anomalous character of Abyssinian mountains, and appetites which modern discovery has confirmed, after a sceptical age had ridiculed the superior wisdom of the gallant Bruce ;* the fabulous massacres and female body-guards of Ashantee and Dahomey,† and the strange practices of the Kabyles of Atlas,— all worthily keep up the character of Africa as the land of marvels, and point it out as a legitimate field for scientific research.

Happily or unhappily, its extremities are now in the hands of the two most polished nations on the earth ; and the day cannot be far distant when a more familiar intercourse will spring up between the Gaul and the Kabyle, the Briton and the Caffre. Though the introduction has been rude, and ushered in by a running accompaniment of powder and shot,

---

* The reader will find an account of the singular geological formation of Abyssinia, and of the raw steaks constituting a chief ingredient in Abyssinian diet, in Major F. B. Head's Life of Bruce ; whose revelations will be seen there verified by the testimony of other subsequent wanderers : pp. 235, 244.

† For a description of the sanguinary practices and negro amazons of Dahomey, see Commander Fred. E. Forbe's Dahomey and the Dahomans, being the Journals of two Missions to the King of Dahomey and a residence at his capital, in the years 1849-50 : vol. i. p. 23. London, 1851.

there is reason to anticipate that Christianity, commerce, and science, heralded by the bayonet, will carry the blessings of lawsuits, civilisation, and doctors, into the heart of Africa, and make us acquainted with its deepest recesses.

European valour and enterprise have already made rapid strides in advance ; and the northern and southern extremities of this vast hemisphere have been carefully and scientifically examined and explored.

This remark applies especially to North-western Africa, which, particularly when viewed in relation with its past history, is much more worthy of study than the fertile but obscure plains and valleys of the Caffre district.

Barbary, or North-western Africa, is undeniably the finest part of that continent. More accessible to Europeans than any other region, it is also more calculated, by its fertility and temperature, to become once again the theatre of a great people. It combines all the qualities that are most adapted to captivate the imagination of the antiquarian and the scholar, to draw forth the energies of the merchant and the speculator, and to engage the researches of the philosopher and the man of science. Once the granary of the Roman empire, it seems intended by nature, under a more happy administration, to replenish the less-favoured regions of the north with the exuberance of its productiveness. Long oppressed by a barbarous and benighted people, it has been for centuries, like Italy its ancient master, the prey and theatre of injustice and rapine, though lately a brighter day appeared *once more* about to dawn upon its shores, under the happier auspices of republican rule. These observations apply more particularly to Algeria; yet all the Barbary states are so closely connected, that the civilisation of one is certain to become infectious.

Historically speaking, the region now under survey is one of the most interesting on the face of the globe. To give the reader a general idea of this part of the continent, historically and topographically, before descending to details, we shall now lay before him its principal landmarks, and the most striking events that have rendered its name illustrious among the nations.

Barbary, properly speaking, constitutes the whole of North-western Africa, and extends from the frontier of Barca and the Gulf of Sidra on the east, to Cape Nun on the west. This vast territory, which includes the regencies of Tripoli and Tunis, the French vice-royalty of Algeria, and the empire of Morocco, corresponds to ancient Carthage, Numidia, the two Mauritanias, and Gætulia. It is our purpose in the present work to give a minute description of the French possessions in Africa, and a general outline of the other states that constitute Barbary, and are situated in North-western Africa. On the present occasion we confine ourselves to a cursory sketch, historical and geographical, of the whole district, chiefly with the view of directing the reader's attention to its interest and importance. The tract under consideration embraces little less than 2000 miles of coast; but its breadth varies greatly, according to the proximity of the sandy waste that occupies the heart of Africa. It is intersected by the great Atlas chain, which, under different names and in different branches, runs east and west through the whole region, generally parallel with the coast, and reaching from the western ocean to the borders of Egypt. Its rivers are mostly insignificant, the distance between the Atlas and the sea not admitting of the formation of a large volume of water.

The father of history has correctly divided this territory into three distinct zones, naturally formed by the character of the soil, and corresponding very exactly with the modern divisions. The first zone, bordering on the coast, and forming the *Tell* of the modern Arabs, he calls the inhabited land; the second zone he styles the wild-beast country,—this region represents the pastoral uplands now called Sahara, a name inaccurately extended to the Desert; and his third division consists in the sandy waste which is the Desert proper of all ages. The second or pastoral zone, the Sahara of the present day, corresponds in part to the ancient Gætulia, and is situated south of the Atlas, between the 30th and 34th to 35th degrees of N. latitude.

The Great Desert occupies the entire breadth of Africa, and stretches through Arabia and Persia into Northern India. Its width varies, being greatest between Morocco and Soudan, and narrowest between Tripoli and Bornou, the route followed by Denham and Clapperton.

Having thus given the reader a faint outline of this interesting region, we shall endeavour to present to him, in a series of brief sketches, the numerous remarkable social and political revolutions that it has undergone.

The history and geography of North-western Africa present the image of a vast archipelago, containing to the north steep and verdant islands, and to the south flat and sandy islands separated by long intervals, and the sea that severs them has risen and fallen in successive tides, encroaching on them at high-water, and losing ground during the ebb. Occasionally during the flood the waves have covered the tops of some of the lower islands, whereas at low-water some of the space separating them has been left dry, and the waters receding even below the lowest gorges, the islands have lost their character, and the archipelago has become a continent. Yet some of the sandy and rocky summits have never been reached by this stormy sea. Such has been the picture presented by Northern Africa in its historical and geological development through the phases of time and the fields of space; the physical characteristics of the soil accurately corresponding to the social phenomena that they represent. The steep islands are the mountainous ridges; the flat islands are the oases; the secular tides are the invasions. All these islands representing groups of the same nation, whilst the flood that sweeps round them is in its turn Phœnician, Roman, Vandal, Greek, Arab, Turk, and French.

Successive tides of humanity have thus flooded the plains of North Africa, each leaving deposits behind; and the mountain-chains, as usual, have been the refuge of the oldest and most conservative hordes. Thus the present Kabyles, or Djebalis (highlanders), of Algeria are to all intents and purposes the same people as the primitive Numidians of the time of Sallust and Polybius. The most important element among the different nationalities represented in Northern Africa is undoubtedly the Semitic, which forms the staple of its population ; and it is probable that the aboriginal Numidians of tradition, the Carthaginians, Arabs, Moors, and perhaps the modern Kabyles, all belong to that remarkable family of the human race.

The Mediterranean cruiser that sails along the coasts of Mauritania and Numidia hails the classic kingdoms of Iarba, of Dido, of Juba, of Jugurtha, of Siphax, and of Massinissa. The traveller while pacing its sunny shore recalls the glories and the heresies of the North-African church ; its Cyprian, its Augustine, its Hippo Regius, and its Cirta. Passing the supposed site of the ruins of Utica, his mind dwells on the heroic death of Cato, the last republican, whose lofty spirit preferred a violent death, rather than bend to the general oppression of the empire ; standing on the ruins of Carthage, he reflects on the revolutions of empires, the Scipios, Hannibal, and Regulus. The image of the gentle, saintly king of France floats before him, as he lies on his couch of ashes on that pestilential shore.[*] Crossing to Goletta, the fort of Tunis, he sees the walls and towers that bear witness to the Christian zeal and valour of a Spanish emperor and a British admiral.[†] In one place he crosses a river in whose turbid stream the veteran Massinissa found his last home ; farther on, he reaches the spot where Genseric and his Vandal host, descending from Spain on the devoted land, proceeded to convert the granary of Rome into a howling wilderness. Not far hence he views the plain where the Greek army of the gallant Belisarius levelled the Vandal pride with the dust. Or if he visits the crumbling battlements of Kairwan, Tlemsen, or Fez, his mind reverts to the days of Arab glory, when the gallant band of Islam flashed like a meteor over the valleys and plains of Mauritania, and plunging on their fiery chargers into the Western Ocean, threatened to reduce the stormy sea into subjection to the Crescent. A melancholy grandeur hovers over this historical land, and the shades of mighty hosts and nations long since gathered to their fathers seem still to linger and haunt its spectral cities.

> " Giace l' alta Cartago ; appena i segni
> Dell' alte sue ruine il lido serba.

---

[*] St. Louis, A.D. 1270. [†] Charles V. in 1541, and Blake in 1655.

Muoiono le città, muoiono i regni ;
Copre i fasti e le pompe arena ed erba:
E l' uom d' esser mortal par che si sdegni :
O nostra mente cupida e superba."
Tasso's Gerusalemme, 1. 15, c. 20.*

Taking a broad survey of the chronology of North-west Africa, we have first the primitive immigration of the Berbers or ancient Libyans, assuming that people, according to its oldest traditions, to have come originally from the Semitic corner of Asia. These are followed by Phœnician colonists, the founders of Carthage, who still belong to the Semitic, and are eventually subdued by the Romans belonging to the Indo-Germanic stock. The Vandal invasion brings in a new branch of the latter variety, constituting a part of the great Gothic family that swept over Europe at the fall of the Roman empire; but after a short triumph, they shared the fate of their predecessors, and were forced to submit to Justinian and the Byzantine Romans, who once more regain the supremacy on the African shore. From the fall of Carthage to this period, from B.C. 146, to about the latter half of the seventh century of our era, different families of the Indo-European variety had held sway in Northwestern Africa; but about the beginning of the eighth century a flood of the Semitic tide once again deluged the land under the name of Saracens and under the crescent of Mahomet, which brought the cross into subjection and extinction on those shores, after it had reigned there about five hundred years. This Arab or Saracen family of the Semitic variety held sway in Barbary from the eighth to the tenth century, when a band of daring desperadoes, belonging to the Turkish branch of the Mogul variety, reduced the Algerine portion to subjection, and ruled it with a rod of iron, till the French conquest in 1830 restored the cross and the supremacy of the Indo-Europeans. Morocco has invariably, Tripoli and Tunis have generally, continued under Arab or Semitic rule since the conquest in the eighth century, though the two latter regencies have been nominally subject to the Sultan of Turkey for a long course of years.

This cursory view of the history of Barbary will shew that it has been the theatre of numerous important and violent revolutions, and will serve to fix the attention and engage the interest of the intelligent reader. The minuter details of its history are reserved for a future chapter.

There is another consideration that increases the interest which surrounds a study of this remarkable country ; I mean, the present position

* " Ill-fated Carthage ! scarce, amid the plains,
   A trace of all her ruined pomp remains !
   Proud cities vanish, states and realms decay,
   The world's unstable glories fade away !
   Yet mortals dare of certain fate complain.
   O impious folly of presuming man !"
   Hoole's Tasso, 1. 14-6, b. 15.

and future prospects of the French power in Africa. If we examine the causes of the disputes and struggles between nations, it is probable that a large proportion will be found to originate in misunderstandings and ignorance. A more accurate survey of, and a closer acquaintance with, the position and power ·of our neighbours would generally or frequently anticipate and prevent the deplorable results to which we have alluded, by enabling us to arrive at a correct comparative estimate of our strength and resources, and by teaching us what we have to expect.

The progress of the French power in Africa is an instructive example of the aggressive and invasive spirit and propensities of our neighbours; and it is important to remember that the French government has within call a powerful army of above 100,000 veterans, inured to hardships and war, and officered by men who have grown grey in camps.

The observations of an eminent writer on the fall of Carthage are moreover especially applicable to this country and to the present situation of the continent. " The fall of Carthage," he remarks, " has been ascribed to that neglect of her maritime forces which was manifested during the last Punic war. When Scipio crossed from Sicily to Africa, there was not a fleet to oppose him. But the principal cause of her decline and ultimate overthrow was the fierce hostility of rival factions within her own walls. * * * In the fate of Carthage was exemplified the usual result of a popular government and of civic contention ; the voice of clamour is silenced only by the shouts of a triumphant foe, who puts an end to the rivalry of parties by treading all distinctions under foot."*

A memorable instance of this truth was afforded in the *coup d'état* of December 1851. Let us hope that the ruins of Carthage and the present slavery of France will have a warning voice for England, and teach us to avoid the abuses that led to these catastrophes.†

From the preceding remarks the reader will perceive that the past and present history and position of French Africa and its borders are an instructive study for the philosopher, the statesman, and the patriot; and though our limits have necessarily prevented us from dwelling on the mysterious valley of the Nile and the past glories of Cyrene, the antiquarian and the politician will be amply rewarded if they extend their minute survey to the north-eastern part of this land of the sun. The wonders of early Egyptian culture, the wealth, luxury, and learning of the Pentapolis and Alexandria, and the Mameluke beyliks of Kahira,

---

* Dr. Russel's Barbary States, p. 83.

† Some of the more inflammable spirits of *la jeune France* have been ready to anticipate as one of the results of the new France now occupied on the north coast of Africa, that the classic Mediterranean will be shortly converted into a French lake. We confess our inability to do justice to this conclusion while our batteries of Gibraltar frown on the Straits, and unless our modern vikings are sadly degenerated from their sires.

are calculated to command the reverence and dazzle the imagination of the ages.

Having thus given a view of what he has to expect among the plains and valleys of old Atlas, we shall transport our reader on board one of the numerous steamships that plough the waters of the Mediterranean, and after a rapid and easy passage deposit him on the quay of Algiers.

CHAPTER II.

Physical Geography.

THE ZONES — TELL AND SAHARA — OROLOGY — THE ATLAS CHAINS — THE AOURES — POTAMOLOGY — PRIMARY BASINS — THE SHELLIF — LAKE MELRIR — EL H'OD'NA AND THE OUAD MZAB — SECONDARY BASINS — NATURAL HYDRAULICS — THE LAKES — TERTIARY BASINS — GENERAL ORGANIC LAWS.

AFRICA, from the north to its centre, is divided by nature into three distinct regions. The first, to which the name of Tell, or the corn-country, has been applied, ascends by a gradual slope to the region of high table-lands. The latter, forming the second region, extends, under the name of Sahara, from the Tell to the Desert, which is nearly on a level with the sea. The high table-lands of the Sahara afford pasture for numerous flocks of sheep; and at intervals you meet with oases containing fortified towns, forming depôts for the corn and merchandise of the nomadic tribes. To the eastward of the oases of the province of Oran, in Algeria, begins the country of the Beni-Mzab,* which contains seven important towns, forming emporiums for the whole commerce of the south, and peopled, according to tradition, by the descendants of the Moabites. The fact is, that almost all of them have blue eyes and fair hair, whilst their language also differs from the Arabic. They are, moreover, schismatics, because they do not belong to any of the four authorised Mussulman sects. But the severity of their morals, their union, and their honesty, have given them a high reputation; and their active character has centered in their own hands most of the barter trade between the Tell and the Desert.

To the south of these table-lands of the Sahara, parallel to the Tell and to the sea, begins the third region of Africa, consisting of the Desert; but not such a desert as is pictured by a European imagination—sand, and nothing but sand to the end of the chapter. The desert is in reality composed of immense plains, analogous to the steppes of Russia, the pusztas of Hungary, and the llanos and pampas of South America, with this essential difference, that they have no wood or vegetation, and very little water, which is confined to certain favoured spots few and far between, that become the necessary halting-places of the traveller.† It is true that tracts of sand frequently occur, which have been spread over its surface by the action of the winds ; and the natives often apply to them very singular

---

* بنى مزاب *Beni-Mzab.*   † Humboldt's Views of Nature, pp. 2-3.

appellations, such as *veins* or *nets*, according to the shape given to them by the caprice of the winds. But the desert contains in like manner oases, and whole countries clothed with vegetation and inhabited by a numerous population, such as the 'Great Oasis of Touat. Beyond these vast plains rises a chain of mountains, rivalling the Atlas in verdure and vegetation, and forming the country of the Tonaregs, who are the buccaneers of the desert. Lastly, to the southward of these mountains, you reach the land of Soudan, the Negroland, the chosen home of the marvellous, and the seat of fabulous realities. A straight line drawn from Algiers to Kachna, at the distance of more than 800 leagues (2200 miles) from the coast, passes through the three regions that we have just described ; and there is every reason to believe that the whole of Northern Africa presents similar characteristics and divisions. The kingdom of Haoussa, of which Kachna is the metropolis, was conquered about thirty years ago by a white Mussulman race called the Foulanes ; and thus, by a singular chance, whilst a Christian power was establishing its dominion in Northern Africa, Islam was imposing her arms and her creed on the centre of that continent.

Between the Tell and the Sahara are vast undulations of ground, celebrated for their pasturages, and called the Sersous.* This district is the residence of wandering tribes and vast flocks of sheep, which constitute their sole wealth, as they abstain from all agricultural pursuits.

Nominal Algeria,† that is to say, the old regency, is divided by a line running nearly east and west into two distinct zones, called by the natives *Tell* and *Sahara*. The Tell, according to some authorities, takes its name from the Latin *tellus* (cultivable land) ; it constitutes the zone bordering on the Mediterranean, and is the land of harvests and agriculture. The Sahara stretches to the south of the Tell, and forms the region of pastures and fruit. Hence the inhabitants of the Tell are agriculturists, and those of the Sahara are shepherds and gardeners. The Tell is formed of a series of fertile basins yielding almost exclusively different kinds of corn, especially wheat and barley ; and its flattest parts compose one of the richest countries in the world, but at the same time one of the most uniform. The chains separating the basins are clothed with timber, but being peopled by Berbers are inaccessible to the Arabs.

The Sahara was long a fabulous land, being called by some the Great Desert, and by others the Country of Dates,—contradictory appellations resulting from the confusion and imperfection of geographical knowledge previous to the French conquest. It was very generally supposed that from the mountains of the Tell to Nigritia there stretched one continued

---

* See note, p. 110, of Marshal de Castellane's Souvenirs de la Vie militaire en Afrique: Paris, 1852.

† See the Exploration scientifique de l'Algérie ; Study of the Roads followed by the Arabs, by E. Carette, introduction.

plain of sand, a wilderness infested by savages. Such is not, however, the true aspect of the Sahara, which consists of a vast archipelago of oases, each offering an animated group of towns and villages. A large belt of fruit-trees surrounds each of these villages, among which the palm rules supreme from its height and value, though you have also pomegranates, figs, apricots, peaches, and vines. This massive verdure, with its profusion of fruit and shade, may give the reader some idea of the strong love entertained by the people of the Sahara for their country, which must not be regarded in the light of a desert till you have advanced a great distance beyond the southern limits of the regency.

The Sahara also stretches to the south of Tunis and Morocco, the northern zones of those countries being likewise styled Tell. The Algerian Sahara is comprised between the Tunis Sahara to the east, the Algerian Tell to the north, the Desert proper to the south, and the Morocco Sahara to the west.

Considered orologically, Algeria consists principally of the assemblage of several chains of mountains running parallel to the sea-shore, i.e. in an east-north-easterly direction, and intersected in their eastern extremities by other transverse chains running east-south-east. It results from this conformation that Algeria is divided naturally into two parts: one western, where the accidents of the ground are very simple, and almost all subject to the same direction; the other eastern, presenting frequent crossings or breaks, and for that reason displaying the loftiest points. The north of Africa presents, as I shall shew, three directions of mountain-chains: one parallel to the Mediterranean, running in an east-north-east direction, and constituting the dominant ridge; a second chain running in a north-north-east direction, and determining the general direction of the coasts of Morocco on the Atlantic Ocean, and also that of the Tunis coast—the third direction is east-south-east, and presents itself distinctly in the ridges of the province of Constantina and the regency of Tripoli; it determines the direction of the sea-board in the latter country.

This compound chain has its highest point in Morocco, where the mountain named Miltsin, near the capital, attains an elevation of 3475 metres (11,398 feet) above the sea. The ridge sinks rapidly in the vicinity of Mlonia, and its lowest points are about the meridian of Mostaganem, Mascara, and Saida, i. e. about the second degree west longitude from Paris, where its greatest elevation does not exceed about 700 metres (2296 feet). Farther east the mountains rise again as far as the Chellia, the culminating point of the Aouress ridge, situated 108 kilometres (67 miles) south of Constantina, and rising to the height of 2312 metres (7583·36 feet).

The Aouress mountain is the highest summit in Algeria and in the whole country that lies behind Morocco and Abyssinia. To the east of the Aouress the mountains rapidly sink to the Halouk-el-Mkhiba, 110 kilometres (68·4 miles) east of Tebessa; this mountain is 1445 metres (4739·60 feet) high, and its summit seems to command the whole regency. The Rarian, which is almost the only chain in Tripoli, does not appear to rise to a greater elevation than 800 or 1000 metres (2624 or 3280 feet).*

Algeria, at the time of its sovereign Hussein Pasha, comprised a great part of the northern shores of the continent of Africa. Its territory at that period extended from the fourth degree west to the sixth east longitude of the meridian of Paris,† and from the thirty-fourth to the thirty-seventh degree of north latitude. The Atlas chain runs through this territory, forming a segment of a circle, of which the extremities approach the sea, while the centre departs from it and approaches the desert. This great chain of mountains must be divided into three zones, which extend east and west in nearly parallel lines; and which may be appropriately styled the Great Atlas, the Middle Atlas, and the Little Atlas. Each of these zones presents almost similar sinuosities. Between the sea and the Great Atlas, which approaches the desert, is the Middle Atlas, a secondary chain, cut by another longitudinal chain, which from east to west approaches more and more to the shore; this latter ridge is the Little Atlas. A number of smaller chains lie between the principal ones and the sea, forming so many ascending steps or degrees. The most northerly point of the Great Atlas is about 15 leagues (37 miles) from Setif, not far from the source of the Ksour and Bousellam. To the west of Tlemsen, in the province of Oran, the Middle Atlas is linked to the Little Atlas,‡ which latter range runs over a space of about 100 leagues (250 miles), and reaches the Shellif, which breaks through it at six leagues (15 miles) distance from Medeah. This chain, which forms an elbow to the east by another branch (or spurt), appears to advance south to join the Great Atlas; while on the other hand, by the Bibans, it follows an easterly direction to form a northern angle at Constantina. It reappears on the right bank of the Seybouse as far as the frontier of Tunis.

At six leagues from the sea the Tafna cuts the Little Atlas, and the latter commands successively the right bank of the Isser and the left of the Sig, which it crosses as well as the Habrah. Then it draws near the shore, which it follows almost in a parallel line for 60 leagues (150 miles) till it abuts in the Col de Mouzaia. Soon after having passed this point, under the name of Djordjora, it dominates the Adouse, and for a moment disappears at Bugia; but a little distance farther on it is again seen drawing near the Middle Atlas.

The Atlas thus presents groups of parallel mountains intersected by and containing a series of basins furrowed by streams in different di-

---

* Exploration scientifique; M. Carette's Géographie et Commerce de l'Algérie méridionale.

† From 1° 39′ 45″ W. to 8° 19′ 15″ E. of Greenwich.

‡ Many geographers regard the Middle Atlas merely as a branch of the Little Atlas.

rections. Those rising near the sea, having but a short course and a very rapid descent, are at certain seasons furious torrents, and at others dry beds. Those, on the contrary, that come from farther inland have to pierce a channel through the] transverse ranges. Such are the Ouad-Rummel and the Shellif, which have to break the barrier of the Lesser Atlas.*

In looking on the map of Algeria,† it may be seen that this country, which extends in length between the Great Atlas and the sea about 250 leagues (625 miles), with a mean breadth of 120 leagues or 300 miles, is divided from one extremity to the other into two regions by the chain of the Little Atlas, the superior region lying between the Great and Little Atlas, and the inferior or maritime between the Little Atlas and the sea-coast. If you seek for the communication that nature has effected between these two regions, you will find dark and hilly defiles, by which at three or four points the waters of the first region find their way to the sea. These issues, opened by the force of the waters, are also occupied by it. Man can hardly venture among them; and thus the two regions which these issues were intended to unite are still left isolated. The division does not stop there. From the intermediary chain of the Little Atlas numerous branch ranges are thrown out to the north and to the south towards the Great Atlas on one side, and towards the sea on the other. These two regions are thus divided into a multitude of valleys, with no common communication between them; so that the country, divided into two long halves by the Little Atlas, and subdivided into numerous fractions by these branch ranges, somewhat resembles a chess-board depicted by the mountains, natural barriers being thus offered to the communication of the population inhabiting it. You may search in vain for a natural centre to the broken country; nature has refused it. Neither are secondary centres to be found; all the maritime region is composed of narrow valleys running to the sea, and these being ranged parallel to each other, resemble the stalls of a stable. Each glen has its river, or more correctly its torrent, flowing from the far end, and following a direct line to the coast. The valleys of the superior regions are more extensive by reason of the waters, which, kept back by the barrier of the Little Atlas, have formed vast basins. But they do not communicate one with another; each is a little world in itself; and to command two contiguous basins, it is necessary to take up a position on the chain dividing them. We have previously seen that from Algiers to 40 or 50 leagues (125 miles) inland is called the Tell, and presents a surface of about 16 millions of hectares.‡

* M. Berbrugger's Algérie historique, pittoresque, et monumentale,—Introduction, by M. De la Haye, editeur : Paris, 1843.

† This excellent description of the physical characteristics of Algeria is derived from an article by the late M. Jouffroy in the *Revue des deux Mondes* for June 1838.

‡ It will be convenient for the reader to remember that there are about two and a half English acres to a French hectare. 16,000,000 hectares = 40,000,000 acres.

C

The principal river of Algeria is called the Shellif, after which we shall enumerate the others in what appears to us the natural order of priority. These are the Seybouse, the Summam or river of Bugia, the Habra, the Tafna, and the Rummel; the three latter being nearly equal in velocity. The Shellif has the same length of course as the Garonne and the Seine, but its basin is not equal to one-half of theirs. It appears about equal to the Marne. In the state of Tunis the Medjerda has only two-thirds the length of the Shellif, and it is also probable that it rolls along a smaller volume of water. In the empire of Morocco, on the contrary, the Omm-er-Rbia and the Tensift, though somewhat less in length of course than the Shellif, seem to contain a more considerable body of water. The Ouad-Sbour, which passes near Fas, is also an important river. These are the principal water-courses in the north of Africa, and the mountains from which they rise are the highest.*

Before, however, we proceed any further in individualising the rivers of Algeria, we shall lay before the reader the intimate connexion between the highest lands and largest water-courses of Barbary;† thus generalising the relation between the orology and potamology of the district under survey.

There are four great water-courses in Algeria, and in all Barbary, forming the four great arteries of the country. To the east the valley of Lake Melr'ir' exceeds the eastern frontier of Algeria, and crosses in all its length almost the whole regency of Tunis. To the north the valley of the Shellif reaches the northern frontier of the region, which is the Mediterranean. To the west the Ouad-Seggar reaches and passes the western frontier to enter the empire of Morocco. To the south, the Ouad-Mzab reaches the borders of the desert, which is the southern limit of Barbary.

The region whence all the streams flow is a platform commanding all the low lands of Algeria. The sources of the four rivers lie near together; thus the Melr'ir' lake and the Ouad-Seggar are supplied by the southern and western slopes of the Djebel-Amour. This mountain, which commands all the plateaux of the four rivers, must be one of the highest in Algeria; and hence the peak of El Ga'da, which separates three of the four great basins, and is held by the natives to be the top of Djebel-Amour, is one of the highest summits of French Africa.

The seven basins, of which the east and centre part of the Algerian Sahara consists, may be divided into two distinct groups: 1. to the north, the basins of the Upper Shellif, the Zâr'ez, and the H'od'na; 2. to the south, the basins of the Ouad-Mzab, the Ouad-Rîr, the Ouad-Souf, and the Melr'ir'. Except the Shellif, all these basins are shut; they

---

* Exploration scientifique;' Géologie par M. Renou, Par. I, Géographie physique.
† See Exploration scientifique de l'Algérie; Recherches sur la Géographie et le Commerce de l'Algérie meridionale, par E. Carette, Capitaine de Génie, p. 70.

also correspond in a general relation of direction; thus the bottoms of the north basins and the bottoms of the south basins are situated in two parallel lines, in a N.N.E. direction, and distant about 250 kilometres (155·4 miles).

In this interval the lines of the ridges, like the lines of partial bottoms, also obey the direction of the extreme lines. If this E.N.E. direction, which especially determines the configuration of Western Algeria, prevailed also in the east of the regency, the basin of El H'od'na and that of the Melr'ir' would send their waters, like the Shellif, to the Mediterranean. But another chain running E.S.E. extends without interruption from the Djebel Dira, under the meridian of Hamza, to the meridian of Tebessa for about 400 kilometres (248·54 miles), and bars them effectually. The chief rings of this chain are the Djebel Dira, the Ouennour'a, the Bou-T'âleb, the Mest'âoua, the Aoûress, and lastly the mountains of Amâmra and of the Nememcha. This first chain forms the basin of El H'od'na; another parallel chain rises to the southward, which bars the basin of Lake Melr'ir', and determines the direction of the inferior branch of the Ouad-ed-Djedi. It is at the Djebel Metlîli that this second fold is knotted on to the E.N.E. chain, and it is at the foot of their southern slopes that the Sahara ends. The rivers that descend from the Djebel Aoûress all cross this second chain, to lose themselves in the Ouad-ed-Djedi. This circumstance, joined to the great quantity of snow that feeds them, and to the great degree of cold prevailing in these regions, is an evidence of the elevation of the Aoûress group, which is the highest mountain-ridge in the eastern part of Algeria. Thus the dominant masses of the Sahara, and perhaps of all Algeria, are the Djebel Aoûress and the Djebel Amour; they determine the direction of the greatest valleys, the one to the east and the other to the west; and the Ouad-ed-Djedi forms the link that unites them, since it receives the southern waters of both.

Having thus endeavoured to analyse the great organic laws of the orology and potamology of Algeria, we shall proceed at once to individualise the characteristics of the principal streams of the region under survey.

The doctrine of basins may be styled the philosophy of topography, giving at once the key to the physical geography of a country. We have seen that Algeria contains four primary basins, and that it is subdivided into a number of secondary and tertiary basins.

The primary basins are the channels of, 1st, the Shellif; 2d, the Ouad-ed-Djedi and Lake Melr'ir'; 3d, the El H'od'na and Chott-es-Saida; and 4th, the Ouad-Mzab and Lake Ngouça.

The secondary and tertiary basins contain a series of salt lakes or rivers; the latter, or tertiary, generally situated between the Little Atlas and the sea. We shall first give a brief survey of the hydrography of the

primary basins, subsequently noticing the others; and we shall begin with
the basin of the Shellif, the only Algerian river that finds its way from
the Sahara to the sea, because it flows through the only open primary
basin.

The Shellif rises in the north slopes of the Djebel Amour, 300 kilometres (186·41 miles) in a straight line from its mouth, but including its
bends 600 kilometres (372·82 miles). Its two chief upper tributaries
are the Ouad-Sebgag and the Ouad-el-Beida. The former, issuing from
the rocks of El-Khiar, falls into the Ouad-el-Beida, which, after traversing
the plain of that name, crosses the plain of Seresso under the appellation
of Ouad-el-Touil, receiving a number of small tributaries before reaching
the Shellif.

The Ouad-ed-Djedi is the chief tributary of Lake Melrir, the first basin,
flows 300 kilometres (186·41 miles) between the cultivable lands on one
bank and the sand on the other, and is often nearly dry, but after rain
a mighty sheet of water. Its name is thought to be derived from the
Berber, *Idjdi*, sand; *Irzer Idjdi*, the river of sand, corrupted in Arabic
into *Ouad Djedi*, the river of the goat. When the arable land forms
both its banks near El-Ar'ouat', it changes its name to Ouad-Mzi, a Berber term. It rises in the Amour, and is formed by several streams, the
Ouad-el-Richa being the principal, rising by one of the highest summits
of the Amour. The Ouad-ed-Djedi is formed by the union of the Ouad-
Mzi and the Ouad-Msaad coming west, the confluence being a little south
of El Arouat ; it receives afterwards the Ouad-Bedjran, the Ouad-Mlili, a
river of fabulous size, owing to its vast channel, and a number of other
streams near the Aouress; all these flow in on the left, coming from minor
basins. The right bank presents few tributaries save the Ouadi-et-Tell,
a valley 130 kilometres (80·77 miles) long and 25 kilometres (15·53
miles) wide. This channel is generally dry on the surface, with water
underneath.

The basin of El Hodna is occupied by the salt lake or Sebkha Msila,
commonly called Chott-es-Saida, the bank of the Saida, which is, like the
Melrir, a vast salt-marsh. The Ouad-Msila rises on the north slope of a
mountain, and flows round to the south of it. A number of other Ouads
flow into the marsh, the chief to the north being the Ouad-Msila ; to the
east the Ouad-Metkaouk, of the same rapidity as the Remel at Constantina, and never dry. The chief stream on the west bank of the Chott is
the Ouad-ech-Chelâl, which changes its name several times, and receives
many tributaries from the Djebel Dira ; on the south we have the Ouad-
bou-Sada and the Ouad-ech-Chair.

The Ouad-Mzab is the largest valley that pours its waters into the
salt lake Ngouça, a *bas fond*, or marsh without an outlet. Its chief tributary, the Ouad-Metlili, rises at the west part of a plateau called El Ferad,
forming Djebel Mahiguen, a day's journey south of El Arouat, and parallel

to Djebel Amour. First it bears the name of Ouad-Mahiguen, then of Ouad-Metlili near that town, and soon after reaches the Ouad-Mzab, which also comes from the Mahiguen, and first bears the name of Ouad-el-Abied, *the white river*. After passing through the oases, it falls into the Ouad-Noumrat. A number of other streams swelling the current, it falls, under the name of Ouad-Mia, into the lake Ngouça. All the streams of this basin dry up, and deluge the country after rain. Notice is given by horsemen directly the northern horizon blackens, gun-shots are fired as soon as the torrent appears, all objects are removed, and soon, with a terrible noise, the flood rolls on, and the Saharian city stands by magic on the banks of the waters, which rise to the palm-tufts; but a few days only elapse ere all disappears.[*]

The rivers of Algeria divide it into a great number of basins, of which the following is a rough estimate:

	square miles.	square myriametres.
Basin of the Shellif	17,325 . .	450
„ „ Habra and Sig	5005 . .	130
„ „ river of Bugia	3850 . .	100
„ „ Tafna	2887 . .	75
„ „ Rummel, or Remel . . .	2502 . .	65
„ „ Seybouse	2310 . .	60

The shut basins contain a much larger surface; the chief being those of Melrir and Ouaregla; those of the Tell of Constantina may embrace about 125 square myriametres (4424·5 square miles). The following table gives the inclination of the chief rivers:

	inches.	metres.
Chiffa, in the Mtidja	·0031 . .	·0008
Seybouse, plain of Bona	·1053 . .	0,0027
Rummel or Remel, from Constantina to the sea	·0975 . .	0,0025
Tafna	·0975 . .	0,0025
Mazafran, from Kolea to the sea . . .	·0507 . .	0,0013
Harrash, from the middle of the Mtidja to the Maison carrée	·0250 . .	0,0010

The inclination of $\frac{1}{400} = ·0025$ (about 1 inch in 390) is very common in Algeria; it is ten times greater than that of the Loire between Orleans and Tours, and twice that of the Meurthe between Saint Dié and Nancy. The great mountain districts of Europe alone present similar inclinations.

These remarks apply also only to the lower part of the course of Algerian rivers. Cascades are frequent, the most remarkable being that of the Remel at Constantina, where it falls 70 metres (219·60 feet) in one leap.[†]

---

[*] Exploration scientifique; Géologie, par M. Renou, Par. I, Géographie physique, 1-15.

[†] Height of the rivers at different points in their course:

	feet.	metres.
Ouad bou Sellam, near Setif	3280 . .	1000
Remel above the cascade	1577·68 . .	481

The surface of the lakes alone is considerable; the whole south-east of Algeria presenting a country partly occupied by Sebkhas, and covering 500 square myriametres. The following estimate of surface may be depended on:

Sebkha, or Salt Lake, dry in summer.

	hectares.	acres.
The Great Chotts of the province of Oran together	255,000	. . 612,000
Chott-el-Hodna, or the lake of Msila . .	150,000	. . 360,000
The East (Chergui) Zarez	56,500	. . 135,600
The West (Gharbi) Zarez	28,300	. . 67,920
Sebkha of Oran	31,250	. . 74,500
Sebkhas of the plateaux of Constantina .	40,000	. . 96,000

Salt Lakes, never dry.

Fzara, near Bona	14,300	. . 34,320
El Maleh	867	. . 20,808

Fresh Water.

El Houbeirat, near La Calle . . .	2848	. . 6635·2
El Hour ,, . . .	2367	. . 5680·8*

The limit of the Mediterranean basin is formed by a sinuous line generally parallel to the main line (consisting of the long chain running east-north-east, and containing the Djebel Amour), except at the centre, where it reaches the summits of the Djebel Amour. It has a surface of about 1300 square myriametres (52,050 square miles); but it is decomposed into a number of closed basins, such as

	hectares.	acres.
Basins of the salt lakes of Oran and Arzeu .	248,000	. . 595,200
Sebkha of the plain of the Mina . . .	31,250	. . 75,000
Basin of the Fzara, near Bona . . .	322,000	. . 772,800

or about 32 square myriametres (123·2 square miles), which reduces to 1268 square myriametres (48,818 square miles) the surface of the basin that sends its waters to the sea.

The salt lakes, or Sebkhas, of the north slope present very clearly the two directions that prevail in Algeria; the seven principal lakes, lengthened out to an extent of about 750 kilometres (466·03 miles), describing an east-north-east direction; whilst those of the province of Constantina follow an east-south-east direction. The former present a total surface of 725,000 hectares (1,740,000 acres); the others, about twelve in number, may have about 35,000 to 40,000 (84,000 to 96,000 acres).

---

	feet	metres.
Remel below the cascade	1348·08 .	. 411
Seybouse at the confluence of the Ouad Cherf and Ouad Zenati	918·40 .	. 280
Chiffa, issuing from the cutting	492	. . 150

* The last two are each of them three-fourths of the lakes of Thun and Brienz in Switzerland.

The aspect of Algeria is uniform; the existence or absence of forests being the greatest feature. The country at Constantina is bare, at La Calle woody. Drawing near the coast, you first see the higher summits; but soon you come under a lower ridge of 1000 or 1200 metres (3280 or 3936 feet), almost always green, from Tunis to Tangiers, though there are some breaks in this ridge at the towns, especially at Tenes and Oran.*

The three prevailing directions of the mountains are not always clearly perceived; at Constantina they look like a chaos. The highest points seen from that town are the Guerioun, 1727 metres (5664·56 feet), and the Nif-en-Nicer, 1534 metres (5031·52 feet) in height; from the Chet'tba, eight kilometres (4·34 miles) from Constantina, you see the Aoûress; the Djorjora is seen eight kilometres from Algiers; and the Ouanseris at an immense distance.

The Sahara is divided into two regions: the northern mountainous, more populous, and better watered; the southern lower, less peopled, and consisting of oases, and containing 253,000 square kilometres, or 97,405 square miles.

The separation of Tell and Sahara is more simple to the west, but complicated to the east, where the line of separation descends to the south face of the Aoûress. This results from the greater height of the mountains, for the Chot't region in the west corresponds exactly to the Sbakt of the province of Constantina; but the lowness of the western mountains makes the land sooner arid, there being fewer streams; while in the east the lands around the salt lakes are often very fertile; hence the Tell is broader there.

The division of the two parts of the Sahara is very simple; it is the foot of the mountains, forming an obtuse angle, the west side parallel to the great ridge or watershed, and the east side parallel to the E.S.E. chain; the same angle is described by the great salt lakes. This limit is dotted with a line of k'sour, or walled villages, starting from Figuig to the west, and joins the frontier of Tunis north of Nefta. This is a great channel for the Mecca caravan.

The Ouad-ed-Djedi and the Lake Melrir indicate the same limit; the west country is not quite so well known. This north zone has every where a breadth of about 300 kilometres (186·34 miles).

The Sahara has a west and east slope, traceable, but not so clearly, into the Tell. Their line of demarcation runs a little east of Algiers, passes by the Djebel Amour, then near Stiten, and on the east limit of the oasis of Touat, where it cuts the meridian of Paris in lat. 27°. A low chain of hills coasting the road from Algiers to Timbuctoo separates the two slopes. Dividing Algeria into two slopes, north and south, or east and west, the Djebel Amour is the pivot and focus of its physical geography. The water-

---

* Exploration scientifique, sciences physiques; Géologie de l'Algérie, par M. E. Renou 1ere partie, Géographie physique, pp. 1-14.

courses, of which the Shellif and Ouad-Djedi are the chief, irradiate from this centre, whose height is about 1600 metres (5248 feet).

The oases depend entirely on the orography of the country, the mountains supplying them with water, and giving them life.

Metlili and the Ouad-Mzab towns alone occupy valleys where streams run beyond them. In all the other oases the rivers come to an end. The angle and height of the mountains near Biskara explain the knot of oases there. Ouaregla receives not only the north waters, but an immense torrent, the Ouad-Mia (100 streams), from Insalah, in the oasis of Touat. Other oases, like Ouad-Souf, have a knot of sand-hills instead of a flat bottom ; this makes them salubrious.

All the south-east of Algeria is a flat uniform country, consisting almost entirely of one sebkha, and embracing, with a part of the Tunis Sahara, 500 square myriametres (19,250 square miles), looking like the sea.

Including the villages of El Goha and Ocdan, and all the tribe of the Chamba, which would extend its south limit to the 30th degree of N. latitude under the meridian of Paris, Algeria would have a surface of 4700 square myriametres, (it has 390,900 square kilometres, according to the *Tableau de la Situation*, i.e. 150,496·5 square miles), only one-tenth less than France. The centre would then fall about the 34° 7′ lat., and 1° 4′ east long. of Paris (3° 23′ E. of Greenwich), *i. e.* between Demmer and K'sir-el-H'îran.*

The division of Algeria into Tell and Sahara resulting from orology and potamology, or what we may call natural hydraulics, depends on geological and meteorological causes, to be determined by the quadrant, the anemometer, and general scientific analysis and synthesis. It cannot be doubted that a great icy chain of 5000 or 6000 metres (16,000 or 19,000 feet) elevation in Central Africa would convert the Sahara into a Brazil or Hindustan.

Heights of the Plains, Lakes, and Marshes.

	metres.		feet.
Medjana, south of Setif	1000	.	3280
Hachem Reris (plain of Mascara)	350	.	1148
Mitidja (at Mered Blockhaus)	148	.	485·44
Salt lake of Oran	60	.	196·8
Marsh of Bou Farik	43	.	141·04
Marabout of Sidi Denden, on a hillock in the plain of Bona	38	.	124·54
Lake Houbeira, La Calle	30	.	98·40
Lake Fzara, Bona	0	.	0
Plain of Bona	0	.	0.

* The length of Algeria between Tunis and Morocco, *i. e.* the mouths of the Zena and Adjeroud, is in a straight line 974 kilometres (605·23 miles). This estimate is less than M. Jouffroy's ; but it is that of the Exploration scientifique, and is probably the most exact.

Slopes of Plains.

	inches.	metres.
From camp of Ouad Khmris to the sea in the Mitidja .	·4875	·0125
Plain of Oran towards the south, between the town and the salt lake	·234	·006
Plain of Tlemsen towards the north	·6084	·0156

Central Asia has much analogy with the Sahara, especially in climate : the distance from the sea occasioning extremes, at 45° N. lat. in Asia, you have the cold of Iceland and the heat of the Gambia.

The undertaking of the French to reclaim the *landes* of Gironde by planting is not impossible in the desert. Many shrubs live with little water, and might attract rains and give birth to springs. The attempt is somewhat problematical and hypothetical, yet experience can alone establish its practicability or impracticability ; nor should we be too ready to pronounce innovations Utopian in this age of wonders.

The height of the Ouanseris, whose name has been so disfigured, has now been ascertained. It can be easily seen from Medeah, 106 kilometres (65 miles) off, and from the Plateau des Santons above Oran, though 220 kilometres (136·64 miles) off. In January 1842 it was all white with snow. The Aoûress presents gentle slopes, the Djordjora steep ones, with sharp needles, and some points covered with snow the whole year. The Dolomite mountains, near Tlemsen and Ouchda, are the steepest points in the west.

As regards the slopes of the plains in the Tell, those of Bona and of the Habra are the flattest; in the south-east Sahara, as previously observed, you find immense flat plains, little raised above the sea.

Heights of Algerian Mountains.

	kilometres.		metres.	feet.
Chellia (Aoûress) .	108 (66½ miles) S. of Constantina .	. .	2312 .	7583·36
Djerdjera . .	94 (58 miles) E.S.S 24° of Aly .	. .	2126 .	6963·28
Ouânserîs . .	78 (48 miles) S.S.E. of Tenes .	. .	1800 .	5904
Amour . . .	155 (96·31 miles) S.E. of Tiaret .	. .	1600 .	5248
Mouzaia . . .	157 (96 miles) S.E. of Blidah . .	. .	1597 .	5237·16
Zakkâr . . .	7 (4 miles) N.E. of Milianah .	. .	1534 .	5031·52
Gonfi . . .	47 (29 miles) W. of Philippeville (7 capes) .		1096 .	3587·88
Edough (Idour) .	10 (6·20 miles) W. of Bona . .	. .	972 .	3188·16
Kahar, Mountain of Lions	15 (9 miles) N.E. of Oran .	. .	615 .	2017·20
Bouzareáh . .	4 (2·48 miles) W. of Algiers .	. .	402 .	1318·50

The mean slopes of the mountains are as follows :

	metres.	inches.
The Santons, at Oran, S. side . .	·0218	·8502
Gouraia (sea-face) Bugia . . .	·053	2·067
Zakkar, near Miliana	·021	·819
Bouzareáh, near Algiers, N.E. slope .	·020	·780
Mouzaia, near Blidah	·018	·702

It may not prove unacceptable to the reader to be presented here with a table of the elevation of the chief towns in Algeria.

	metres.	feet.
Telegraph of Djemadra above Blidah . .	1400	. 4592
Setif	1100	. 3608
Betna	1100	. 3608
Medeah	920	. 3017·60
Milianah	900	. 2952
Fort Gouraia, 2 kilometres N. of Bugia .	671	. 2200·88
Constantina	656	. 1851·68
Tlemsen	500	. 1640
Mascara	400	. 1312
Guelma	286	. 938·08
Emperor's Fort, Algiers	210	. 688·80
Casbah Algiers	124	. 416·72
Casbah of Bona	105	. 344·40
Oran, top of the town	100	. 328
Scherschel	20	. 65·60
Algiers, Place du Gouvernement, lowest part of the town	20	. 65·60

Such are the broad features stamped by the hand of nature on this country, which, like all other inhabited lands, has been arbitrarily decomposed by man, according to the whim of despots or the sway of races. The political divisions of Algeria will be enumerated and analysed in the following chapter.

CHAPTER III.

𝔓𝔬𝔩𝔦𝔱𝔦𝔠𝔞𝔩 𝔊𝔢𝔬𝔤𝔯𝔞𝔭𝔥𝔶.

POLITICAL DIVISIONS OF NORTH-WEST AFRICA — POLITICAL DIVISIONS OF ALGERIA — LATITUDES AND LONGITUDES — ARAB MENSURATION — TURKISH DIVISIONS AND SUBDIVISIONS — SCIENTIFIC FRENCH DIVISION — SIX DISTRICTS — DISTINCTION OF TELL AND SAHARA — ESOTERIC ANALYSIS — EXOTERIC DELIMITATION — SURFACE — ARAB APPELLATIONS — ZONES AND DEPARTMENTS.

NORTH-WESTERN Africa has been variously divided at sundry epochs, according to the predominance of races and dynasties. The territory of the Republic of Carthage appears to have corresponded in a great measure with the present regency of Tunis ; but its influence extended over a much wider surface, embracing the greater part of Northern Africa, and comprehending a great multitude of tributary hordes, who, like the present Arabs, led a nomadic life. To this class belonged the Numidians, with many tribes of Libyans, including possibly the Gætulians. The territory of these tribes was naturally fluctuating, in consequence of the roving mode of life of its population ; but as soon as the Numidians stand forth as a free people, and assert their right to distinct individual nationality, their territory seems to have answered with tolerable accuracy to the present province of Constantina in Algeria. The two Mauritanias, as they were afterwards called by the Romans, comprehending the remaining portion of Algeria and the empire of Morocco, were brought into a state of partial and nominal dependence on Carthage by Hamilcar, the father of Hannibal. But it is difficult to assign any definite limits to the lands occupied by those nomades at this early period of history. The contest between Syphax and Massinissa, and the tragedy of Sophonisba, attest the uncertain sway of Carthage over her turbulent neighbours.

After the Roman conquest we arrive at more precise territorial notions respecting North-western Africa. The immediate district dependent on Carthage received henceforth the name of the Province of Africa.* Numidia was made tributary to Rome under native princes, and its capital retained the name of Cirta, till the Emperor Constantine conferred upon it his own title, which it has retained to the present day, though corrupted

---

* Michelet's History of the Roman Republic, ch. iii. iv. v. ; Herder's Philosophie der Geschichte, b. xii. sec. 4 ; Dr. Russel's Barbary States, ch. i. Montesquieu, Grandeur et Décadence des Romains, ch. 4.

by the Arabs into Cossantina.* Mauritania was divided into two provinces, Mauritania Cæsariensis, extending from Numidia to the River Mulucha, and Mauritania Tingitana, from the latter stream to an indefinite limit, corresponding to the present southern border of Morocco. The first of the provinces in question answers pretty accurately in length and breadth to modern Algeria, the second is represented by the empire of Morocco. The country to the south of these provinces, known by the name of Gætulia, was in a great measure independent of the Roman sway, and embraced a considerable portion of the three Saharas of Tunis, Algeria, and Morocco, besides an unlimited stretch of the true desert. Further details respecting these divisions will be found in the chapter on archæology.

The political geography of North-western Africa in the middle ages is an obscure and intricate matter, as shifting and transitory in its demarcations as the Saracen dynasties that ruled it.

After the Arab conquest, the capital of North-western Africa, while it remained subject to the Asiatic caliphs, was placed at Kairouan or Kairwan, a city which they erected in the province of Africa, or the territory of Tunis, fifty miles south of the latter town, and twelve miles from the sea. Under the African Khalifs† the capital was at Mehadia. After the yoke of the Fatimites had been thrown off by the Sanhadja Berbers, the first branch placed their capital at Achir, on the road from Bou-Sada to Bugia, and afterwards restored it to Kairouan; the second branch placed it at Bugia, in the province of Constantine of modern Algeria. The Almoravides, another independent dynasty of Moorish sovereigns, made Morocco their capital; and the Almohades, who succeeded them, followed their example till the division of their empire. Then the branch of the Beni Mrin made Morocco and Fez their metropolis; that of the Beni-Zeian settled at Tlemsen in the province of Oran in modern Algeria, and that of the Beni-Hafes at Tunis.

The reader will perceive from this outline that the political divisions of Barbary during the middle ages were as confused and intricate as those of our European sires. At the period of the Turkish conquest in 1515, Algeria in particular had been parcelled out into a multitude of petty states, each governed by a petty sovereign, and all independent of each other. But leaving these insignificant divisions, which topographically and ethnologically are of no more importance than some of the smaller counties of England, we shall proceed to lay before the reader a compendious sketch of the political divisions of North-west Africa since they have received a permanent and definite seal by the Turkish conquest. After the brothers Barbarossa had reduced the territory of modern Algeria to subjection, they distinguished it from the Empire of Morocco to the west by the mountains of Trara in the province of Oran, and from Tunis to the

---

* قسنطينة Qosanthina.   † The Fatimite dynasty in Egypt.

east by the Ouad-el-Zaine, a river near La Calle. The breadth of the regency has always been somewhat fluctuating, owing to the sandy border that forms its southern limit ; but during the Turkish sway the tribes of the oases of Zab and the Mozabites inhabiting the Beni-Mzab district were partially and nominally subject to the Janissaries, who maintained a garrison at Biskara. The empire of Morocco, since it came under the sway of the present dynasty in 1519, has been confined to the limits of the ancient Tingitanian Mauritania, extending from the river Mulvia on the east to Tafilet in the south, and comprised between the Atlas and the ocean. Tunis, since 1520, has corresponded in most respects to the ancient territory of Carthage and the Roman province of Africa, the Zaine river forming its west limit towards Algeria, and the island of Jerba its east limit on the side of Tripoli. The breadth of this regency varies from 100 to 200 miles.

Having given this rough outline of the political divisions of Barbary down to the French conquest in 1830, we shall proceed to fill up the canvass with minuter details as regards the regency of Algeria, the special subject of the present work. And first, as to the territorial subdivisions of Algeria under the Turks, it may be desirable to state here that the regency under the Ottoman rule was governed by a despotic sovereign nominally dependent on the Sultan of Turkey, and named the Dey. The seat of his residence was Algiers, which was regarded as the metropolis of the whole regency, which comprehended four provinces or beyliks. These were governed by three beys, who were officers nominally subject to, but virtually independent of, the dey. The beylik or province of Algiers, being immediately dependent on the dey, did not stand in need of a special bey ; consequently, though there were four beyliks, there were only three beys. The other beyliks, after Algiers, were Oran, the western, capital Oran ; Tittery, the southern, capital Medeah, 73·32 miles from Algiers ; Constantina, the eastern, capital Constantina. Since 1830 the province of Tittery has been added to that of Algiers, and hence the present viceroyalty of Algeria contains three provinces : 1. Algiers to the centre ; 2. Oran to the west ; 3. Constantina to the east. Of these the last is much the largest. Proceeding to analyse the individual provinces, we find that the distance of the city of Algiers from the nearest and principal points in France is as follows.* The pharos of Algiers is 758 kilometres (471 miles) from the bottom of the port of Marseilles, which represents about the centre of the town. Algiers itself lies 7° 26′ south and 2° 34′ west of Marseilles. The distance from Algiers to Paris, measured between the centres of the two towns, is 1342 kilometres (833·88 miles). The distances of Algiers from the extremities of France are, that from Port Vendres 645 (400·79 miles), and from Dunkirk 1585 (984·88 miles). All these

---

* The latitude of Algiers is 36° 49′ 30″ N.

distances have been obtained by mathematical calculation. The town of
Oran is in 35° 45′ 57″ N. lat. according to French observations (35° 58′
English observation), and in 2° 4′ 52″ W. long. (24′ W. of Greenwich) ;
and its distance from Algiers is 66 leagues (165 miles) west, and fifteen
hours' sail from Carthagena in Spain.

The town of Constantina is in 36° 22′ 21″ N. lat. (36° 28′ in the Cabinet Atlas), and 4° 16′ 36″ E. long. (6° 26′ E. of Greenwich), and is 320
kilometres (198·84 miles) E., and 7° 17′ S. from Algiers, as the bird
flies.* In estimating land-distances in Algeria, it is very essential to be
careful in making the statements of the natives an authority. Arab measures are always uncertain, and often incorrect. Their principal distinctions in mensuration are : 1. the day's march ; 2. the hour's march ;
3. the mile ; 4. the farsekt.† The day's march is necessarily very variable, owing to what may be called subjective and objective circumstances ;
e. g. the motive of the traveller, and the nature of his vehicle, or the
country over which he journeys. The only divisions of the day known
to the Arabs are the times of prayer, or the position of the sun : these
are—El fedjer, daybreak ; Es s'bah', sunrise ; El oul, 10 A.M. ; El alem,
mid-day ; Ed dohor, 1 o'clock, P.M. ; El acer, 3 or 4 o'clock, P.M. ; El
mor'reb, sunset ; and El lîl, nightfall. The term ' mile' when used by
the natives in Africa is also a variable and optional distance. By the
French, however, the distances throughout Algeria have been ascertained
with their usual mathematical accuracy ; and it has been found that
the actual extent of the whole vice-royalty from east to west, including
Great Kabylia, is between 240 and 250 French leagues (625 miles). This
estimate agrees imperfectly with that of Dr. Shaw, about 100 years ago,
who gave the regency a length of 480 miles. Its breadth from north
to south, that is to say, from the Mediterranean to the true desert, varies
from 60 to 200 leagues (120 to 500 miles), containing, according to the
computation of Marshal Bugeaud, an Arab population of from three to
four millions, though other authorities represent it as much less or
greater.‡ About two-thirds of this territory presents a surface of rugged
and wild mountains, intersected, however, as we have previously seen, by
numerous fertile valleys in many parts.§ It was in 1843 that, according to the division of the Minister of War, French Africa was divided
into three provinces, Algiers, Oran, and Constantina, each of which were
made to contain several subdivisions. Thus Algiers was divided into Al-

---

* These diverging mensurations are from the Exploration scientifique, and the Cabinet
Atlas and Universal Gazetteer.

† The farsekt is probably derived from the Persian mile, farsang, παρασαγγης, consisting, according to Passow, of 30 stadia, or 3750 paces, three-fourths of a German mile,
or nearly four English miles.

‡ For further particulars on the population the reader is referred to the chapters on
the native races and statistics.

§ Dawson Borrer's Campaign in the Kabylie (Longmans, 1847), p. 233.

giers and Tittery; Oran into four, namely, Oran, Mascara, Mostaganem, and Tlemsen; and Constantina into two, Bona and Setif.*

The old province of Algiers was bounded to the east by the river Booberak, to the west by the Massafran, and was much smaller than the two others (the Tell), being scarcely sixty miles in length and breadth. Under the Turkish sway, as previously observed, the territory or province of Algiers Proper was independent of the other beys ; and its kaids or mayors were immediately under the dey, whose direct authority thus extended over a circuit of six square German miles† (120 English square miles). It is proper to add, that the limits of this territory were very fluctuating, owing to the caprice of the deys, who found it frequently convenient to extend their direct authority by encroaching on the territory of the refractory or obnoxious beys. Thus Blidah, which properly belonged to the province of Oran, and the plain of Hamza to the iron gates (a mountain pass), were administered by the aga of the Arabs, who had the direction of the province of Algiers.‡

The Turkish province of Tittery, which has now been swallowed up in that of Algiers, was much smaller than those of Oran and Constantina; and its name has been derived by some from the Arabic *iteri*,§ cold, because it contains some snowy mountains. The four chief divisions of Algeria under the Turkish rule were frequently classified as follows : 1. the western province, or Mascara ; 2. the territory of Algiers ; 3. the middle or southern province of Tittery ; 4. the eastern province, or Constantina.

The western province was that of Mascara, now called the province of Oran.

This province embraces now a surface of 102,000 square kilometres (39,270 square miles), with a population of 600,000. The present province of Algiers contains a surface of 113,000 square kilometres (43,505 square miles), with a population of 900,000 persons.||

The province of Constantina lies between the meridians of the rivers Booberak and Zaine, and is nearly equal to the other two in extent, being upwards of 230 miles long, and more than 100 broad. This province has a surface of 175,900 square kilometres (67,721·5 square miles), with 1,300,000 inhabitants,¶ and includes the remarkable district of Algeria known by the name of Great Kabylia, which has long been celebrated for the sturdy independence of its mountaineers, and has lately become the theatre of some of the boldest French exploits in Africa. As we

---

* Dawson Borrer, c. 16, on the Arab tribes.
† Nachrichten und Bemerkungen, &c.
‡ Adr. Berbrugger's Algérie historique, pittoresque, et monumentale: folio, Paris, 1843, p. 27. § Blofeld.
|| Tableau de la Situation des Etablissements français en Algérie, 1850, p. 719.
¶ Ib. p. 719.

propose devoting a special chapter to this interesting region, we shall
confine ourselves on the present occasion to a few brief statements respecting Great Kabylia, which contains a surface of 7800 square kilometres (3003 square miles), with a population of 370,000, and an average
number of 80,000 fighting men, presenting a sea-face of 146 kilometres
(90·72 miles) on the Mediterranean, between Dellys and Bugia.

Previous to the French conquest and exploration of Algeria, there cannot be said to have been any proper or accurate political divisions in the
country. It is only lately that they have been methodically established
for the sake of convenience ; and we here introduce those approved and
suggested by the scientific exploration of the French government.*

1. The Algerian Sahara is intimately bound to the Tell ; and the
union of the two regions constitutes Algeria. 2. All the partial threads
that compose this web are divided into three distinct parcels. Thus some
are found in the east Tell (Bona, Constantina, Setif) ; others in the centre
Tell (Algiers, Medeah, Milianah) ; others, again, in the west Tell (Oran,
Mascara, Tlemsen).

The Sahara is further divided into three parts : 1. the east Sahara ;
2. the central Sahara ; 3. the west Sahara. Thus Algeria, besides two
transverse zones, is decomposed into three meridian segments, formed of
the corresponding parts of the Tell and Sahara. We shall henceforth
adopt this classification.

Algeria, politically regarded, means all the territory comprised, really
or nominally, in the old pashalik. This territory is divided by the commercial habits of its population into three meridian segments, called, 1st,
East Province ; 2d, Centre Province ; 3d, West Province. These correspond to what the natives call Beilik-ech-Cherguiia, Beilik-el-Oustaniia,
Beilik-el-R'arbiia. Each province is divided into two regions—1st, north,
2d, south—essentially different, and belonging to the Tell and Sahara.

Hence Algeria is divided into six distinct regions, called thus :

FOR EUROPEANS.		FOR NATIVES.
North.		*Tell.*
East Tell	. . . . . .	Tell-ech-Chergui.
Centre Tell	. . . . . .	Tell-el-Oust'âni.
West Tell	. . . . . .	Tell-el-R'arbi.
South.		
East Sahara	. . . . . .	S'ah'ret-ech-Cherguiia.
Centre Sahara	. . . . .	S'ah'ret-el-Oust'âniia.
West Sahara	. . . . . .	S'ah'ret-el-R'arbiia.

It will be seen that in the political as well as in the physical geography of Algeria the great characteristic distinction is that of the Tell

* See page 81. part ii. of E. Carette's Recherches sur la Géographie et le Commerce de
l'Algérie méridionale, in the Exploration scientifique : 4to, Paris, 1844.

and Sahara. Before we proceed to determine more accurately the frontiers of the viceroyalty, we shall pause for a short time to consider the most striking natural and social features of these regions. By determining the northern border of the Sahara or southern zone, we shall be able at once to determine the outline of the Tell.

The limits of the Tell and Sahara* are determined by their produce. There are, however, transitional, hermaphrodite regions or zones, where the date and the ear of wheat equally ripen ; and there are others again which produce neither : these latter zones, being unenclosed and unfit for culture, come under the head of Sahara. The natives distinguish the zones thus : the country where corn is the rule belongs to the Tell ; the country where corn is the exception belongs to the Sahara.

The Ouad-R'is'rân divides Algeria and Tunis throughout its course. At the point where it enters the plain of El Mîtli there are ruins also called R'is'rân. Here the limit of the Sahara touches the frontier of Tunis. These ruins are at the foot of a chain of mountains which is prolonged without interruption east to the Djebel H'adîfa, near Gabes, in Tunis, west to the Djebel Metlîli, near El Gant'ra. The edge of the Sahara follows the foot of these mountains. Leaving the ruins of R'is'rân, the limit of the Sahara of Algeria, all through the countries that we have studied, may be divided into three parts : the first extends from R'is'rân to the Djebel Metlîli, and remains constantly in the basin of the lake Melr'îr'; the second extends from the Djebel Metlîli to the peak known as the Grîn-el-Adaorâ (the little horn of the Adaorâ), and follows constantly the basin of El H'od'na; the third extends from Adaorâ to the village of Frenda, and remains throughout in the basin of the Upper Shellif.

It follows that the Algerian Sahara does not advance so far north in the eastern as it does in the western part of the viceroyalty. R'isr'ân is in lat. 34° 20', and 140 kilometres (86·99 miles) from the coast. In the meridian of Dellys it comes to lat. 36°, and only 80 or 90 kilometres (49·06 or 50 92 miles) from the coast. Thus in the east and centre the Tell or corn-country passes beyond the limits of the basin of the Mediterranean ; in the west it does not reach those limits. The valleys of the Ouad-ed-Djedi and Ouad-el-Arab produce in their lower parts dates and grains, and are thus of a hermaphrodite nature. To the west the upper basin of the Shellif only produces dates. Hence on the limits of the Sahara there are doubtful districts, to the eastward doubly productive valleys, to the west immense ungrateful steppes. These intermediate zones present three basins : to the east, double culture, that of dates and corn ; to the centre, double culture intermixed with pasture ; to the west, pasture only.

---

* See chap. iii. of E. Carette's Recherches sur la Géographie, &c , ubi supra.

We have seen that the Algerian Sahara is divided into basins : 1st, that of the Ouad-Mzab ; 2d, that of the Ouad-Zargoun ; 3d, that of the Upper Shellif. The Ouad-Zargoun only enters partially into this territory, which may more correctly be analysed into four primary basins : 1st, the Lake Melr'ir' ; 2d, the H'od'na ; 3d, the Upper Shellif ; 4th, the Ouad-Mzab : and into three secondary basins ; 1st, the Zâr'es ; 2d, the Ouad-Rîr' ; 3d, the Ouad-Souf. It is proper to add, that the inhabitants of the Sahara know no other division of the country than that into oases and tribes.

The contrast between the Tell and the Sahara and their populations may be summed up as follows :

" The knowledge of the solar months, though necessary in agriculture, is less spread in the Tell than in the Sahara. In the Tell the marabouts give the signal for tilling and harvest. In the Sahara, where the labour is more individual, each proprietor regulates himself the order of his work. In the Tell there is great ignorance and apathy when epidemics prevail or approach ; in the Sahara, on the contrary, there is much foresight. The Sahara contains a great many towns and villages, whose construction does not imply any great skill, but much more than a tent, the usual dwelling of the Tellians, excepting the mountaineers of Kabylia, who live in houses. The Tellian only knows his neighbour ; the Saharian is a great traveller. The first only knows the day's march as a measure of time ; the Saharian knows the Roman mile. The Saharian believes in labour, and seeks it—he is strong, active, and clever ; the Tellian lazy and awkward. The first men who greet you on landing at Algiers are Saharians, who constitute the porters and carriers of the capital. The question then arises, is there more civilisation in the north or south of Algeria among the natives ? Except the Kabyles, who inhabit the mountains of the Tell, there is decidedly more civilisation in the south, and even the Kabyles themselves are greatly inferior to the Saharians in sociability, though equal in industry. The Saharians have a loftier mind and a more lively imagination ; allegory is common among them, and some know even how to paint. They are the only population in French or all north-western Africa who shew a little vein of culture. If European civilisation penetrates Algeria, industrial arts will go to Kabylia, but letters and sciences to the Sahara."*

Having now analysed the chief features of Algeria esoterically, we shall proceed to determine its limits more clearly in an exoteric point of view.

We have said that the Trara mountains have been generally regarded† as the western limit of Algeria, and stretch a considerable distance from

* E. Carette's Recherches, &c. p. 236.
† Dr. Shaw's Travels in Barbary ; Nachrichten über d. Alg. Staat. 1798, 1800, 3 th 1 th. p. 19 ; Dr. Russel's Barbary States, p. 315.

north to south, the northern point constituting the promontory known by the name of Cape Hone. Some writers have represented the river Mulloviha or Malva to be the limit, which may have proceeded from the circumstance that the district between the Trara mountains and the Malva river is almost a desert, and a kind of neutral ground in the possession of roving tribes independent of Morocco and Algeria. The distance from the Trara mountains to the Ouad-Zaine, the east limit of Algeria, is from 1° 40′ W. to 9° 15′ E. of Greenwich (4° 39′ W. to 6° 54′ E. of Paris).

A short distance to the west of Cape Hone is Twunt, which, with the Trara mountains, is, according to Blofeld, the west end of the province of Oran and of Algeria.

The natural frontier of the Algerian Sahara to the south was long a doubtful matter ; nevertheless it has one which consists in a chain of oases in Algeria. These are cut off sharp from the south by an abyss of sand; and proceeding from east to west, they occur in the following order : the Ouad-Souf, the Ouad-Rîr, the Temaim, Ouaregla, the Ouad-Mzab, El Abied, and Sidi Cheikh. Beyond this chain of oases, sands and droughts are effectual barriers to the advance of ambition and commerce. This desert is also the southern limit of Tunis and Morocco ; and North Africa obtains in this manner the character of an island, whose clear limits are the ocean, the Mediterranean, and the desert.

We have previously observed that very false notions have long prevailed respecting the great southern waste that occupies so large a portion of the surface of Northern and Central Africa. Sand is the smallest component element of this district, and only extends a few days' march from the coast, and then you reach a stony and arid table-land cut up into immense valleys of 50 or 60 metres (196·8 feet). This plateau abuts on a mountain-chain running from Cape Bojador to an unknown limit to the eastward ; but to the northward these mountains touch on Morocco, and are clothed with forests. Sand is only met with in the lowest places, where you also find well-water, whereas the hills and plateaux have none.

The oasis of Touat is surrounded at some distance by mountains to the westward and north-west ; the country that separates it from Morocco is scattered with them, but we know nothing of their distribution. Between Morocco, Algeria, and Touat lies an uninhabited desert without any water, and south-east of Algeria exists a like country stretching to R'dâmes ; but between the two, near Ouad-Mzab, there exists a mountainous country which extends only a little way east and west, and appears to end a little before El Goh'a. The whole road from Algiers to Touat only presents sand around El Goh'a, which stands about half-way from South Algeria to the oasis. The desert resembles many other countries topographically, but it is distinguished by a number of great shut basins with a sandy bottom, flat, and more or less salt, containing brackish water a little underground. The Arabs call these plains, which have beds of salt, R'oût.*

Passing from the southern frontier of Algeria to the east, we find that the Algerian oasis nearest to the regency of Tunis is the Ouad-Souf; and the Tunis oasis nearest to the regency of Algeria is the Belad-el-Djerid, of which Neft'a and Tôzer have an almost equal right to be called the capitals. The frontier-line is not accurately determined, but falls near the sand-mountain Bou-Nab, belonging to the Algerian tribe Rbeia; and the wells El Asli, belonging to the Tunis tribe Neft'a. There is a large space of neutral ground between the two territories to the north of these oases in the vast basin of the Lake Melr'ir'.

Negotiations have taken place between the French government and that of Tunis, in relation to certain points, within the last few years, since when the border-line has been more accurately determined. The limits of the two states in their southern part are, the wells of Bou-Nab, the sand-hills around the Ouad-Souf, the plain El Mîta and Ouad-R'isr'ân, the course of the Ouad-Helâl, the defile of Bekkaiia, the ruins of H'idra, the course of the Ouad-H'idra, the Ouad-Serrat, and the Ouad-Malay.

The reader should bear in mind that there are many neutral grounds in Algeria, occasioned by the hostilities of tribes, some of them being 78 or 80 kilometres (50 miles) in width.

A few years ago there were but two practicable roads from Algeria to Tunis, that along the shore, and that of the Sahara ; every where else the traveller was murdered; and you could only follow the first-mentioned route by paying the tribe of R'ezoân a duty of 25 fr. per mule.†

We have seen (p. 49) that the Ouad-R'isr'ân divides Algeria and Tunis throughout its course, and at the point where it enters the plain of El Mîti are the ruins of R'isr'ân, where the east end of the Algerian Sahara touches the frontier of Tunis.

The French documents on the limits of Algeria and Tunis near the coast are somewhat contradictory.‡ Thus the maps prepared at the Dépôt de la Guerre have successively placed it at the ruisseau of Saint Martin, near La Calle, and at the Ouad-el-Zaine, two leagues farther east. According to M. Berard, it ought to be the channel leading from the lake of Tonegue at one league and a half east of La Calle.

Marmol§ and Gramaye‖ include the island of Tabarca in the province of Constantina ; Pierre Dan¶ also places the limit of Algeria towards Ta-

---

* Notice géographique sur l'Afrique septentrionale by Renou, in the Exploration scientifique, p. 332.

† Recherches, &c. of E. Carette, in the Exploration scientifique, p. 17.

‡ Baron Baude's Algérie, i. p. 269, appendix, note.

§ Africa of Marmol, b. vi. c. 2.

‖ Gramaye's Africa Illustrata, l. 10.

¶ Pierre Dan, Histoire de Barbarie et de ses Corsaires : 4to, 1637, liv. ii. c. 1.

barca. Dapper* places Tabarca in the province of Bona; and he fixes the western limit of Tunis at the Ouad-el-Burbar and El Zaine, the ancient Tusca. Peyssonel,† about 1724-5, places the limit of the two regencies at Cape Roux.

Dr. Shaw says (1732) that the Ouad-el-Erg was for many years the limit of the two regencies ; this stream flows from the lake of the Nadis (of Tonegue) five leagues east of La Calle.‡   But as the territory between the Ouad-el-Erg and the Zaine was often put under contribution, Shaw places the frontier at the Zaine, four leagues farther east.

Shaler, the United States consul at Algiers (in 1826), places the limits at Tabarca, at the mouth of the Zaine, in 9° 16' E. long.

Numidia and the territory of Carthage were in like manner separated by the Tusca, now the Ouad-el-Zaine ; Tabarca and Vacca were Numidian towns.   In 1741, the Lomellini of Genoa paid 25,260 livres to the government of Algiers, and 15,285 to Tunis, for the island of Tabarca ; hence it is evident that Algiers must then have laid claim to the left bank of the Zaine, because to the west of La Calle the commerce of the coast belonged at that time entirely to the French.

Half-way between La Calle and Tabarca, and at the distance of three leagues from each, Cape Roux advances into the sea ; and Mount Khoumir, whereof the cape is a prolongation, rises in sharp peaks to an elevation of 1000 metres (3280 feet).   Its almost inaccessible ridge bisects the contested territory, and has been placed as a limit between the two regencies by the hand of nature; hence the Algerines and the men of Tunis have never attempted to establish themselves permanently on the opposite side of this cape to their own country, without the aggression leading to discord and strife.

Algeria, limited to the oasis of Metlili and of Ouai-regla, presents the following surface :

Tell	.	.	.	.	1480 square myriametres.	56,980 square miles.
Sahara,	(North zone			1400 ,, ,,	53,900 ,, ,,
or S'ah'ra	(Oases		.	1320 ,, ,,	50,820 ,, ,,
				4200	160,700

We have already seen (p. 41) that, comprising the villages of El Goha and Ocdan, and all the tribes of Chamba, which would extend the southern limit to the thirtieth degree of latitude north of the meridian of Paris, Algeria would have a surface of 4700 square myriametres (180,950 square miles), or only one-tenth less than France.   The centre would then fall about 34° 7' N. lat. and 1° 4' long. E. of Paris, or 3° 23' E. of Greenwich ; or in other words, between Demmel and Ksir-el-Hiran.

* Dapper, Description de l'Afrique : Amsterdam, pp. 188, 189, 199.   1686.

† Peysonnel, Voyage dans les Régences d'Alger et de Tunis : 2 vols. 8vo, Paris ; published first in 1838.

‡ Berard says 1½ leagues (3¼ miles) E. of La Calle, which is probably nearer the truth.

The length of Algeria between the frontiers of Tunis and Morocco, *i. e.* between the mouths of the rivers Zena and Adjeioud, is in a straight line 974 kilometres (605·23 miles). This distance is about the same as that which separates the Point of Raz, in Cape Finisterre, from the mouth of the Lauter in the Rhine ; the direction is about the same, and the eastern extremities fall under the same meridian ; but the Point of Raz exceeds the extreme west meridian of Algeria, because of the difference of the length in the degrees of longitude.[*]

The etymology of the word *Tell* is doubtful. The talebs (طالب), who are the Arab *savans*, call *seheur* the inappreciable moment that precedes daybreak, when night is no longer night, and day is not yet day ; at the period of the Rhamadan, as soon as you can distinguish a white thread from a black one, abstinence is incumbent on all true Mussulmans. The seheur precedes that instant, and it is more easily appreciable in a country with a wide horizon ; hence, according to these sages, the name of Sahara has been given to this region of lofty plateaux which comes after the Tell, of which the etymology, according to some authorities, is not the Latin word *tellus,* but the Arabic word *tal* (طَالَ *to tarry;* طُول (toul), *length*), which means ' to be last,' because the seheur is only seen there later. Whatever may be the true history of these etymologies, the French understand by Tell the land that yields grain ; and by Sahara the land of flocks and pastures. As an Arab named Mohamed Legras once expressed it to Marshal Castellane, " The Tell is our father ; she who married it is our mother ;" or according to the saying of the nomadic tribes, " We cannot be either Mussulmans, Jews, or Christians; we are the friends of our bellies."[†]

The Arabs themselves sometimes style the people of the divisions of Barbary, including Algeria, by their productions. Thus they call the inhabitants of the towns the *gold people ;* the inhabitants of the Tell, the *silver people ;* and the inhabitants of the Sahara, the *camel people.*[‡]

A name commonly applied by the Arabs themselves to the Sahara is Blad-el-Djerid (the country of dates) ; an epithet that older European geographers caused to supersede the more correct appellation of Sahara, which they erroneously transferred to the Great Desert. The Arabs of the Sahara, in familiar conversation, frequently style themselves *Djeridi,* which might be rendered *palmers.*[§]

We have previously stated that the first plateaux of the Sahara are named *Serssous,* and form a succession of *mamelons* or mounds of almost equal elevation, following each other in succession for an immense distance ; you would take it to be the swell of the sea magically stayed and petrified

---

[*] Exploration scientifique.

[†] Souvenirs de la Vie militaire, &c. pp. 253-4.

[‡] Le Grand Désert ; itinéraire d'une caravane au pays des Negres, by General Daumas : 1850, p. 34.

[§] Ibid.

by some invisible hand. Amidst each of these inundations are found springs of fresh water; and fertile pastures, with short and thick grass, stretch away, supporting and nourishing those sheep so famous for their delicious meat and valuable wool. Farther on and beyond the first horizon of mountains, at twenty leagues distance from the mountains of the Tell, begins the real Sahara : there Count de Castellane was informed that the traveller meets vast, empty, and naked plains and mountains ; oases with tapering palm-trees, and other lands where towards the spring and during the winter you can still find pasturage for the flocks ; and farther still, at a great distance, you come to the mysterious world of sand.*

The surface of Algeria, including the Tell and the Sahara, is reckoned at 390,900 square kilometres (150,496·5 square miles), which amounts to about four-fifths of the superficies of the eighty-six French departments.† This territory contains 1145 tribes, with a population of 3,000,000; to which if we add the population of the towns, we shall obtain a grand total of 3,196,140. Except some Kabyle districts between Dellys and Philippeville, and a few tribes on the borders of Tunis, the whole Algerian Tell (137,900 square kilometres, or 53,091·5 square miles) may be regarded as entirely subdued by France. The Sahara, embracing 253,000 square kilometres (97,405 square miles), also acknowledges the French authority; but its population is much thinner and more scattered than that of the Tell, and the French troops only occupy a few detached posts in it. The influence of the tricolor has now penetrated to the southern limits of the Sahara, especially since the capture of Zaatcha and Laghouat, and the French authorities have representatives in the whole zone of the oases.‡

The esoteric political divisions of Algeria have undergone considerable modifications since the organic decree of the 15th of April, 1845, which maintained the old division of the regency into three provinces. In the first place, the territory of each province was subdivided into three zones : i. e. the civil zone, under the administration of the common law as decreed by the legislature of Algiers, and under the direction of the civil power, save in the case of certain restrictions applicable to natives. 2dly, the mixed zone, where the European population being thinner was placed under an exceptional *régime*, all the administrative, civil, and judicial functions being performed by military men. 3dly, the Arab zone, which was administered by martial law.

---

* Souvenirs, &c. p. 255. 1852.

† The Tableau de la Situation and the Exploration scientifique differ slightly in their estimate of the surface of Algeria, the former reducing it to 150,496·5 square miles, and the latter extending it to 160,700, making a difference of 10,204 square miles. The tendency of all colonial governments in general, and of the French in particular, to extend their limits, easily accounts for the inclination shewn by our neighbours to encroach on the sands of the deserts, ultimately embracing a surface of 180,950 square miles, and reaching the 30th degree of N. lat. See page 53, and Le Sahara Algérien, by General Daumas.

‡ Tableau de la Situation, &c. 1850, pp. 77-79.

The particulars relating to the administration of Algeria being minutely described in another chapter, we shall here confine ourselves to changes in the territorial divisions ; one of the most important of which was that which, by a decree of the executive power of August 16th, 1848, decided that the colony should be subdivided into parishes. By the decree of the 9th of December, 1848, the old division of Algeria into three provinces was still preserved ; but the distinction between the civil, mixed, and Arab zones was suppressed, and Algeria was simply divided into civil territories or departments, and into military territories, whose limits were fixed by the executive power. The civil territories have been erected into three departments, taking the names of the three provinces.*

Before concluding our sketch of the political geography of Algeria, it is well to describe a few divisions of the territory peculiar to the natives, and which we have hitherto omitted.

A general and wholesale division applied by the Arabs to the whole of north-western Africa is that according to the cardinal points. The south, a vague and indefinite term applying to the Great Desert and Soudan, is the Guebla. The west, including Morocco, and if you confine yourself to Algeria, the province of Oran, is El Garb, or J'harb, whence the native name for the empire of Morocco is Moghreb, and its people are styled Moghrebins. The east is described by the word Cherg, and admits of an unlimited extension : in its narrowest sense it may mean the province of Constantina in Algeria; in its widest sense it may embrace Egypt, Arabia, and the Levant.

Exoterically the Arabs call all other countries Beurr-el-Adjem (except the Berber districts) where the Arabic tongue is not spoken, even if the inhabitants should be Mussulmans. The spelling of Adjem is the same as that of the word *adjem,* meaning ' ox;' and we are disposed to think that the Arabs in their pride compare all who do not speak their tongue to beeves ; adjel (عِجْل) in the singular signifying the ox that has not been broken into the yoke, *i.e.* a calf.†

Empires depart, races dissolve, religions change phases, form, and substance; but the handwriting of the Almighty on the trackless sands and the everlasting hills remains the same yesterday, to-day, and for ever. Carthage has become desolate, and the royal Hippo a habitation for dragons ; but the three zones of Herodotus still remain as fresh and dry as ever, whilst old Atlas cuts the blue vault with his peaks, and the graceful palm still nods its crest unchanged over the waving murmuring oasis.

* Tableau de la Situation, &c. 1850, pp. 77-79.
† Le Grand Désert, &c., by General Daumas, p. 161.

CHAPTER IV.

Algiers.

PRINCIPAL FEATURES OF THE PROVINCE OF ALGIERS—THE SHELLIFF—THE HARATCH, THE MASSAFRAN, AND THE ISSER—THE MITIDJA—THE SAHEL—SIDI-FERRUCH—CAPE MATIFOU—ALGIERS—THE OLD PORT—THE NEW PORT—STREETS—HOUSES—BAZAARS—THE CASBAH—THE FAUBOURGS.

HAVING given a general outline of the physical and political character-istics of Algeria, it is our purpose to launch forthwith into a minuter topographical analysis of the regency. And in order somewhat to diminish the dulness of dry details, we propose to interlard our pages with copious and apposite extracts from the most recent visitors in Algeria, illustrative of the scenery and topography of French Africa.

Before analysing the province of Algiers, we shall begin, as in our larger survey of the whole regency, by a broad outline of its natural features.

This province comprises, like the other two, its Tell and Sahara, and is bisected twice by the two Atlas chains. The Djebel Amour towers aloft in the southern part of the Sahara, which is watered by the Ouad-el-Djedi, which passing the town of El Agrouat or Laghouat, flows east into the province of Constantina. Farther north we find the two Zarhez lakes, called Chergui and Gherbi, east and west.

In the east of the province the chief feature is the Djorjora range of the Atlas in Great Kabylia, which will be described in another place; and near the sea we have the Great Mitidja plain and the Sahel coast-ridge, of which more presently. The chief river is the Shelliff,* rising at the Djebel Amour, at a place called Sebbeine-Ain, the 'seventy fountains.' Its first tributary is the Nahar-wassal, from the west. Running N.E. it flows past Boghar, near the sanctuary of Sidi-ben-Tyba, a little below Medeah; then passing close to Millianah, it flows west, washing the walls of Orleansville, near which town it enters the province of Oran. It receives large contri-butions the whole way, especially the Ouad-Midremme, the Ouad-Aradji, and the Ouad-Foddha. The river Haratch is the Savus of the ancients, and about one hours' march to the east of Algiers. It is a considerable

---

* Blofeld's Algeria Past and Present. 1844. Blofeld asserts that the whole course of the Shelliff from the Sebbeine Ain to Djebel Diss, *i. e.* the mountain of Spartum, or reedy grass, is little short of 200 miles.

stream, which takes its rise in the mountains of the Little Atlas to the S.E. of Blidah, a French post and small Arab town situated about 10 leagues (25 miles) almost direct south of Algiers. The Haratch traverses the Metidja plain, where it is about 11 leagues broad (29¼ miles), and falls into the bay of Algiers 3 or 4 miles to the east of the metropolis. The water of the river is muddy and brackish, and in winter it is subject to great inundations. Its principal ford is called the *Gué de Constantine;* and when Mr. Borrer visited the regency, the French wooden bridge was carried away, in November 1846, during the prevalence of an unusually wet season, which occasioned extensive and disastrous floods in the Mitidja plain. The wooden bridge in question sailed down the torrent on that occasion, and went to pay a visit to the strong Turkish bridge which is built five miles lower down.*

The Isser's chief source (according to the French map of the province of Algiers, drawn at the Dépôt général de la Guerre, for 1846) seems to be near Berouaguia, about 15 miles S.S.E. of Medeah, and in the territory of the Beni-Hassan. From thence flowing under different names in a N.E. direction for about 45 miles, it suddenly turns in the territory of the Beni-Djaad almost direct north, and flows into the sea some 5 miles to the west of Cape Djinet, a promontory situated about 45 miles to the east of Algiers.

We shall pause for a moment to remind the reader of the present political division of the province of Algiers, which, as has been previously observed, contains at the present time the territory attached to the metropolis and the province of Tittery, according to the divisions under the Turks. This division of the viceroyalty is still much smaller than the other two constituting the provinces of Oran and Constantina, from the former of which it was till lately separated by the river Massafran, and from the latter by the river Booberak. The province of Algiers is analysed into two subdivisions, which are those of Algiers and Tittery; and contains only 113,000 square kilometres (43,505 square miles).† Nor is it in general so mountainous as the other provinces. The sea-coast to the breadth of five or six leagues (12 or 15 miles) consists principally of rich champagne grounds, behind which are a range of rugged mountains composing part of the Little Atlas chain, running almost straight and parallel to the coast. Beyond this range, and particularly in the neighbourhood of Medeah, Titterie Dosh, and Hamza, the ancient territory of the Tulansii and Banniri, are extensive plains, though none of them equal to that of Metijiah.

The latter plain, sometimes written Mitidja, together with the range of hills called in Arabic Sahel, on which the metropolis is built, constitute the most important features of this province. The Mitidja is a vast level,

---

* Dawson Borrer's Campaign in the Kabylie, p. 16.
† Tableau de la Situation des Etablissements, &c. 1849-50, p. 719.

situated between the north slope of the Lesser Atlas and the Sahel, and bounded to the east by the lofty mountain-range of Kabylia, in the province of Constantina. It is watered by two rivers, the Haratch and the Massafran, and is as flat as a billiard-table over its whole superficies. It varies from three to five (and some say eleven) leagues in breadth, forms a semicircle of about fifteen leagues, and touches the sea in two places,— at the Fort of Maison Carrée, a little to the east of Algiers ; and just below Scherschell, formerly in the province of Oran. The Mitidja entirely differs from the Sahel, or as it is sometimes called the Massif, or chain of Algiers. It has been in turn noted for its fertility, for its barren sands, and for its unhealthy marshes. All these statements are true, though apparently contradictory, as the plain contains all these differences in its ample embrace. Several Roman roads used to cross it; the most important of them, following the coast, can be traced to the eastward and west of the metropolis in the direction of Dellys and Scherschell.[*]

This plain is represented by eye-witnesses as a perfect desert now, compared with what it was in 1830 and previous to the French conquest, when upwards of twenty thousand Moorish villas and farms are stated to have dotted its verdant surface.[†]

The Mitidja is a fine valley, eighteen leagues long and six or seven broad (45 miles long and 14 or 15 wide); it is only slightly undulated even at the water-shed separating the basins of the Haratch and Hamiz from that of the Massafran.[‡] The Atlas and the Massif or chain of Algiers, which limit this plain, rise almost suddenly from it without any slopes. The Mitidja to the west is bordered by the Sahel hills, which do not attain any very considerable elevation, and are cut through by the river Massafran in order that it may reach the sea; and to the north-east its boundary is formed by the sand-hills that the Haratch and Hamiz cross at their mouth. It is well cultivated near the mountains, and marshy in its lower parts; its aspect is generally bare : yet in some parts you see, especially to the south, agricultural establishments and Arab hamlets surrounded with impenetrable hedges of Barbary figs, and with plantations of olives, carob, jujube-trees, and some elms. The northern slope of the Little Atlas is covered with brushwood, chiefly oaks and lentisks, and is cut by great valleys, from which issue the streams that water the plains.

Having completed our description of the Mitidja plain, we turn next to the Sahel range, also known by the name of the Boujareah. This hilly district, containing a superficies of twenty-five square leagues (125 square miles), is washed to the north by the sea; to the east by the Haratch; to the west by the Massafran; and the south descends abruptly into the plain.

---

[*] Baron Baude's Algérie, 1841. The French in Algiers, by Lady Duff Gordon, 1845.
[†] St. Marie's Visit to Algeria ; D. Borrer's Campaign, &c. p. 16; Pananti's Avventure, 1817.
[‡] Berbrugger, Introduction, p. 6.

It is intersected by numerous valleys, which are well watered in winter, but dry in summer. The Sahel, which constitutes an isolated range, occupies in front of the Mitidja an almost elliptic area of 33,000 hectares (82,000 acres); the sea bathes its northern hemicycle, and Algiers is built on its side exposed to the Levant or east. The soil of the Boujareah is in general strong and good : the thickets that cover a large part of its surface consist principally of carob, lentisks, wild olives, &c., which are greatly injured by the cattle that are suffered to wander over the country. Here and there, however, you meet Jerusalem pines, whose vigour shews the nature of the soil to be adapted for the growth of wood. The Sahel hills are the last slope of this range to the south, and rise suddenly from the plain to the height of 150 metres (487 feet). The Boujareah has lost many of the sources that it once had, which supplied in the time of Père Dan one hundred fountains in Algiers. The Ouad-Kniss, called by Nicholas de Nicolai (1587) the Savo, used to be a large stream, and is now only a thread. It contains, however, many dry springs, the drying up having resulted in all probability from the stripping of the woods.*

The ridge of Algiers presents a very regular system of gradually ascending hills, cut by numerous gullies ; it sheds its waters to the south into the plain, to the north they fall directly into the Mediterranean. The culminating point of the Boujareah is 400 metres (1300 feet) above the sea. This massif or ridge is covered in the neighbourhood of the town with agreeable habitations, where abundant springs keep up perpetual freshness and vegetation; but it does not present a pleasant appearance on the top : the land there is dry, stony, and covered with short shrubs ; but the ravines when watered are woody, and capable of great cultivation.†

In individualising the minuter features of this province we shall begin with a description of the sea-coast, and deposit the reader at first on the peninsula of Sidi-Ferruch, where the French army landed in the invasion of 1830. After leaving the river Massafran, the western limit of the province under the Turks, the first object that meets the eye is a small tower upon a rocky cape or isthmus, stretching about a furlong into the sea.‡ This is the marabout, or tomb and sanctuary, of Seedy or Sidi-Ferdje, or Feredje, corrupted by the French into Sidi-Ferruch. This building stands on the extremity of the peninsula, which is situated about half-way between Scherschell (Julia Cæsarea) and Algiers (Icosium), and advances about one-third of a league (two-thirds of a mile)§ into the sea, with a breadth of 8000 metres (26,240 feet). The isthmus leaves two bays, one to the eastward and the other to the westward, or to the right and left, bordered with wide beaches and sand-hills. The ground of the peninsula

---

* Baron Baude's Algérie, 1841, i. pp. 78-81.   † Idem.
‡ One-eighth of a mile, according to Blofeld.   § According to Berbrugger.

is mostly low and sandy, but it rises to the extremity and forms a rocky eminence with several constructions. The chief among them is the marabout above mentioned, with a minaret or square tower, to which the Spaniards have attached the name of Torre-chica. The Arab name is Sidi-Feredje, the latter being the name of the native saint buried there, and the word Sidi being an Arab title corresponding to our *lord*. The creeks of Sidi-Ferruch offer at present a refuge and shelter in stormy weather to the sandals of the country and other small craft. They anchor, according to the wind, to the east or west of the peninsula. The natural jetties of rocks by which these creeks are protected might easily be converted into moles.*

The peninsula can boast of five wells of brackish and one stream of good water; and at the distance of about nine miles to the north-east begins the high chain of Boujareah, here called Sahel, a word meaning coast, shore.† Between Sidi-Feredje and the Sahel are some plains, on one of which, bearing the name of *Staoueli*, an engagement took place between the French army and the Turks in 1830.

Before Khaireddin Barbarossa had made a port at Algiers, Sidi-Feredje and Cape Matifou were frequented by the merchant-ships that resorted to the capital. After this change it was still preserved from total neglect by the veneration attached to the marabout, whose name, according to Baron Baude, ought to be spelt Esseid-Efroudj, an epithet corresponding to the Catholic appellation *mon sieur St. Denis*. The Mussulman population have long been firmly persuaded that miracles are performed on this spot by the supernatural power attributed to the saint; and a marvellous legend records how a Spanish captain who had offended the saint had his ship three times enchanted back to the isthmus because he had some portion of the Sidi's property on board. The third miracle operated, of course, conclusively on the mind of the obdurate Spaniard, who forthwith underwent circumcision. It is somewhat remarkable, however, that long before the French conquest a tradition was current in the country to the effect that the French would enter by Sidi-Feredje, and leave it by the Isser.‡ The surface of the peninsula is about eighty hectares, and the marabout on the top of the rock is not wanting in elegance. The promontory terminates in the shape of a T, created by a bank of high rocks which is prolonged by islets, and forms on its sides two little shelters sufficiently valuable on this exposed coast. On the platform of the marabout, on the 14th of June, 1830, the lily flag of the Bourbons was hoisted by Jean Sion, captain of the maintop of the *Thetis*, and by François Louis Beunon, a sailor on board the Surveillante, who were the two first Frenchmen of the expedition that landed on the African shore.§

The marabout of Sidi-Feredje has long been a noted landmark for

---

* Baron Baude, ii. p. 56.   † ساحل

‡ Baron Baude, i. 55.   § Idem, p. 54.

sailors, who generally know it by the name of Torre-chica, a term meaning in Spanish 'the square tower;' and the peninsula is avowedly one of the most convenient landing-places on the coast of Algeria ; hence its great importance to the power possessing or invading Algeria, a fact ascertained by the French in 1830. If a fort were built on the rock of the marabout, the landing would be rendered almost impossible, and elsewhere it would have been attended with great risk. The *genius loci* and the fortified lines traced in 1830 would put an establishment in perfect safety from all attacks of the Arabs on the land side.

To the westward of Sidi-Feredje, between the point of Scherschell and Cape Aqnathir, are every where scattered the remains of ancient cities. Scherschell, which we shall describe more minutely in another place, is a little town of potters and corn-merchants formerly included in the province of Oran, and is thought to stand on or near the ruins of Julia Cæsarea, the capital of Cæsarean Mauritania, and the royal town of Juba II. under the protection of Rome.

As we propose to devote a special chapter to the archæology of Algeria, we shall avoid any farther details of antiquities on the present occasion. Not far from the mouth of the Massafran, and below the town of Coleah, is another marabout named Sidi-Fouqua ;* and between Sidi-Feredje, Ras-Accon-Natter, and Algiers is the tomb of Sidi Halliff, another marabout about half-way between the peninsula and Algiers. Half a league W.N.W. of Sidi-Halliff is the Ras-Accon-Natter or Cape Caxines, beyond which and about three miles to the south-east is the harbour of Algiers.†

As the port of Algiers is described in another place, the present observations apply to the bay. Pointe Pescade, one league and a half north-west from Algiers, is the most advanced portion of the chain of Boujareah. Proceeding thither from the capital, you coast along a beach of about 800 metres, shut in between the point of Sidi-Kettani and that of the Salpetrière. A little farther on, two sources flow from the hollow of the rock into the sea; and Moorish women, with their attendant negresses, are reported still to frequent them, as in the days of Henri Quatre, performing various ceremonies savouring of sorcery and fetichism, such as burning incense and myrrh, and cutting off cocks' heads.‡ The road from the capital to Pointe Pescade crosses several ravines, which are dry six months in the year, and is bordered in some parts on one side by the sea and dangerous precipices, while on the left it is flanked by steep slopes. The soil consists of argilo-calcareous earth mixed with stones.§

Nine hundred metres (2952 feet) north-west of the jetty of Khaireddin, the point of Sidi-Kettani projects to the E.N.E. towards the high sea, by a reef of submarine rocks, which ends in the rock Mhatem at 460 metres

---

* Berbrugger's Algeria, 1843, p. 2.   † Blofeld, p. 30.
‡ Baron Baude, ii. 57 and following.   § Idem, i. p. 117.

(1508·80 feet) from the land. The latter islet is only covered by forty centimetres (15·16 inches) of water.*

To the south of the capital† the coast forms a small creek, where it might be supposed that vessels could safely find shelter; but during the north winds there is a very dangerous surf. The European merchant-ships used to be obliged to anchor in the bad creek called *of the palm-trees*, situated towards the middle of the faubourg Bab-azoun, beyond Ras-Tafourah; and they were in constant danger there, as the least wind raised a heavy swell, from which they had no protection.‡ The rock continues to the opening of a deep ravine, which discharges the rains from the neighbouring heights into the sea; beyond this an extensive beach presents itself, which insensibly curves northward to the river Hamiz, forming thus the greatest part of the circuit of the bay. This beach is generally very wide, and when the sea-breeze sets in, the waves break continually over it, even in fine weather; viewed from the hills by the Fort of the Emperor, it presents a wide border of foam. The eastern part of this bay is closed by a steep and precipitous shore, which rises gradually to Cape Matifou. At this extremity there is a very good anchorage upon a bottom of sand and mud, and sheltered from the east winds. Crossing the Hamiz, another considerable stream, you arrive at Temendfuse, corrupted by the Franks into Matifou, a low cape with a table-land in the middle of it, and a small castle built by the Turks to defend the adjacent roads, which once constituted the chief station of their navy.§

Cape Matifou was the station of the Turkish galleys that used to bring a new pasha to the Algerines from Turkey every three years, and his arrival was always notified to the city by a gun-shot.‖ There exist several remains of an ancient city named Rusguniæ at Matifou, which will be noticed in the chapter on Archæology. Cape Matifou forms the eastern limit of the gulf of Algiers. Between the mouth of the Hamiz and the northern slope of the cape there stretches a mile of highland, and this spot would be healthy were it not for the vicinity of the marshes. After the disastrous tempest which scattered the fleet and hopes of Charles V. in 1541, he was forced to march from Algiers to Cape Matifou in order to embark his troops. He embarked from the ruins of Rusguniæ, of which there existed at that time more remains than appear at present. His army marched on the 27th of October from the suburb of

	kilometres.	miles.
Babazoun to the Haratch	9	5·59
The 28th from the Haratch to the Hamiz . .	12	7·45
The 29th from the Hamiz to the ruins of Rusguniæ .	3	1·86
Distance by land from Algiers to Cape Matifou .	24	14·90

* Baude, p. 30.
† Described more minutely in the following chapter.
‡ Berbrugger, p. 27.   § Blofeld, p. 30 and following.   ‖ Berbrugger, p. 27.

The emperor embarked on the first of November on the fleet of Andrew Doria, which weathered the Cape Matifou after unheard-of difficulty.* Further details of this interesting expedition will be found in the chapter on History.

Cape Matifou is a very important strategical position to the power in possession of Algiers; for it is evident that at the spot where Charles V. embarked a discomfited army in a stormy season, others more fortunate might accomplish a successful landing ; and the disposition of the ground would enable an enemy to establish himself strongly thus near to the capital. These reflections led Baron Baude to perceive and suggest the importance of building a fort on the cape.†

Thus it appears that the gulf of Algiers forms a semicircular indenture in the coast, three leagues in diameter and open to the north.‡ Its shores are mostly desert, and the bottom of the bay is bordered by sand-hills, which though not exceeding an elevation of forty metres (130 feet), yet effectually stop the waters from the plain of Mitidja, in such wise that even the Haratch and Hamiz can hardly get through. Hence there results a zone of marshes one league in depth, which at a rough estimate presents a surface of 1200 hectares (3000 acres) to be drained.

Beyond the rivers Regya, Budwowe, Corsoe, Merdass, and Isser, which run not far from each other and descend from the Atlas, is the little port of Djinet, where a quantity of corn is annually shipped for Europe. Djinet is a small creek with a tolerable anchorage before it. The sea-shore, which from Algiers to Temendfuse, and thence to this place, had few rocks and precipices, begins here to be rugged and mountainous ; and among these hills, three leagues further east, is the mouth of the Booberak, which formed the east boundary of·the province and separated it from that of Constantina till recently.§

Before we make a tour into the interior of the province of Algiers, we shall transport the reader, with his kindly permission, to the busy quays and streets of the capital, and make him familiar with its scenery and population.

The distance from Algiers to Port Mahon in Majorca is 64 maritime leagues.||
To Palma 57   ,,
To Iviça 58   ,,

We have already seen that Algiers is built on the northern slope of the Boujareah range, whose highest point is 1312 feet above the level of the sea, and which has a circuit of 90 kilometres (55·92 miles). The sea defends 44 kilometres of this ridge (27·34 miles), and the Haratch

---

* Baron Baude, i. 73.   † Idem, p. 76.
‡ Blofeld represents it 8 to 9 miles wide and 4 deep.
§ Blofeld, p. 30 and following. || Twenty-four hours' voyage.

and Massafran 10 (6·20 miles); thus leaving only 25 (14·53 miles) to be defended by art to protect this whole district.* The name of Algiers comes from the four islands which are situated out at sea in front of the town. These were called in Arabic Ed-Djezair (the islands), contracted into Djair.† The metropolis stands in 36° 49′ 30″ N. lat., and in 3° 28′ E. long. of Greenwich. The present lighthouse is built on the foundations of the fortress erected by Peter of Navarre on the largest of the four islets, whence it was called Peñon, the augmentative of the Spanish word *pena*, rock.

The present metropolis of Algeria must be divided into two parts: the new town, which is entirely French in its character, and is built on the lower part of the slope and along the sea-shore; and the old town, which occupies the higher region, and is crowned by the casbah or castle, the former residence of the Dey.‡ The suburb of Bab-el-ouad, or the water-gate, almost entirely in the hands of the Europeans, stretches along the sea-shore to the north-west, and that of Bab-azoun to the south-east of the town.

The town of Algiers is a mile in length in front of the sea. The streets of Bab-azoun, and Bab-el-ouad to the northward of the former, both run north and south 3083 feet across the city. The Casbah street, old, tortuous, and steep, leads down from the castle, and the old town to the lower town and the port.§ The Place des Victoires is situated at the foot of the Casbah, and the street of Porte Neuve or Bab-Edjedid terminates at one end of the former place, and at the other leads to the gate south of the Casbah. The Place du Gouvernement is a large oblong space planted with orange-trees, and surrounded with houses built in the European style; and all persons going from one end of the town to the other are obliged to pass through it, which makes it the centre of bustle and activity, presenting a motley crowd of Arabs, Moors, Jews, Frenchmen, Spaniards, Maltese, Germans, and Italians.‖ Along two sides of the Place du Gouvernement are ranges of houses in the European style, four stories high, and fronted with arcades and balconies. When visited by Count St. Marie in 1845, some Moorish houses situated to the right, recently burned down, had been replaced by some wooden barracks; the only ancient structure then remaining on that spot being the remains of a

---

* Baron Baude, vol. i. p. 53.   † جزائر *Djezair*. Berbrugger, p. 27.

‡ The new town is called وظا (*Outa*), the plain; and the old town جبل (*Djebel*), the mountain.

§ Count St. Marie's Visit to Algeria, 1846, pp. 4, 5.

‖ The well-known Casbah Street is a long and very steep street, interrupted occasionally by steps on account of the steepness of the acclivity. Its shops were all lighted and open when Count St. Marie passed through it in the evening; and on all sides were to be heard instruments of music, Moorish, French, and Spanish, with a great noise of bawling, singing, &c. He also observed much drinking in the shops. *Visit*, p. 36.

E

tower called the Janina, surmounted by a dial. To the left the Place is closed by a balustrade breast-high, behind which is a battery of eight guns ; and farther down are seen the quay, the port, the vessels lying at anchor, and the high sea.

The street of Bab-azoun has two rows of houses built on the same plan as those on the Place ; and the Bab-el-ouad Street,* as previously remarked, is built exactly like the former, and parallel to the shore. The Marine Street runs to the right of the Place du Gouvernement, and in it are situated the old baths or hammams of the Deys. You descend to the port through the Marine Gate, passing by the balustrade of a spacious terrace adjoining the Admiralty, and after emerging from the arches of the latter edifice you find yourself in the rear of the lighthouse.

The three streets of the Marine, of Bab-el-ouad, and of Bab-azoun abut in the great central Place. The two last form in reality only one, following the slope of the hill from north to south.†

Leaving Algiers by the Gate of France, which was close to the sea during the existence of the old port, you crossed a mole, about 300 paces in length, to a small island (the Peñon) almost parallel with the walls of the city. This island is about 180 paces long and 60 broad, and at that time it was entirely covered to the height of 12 feet with masonry, laid on a foundation of reeds and sand. Upon this stone platform were erected strong fortifications and arsenals, with a lighthouse in the centre. Thus the port appears as an irregular square bounded on three° sides by the city, the mole of Khaireddin, and the islet. On the arrival of the French at Algiers, this port, which had originally been constructed by the labour of 30,000 Christian slaves, under the direction of the celebrated Barbarossa, was in danger of destruction in spite of the immense works, the only occupation of thousands of captives. The foundations were undermined and contained numerous cavities, while the upper parts were decaying and full of fissures ; in short, it would soon have become so ruinous that a violent sea, so frequent and terrible in these offings, would easily have completed its demolition.

The French, however, soon turned their attention to the port, and threw in by the jetty enormous blocks of granite and marble. The experience of a few years, observes Mr. Blofeld, has proved the efficacy of this plan ; but they had still to adopt means to save the mole, which, built upon moving sand, isolated and projecting, and upon which the waves broke with violence, was partly washed away, and required new foundations.‡ The French therefore formed a pile of blocks of marble all round the mole ; this pile, however, sank below the water the following winter, but its overthrow consolidated a base upon which it became more easy to establish other works. These embankments were fortunately disposed

---

* Count St. Marie, p. 27.   † Baron Baude, i. p. 102.
‡ Blofeld, p. 27.

by the sea much better than art would have done. They formed an inclined plane, which blunted the force of the waves and presented a strong foundation on which were erected other works, that not only protected the ancient ones, but added to the extent of the port. The latter was, however, always much exposed to the north winds, and even within it vessels have been destroyed by the swell of the tempests. It is true that the works undertaken since 1836 made an improvement, and the most recent additions and alterations, as will be seen farther on, have rendered the anchorage quite secure. During fine weather vessels anchor within a mile or a mile and a half of the coast, as at that distance there are from sixteen to thirty fathoms water, with a bottom of soft mud ; but it is advisable to use chain-cables. Ships never anchor to the north of the lighthouse, as all that part of the coast is rocky : they might, perhaps, do so opposite the flat shore of Bab-el-ouad, and in front of a valley you meet there ; but there are rocks in the environs, and they could not remain at their moorings during east winds. The old defences of the port, as encountered by Lord Exmouth, and found by the French in 1830, consisted on the Mole and Peñon of, 1st, the lighthouse battery of fifty guns ; and 2dly, another strong battery north-west of the former and east of the port, with seven mortars. Several heavy guns surmounted the gateway that commanded the mole, and 12 batteries of heavy guns were placed at different distances at the waterside, in front of the town. They were left much in the same state for some years after the French conquest ; and Capt. Rozet* remarked during his visit, that the finest work after the Casbah was the united buildings of the mole and marine forts, which were mounted with 237 guns under the Turkish administration, and were the strongest defence of Algiers. Further particulars respecting the topography and history of the port are furnished by M. Berbrugger, who observes that nature had placed before Algiers the elements of a port of middling extent. A chain of reefs starting from the shore, and following a south-east and north-east direction, runs out and joins, at the distance of about 230 metres (754·40 feet) towards the open sea, four islets arranged side by side, from north to south. This reef has a shape similar to that of the letter T ; and it is very likely that at a distant period it afforded very good shelter, but that the effect of the waves on its schistous masses has loosened considerable pieces, and made breaches which were noticed even in the sixteenth century. However this may have been, the present port is the same as that of the Romans, as is proved by the remains and direction of the Roman *via* in several points of the Rue de la Marine. It was also the same under the Arab chiefs ; and as fast as blocks were washed away by the sea, the Turks substituted others. The French at first did the same, but soon found that

* Voyage dans la Régence d'Algérie, par Rozet, 2 vols. 8vo, 1833.

it was an endless because an imperfect process. In 1831, M. Noll, engineer of the hydraulic works at Toulon, was charged with the duty of remedying this, and succeeded as well as he could ; but for want of a foundation, he could not restore the basis of the jetty at the same time that he had restored its body, so that the breaches extended, and it was necessary to have an additional defence of hydraulic lime and gravel to stop them effectually.

The mole, whose direction is almost perpendicular to that of the winds that blow the strongest into the roads, is much more exposed to injury than the jetty. The projection of the pier-head was repaired in 1831, but destroyed in the winter of 1832. Subsequent efforts to repair the jetty and mole did little good, when violent winter-storms in 1833-4 shewed that the system of loose stones piled round the mole might encumber the harbour with dangerous shoals. M. Poirel then suggested the Roman plan of using artificial blocks of hydraulic lime with gravel : this system was employed at the bridge of Caligula at Pozzuoli (Puteoli). A number of plans were now suggested ; but of these, two projects became the favourites, called the great and little projects, or the *Projet Raffineau de l'Isle* and the *Projet Poirel*. Nothing was decided in 1842,* and the matter seems to have remained in abeyance ever since ; but we learn that the energetic government of Louis Napoleon is seriously engaged in making a great harbour at Algiers.

Writing on this subject in 1841, Baron Baude considers three projects for the improvement of the port most deserving of attention ; 1st, that of M. Berard, author of the *Nautical Description of the Coast of Algeria ;* 2d, that of M. Sander Rang, captain of a corvette, and that of M. Poirel, civil engineer and inspector of bridges and highways in Africa.

M. Berard suggested a circular jetty uniting the north end of the batteries of the mole with the land, and leaving a space of about nine hectares (22 acres) between it and the jetty of Khaireddin, which would have to be opened in the middle, and the present port would then answer the purpose of outer port. M. Sander Rang and M. Poirel both propose to make opposite the quarter of Bab-azoun a large port, of which the present one would constitute the bottom. Several serious objections are made to these plans ; and Baron Baude suggests with reason the propriety of making the new port opposite the Bab-el-ouad suburb, north-west of the town, the only side where there is room for the accommodation of an increasing population and commerce. He proposes to run a jetty from the Sidi Kettani point to the Mhatem rock, thence bending south-east towards the lighthouse rock ; another small mole would run out towards it from the Peñon rock ; and between these two would be the entrance of the new harbour, which would contain twenty-four hectares (60 acres),

* Borbrugger, part i.

only eight (20 acres) less than Marseilles, and the shore would offer an admirable site for warehouses.*

With regard to the improvements of the harbour projected by the present government of France, we find they are now in operation and partially completed. It appears, moreover, that although their improvements are by no means finished, the government of the Prince President recently ordered Vice-Admiral Baron de la Susse, commander of the squadron of evolutions, to ascertain from practical experience whether the means of causing a fleet to enter and anchor in the port are satisfactory. From a report of the vice-admiral, the substance of which was published in the *Moniteur*, it seems that five men-of-war, towed by steamers, severally entered the port, and cast anchor at a cable's length from one another, near a place indicated by the naval authorities. A sixth man-of-war also entered, and anchored on the line set apart for steamers. The steamers of the squadron afterwards anchored, as did also those of the local service. All these ships did not encroach on the space reserved for merchant-vessels, and three men-of-war and steam-frigates in addition might also have been placed without inconvenience. According to the observations made by the admiral, the removal of a rock called *Roche sans Nom*, situated about the middle of the port, would allow a fleet of at least twelve men-of-war and as many frigates to anchor, in addition to the mercantile vessels. Orders have been given to have the said rock removed forthwith; and the port, when completed according to the plan definitively adopted in 1848, will be surrounded on the northern side by a breakwater 700 yards long, on the south by one 1200 yards in length, and the entrance will be 350 yards wide. Each side of the entrance is to be defended by a strong battery.†

The old mole, uniting the island to the town, was 600 paces long, and the phare or lighthouse was 35 fathoms in height.

Seeing the importance of the subject, it has appeared desirable to complete our description of the latest improvements accomplished or projected in the port of Algiers, as described by the French official documents, which rightly observe that the maritime constructions are of the first importance in Algeria, by securing the protection and supplies of the colony.

From 1842 to 1846 various alterations were made in accordance with the project of Mr. Bernard, inspector-general of woods and forests (*ponts et chaussées*); but they only admitted a sheet of water containing 56 hectares (140 acres) as the military and commercial harbour of Algiers, without providing any roadstead.

It was only on the 26th August, 1848, that a distinct project was adopted for its serious improvement and enlargement, in consequence of deliberations of the mixed and nautical commissions of Algeria, of the

* Baron Baude, vol. i.
† See the article of the Paris correspondent of the *Times* of Wednesday, Nov. 25th, 1852.

superior administrative council, of the council of admiralty, and of the general council of woods and forests.

The project adopted proposed to make of Algiers a good harbour for the military and commercial navy, and to prepare a roadstead in front of the port. The means devised were as follows:

1. A jetty called the north jetty, length above water 700 metres (2296 feet), including the pier-head; 2. another jetty, called the jetty of enclosure (*d'enceinte*), to measure with its pier-head a length of 1205 metres (3952·40 feet), and which may be named the jetty Bab-azoun, because it takes root at the foot of Fort Bab-azoun; 3. an internal jetty, called Algefna, which will answer both as a landing-place and a store (*parc*) for coals.

The two great jetties will be separated by a passage of 350 metres (1148 feet), and the sheet of water contained between them will embrace about 90 hectares (222·30 acres) of surface.

The roadstead will be protected, 1. by a breakwater presenting a development of 1200 metres (3936 feet); 2. by a south jetty, also 1200 metres long. The space devoted to the roadstead would amount to about 700 hectares (1680 acres).

The north jetty, which it was most essential to build at once, was begun the first. In August 1842 its length was 180 metres (590·40 feet); on the 31st of the following December, 220 (721·6 feet); at the end of 1843, 256 (839·68 feet); at the end of 1844, 367 (1203·76 feet); at the end of 1845, 409 (1341·52 feet); 502 (1646·56 feet) at the end of 1846; 600 (1968 feet) at the end of 1847; 659 (2161·52 feet) at the end of 1848; and 728 (2387·84 feet), including the shelving slope at the end, on the 31st of December, 1849. This length of 728 metres, composed of 530 metres (1738·40 feet), raised 2 metres and 50 centimetres (8·20 feet) above the level of the sea and finished, and of 112 metres (367·36 feet) raised 2 metres and 50 centimetres (8·20 feet) above the sea and unfinished, and of a submarine part of 86 metres (282·08 feet), sheltered a surface of 78 hectares (195 acres). At the end of 1850 this jetty had reached its entire development, and had been carried out to its pier-head with a depth of 25 metres (82 feet). The sheltered surface already embraced above 80 hectares (200 acres). In 1850 they were engaged in finishing the pier-head, on which it was intended to build a fort with a double row of batteries. A powerful battery was built as early as 1848 at the foot of the same jetty.

The head of the jetty of Algefna was built at the same period, having a length of 81 metres (265·68 feet) and a breadth of 32 (72·36 feet), in order to establish a battery. The jetty of Bab-azoun was in process of execution in 1850 ; a pile of timber caulked with oakum, 70 metres in length, having been established in 1848-49.

Up to the year 1846 artificial blocks of hydraulic lime of from 10 to 15 cubic metres (352·87552 and 529·31328 cubic feet, or 13·06946 and 19·60419 cubic yards) were used for the maritime works at Algiers. In

1846 a mixed system was adopted, which produced a remarkable economy in the expense of building. This system consisted in employing rough blocks of stone as a foundation to within 12 metres (39·36 feet) of the surface of the water on the exterior side exposed to the action of the sea, and to 8 metres (26·24 feet) from the surface on the interior, and in building all above this with artificial blocks. The pier-heads, in the whole of their circumference, are considered as an external facing.

The different works that remained to be done in 1850 to complete the new port and roadstead came under the following heads:

1. The completion of the north jetty, the building of its pier-head and of the fort to crown it; 2. the construction of the jetty of Algefna; 3. the building of the quays going from north to south; 4. the construction of the first branch of the jetty of Bab-azoun, giving it all the length necessary to found the platform intended for the establishments of the navy, and to diminish the swell within the port; 5. the establishment of one of the stairs of communication between the quays and the town; 6. the scarping of a rock existing within the harbour, known by the name of the *roche sans nom* (nameless rock); 7. construction of the establishments of the navy; 8. completion of the jetty of Bab-azoun, the construction of its pier-head and of the fort intended to crown it; and 9. second stairs of communication between the port and the town.

The whole expense necessary to complete the port, without including the roads, which are postponed, is estimated at 41,592,000 fr. (1,663,608*l.*) Up to Dec. 31st, 1849, 14,600,000 fr. (584,000*l.*) had been spent. Hence there remains to be spent in additional work the sum of 26,992,000 fr. (1,079,680*l.*)*

A powder-magazine in rear of the lighthouse exploded not long before Count St. Marie visited Algeria in 1845, reducing the surrounding buildings to complete ruin; but the damage has been since repaired. The mole was at that period 2000 paces long and 6 above the sea, wholly constructed of enormous artificial blocks of hydraulic lime and gravel, and the works then in progress were not completed at the time of the count's visit. It formed an inward curve, contracting the mouth of the port.

We trust that the previous remarks will have made the reader familiar with the port of Algiers ancient and modern; and we propose now to notice the chief buildings and the style of architecture observed in the capital. In 1830 the narrow winding streets of the town underwent a rapid change under the management of the conquerors. The greater part of them had no written name, and none of the houses were numbered, which rendered it impossible to make out any general direction, without having a sort of general plan of the distances between the principal objects in the city. It was found necessary to widen those streets, to adapt them to the convenience of their European inhabitants, and to give them that

* Tableau de la Situation, 1850, pp. 314, 315.

straight form so necessary to all who estimate the value of time. The speculators who travelled in the rear of the army lost no time in erecting houses five stories high, which certainly have a very fine effect; several streets with arcades have been built ; and, in short, all has been done to constrain the natural orientalism of Algiers into a Parisian shape. A rich Moor, a man of great experience and good sense, observed to Madame Prus in 1850, shaking his head sadly at the sight of one of these lofty habitations, the numerous apartments of which accommodated a host of lodgers : "They seem little aware that this is a country subject to earth-quakes ; for here they are building away as they might do in France, while at no great distance from hence the ruins of Oran and Blidah are evident proofs of the danger they incur. Let them look at our Moorish houses and observe how low they are built, and with what care they are propped up on beams, and made so as to support each other even on opposite sides of the street. Then let them ask, why have the natives fixed on this mode of construction ? and I will answer them, that in 1717 an earthquake was felt for nine months, which destroyed three-fourths of the town, while the population lay encamped in the fields, and only returned when all symptoms of the calamity were over. In 1825 another convulsion threw down the walls of Oran and Blidah, and crushed many of the inhabitants under the ruins. Algiers at the same time felt fifty-three shocks in a fortnight. Another took place in 1839 ; and even worse consequences might have ensued but for the manner of building adopted since 1717."

Before this precaution was used, no other remedy against the disaster was known but that of strangling the reigning Dey. Though European fraternity prefers to strangle saints and heroes rather than despots, it would at least be wise in the French if they were to conform to the custom resulting from this dear-bought experience, and sacrifice elegance to security.

In visiting the different quarters of the city and becoming familiar with its architecture, we shall accompany some of many Europeans who have described its curiosities. Count St. Marie informs us* that the Fisher-man's Quay is at the foot of the Government Terrace, which is ascended by a few steps and a sloping path. All the men whom you meet there selling fish are Maltese ; the best fish being the tunny : oysters are rare, and different in form and colour from those of Europe. Leaving the Place du Gouvernement, the party whom we accompany reached in about half an hour the boundary of the city at the Bab-azoun gate, which consisted then of double arches connected by a sort of bridge crossing a ditch, which runs along the foot of the city wall. The principal gates of old Algiers were the following : 1. the new one, Babed-Djedid, on the top of the hill near the Casbah; 2. the Gate of Bab-azoun, through which you pass into the Mitidja plain; 3. the Gate of Bab-el-ouad, to the west of the town ; 4. the Marine Gate, leading to the arsenal and the mole ; and

* p. 31.

5. that of Fishermen. On the right of the Gate of Bab-azoun, and within the city, stands a small marabout, the grated door of which is always open. This building is said to be the burial-place of the *Emperor** Barbarossa, as St. Marie curiously styles him, evidently meaning the pirate Khaireddin : it is held in great veneration by the Arabs.†

We shall next accompany our friends into the interior of a Moorish house in the Bab-azoun Street. After ascending a few steps, they entered a large court with flags of white marble, having in the centre a basin of water with orange-trees about it. Along four sides of this court ran two galleries, one above the other, fronted with beautiful carved wood, and supported on marble columns. One side of the house in question contained the city Museum, which possesses a collection of animals, minerals, Roman and Carthaginian tumular stones, and old arms. Within the same building you find moreover a library, also in other parts a college called royal. All the houses of the Moors in Algiers are like the one now described. They are massive square buildings, and have no windows towards the street, the entrance-doors being low and small. The ceiling consists of carved wood gilt, and the walls are pierced on the inside with small dormer-windows. The walls of the apartments are hung with flags and draperies, and faced with Dutch tiles or varnished bricks with passages of the Koran inscribed on them, and gilt or coloured ornaments. On the floors are spread in the better class of houses costly carpets and cushions of cloth-of-gold. The ground-floor is appropriated to the slaves, and a narrow winding staircase leads up thence to the first-floor, which is occupied by the family ; the flat terrace on the roof being used as a promenade. The architecture of the Moorish country houses is similar to that of their town residences, save that they are surrounded with walls two feet high, and almost impenetrable plantations of thorny figs and aloes.

Before we proceed any farther on our fatiguing round of sight-seeing, we will seat the reader in a fiacre, and drive to the most prominent objects of curiosity. These fiacres resemble a basket made of wood, and hung round with curtains of various colours. The drivers are frequently Spaniards, with a small Spanish hat adorned with streamers of velvet. Proceeding to the old town, we find the narrow streets almost roofed over by projecting houses, the fronts of which nearly touch each other from the first story to the terrace on the top. The streets in this part of Algiers are paved with round uneven stones ; and at this quarter is the Gate of Victory, on one side of which is a fountain of white marble, constructed among the ruins of an ancient Roman aqueduct.‡ Algiers is built in the form of an amphitheatre, and is commanded by the Casbah ; but the moats and ditches

---

* Count St. Marie, always more remarkable for the facility of his style than for the solidity of his facts, is a good specimen of the literary discrimination of authors and readers in this veracious age. Examples : Baba-Aroudj is converted into Barbarossa, the amiable German oppressor of Milan ; and the Vandal invasion is placed in the seventh century.

† St. Marie, pp. 4-7.   ‡ Ibid. p. 16.

which run alongside the walls of the city on the right and left used not to extend to the walls and bastions of this ancient abode of the deys. The Casbah* is hardly recognisable even by the Arabs, from the changes that have been made in it by the French, the little kiosk where Deval the French consul was insulted by the blow of a fan of the dey (1827) remaining alone unaltered : the walls of this pavilion are lined with porcelain. "From the courtyard," continues Count St. Marie, " we descended into some vast caverns divided into chambers, where the French found numerous treasures amassed in 1830 ; but previous to that date their approach was rendered impossible by a number of tigers† and hyenas being chained near

GARDENS OF THE DEY AT THE CASBAH.

to guard them. All other parts of the place are entirely changed, and— *proh pudor !*—the women's apartments and the harem are converted into quarters of artillery ; almost an equal sacrilege to that of converting the marabout of Sidi-Djemyah into a station-house for gendarmes. In a beautiful little kiosk attached to the Casbah, commanding a magnificent view of the sea, the city, and the country, there is now an ambulance or mili-

---

* قَصَبَة. The word Casbah means literally 'reed.' *Cours d'Arabe vulgaire*, par A. Gorguos, vol. i. p. 189.

† This must be an error of the count, as there are no tigers, but only animals of the leopard tribe in Algeria.

tary hospital. Near this spot are fountains of fresh clear water, and marble reservoirs in which the soldiers now wash their linen ; and a small mosque at a little distance has been converted into a Catholic chapel, surmounted with a cross. The French, on taking possession, guaranteed to the Arabs the free enjoyment of their religion; but they have turned their mosques into Catholic chapels. The Protestants have purchased ground for chapels, and the Jews have converted certain houses into synagogues."* The Casbah commands the whole town, and the hill on which it stands is 500 feet above the level of the sea. Gloomy battlements surround the castle, which is capable of accommodating two battalions, but is itself commanded by the Fort de l'Empereur on the road to Douera in the Sahel, of which more anon.

" Algiers," observes our friend St. Marie, " is the only town in the regency which, by the erection of new buildings and the accumulation of French inhabitants, presents the aspect of a rising colony. All the other towns which surround Algiers preserve for the most part their primitive aspect, with the exception of some large buildings erected here and there by the French for barracks and hospitals." Descending once more to the lower town, we pass from the middle ages to our high-pressure civilisation, and fancy ourselves in the handsome streets of a European capital. Those of Bab-azoun and of the Marine are spacious and elegant, and contain some good shops. The bazaars are constructed in the Moorish style, and in general are most curious.† That in the Rue du Divan is principally occupied by Moors employed in various embroideries on leather and silk, for which the capital is famous, such, for instance, as ladies' slippers, purses, portfolios, &c. Farther on are venders of essence of roses, jasmine, and other perfumes ; and in the shops are displayed *chachias*, or leathern caps, such as are made at Tunis, silk scarfs or fotas, and many articles of the same description. The della or auctioneer walks about laden with burnouses, "djaba dolis," or men's vests, rhlilahs or women's tunics, and frimlahs, a sort of spencer worn by ladies. His fingers glitter with diamonds, and his hands are hardly able to grasp all the insaias (anklets), rdites, (bracelets), sarmas (ornaments worn by married women), and other articles of value, which he is employed to dispose of for the benefit of Moorish ladies pressed for want of money.‡ Some immense works, observes Mr. Blofeld, have been made in the Place du Gouvernement, and in the streets *de la Marine*, of Bab-azoun, and of Bab-el-ouad ; these have a handsome appearance, with their long galleries, their shops, and the crowds which animate them. In the street of Bab-el-ouad the passengers are more numerous than those in the Strand in London.§ In these places, excepting in some parts of the Rue Bab-el-ouad, there are no longer any

---

* St. Marie, p. 16.   † Madame Prus, 1850, p. 216.   ‡ Idem, p. 216.
§ Blofeld, p. 13.

Moorish houses ; all is changed ; and were it not for the throng of Turks, Moors, Arabs, Negroes, &c., the stranger might fancy himself in one of the principal French cities.

While on the subject of the shops and bazaars, it is well to remark that the shops of Algiers contain now the luxuries, comforts, and fashions of Paris, bronzes, porcelain, glass, rich shawls, embroideries, woollen stuffs, silks, cottons, &c.

On leaving his hotel in the Place du Gouvernement, Count St. Marie passed through the Janina arch and saw the governor's palace, an old Moorish house faced with marble and adorned with marble columns. In front of it is the bishopric, a miserable place as to its exterior, which is, however, better inside the walls. M. Dupuch, who was bishop in 1845, had been previously a counsellor at Paris, subsequently became a Carthusian monk, and ultimately a prelate. Soon after they reached an Arab bazaar, consisting of a spacious gallery, newly built and of curious construction, containing ranges of arches, each forming a separate shop for the sale of various merchandise. In one of them the count saw, as at Stamboul, attar of roses, fragrant pastilles, silk fillets of various hues ornamented with gold and silver, bracelets of plaited silk, intermingled with coral beads, hose, red trousers, girdles or scarfs of gold, and little pots of colours—blue for the eyebrows, red for the cheeks, and yellow for the nails.

The barbers in these bazaars are mostly Koulouglis, or sons of Turks by Moorish women. In the centre of the bazaar is a little rotunda for sales by auction.* The bazaar of the Fig-tree, a small open space, contains the shops of the richest tradesmen.

The principal streets of the capital are twenty feet wide, most of the others being just wide enough to admit of three persons walking abreast ;† and though the lower town is quite European in its character, the upper or Moorish quarter resembles most other cities of the East, containing narrow winding streets, obscured by projecting stories and overhanging roofs.‡

"The lower part of the town which surrounds the port," observes M. Lamping, "has already acquired a completely European character. The streets of Bab-azoun and of the Marine are as handsome and as elegant as the boulevards of Paris ; but the upper town retains its Arab appearance, and is almost exclusively inhabited by Moors and Jews. The streets are there so narrow, that two horses cannot pass without difficulty. The Arabs have no notion of carriages." A motley crowd fills most of the great thoroughfares, consisting of various races. Next door to an elegant French milliner, an Arab barber was shaving the head of a Mussulman ; and an Italian restaurateur, who extolled his maccaroni to every passer-by, was

---

* St. Marie, p. 43.   † Blofeld, p. 3.
‡ The Foreign Legion : 1st Part of Lady Duff Gordon's French in Algiers, p. 16.

the neighbour of a Moorish slipper-maker.   Every thing, moreover, in the
capital wore a martial aspect.*

In the streets of old Algiers the windowless houses scarcely leave an
interval of two metres (6·56 feet) between them, and the salient eaves
overhead belonging to the upper stories hardly suffer the passenger to see
the sky.   The narrowness and obscurity of these lanes at first shock the
European, but the coolness resulting from the same cause speedily recon-
ciles him to these drawbacks.   The only things wanting in the Moorish

STREET IN ALGIERS.

houses are exterior openings to ventilate them.   They are in other respects more picturesque and better adapted to the climate than our architecture.   The inside of the Bourse of Paris, reduced in scale and with the
African sky overhead, gives a good idea of the interior of a Moorish house.†

"I toiled through the narrow streets," says M. Lamping, "up to the
Casbah, the former residence of the Dey, the road to which is so steep

* The Foreign Legion, p. 15.   † Baron Baude, vol. i. pp. 50-52.

that steps had to be cut to form it. As I did not know the shortest path, it was at least two hours before I reached the top."* Algiers itself is built in the shape of an amphitheatre on the declivity of the Sahel hills, and when seen from a distance looks like a huge white pyramid, for the town forms a triangle, the highest point of which is crowned by the Casbah.†

Before we pass through the former gates of the capital to visit the suburbs, we shall present the reader with the following sketch of the appearance of Algiers on landing, from the pen of Marshal de Castellane, the latest authority on the subject.

" On approaching Algiers from the sea, it presents the appearance of a town tranquilly and lazily reposing along the slope of a hill, surrounded by a fresh and verdant country. On penetrating into its precincts, however, European bustle and activity belie the indolent exterior of the city. The fact is, that Mussulman Algiers is at an end, and is making room daily for its Gallican successor. On first landing, the visitor is greatly struck with the strange and motley crowd in its streets, where every one seems to run rather than walk. A novel display of various costumes attracts the eye on all sides. One moment you meet some Biscris moving along with a rapid and cadenced step, carrying a heavy load on a long pole; presently an Arab appears in his bournous, then a Turk still sporting the graceful turban, a Jew with his sombre attire and cautious look, the oil-carrier with his goat-skin pitchers, and to crown the tumult thousands of asses and their negro drivers, curricles with two or three horses, baggage-mules‡ proceeding in long files with provisions for the military storehouses, horsemen galloping full tilt contrary to the police regulations, colonists with white hats and broad brims, or glittering officers lording it over every one in conscious self-importance. In short, you have the confusion and agitation of an ant-hill; every where energy, hope, and its offspring, steady and active labour.

" The lower town, by the port, is the seat and scene of this activity, and presents a great contrast to the silence and repose of the higher part of the town, which is the refuge and head-quarters of Mussulman gravity, and offers a labyrinth of narrow and winding lanes where two men can hardly walk abreast. Occasionally a white phantom glides past you as you thread your way through its narrow streets, a door is seen to open

---

* The Foreign Legion, p. 38.   † Ibid. p. 15.

‡ The baggage-mules are always styled ministers in Algeria ; and if you ask the soldiers why, they will answer you, because these beasts are charged with the affairs of the state, or because they have the telegraph at command, pointing to their long movable ears. It happened once upon a time that a real minister, M. de Salvandy, visiting the province of Constantina, was escorted from Philippeville to Constantina by soldiers of the waggon-train. On climbing a hill his ears were suddenly offended by hearing the word 'Minister!' shouted out on all hands, amidst a shower of imprecations and blows. Astonished, he asked what it all meant ; and when informed, he laughed as heartily as any one at the joke.— *Castellane's Souvenirs*, p. 11.

mysteriously, and the apparition vanishes. It was a Moorish lady. The old despotic spirit of the Deys seems still to brood over this part of Algiers, though the French *tricolore* has long waved over the Casbah."*

Having completed our description of Algiers within the old walls, we shall transport the reader to the old faubourg of Bab-el-ouad, to the west of the metropolis, passing through the gate of that name. The first object that here claims our attention is the Fort of Twenty-four Hours, called by the natives Bordj Sitti-Takelits (the fort of Madam Takelits), because it was built near a marabout of a holy woman of that name. It is situated at a few fathoms from the sea, behind the Tophana† or battery of Sidi-

TERRACES OF ALGIERS.

Kettani, another saint honoured here. It is an oblong square with irregular sides, without lower embrasures, commanded to the west by heights, which could not hold out after the town was taken. The French have laboured recently to scarp it by cutting down the limestone rock on which it is built level with the esplanade of Bab-el-ouad. This fort was built in the 18th century, and is now occupied by *disciplinaires*.

* Souvenirs de la Vie militaire en Afrique, par le Comte P. de Castellane (now a Marshal), p. 1. 1852.

† Tophana is a Turkish compound word meaning 'gun-wharf:' *top*, gun; *hana*, wharf.

The great changes that have been recently made in Algiers have enclosed the Forts Neuf and that of Twenty-four Hours within the present walls, which also embrace the old faubourg of Bab-el-ouad. The present gate of Bab-el-ouad is opposite the point of Sidi-Kettani ; and the new faubourg of Bab-el-ouad stands opposite the *anse* (or cove) *de la Salpetrière,* and underneath the hospital of that name.

Between this Fort of Twenty-four Hours and the road of the Jardin du Dey (Dey's Garden) you see an isolated structure on a chain of rocks, with some luxurious trees rising above its walls. The cupola of a marabout announces it to be a saint's tomb consecrated to Sidi-Djemyah, but he has been unceremoniously thrust out by a post of gendarmes. The garden of the convicts is on the other side of the road, in which you see the elegant agave-flower, while Mount Boujareah forms a background to the prospect.*

"Nearly facing the Fort of Twenty-four Hours," says Count St. Marie,† "we entered a garden called the Jardin Marengo. It is a pretty place, belonging to Colonel Marengo, formerly the commander of the citadel of Algiers. The garden has been cultivated by condemned soldiers, to whom it must be a severe punishment, owing to the great heat. Scarcely a day elapses without some of them experiencing *coups de soleil* and other accidents, occasioned by exposure to the sun, whose ardent rays destroy the freshness of vegetation; and though much care is bestowed on the cultivation, it is not so beautiful as it would be in a more favourable locality. In this garden is situated an old marabout, the walls of which were faced externally with white, blue, and green porcelain. This little temple has been surrounded by flags, and has a very pretty effect. The real name of Colonel Marengo was Capon ; and his father, who distinguished himself at the battle of that name, received in jest from Buonaparte the appellation, which is still retained by his son, though it is said that he has not much military talent."

The Fort Neuf (Bordj-el-zoubia) is situated at the northern angle of Algiers, and was so called by the Europeans because it was a recent erection, having been hardly completed in 1806. It was one of the first places that occupied the attention of the French after the conquest, and additional works were constructed to put it in a state of defence. They began to surround it with a moat, and to make *revêtements* and masonry escarps; which, with other improvements, enabled 1200 men to find accommodation in its vaults. It was, however, afterwards given up to the military convicts under Lieut.-Col. Marengo.‡

Baron Baude observes that the convicts have been usefully employed on many works in the port, and that they have formed a good garden at the barracks, besides an excellent *champ de manœuvres.* The system of convict-labour has worked well, and they have improved morally and ma-

---

* Berbrugger, part i. p. 39.   † p. 27.   ‡ Berbrugger, p. 39.

terially under the treatment they have experienced.* M. Berbrugger, who examined into the condition of the military convicts, bears witness to the cheering results presented by the instruction and discipline to which they have been subjected. The men have been taught general elementary knowledge, and, what is still more important, self-respect and esteem for their superiors. There is some good element in most criminals, even the most obdurate; and by touching the right chord, they can generally be reclaimed. Vice is much more circumstantial than inherent in man. After the expiration of their term of servitude, the men have returned to their regiments, where they have almost universally behaved well.

The old gate of Bab-el-ouad† opened to the north of Algiers, on a plain where there is more room for building than on any other side of the town. Nevertheless the pirates preferred the hill, thinking that it would place them in a safer position; but since 1830 there has been a determination of population towards this plain, and most of the public establishments belonging to the colonial government have been erected on that side. The space contained between the old gate of the town and the sea is filled by the Fort Neuf, which, as previously stated, is a prison-barrack of military convicts.‡

The Fort of Twenty-four Hours was built in the oldest part of the vast Mussulman cemetery stretching from Bab-azoun to Bab-el-ouad, along and outside the walls. A new zone of tombs began beyond this circle, forming that of the Christians; then beyond that, and in the direction of the ravines of the Boujareah, lay that of the Jews. The tombs of several of the deys, such as those of Mustapha, Moussa, &c. were situated in that part of the Mussulman cemetery lying between the Fort Neuf and that of Twenty-four Hours. They were shaped like marabouts, of a square form, with a cupola at top; but were destroyed by the French in 1830. The ground of these cemeteries has been greatly encroached upon by French settlers; and it is anticipated that all traces of them will gradually disappear as the European town stretches out on the road to Pointe Peseade, beyond the old Bab-el-Ouad gate.§

Proceeding to the other extremity of Algiers, we pass through the old gate of Bab-azoun into the old suburb that bears the same name. The faubourg of Bab-azoun only exists in the memory of the first-comers to Algiers, most of the buildings having been knocked down to enlarge and open up the approaches to the town. Very little remains of this picturesque quarter, except Le Quartier des Spahis and some little Moorish shops, where the Arabs come to buy rope, and straw mats, iron, pottery, &c. But the population frequenting this district has much changed of late; and on coming from the steamboat you see there in a few minutes specimens of all the Algerian races. The Rue Bab-azoun passes through

---

* Baron Baude, vol. ii. p. 57. † Ibid.
‡ Ibid. vol. i. § Berbrugger,, p. 1.

F

the gate of that name to the country.   It was at that gate (the old one
now destroyed), that during the disastrous expedition of Charles V., Ponce
de Balagner, dit de Savignac, knight of the Temple, plunged his dagger
into the gate, and fell a victim to his daring gallantry.   The walls used to
be lined with heads of the innocent and guilty; and on to the iron hooks
that projected from their sides, criminals, imaginary or real, used to be
precipitated from above, and remained suspended in agony till death put

STREET IN ALGIERS.

an end to their sufferings.   A square planted with trees, and having in its
centre a basin, is the place of execution; and at the foot of an escarp on
the right is a row of curtained carriages (*voitures tapisseés*), to take the
travellers about the environs.   A little farther on, you probably meet with
a native band, whose music being rather more remarkable for noise than
melody, speedily puts to flight all who have any pretensions to an ear.
Sometimes you may also meet in this locality serpent-charmers from
Morocco, who display their mesmeric influence over the tribe of creeping

things. Above this spot is situated the wood and charcoal market, containing tattered tents, camels, and dirty Bedouins proudly wrapped in rags. A little to the right are the barracks of the spahis;* while to the left is a fine high building, which is the Caserne du Train des Equipages, or the barracks of the waggon-train.† Most of the old structures in this vicinity are demolished or condemned, and handsome streets and public buildings will shortly meet the eye of the visitor on passing through the new gate of Bab-azoun. Between 1841 and 1845 the new faubourg of Bab-azoun was created, and considerable expense incurred in levelling and paving. The whole district is now within the new walls.‡

* Native troops. See the chapter on the French army in Algeria.
† Berbrugger, p. 6.   ‡ Tableau de la Situation.

PART OF ALGIERS AND MOSQUE OF ABD-ER-RAHMAN-EL-TSALEBI.

CHAPTER V.

Statistics of Algiers.

RELIGIOUS EDIFICES—BATHS—FOUNTAINS—DRAINS—NEW CIVIL EDIFICES—
HISTORICAL STATISTICS OF ALGIERS—THE POETRY OF EASTERN LIFE—AN-
TAGONISM OF THE SOCIAL STATES OF EUROPE AND AFRICA—NEW MILITARY
EDIFICES AND DEFENCES.

IN 1833 Algiers contained 120 mosques and marabouts, fourteen synagogues, and one Roman Catholic chapel. Three of the mosques had
in 1843 been turned into Catholic places of worship, and one of them is
now the French cathedral church, and has some very beautiful arabesques
on the walls and ceiling, and the doors have flowers carved upon them
in a style not excelled by Grindling Gibbons.

As regards the native sacred edifices, they are commonly divided into
three classes: 1st, the djamas, which are the principal mosques; 2d, the
mesjids, called in Egypt mesguid, whence came the Spanish term 'mez-
quita,' and our *mosque*. The khotbah or public prayer is offered up in the

djamas on Friday, the Mussulman Sunday. The third class consists of marabouts, which are the tombs and sanctuaries of saints; of this class more anon. The Algerian mesjids are somewhat like our Gothic churches in their interiors, but instead of seats and benches, they strew the floor with mats, upon which they perform the several stations, sittings, and prostrations that are enjoined in the ceremonies of their religion, and which are so accurately represented in Lane's *Modern Egypt*. Near the middle of the mesjid, or more especially of the djama (the great), is a huge pulpit, balustraded all round with a few large steps leading to it. In this the mufti, or one of the imams,* places himself every Friday, and explains some parts of the Koran, and exhorts the people to piety and good works. The wall of the mosque on the side towards Mecca is called kibla, in which is a niche representing the presence and the invisibility of the Deity. A minaret rises commonly at the opposite end of the mosque, having a flag-staff at the top. The mesjids, sanctuaries of marabouts, the muftis, imams, and other dignitaries attached to them, are supported by revenues of houses and lands bequeathed by will, or appropriated by the public for this purpose.†

A good specimen of a mosque of the second class is presented by that of Sidi-abd-er-Rahman-el-Tsâlebi, situated between the marabout of Sidi-Sadi and the west rampart. It is a charming edifice, held in high vene-ration by the Mussulman population on account of the saint buried there. The flags of the Turkish troops used to be kept in it, and the following inscription in Arabic was over the door : "In the name of the gracious and merciful God : may God shed his mercies on our lord Mahomet! The building has been finished, with the divine help, by the hand of our emir, the very powerful and generous El-Hadj Ahmed-ben-el-Hadj Massli. May God direct him towards grace by the merits of Zerroug and those of the sincere Abou-Beker. Its date, O thou who inquirest, is in the words qad djaaltouhou min sabiquin (I have formerly established it)." This implies 1108 of the Hegira, as letters in Arabic have a numerical value. A new inscription shews the edifice to have been built in 873; and it ap-pears that the marabout Sidi-abd-er-Rahman was born at Tsaallah in the province of Constantina, as his name implies.

We add a list of the duties and obligations attached to this establish-ment. 1st, the distribution of alms and aid ; 2d, the repair of fixtures ; 3d, daily expense in giving food to the natives who resort to it ; 4th, re-ligious expenses of all kinds. Offerings daily placed on the tomb of the marabout, and the rent of certain endowments, make up an income of 8000 boudjous (1 boudjou=1 fr. 80 cents=1s. 6d.). The expenses amount to 6,500, leaving an excess of 1,500 boudjous. The officers of the establish-

---

* Different ecclesiastical officers of the Moslem hierarchy, of whom more anon, in the chapter on Religion and the Law-tribunals, Part II.

† Blofeld, p. 136.

ment consist of an oukil or administrator, three imams, a chaouch or beadle, three heuzzabins or readers; and one woman to sweep it.*

In many of the towns of Algeria, especially the capital, since the French conquest, the number of mosques being found excessive, several of them have been converted into hospitals, warehouses, and even Catholic churches. Thus at Algiers two mosques have been turned into the cathedral and the church of Notre Dame des Victoires. A sufficient number of mosques, however, have been preserved and repaired to meet the wants of the Mussulman population.

The French official documents† divide all the mosques in Algeria into five classes, save the great mosque of the capital: 1st, the mosques with great minarets; 2d, those with a pulpit for the khosbah; 3d, the mosques with less important pulpits; 4th, the mosques without pulpits; 5th, the small chapels. Of the 1st class, Algiers has 3; of the 2d class none; of the 3d class none; of the 4th class 4; and of the 5th class 12.

Thus Algiers, including the great mosque, has twenty Mussulman temples, whose ecclesiastics will be enumerated in another place.

The Jews have twenty-five synagogues at Algiers; the Catholics have two churches and one chapel; and the Protestants one place of worship at Algiers, and one at Douera (a neighbouring colony).

Next to fresh air, good water is the first necessary and greatest luxury of life. Without plunging into the excesses of hydropathy or teetotalism, it may be readily admitted that ἄριστον μὲν ὕδωρ, and that *aqua fresca* is equally valuable with the Promethean fire, especially in the realms of the sun, where, if any where, cleanliness is next to godliness. Drains, baths, and aqueducts were the first care of the Romans, and their vestiges may be traced throughout Algeria.

The capital used to be well supplied with the crystal liquid from the Boujareah under the earlier deys; but Turkish improvidence [neglecting the plantations has caused many springs to dry up. The French seem at length aroused to a sense of the importance of a good supply, and active measures are taking to secure it. In many instances the old Roman, and sometimes the more recent Turkish, conduits and channels have been repaired and employed; 1400 years not having sufficed to ruin the cyclopean structures of the masters of the world.

Between the years 1840 and 1847, the French government has completed the erection of nineteen fountains in the capital of the colony and its precincts. These works have cost the sum of 141,446f. 22 cents (5657*l.* 17*s.* 8*d.*), and have been erected in the following localities:

Rue de Chartres, at the angle of Rue Porte Neuve . . .	1
Rue de Chartres, at the angle of Rue Bruce	1
On the Place de Chartres	1

---

* Berbrugger, part i. p. 34.　　　　　† Tableau des Etablissements, &c. 1850, p. 362.

On the Casbah hill 1
Rue du Palmier 1
At the corner of Rues Reynard and Regard 1
Rue de la Revolution 1
Rue de l'Intendance 1
Rue de Nemours 1
Rue Bruce 1
A l'Agha 1
Rue de la Giraffe 1
Rue de Chartres, corner of Rue du Chêne 1
Bottom of Rue de la Casbah 1
Rue d'Annibal 1
Rue de Navarin 1
Corner of Rues du Chat and du Locqdor 1
Rue de Staouëli 1
Mustapha barracks of waggon-train 1
                                                    ——
                                                    19

Draining is another subject to which the French government has devoted a good deal of attention; and it is somewhat mortifying to reflect, that those great nations of the West who boast of their enlightened polity and humanising civilisation, should be still distanced in undertakings of public spirit by the old-fashioned men of the Augustan age. So evident is it that our progress has been very onesided and revolutionary in its character. Stern necessity, the cholera, and the footprints of Rome have at length roused the French to purge their cities and span the colony with the mileage of high roads.

The following are the larger-sized drains which the French call *de grande* section :

	Length.	Expense.
Drains in the Rue de Chartres	315 metres	37,894 fr. 26 cents.
Drains in the Rues Doria, Des Trois Couleurs, Mahon, Duquesne, de la Marine, du Marteau, &c.	840 ,,	. . 45,399 ,, 04 ,,

(Middling size.)

Drains in 45 streets (1842-5)	1279)	
Repairs in 64 streets	1277) ,,	. . 142,861 ,, 10 ,,
New drains (1846-9)	2800 ,,	. . 121,000 ,, 0 ,,
	6647 metres	. . 387,154 fr. 40 cents.*

A plan has been started for building a great drain, destined to carry off all the filth of the town beyond the port, so as to avoid the stagnation and effluvia that result, as at Marseilles, from imperfect sewerage. It is to be hoped that this project, which smacks somewhat of colonial grandeur, may receive the sanction of the government.

While on the subject of sanitary measures, the following regulations of the French authorities to preserve the cleanliness of the town are deserving of notice.

* Total of drains, 21,802·16 feet ; expense, 15,486*l.* 3*s.* 4*d.*

By an arrêté, or decree of government, of the 26th July, 1843, every resident is obliged to have swept that part of the way contiguous to his house or other premises, and to clear away the mud opposite his dwelling as far as the middle of the street. All rubbish is to be heaped up and carried away by the scavengers. All glass, &c. to be thrown aside separately, where it cannot inflict wounds. No fires are to be lighted in the streets, nor is it allowed to throw any thing out of the windows. From the 1st of June to the 1st of October, all the inhabitants are required to water the streets twice a day ; for which purpose the water is to be obtained from the public cisterns only.

The next subject that claims our attention is the historical statistics of the public streets in the capital since the French occupation. The Rue de Chartres was paved with lava in 1841-42 ; and in the last-named year the names of the streets and squares were put up. From 1840-1842 the squares were planted ; and between 1842-44, the Place Royale, Place Mahon, and Place de Chartres were paved ; besides which the streets of Joinville, Tanger, and des Mulets were opened. In 1844 the square of Isly and the streets of Mogador, Isly, Joinville, &c. were paved ; and from 1845-46, foot-pavements and various plantations were made.

We now pass to the new civil and military edifices, the former of which we shall classify under the following heads : 1. public justice ; 2. education ; 3. divine worship ; 4. general administration ; 5. finance ; 6. municipal ; 7. hospitals ; 8. archæological and literary.

Justice. The court of appeal and the tribunal have been established in two vast houses of Moorish construction. The central prison of Algiers, built on a half-cellular system, is not quite finished, but it is already opened for the reception of prisoners. It is, however, only a departmental prison, and not a house of detention ; and they still send to France prisoners condemned to a longer space of confinement than one year. The expense is estimated at 744,000f. (30,160*l.*)

Education. A lyceum was founded and built at Algiers between 1847 and 1849, costing 51,500f. (2060*l.*); and a (mutual) school between 1840-49, costing 10,330f. 70 cents. (413*l.* 4*s.* 2*d.*)

Worship. The cathedral at the metropolis is a vast building, begun in 1840, and though not yet completed, is already partially consecrated and devoted to divine worship. A good deal remains to be done before it will be finished. The expense, up to December 1849, amounted to 730,215f. (29,208*l.* 12*s.* 6*d.*) A handsome Moorish house, suitably repaired, has been converted into the bishop's palace. A great seminary has also been established in the old camp of Koubah : Notre Dame des Victoires and the chapel of Bab-azoun were formerly mosques, and have been previously noticed.

Administration. The hotel of the Prefecture for the general direction of civil affairs was begun in 1845, and finished in 1849 ; expense

A STREET IN ALGIERS.

200,000f. 50 cents. (8000*l*. 0*s*. 5*d*.) The central police-station was finished in 1847, costing 4729f. 40 cents. (189*l*. 3*s*. 8*d*.)

Municipal service. When municipalities were established in Algeria, in 1847, a new mairy was placed in the new building adjacent to the old *direction de l'intérieur* (colonial office), now converted into the general secretaryship of the government.

Hospitals. Mediterranean usage entailed on Algiers the necessity of building a lazaretto from 1841-42, at an expense of 461,922f. 92 cents. (18,476*l*. 19*s*. 10*d*.)

The hôpital civil has been established in the old barracks of Janissaries at Bab-azoun, the repair of which cost 92,999f. (3720*l*.) The orphan asylum at Mustapha has cost 42,415f. (1696*l*. 12*s*. 6*d*.), and the house of the sisters of mercy (1848) 30,634f. 66 cents. (1225*l*. 7*s*. 6*d*.)

Museum and Libraries. The library and museum were removed in 1845-6 from the college of Algiers to a house in the Rue des Lotophages.*

This house, the first story of which is appropriated to the museum,* and the second to the library, was built about sixty years ago by El-Hadj Omar, grandson of Hassan Pasha, on some rocks by the sea-shore, bathed by the waves on two sides. It is a splendid Moorish dwelling, and one of the most complete and curious models of that native architecture which has almost entirely disappeared at Algiers. In this respect the house itself may be regarded as a museum. The library, placed, as we have said, on the first story, comprises four halls (*salles*) opening on a pretty gallery paved with squares of porcelain.

We shall pause awhile to dwell on the literary monuments of a patri-archal race and of a waning religion contained in this edifice. The first room contains works of theology and of philosophy, maps, and stamps. The second, archives, books of natural history, of astronomy, of mathematics, of physical science, of chemistry, of architecture, of medical science, of agriculture, of history, and of what relates to war, marine, and belles-let-tres. The third compartment contains two reading-rooms ; one for Eu-ropeans, the other for natives. In connection with the last is a large glass cabinet, in which the Arabic MSS. are deposited.

In the European reading-room you find the works relating to Algeria, and in general those that are in most demand.

Printed books. This collection is already of considerable importance in

---

* The antiquities and specimens contained in the Museum will be noticed in the chapter on Archæology. The only curiosity we shall here record is a discovery by the celebrated naturalist, Bory de St. Vincent, who, gaping for lions in this virgin field of science, was delighted one fine morning to see a singular specimen of natural history brought in by a *sous-officier*. He rewarded the man handsomely, and, enchanted with the novelty, he wrote a learned description of his wonderful variety to the Jardin des Plantes, describing its singular proboscis, resembling an elephant's trunk, and giving it the name of *rat trompe*. Judge of his dismay, after the lapse of a few days, to find that the proboscis consisted of another rat's tail artfully put through the nose of the speci-men ! *St. Marie.*

supplying intellectual food to the metropolis, though it is not large enough to meet the wants of its studious inhabitants. The number of books inscribed in the catalogues amounts at present to above 5,500 volumes, pamphlets, maps, and plans, distributed in 2100 works, classed as follows : 1st, Algeria, all works and documents, &c. on the colony ; 2d, moral sciences, including mental philosophy, geography, philology, and archæology.

Manuscripts. The collection of Arabic MSS. is greater than the wants of the place. The natives hold them in high esteem ; but unhappily there are but few hard workers among them, and this part of the collection will not be justly appreciated till a greater number of Europeans apply themselves seriously to the study of Arabic.

The period of Ramadhan, which ends in fêtes, leads the Mussulmans to extraordinary expenses, and always produces a rich harvest of Arabic MSS. The year 1850 was remarkable for the number it yielded. The library of Algiers has taken advantage of this circumstance in adding to the store a variety of good works, especially a geography of Mohereb,* containing some curious details on the Roman antiquities of each place. The number added to the collection since 1846 amounts to 200 MSS. on every variety of subject.

It will be gathered from these observations that the library of Algiers contains a most remarkable and matchless collection of matter relating to the special literature of Northern Africa.

At a time when the ancient Moslem empire seems about to fall in pieces, when the mysteries of harems and pyramids and mosques are being trodden under foot by the Giaour, and the Crescent begins to pale before the Cross, it is not without pleasure that we hail all strenuous efforts to preserve relics and monuments of that singular race, which, under the impetus of faith, burst like a whirlwind from the desert, sweeping over the plains of Africa and the vales of Spain, till the scimitar flashed on the banks of the Loire, and the muezzin's call reverberated amongst the valleys of the Basques.

The historical statistics of the city of Algiers present us with the following details : In the beginning of the 17th century, Algiers, as described by Jean Baptiste Gramaye, in his *Africa Illustrata,†* contained 13,000 houses, many of which held 30 families. In the Jews' quarter, the house of Jacob Abum had 300 inhabitants, and that of Abraham Ralhin 260. There were 100 mosques, each attended by three marabouts, and some by 30 or 40 ; and there were moreover innumerable oratories. The number of baths was 86 ; and besides superior schools, in which the Koran was interpreted, there were 86 schools in which children were taught to read and write.

* مغرب *Morocco.*

† Jean Baptiste Gramaye was born in 1580, and his *Africa Illustrata* was published in 1622.

Haedo counted 10,000 gardens in the district of the capital, but the registers of the regency made them 14,698 ; and all of them contained two or three, but most of them eight slaves. There were at that time about 35,000 Christian slaves in Algiers and the neighbourhood. Ali Mami had 132, many had 60 and 70, and the Dey's bagnio had 2000.

Haedo, who had lived there, estimates the white Moors at 2500 families, and the black Moors, or Kabyles, at 700 families. Of Arabs and beggars there were 3000 ; and the Modajares, driven from Spain, made up 1000 additional families ; besides which there were 1000 Valencian Moors. There were 1600 Turkish families besides the Janissaries, 6000 renegades, 6000 Janissaries, 136 families of caids or civil authorities, 300 rais or masters of ships, 86 scherriffs,* and 800 hadjis, or men who had made the pilgrimage to Mecca. Each of the three galleys was manned by 80 Turks, the others had about 30 men. The city contained, moreover, 80 black-smiths, 1200 tailors, 3000 weavers, 120 cheesemongers, 300 butchers, and 400 bakers ; it had also 150 Jewish houses, and, according to report, 8000 Jews. De Breves,† ambassador of Henri Quatre to Turkey in 1628, gives Algiers 100,000 inhabitants ; and Pierre Dan,‡ in 1637, ascribed about the same number to it. At the French conquest in 1830 it had about 40,000, though the size of the town in both cases was nearly the same, comprising 50 hectares and 53 centiares (125 acres). Besides this, the jetty contained four hectares 0·9 centiares (10 acres), giving as the general result, 54·62 hectares (136 acres). In the most crowded quarter of Paris, that of the Arcis, you find about 1554 persons per hectare ($2\frac{1}{2}$ acres); this proportion would give 80,000 to Algiers.

In 1841 there were only 16,000 Mussulmans in Algiers ; hence 14,000 must have emigrated since 1830. This result had been caused in part by the increase in prices. In 1830 wheat and barley were sold at 2 fr. the hectolitre (1s. 8d. sterling per 22·009667 gallons, or $2\frac{7}{9}$ths bushels); an ox cost 18 fr. (16s.); a sheep, 2 fr. 50 cents. (2s.); 100 eggs, 1 fr. 20 cents.: and these prices remained almost the same up to 1834 ; but in 1841 provisions had become almost as dear as at Paris.

The population of the capital is by some thought to have amounted to 70,000 before the French invasion. After that date the natives have been reckoned for some years at 30,000, analysed as follows :

Moors	17,000
Jews	5,000
Turks	4,000
Negroes	2,000
Kabyles and Arabs	1,000
Biskris and Mozabites	1,000
	30,000

* Scherriffs are descendants of the Prophet.
† Relation des Voyages de M. de Breves, 1628.
‡ Histoire de Barbarie et de ses Corsaires, 4to, 1637.

To these must be added 30,695 Europeans, which will give a total of 60,695 inhabitants shortly after the conquest.

In 1833 there were 2,920 houses, 148 public fountains, 14 synagogues, one Roman Catholic chapel, 120 mosques and marabouts, and 48 schools for boys and girls. In 1843 the city contained, besides, two theatres, the Grand Théâtre and the Théâtre des Petites Variétés,* good libraries, and two good newspapers; one of which, the *Akbar*, was published twice a week, and contained four pages. Three of the mosques had been converted into churches, one of them constituting the French Catholic cathedral; but many of the numerous fountains were dry, and there was a want of good water.

The European population of Algiers has much fluctuated, as will be seen in another place. It appears to have reached its maximum in 1847, having amounted, on the 31st December of that year, to 42,113 persons, whereas on the 31st December, 1848, it had fallen to 37,572; and at the same date in 1849 it had been reduced to 37,114. Various causes have contributed to this result, especially political agitation, and the greater safety and facility of colonisation in the interior.†

As regards the present statistics of education and public worship at Algiers, the fullest particulars will be given on these points in a future chapter. We shall here simply state, that the number of European pupils of both sexes, public and private, amounted in 1849 to 1178 children.

Many important alterations and improvements have been effected in Algiers since the visits of M. Blofeld and Count St. Marie. If the reader casts his eye over the map of the capital accompanying this work, he will perceive more easily than by any other method the changes that have taken place. First, the old wall and precincts no longer form the boundary of the city, which includes the old faubourgs of Bab-azoun and Bab-el-ouad, the Moorish gates having been destroyed. The city has been surrounded with new walls, ditch, and bastions, and a new citadel is erected, embracing the old Casbah. Several new gates have been constructed, including the Porte Constantine, a little above that of Bab-azoun; the new Porte Bab-azoun, adjoining the fort of that name; and beyond, the Porte du Sahel, east of the citadel, and the Porte Vallée, west of the citadel; besides the new Porte de Bab-el-ouad, close to the point of Sidi-Kettani, and the Fort of Twenty-four Hours. Several new streets have also been formed, that of Bab-azoun being prolonged, and widening all through the ancient faubourg of that name to the Porte Bab-azoun. The new street of Bab-el-ouad is a prolongation of the old one, and passes through the great Place d'Armes, opposite the Jardin du Dey, to the gate of the

---

* Algeria, resolved not to be behindhand in the amenities of civilisation, has completed a magnificent theatre; and an excellent operatic *troupe* has just left Paris to commence operations there (July 1853).

† Tableau de la Situation, 1850, pp. 94-96.

same name. The Rue d'Isly runs from the Porte de Constantine parallel to Bab-azoun Street, passes through the Place d'Isly, and joins the Rue de Bab-azoun at the Place de Garamantes by the Rue de Rovigo. Rue Poudrière runs down from Porte and Place Sahel to the Rue d'Isly. A number of other labyrinthine streets are christened Rue de Rovigo, and the Rue de la Lyre runs from the Place du Gouvernement to the Rue d'Isly; parallel with the Bab-azoun Street several large open spaces have been cleared, especially the Place Nationale, close to the quays, opening into the Rue de la Marine; and a new street, called Rue du Rempart, that runs along the quays eastward. The Place Nationale (formerly Place du Gouvernement)[*] is planted with trees, and is the principal square of Algiers. All the old rampart to the westward from the Casbah to the old Porte Bab-el-ouad has been converted into the Boulevard Vallée. These, and many other minor improvements which have been made, quite alter the character of the city, conforming it to a third-class European capital. It is doubtful if it is so well suited, however, to the climate and country.[†]

After entering Algiers through the new gate, Bab-azoun, it is proposed to erect on the left, opposite Fort Bab-azoun, an *entrepôt de tabac*, or depôt of tobacco, and a *halle au blé et aux huiles*, or corn-exchange; and all the and on the right of the street between the gate and the Mosque of Sibi-Abd-el-Kader is reserved for military constructions.

It is needless to add, that when these alterations and improvements (?) are effected, the whole of the lower part of Algiers will be identified in appearance with most large continental fortified and seaport towns; and a long interval will not elapse ere the old town, the Djebel, will melt away in the embrace of its juvenile successor, the Outa.

As a relief to the somewhat dry details of this statistical chapter, we here present the reader with the impression made by Algiers on the mind of an intelligent French officer, now a Marshal of the Empire.

" The town of Algiers," observes the soldier, " combines the gaiety of Paris and the charm of eastern life; and contains, in particular, one terrace that recalls the enchantment of the Arabian nights. You go there, when the oppressive heat of the day is passed, breathe the refreshing breeze, while you contemplate the sea with its thousand scintillations, above your head hang, apparently suspended, the white walls of the houses; then surveying the bay of Algiers, your eye rests on slopes covered with roses and verdure, and on the mountain outlines that fade and shade into the Jorjora, whose barren ridges cut sharply the blue canopy of the sky."[‡]

Before we take leave of the island city (Djezair), it may not be unprofitable if we offer a few observations on the contrasts of European and oriental social life and architecture, and on the main principles manifested in both.

---

[*] Names in France and its colonies are as fluctuating as dynasties.
[†] Tableau de la Situation. [‡] Castellane, p. 249.

An imperfect idea of this antagonism may be given by saying that eastern life is poetry, and western prose. The fascination of the fabulous and the hues of romance will ever gild the battlements of Damascus, and hover round the minarets of Cairo, casting into a stern shade and pallid twilight the dismal machinery of Teutonic and Scandinavian poetry. To the sunshine of imagination, Saladin, Alraschid, and the Mameluke Beys will ever carry off the palm from Round Tables and the aureole of Roncesvalles. There is a wealth of wonder, a gorgeousness of tint in oriental life and thought, that can never square with doublet, point lace, trunk-hose, or inexpressibles.

Chivalry and gallantry first passed from Saracen tents under the crests of northern barons, and inspired the rugged breasts of steel-clad Goths with gentleness in bravery. Thus, to the airy minaret, the tinkling fountain, the tapering date, and Ali Bey on his barb, belongs the diadem of fancy. Yet the westerns shall have their due, and in the workshops of Manchester and the *ateliers* at Paris, I ween that you shall find miracles that put Aladdin's lamp to the blush. Look, however, to the Vulcan, and your lamp goes out, for you shake hands with ragged socialism and hoarse radicalism.

The mind of man leaves its stamp on his greatest as well as smallest creations, and his clothing, his thatch, in short, all that reflects him, is an image of, and correspondence to, his character, modified by time and space. Hence the social state of a people can be gathered from its architecture and its tailoring, which also give the key to the climate that it inhabits, to its dominant pursuits, and national propensities.

The great contrast of Moorish and European houses is a type of their national antagonism. The latter are impelled by a vague instinct of association to issue from the castellated isolation of families in the dark ages, and to hive together in vast agglomerations of humanity, where the individual and the family become fractions of the social body. Such agglomerations are no doubt without any form or organisation, and only cemented by physical position ; but they form the natural and necessary bridge from the hostile isolation of barbarism to the complete association of humanity, to which all the higher tendencies of modern civilisation are pointing. A Moorish house shews at one glance its great distance from this consummation. Generally small, they can only hold one family ; and whilst our European houses give free admission to the light of heaven through large and numerous windows, the Moor gropes about in a perpetual twilight, his walls presenting the appearance of a prison.

These two facts are symbolical of the great characteristics of eastern and western life. The more progressive race, leading a more public life, required vaster and more comprehensive edifices, embracing numerous groups, who find daily the advantage and amenity of a greater social approximation between the members of society, accepting material as-

sociation in the first instance as a prelude to the general extension of this
great principle to more elevated interests. But in oriental life, where
man has never conceived of a higher association than that of private families in the most imperfect form, through the slavery of woman, no other
dwellings could be expected than houses uniting the character of castle
and dungeon.

MANUFACTURE DE TABAC.

It is natural to infer from their residences that one of these hostile
races is inquisitive, sociable, and accessible, on seeing the number of windows in their houses; nor can we wonder at the Arab captives at Marseilles comparing the French dwellings to large ships pierced with port-
holes. And do not the long bare walls, with a few rare pigeon-holes and
barred openings, announce a people careless about every thing beyond
their family group, disdaining to look abroad, and anxious to hide the
mysteries of the household from the profane crowd? The inquisitive
and restless citizen of the West required the broad daylight and a wide
horizon to look about him, learn the news, and see what was going

on ; but a jealous nation, shut up in individualism, could not endure to lay bare the privacy of its seclusion to neighbours and strangers ; patriarchalism could not brook the fraternising co-operation of our social life. Climate has also much effect in modifying the architecture of the two races, and shews our folly in trying to naturalise our architecture, diet, and tailoring at the Poles or under the Line.

Nature having been sparing of heat and light to the European, he has been forced to exert his ingenuity in making the most of the share allotted to him.　Like the plant growing in the shade, that stretches and inclines towards the glorious sunshine, the European throws open his walls to let in the pale rays of his watery sun.　But in Africa, with its cloudless sky, burning sun,˙and dazzling light, the severest winter is like a fine autumn with us ; and through most of the year shade being the great desideratum, windowless walls, cool arcades, courts, and fountains, are the architecture indicated by nature and followed by man.

As regards the latest military works, the greater part of the fortifications on the land-side were completed in 1850, including bastions, ditch, curtains, &c.　On the Islet of the Marine six batteries had been established, besides a battery for twelve pieces on the rock Algefna, the battery of El Kettani, and the Fort des Anglais.　They have also established three powder-magazines, to contain 300,000 kilogrammes (660,000 lbs.) ; but the sea-defences were incomplete in 1850.*

* Tableau, p. 15.

CHAPTER VI.

Precincts of Algiers.

PRECINCTS OF ALGIERS——THE TWO MUSTAPHAS——JARDIN D'ESSAI——BUFFARICK——
MODEL FARM——MAISON CARRÉE——THE CAFÉ OF HAMMAH——THE CONSULATE
OF SWEDEN——AYOUN BENI MENAD——POINTE PESCADE.

WE propose now to make a few excursions in the environs of Algiers, in the society of some select friends who will act in the capacity of guides.

"Leaving the back gate of the Casbah," says Count St. Marie, "we had before us, on a little eminence, the entrenched camp of Tagarins. It consists of a large square enclosed by wooden palisades, containing eight rows of parallel barracks, with sufficient room between each for the free movement of the troops. The bedding at that period was miserable, the hammocks consisting of canvas without mattresses or covering, and they were strung by ropes to the walls and to poles. The men quartered there in 1845 consisted of the celebrated Chasseurs d'Orleans, now known as the Chasseurs de Vincennes, the first body of troops that were provided with Minié rifles.

"Pursuing our course (the other side of the Tagarins), we came in sight of four rather large hospitals, which, being exposed to all the winds, are in a very unfavourable situation. On the opposite side of a picturesque ravine which lay open before us, we saw two buildings comprising the Hôpitaux du Dey and la Salpetrière, the former of which is very large, and situated nearer to the sea than the latter. The principal room in the Hôpital du Dey is calculated to contain 2000 beds, and was used under the Deys as a receptacle for plundered goods. The surrounding rocks are clothed with plantations of aloes and acacias. This hospital is admirably arranged and conducted, containing clean neat rooms with iron beds, all of good quality and in good order."

On another occasion our friend St. Marie entered the street of la Charte within the walls, which was thronged with people, because the market held in the Place de la Charte was about to open. In the middle of this square is a fountain surrounded by orange-trees; and it presented on this occasion a busy scene, with country people seated in rows displaying the different objects of their cultivation. Various fruits, which are almost unprocurable in Europe at that season (winter), were exhibited in great profusion in this market, which was crowded with negresses, Maltese,

Marseilles flower-girls, &c. &c. A short distance after leaving the market, our party passed a Protestant church of moderate dimensions, which at that time was nearly completed. When they had issued from the gate of Bab-azoun, they turned up an ascending road to the right, where a stone has been placed with an inscription stating that it was traced out by General Berthezene in 1831.

Following this road they reached the Fort de l'Empereur, which commands a magnificent view of the coast and town. Between the fort and the shore the eye plunges into a large ravine thickly studded with houses surrounded by gardens; more to the right is a heap of ruins, which are the only remains of our consul's villa; and looking back you see the commencement of the Sahel and Delhi Ibrahim, a small European village on the road to Douera. The Fort de l'Empereur forms a large square on an eminence completely commanding Algiers; but it is no longer fortified, and is only garrisoned by one company of *disciplinaires*.* This fort was the largest work in the vicinity of the capital under the Turkish government, and was named after Charles the Fifth. It is situated to the right of the town, and commands the approaches from the land side. The hill on which it stands is 1100 metres (3608 feet) south of the Casbah, and 210 (688 feet) above the sea ; and it consisted in 1830 of three bastions with a cavalier in the centre, and used to mount 50 cannon.†

After passing the fort, St. Marie proceeded along a broad road called the Girdle road, which, however, was not in a fit condition for the passage of carts, having on each side hedges of myrtle, hawthorn, and lilac, and on one side a limpid little stream. These features of scenery, added to the view of the roadstead on the left and clusters of shady trees on the right, made this part of the ride most delightful. Soon after, a pathway down a steep declivity brought them to the village of Upper Mustapha, where a terrace in front of the restaurant commands a fine view. To the left appears the city, with the Fort of the Emperor ; further down the village of Lower Mustapha and its cavalry camp ; to the right you see the village of Koubah;‡ and nearer the sea-shore the Jardin d'Essai (experimental garden), the mills of Hussein Dey, and near the end of the curve, the white walls of the Maison Carrée. Broad roads connect these different points, and the picture is enlivened by numerous country-houses and green pleasure-grounds.

The most recent works effected by the military engineers at Mustapha are the construction of a forge and a cart-house, a masonry trough, and

* St. Marie, pp. 20-23 and 47-49.

† See the description of this fort shortly after the French conquest, by Captain Rozet, Voyage, &c. Prince Pückler Muskau's Semilasso in Africa, vol. i. ; and Dr. Russel's Barbary States.

‡ The military engineers have lately established temporary ditches at Koubah, and put in order the buildings of this camp before giving it up to the civil administration (*des domaines*). Tableau, &c. 1850, p. 17.

GREAT COURT, PALACE OF MUSᵗ APHA PASHA.

three aqueducts to bring in a supply of water. Several *bureaux* for the different officials have been added, and some stables built. They have, moreover, lately erected there a store for forage on the side C, two masonry basins containing 300 hectolitres (about 6600 gallons), and a branch from the aqueduct of Hammah to feed them.*

The villages of Upper and Lower Mustapha are built on the slope of a hill; and the latter contains a cavalry camp, which was occupied in 1845 by the first regiment of the Chasseurs d'Afrique, commanded by General Bourgon, a man who has distinguished himself in Africa. The camp is surrounded by wooden palisades, and the stables form one of the four sides of the upper quadrangle. Within its ample precincts are contained a small hospital, veterinary hospital, and surgeon's quarters : it possesses also extensive magazines of forage, and is supplied with water from a neighbouring stream, which requires filtering.†

Upper Mustapha (Mustapha Supérieur) is built on a declivity of the Sahel about a league from Algiers, and is surrounded by most exquisite fruit-gardens. It was formerly the palace of the Dey's son, and boasted a great degree of splendour. The edifice was built round two courts, the smaller of which is adorned with 64 marble columns supporting magnificent rooms that formerly constituted the seraglio.‡ Nevertheless the bland repose of European scenery is wanting here, the lines being rigid and not sufficiently softened off. On the opposite side of the road stands the country house of the Governor, once the Dey's ; near it is an old Moorish house occupied by the Colonel of the 1st Chasseurs ; and further back is the country residence of General Yussuf.§ All these are large buildings, and the gardens surrounding them contain fine fountains and orange-trees.

The cavalry camp of Lower Mustapha is very clean, with flower-beds under the officers' windows, besides a café and restaurant. The privates sleep on iron bedsteads; at the head of the bed are the trappings, arms, and bridle of the horse; at the foot of the bed the saddle is placed ready for use, and in eight minutes the trooper may be mounted. When visited by St. Marie, wild boars, eagles, &c. were roaming about the barracks ; these animals having been tamed by the officers, who are much attached to the chase. At a little distance, we find a Moorish coffee-house called "the Plane-tree,"‖ on one side of which is a marble fountain and a small marabout with fine plane-trees; facing this café is the railing of the Experimental Garden. The road, after passing the café, followed the curved

---

* Tableau de la Situation, &c. 1850, p. 17.   † St. Marie, pp. 47-50.
‡ The Foreign Legion, p. 16.
§ A distinguished cavalry officer in the French service, of uncertain extraction, of whom Prince Pückler Muskau gives an interesting description, and whose romantic adventures will be noticed in the chapter on the French Army, Part II.
‖ Hammah.

line of the sea-shore; and after riding some distance, says St. Marie, we came to some water-mills on a shady little stream. These mills, the rivulet, and a bridge, are all called Hussein-Dey. The bridge is thought by the Count to be of Roman construction, and the banks of the stream are clothed with a luxurious vegetation. The acanthus, with its broad, glossy, dentated leaves, looked at a distance like a Corinthian capital level with the ground. The road in this part is overshadowed by enormous fig-trees, and the wild vine and ivy are seen climbing up the acacias, orange, and lemon trees. The officers of Mustapha barracks were observed in this vicinity, on their road to hunt wild boars on the banks of the Haratch, armed with Cossack lances, and followed by large lion dogs.

The road still descending brought them to the bridge of the Haratch,* at the foot of the hill on which stands the Maison Carrée. This bridge, which is of Moorish construction, consists of ten arches; but when visited by St. Marie water flowed through one of the arches only, whilst in the rainy season all are filled. At the end of the bridge is a post of native tirailleurs, and further on occur a few European houses used as inns. It is a curious fact, that all the houses in the villages of Algeria are places of public accommodation, i. e. drink and other refreshments are every where sold. Beyond this spot they arrived by a zig-zag road at the summit of the hill and before the fort of the Maison Carrée, which is a barrack rather than a fortress. There are embrasures on all sides of the walls, which are twelve feet high; four sides of the interior are occupied by buildings used for the service of the barracks ; and a little square building in the centre of the court contains the officers' apartments, the powder-magazine, and the stables. The fort can contain about 1200 men, and would be the key to the road to Algiers if captured by an armed force coming from Fondoulk or Kabylia. " This fort," says St. Marie, " is of Moorish construction, but I could not learn the particulars of its origin, though the building appears contemporary with the Emperor's Fort. Leaving the Maison Carrée, and turning your back to the sea, you have before you a distant view of the Fondoulk, the Mitidja, and the beginning of the Lesser Atlas chain. A few white spots on the horizon shew the site of Blidah. All around you in this spot is barrenness and stunted vegetation."

Men of sense in Algeria spoke in 1845 (the panic year) of running a jetty round the shore of the bay, which is a quicksand, for a railway leading from Algiers to Blidah, passing the Maison Carrée, and affording communication with some new villages about to be built at the foot of the mountains of Kabylia.

After the Maison Carrée St. Marie and his party came to a plain of sand along the shore, and halted at a sequestered building called the Water Fort, now no longer a military post, but the property of a colonist.

* By an unaccountable oversight St. Marie calls the Haratch the Shelliff which runs west of Miliana.

They then left the sands, plunging into thickets of jujubes by a path leading in the direction of the Rasauta, a secluded farmhouse, which was at that time surrounded by several little encampments of Arabs. It belonged at that period to a Spaniard, and was surmounted by a steeple. The encampments were almost hidden by plantations of fig-trees and aloes, but they perceived that the tents were high and covered with skins of animals, and their approach was guarded by numerous Arab dogs, who have a natural antipathy to Europeans, whom they would worry and devour if they entered a douar (Arab village) without the protection of a native. Continuing their excursion, they passed on the left the little French village of Fondoulk,* situated nine leagues (22½ miles) E. of Algiers, where there is an entrenched camp ; and they proceeded to ford the Haratch,† which in this part is a narrow stream with high banks. Soon after passing the river they came to a road running through an immense ditch, which the colonial government proposed at one period to carry as a vast moat round Algiers in a circuit of ten miles, for the purpose of enclosing and protecting a portion of the Mitidja and the remaining district near the capital.

A short distance further on they reached the *Ferme Modèle* (Model Farm), an establishment formed for the purpose of improving the breed of cattle, and the quality of fruit, vegetables, &c. It has, however, proved an entire failure, owing to its exposed situation, having been plundered by the Arabs, who, moreover, destroyed the crops of corn around it. The fields belonging to it are now used for fodder, and as soon as the grass is mown it is given to the government, as otherwise the Arabs would burn it. When visited by St. Marie, the farm was in a ruinous condition, and he was of opinion that it ought to be abandoned. Pursuing their road, our party came to a Moorish fountain, which some military wag had christened Cabaret du 43me, an inscription then legible on the masonry. Soon after passing this spot the road became monotonous, the vegetation on this part of the Sahel being stunted in its growth, and consisting principally of thickets of brambles. They advanced to a more verdant hill, and to some mills belonging to a Maltese, who arrived almost penniless at Algiers a few years ago, and now by thrift and steadiness has been able to marry his daughters with handsome dowries. In a ravine at their feet they saw from this spot the village of Birkadem,‡ containing some colonists' houses in the European style, a handsome Moorish coffee-house, and a white marble fountain in the Byzantine style. All the houses of this village have pretty gardens with running streams before them, and the neighbouring country presents a delightful pastoral appearance and a

---

* The word Fondouk فندق means literally 'bazaar.' A. Gorguos' *Cours d'Arabe*, &c. vol. i. p. 210. It comes from the Greek πανδοχεῖον.

† Here again St. Marie calls the Haratch the Shellif, p. 54.

‡ بیرخادم, 'Birkadem,' the Well of the Negress.

luxuriant vegetation. They proceeded hence along a well-made road to the village of Birmandreis, which in most respects resembles that of Bir-kadem. After descending another little declivity, they mounted to the summit of the ridge, where stands a monument erected to General Voirol, under whose direction this road was made. Descending the other side of the hill to the capital, they were delighted, as the shades of night came on, with the brilliant phosphorescent appearance of the sea at their feet.*

The whole coast from Algiers to the fortified camp of Kouba was formerly inhabited by the most wealthy Turks and Moors, who spent here in pleasure the prizes they gained in piracy. Many of these villas are still in good repair, and in the hands of French and Spanish pro-prietors; and the soil around them is very productive, owing to the springs which rise in the hills. There still remain many traces of the Roman and Moorish mode of irrigation; but the bold arches built by the former have long been in a state of decay, while the modest pipes laid down by the latter underground are found to be still serviceable. The bay presents the most enchanting scene for a few miles E. of Algiers, the sides of the mountains being crowded with beautiful gardens and villas built in the Moorish style. On the ridge of the Sahel there used to be (1841) a semicircular chain of fortified camps and blockhouses, intended to protect this fruitful district against the Berbers. Many of these still exist.†

Opposite the barracks of the waggon-train previously described, in the faubourg of Bab-azoun, is the road winding up to the Fort de l'Empereur on the top of the mountain. This was the first of the military roads that have now become so common in Algeria; it bears the name of the Chemin Rovigo, having been made in 1832, when the duke of that name was governor-general. After letting off this branch, the street of the faubourg Bab-azoun coasts along the strand, passes between the marabout of Sidi-Abd-el-Kader-el-Djelali (frequented by women who want to have children), and some French guinguettes (drinking-booths) shaded by palm-trees, whose graceful shapes and verdant freshness present a strange contrast to the prosaic wine-bibbers that frequent the pot-houses underneath them. Beyond these European structures the faubourg ends, though you meet a number of habitations with shining whitewashed walls, that contrast well with the verdant country all the way to the Maison Carrée.‡

The square fort to the right of the city, fort Bab-azoun, standing on the sea-shore, which it commands, is situated one-fourth of a league (two-thirds of a mile) from Algiers beyond the Bab-azoun faubourg, and con-sists of a simple rectangle of masonry. It has an elevation of 15 metres (49·20 feet); it presents a fine battery on the side facing the sea; and it

* St. Marie, p. 64.   † The Foreign Legion, pp. 15-18.
‡ Berbrugger, p. 19.

stands S.S.E. of, and distant about one mile from, the mole. Beyond
fort Bab-azoun, one-fourth of a league to the east of Algiers, is a plain
1200 metres (3936 feet) in breadth, extending to the Mitidja, and enclosed
between the sea and the coteau or declivity of Mustapha. The French
have planted what they call a Jardin d'Essai (experimental garden) in this
plain, which is named Hammah. To the east of the Haratch, three miles
beyond fort Bab-azoun, is the Fort de l'Eau, which is an irregular building,
but no longer a military post. There used to be several batteries, and
some still exist, between Algiers and Cape Matifou along the sea-coast,
the distance separating those two points being 24 kilometres (14·29 miles).
The Maison Rouge, *alias* Maison Carrée, is situated on a hill above the
Haratch, and was formerly the Haouch of the Aga, where he kept 2000
men in garrison ; it consists of a square building, each side measuring
85 fathoms.

On a subsequent occasion St. Marie made another excursion, passing
through Lower Mustapha, Birmandreis, and Birkadem,* till he came to a
plateau that commands the Sahel range. Turning to the right from the
Model Farm, and following a narrow road, after proceeding eight leagues
(20 miles) they saw the chain of the Little Atlas before them. Around
them an immense plain, the Mitidja, extended on all sides, with only one
solitary palm-tree visible on its ample surface. At length they arrived at
Buffarick after a four hours' ride : it is surrounded by verdant poplars, in
a delightful situation, and is supplied with plenty of water ; indeed the
supply is somewhat too copious for the salubrity of the place. The streets
of Buffarick are wide and straight, and shaded by rows of poplars and
willows. Many of the houses are built of stone, instead of the poor
wooden ones which used to constitute the Camp of Erlon, part of which
was still in existence in 1845 at one end of the town, and contained
some troops.†

" Buffarick," says M. Lamping, " is another fortified camp and a small
village which stands on the river Haratch, in the middle of the plain of
Mitidja. The soil is here very productive, but the air so unhealthy that
the village has been depopulated more than once." Official documents
add that this place, which was once so unhealthy, has recovered its salubrity
through the extensive system of drainage that has been introduced. Its
buildings are spacious and numerous, it is surrounded by a considerable
extent of cultivated land, and its colonists are in easy circumstances.
Nineteen farms radiate from this town as from a common centre ; and
the authorities are now engaged in extending its territory, as its narrow
limits form the only obstacle to its rapid increase.

The Mitidja plain, says another tourist, in one place cultivated with

* The camps of Birkadem and Beni-siam are to be converted into hospitals. Tableau,
p. 17.
† St. Marie, p. 78 ; the Foreign Legion, p. 41 ; Castellane, pp. 4, 5.

corn, in another stretching out in wide expanses of brushwood and coarse grass or vast marshes, producing forests of lofty reeds, offers in some parts a fine covert for the wild boar and the panther.* At a spot in the plain called Arba there is held once a week one of the greatest markets of the neighbourhood, which is much frequented by Arabs, who bring to it their horses, cattle, and other property. Arba is a pleasant spot. Delightful groves of orange, lemon, and pomegranate trees, with massive clumps of lentisks and wild olives, adorn this portion of the plain ; and at that season (May) the earth was gay with flowers of every hue, whilst the song of the nightingale was heard on all sides ; and what was better still, the travellers' horses were revelling among fine herbage. This position is at the foot of the Djebel Moussa, one of the inferior heights of the Little Atlas. Numerous streams water the plain in the neighbourhood, of which the principal is the Ouad-Arba ; yet though the water they contain is clear, it is not wholesome, being liable to produce diarrhœa.†

Taking the gentle reader by the hand, we shall now lead him to some of the most picturesque and characteristic haunts and lounges of the natives near Algiers.

On a ridge that commands Algiers, at the distance of 150 metres (492 feet), towering over the immense ravine that separates the Boujareah from the hill on which the capital is built, there stand the remains of a fort raised by Hassan Pasha. It was built of a kind of mortar which reminds one of the Roman cement, and is consolidated by corner-stones of strong masonry, which must have given it the strength of Roman buildings, and in this respect it departs widely from most native structures. Enormous facings of wall stand still erect, owing to the great adherence of the materials, and astonish the beholder by their size ; but hanging over a precipice undermined by the action of rain-water, they constantly threaten the demolition of the frail structures scattered over the hill beneath.

It was in these ruins, called by the French the Fort de l'Etoile or des Tagarins, that the batteries were begun, intended to breach the Casbah in 1830, when the people of Algiers, fearing a storm, forced the Dey Hassan to enter into negotiations preliminary to the surrender. The Tagarins was in ruins in 1830, having been blown up by a negress, who, jealous of her master the governor, fired the powder-magazine and perished with him.

The Emperor's Fort was used as a prison for officers in 1843. If the main building rose a little higher above the walls, it would form a tolerably agreeable dwelling, and the delicious view that it commands would be a great compensation for a short captivity within its precincts.‡

The fort is what the French call à cheval, or astride, on the ridge that descends from the culminating point on which towers the Casbah, and commands a view of the road to Blidah, by the Sahel ridge, also of

---

* Dawson Borrer, p. 20.   † Ibid. p. 21.   ‡ Berbrugger, part i.

the road to the same town that passes by Birkadem and the plain, and of a third road that runs along the sea-shore towards the Maison Carrée, where it divides into several branches, some leading to the farms of the territory of Beni-Mouça, whilst another terminates in the camps of Fondouk and of Kara-Mustapha, and a farther branch leads to the solitudes of Cape Matifou.

These different roads, which are continually paced by a population presenting an inconceivably bizarre mixture, and animated by an extraordinary movement and circulation, offer a most attractive spectacle.

Add to this the sea-view, the continual arrival and departure of ships of war and merchant vessels, the appearance of the pretty villas surrounding the fort, some of which, suspended over abrupt precipices, look like pictures hanging to a wall,—and the reader may form a proximate idea of the noble scenery commanded by the Fort de l'Empereur.*

Though the Barbary pirates were no respecters of persons or of nations, the ambassadors of Christian states seem generally to have led a luxurious and easy life at Algiers.

The consulate of Sweden was one of those charming country-seats so numerous near the capital before conquest, war, and military occupation had left fatal traces of their passage in felling most of the noble trees that adorned its gardens. The spot where the consulate stands must be the site of some Roman structure, from the remains that have been lately discovered there ; nor is this strange, as the slopes of Mustapha must always have been a favourite spot for villas, and most of the consuls resided on this side, including those of Holland, Spain, Denmark, and Sweden. The residence of the latter is situated on one of the culminating points of the slope of Mustapha, and the eye embraces both declivities. Few views can equal that which meets the eye from this point ; as you behold in one glance all the details of a richly cultivated landscape, adorned with all the attractions of art, and the wild background of the rugged and precipitous mountains that frown above it. The blue waters of the Mediterranean perpetually breaking against the dark schistous rocks of the coast cover the shores with a circle of white foam, that presents the appearance of a broad silvery band, diversified at night by phosphorescent streams of fire. Mountain, plain, and ocean harmonise beautifully in this graceful view, which the eye is never satiated with beholding.†

Between the Fort Neuf and the Jardin du Dey, to the left of the road of Boujareah, there appear a number of whitewashed tombs, which at a distance look like a flock of sheep in a meadow. Their shape is very like the hull of a ship reversed, and placed on a rectangular base. Some of these monuments are of marble, and almost all contain inscriptions, of

---

* Berbrugger's Algérie, part i.   † Ibid.

which some are very short, only giving a simple enumeration of the names
and qualities of the deceased, whilst others, much longer, cover the stone,
and contain many scriptural extracts. This is the Jewish cemetery.*

Following the road from Algiers to Koubah, the traveller finds at the
foot of the hills, and opposite the Jardin d'Essai, the pretty Café of Hammah, called by Europeans the Café of the Plantain-trees. This name is
derived from the fine trees that shade the native building, whose appear-

CAFE DE HAMMAH.

ance, however, has been greatly changed since the conquest. The pitiless
hand of civilisation has here, as elsewhere, almost demolished the picturesque. The narrow shady path that used to lead there has been replaced by a wide, straight, dusty road, the work of civil engineers. The
formal avenues and regular alleys of the Experimental Garden are the present substitutes for the wild and capacious clumps of trees that used to
separate it from the Mediterranean. Then the noisy French guinguette
(wine-shop) has hung up its symbolical cork alongside the Moorish café,
typifying the contrasts of the two races. Thus, next door to the lively,

* Berbrugger, part i.

gay, and noisy French, adding to their natural excitement the fictitious excitement of fermented liquors, you see the grave and immovable natives sipping Mocha and pure water,—inoffensive tonics that leave the reason clear.

Leaving the broad prose of the wine-shop, let us enter the poetical atmosphere of the Moorish café, realising the dreams of Eastern romance. Several large mats are extended in the shade of the plantains, and the customers may be invariably seen seated there with their legs crossed, or recumbent in the scriptural and classical attitude of John and Alcibiades. The shop of the *qahouadji* قهواجي or coffee-house-keeper faces the centre tree, and contains benches covered with mats ; but it is seldom resorted to, save in bad weather. Near a stove, always containing boiling water, stands the mortar in which the coffee is pounded ; and over it hangs a board destined to receive the names of those customers who are sufficiently well known to obtain credit. Some pipes, a few wooden footstools, and two or three draught-boards, form the rest of the furniture. There is a great distance between this simple establishment and the dazzling luxury of French cafés ; but the situation, architecture, and arrangements of these native Algerian coffee-houses are so picturesque, original, and antique, that they give birth to tranquil and primitive emotions, foreign to the gildings and trappings of the French metropolis. Though frescoes and gilding are wanting, there is nothing to excite the painful reflection of palled appetites and bankrupt competition, as in our princely houses of entertainment.

The *qahouadji* of Hammah, without the dread of failures or rivals, passes his happy days at his stove or among his customers. Armed with a little pair of tongs, he may be seen hurrying to deposit a live coal in the pipe of one customer ; whilst he hands a fendjal, or cup of aromatic coffee, to another, for the modest price of five centimes (a halfpenny). When not engaged in these duties he is always at his post by the stove, concocting the precious liquor that forms the basis of his revenues. When the water boils, he pours in the bruised coffee, stirs it a few minutes, and then after pouring it several times from one pot to another, discharges it at length into very small cups, with copper egg-cups as saucers. The beverage taken in small quantities in hot weather is very wholesome and refreshing, and a happy substitute for those copious libations of debilitating fluids that predispose the system to fever and dysentery.

The natives do not resort to these places only to drink coffee. They play at many games, especially cards, making use of Spanish packs and terms. Thus they call the colours, *oros, copas, espados, bastos,* and the court cards, *rey, dama, sota,* and the others, *cuatro, as, seis,* &c. The frequent intercourse between Barbary and the Peninsula, and the Andalusian origin of many Moors, will explain this fact.

Draughts are also a favourite amusement ; but the squares, instead of being black and white as with us, are hollow or flat alternately. They also substitute for our men two kinds of pieces, whereof one resembles the castle, and the other the pawns in chess. Their mode of play likewise differs somewhat from ours ; *e. g.* no one can be forced to take.

But the entertainments of the Rami, or story-teller, are the great attraction. It is chiefly in the Ramahdan fast that this worthy displays his powers. The Thousand and One Nights are the chief fund on which he draws ; and when he originates the matter, his improvisations have a revolting obscenity to European ears. Some expressions are continually repeated in their discourses, such as قال *qal,* قالت *qalet,* قالو *qalou* (he has said, she has said, they have said), قال في ل مـِثَـل *qal fil matsel* (they say in the story), كَما فولو *kiman quolou* (as they say, &c.), rendering an Exeter-Hall patience necessary to endure such monotonous delivery.

There are some other recreations to which the less rigid Mussulmans addict themselves at the coffee-houses, including a certain description of intoxication, called *kif,* not prohibited in the Koran. Some take âfioun (opium) ;* others munch a kind of bean named bouzaqa, which is reported to kill all animals having the appendage of a tail (zaqa). They also eat an opiate paste, madjoun ; the women are particularly fond of this substance. Boundje is another intoxicating substance that they employ ; but hachich,† or Indian hemp, mashed fine, and smoked in very

---

* From عَافِيَة *aâfya,* health, calm, serenity.

† The botanical features of this plant will be found in the chapter on the Algerian Flora ; but we propose to give in this note the substance of Dr. Lagger's remarks on the mode of preparing and using the plant. Kœmpfer says that the term *kif* is used in Persia to designate all substances that generate intoxicating effects. The principal of these substances are tobacco, the poppy, and hemp. Silvestre de Sacy informs us that the Arabs of Egypt use the term *kief* to designate the stupor into which the use of the hachich throws them.

In Algeria they apply the names of *kif,* of *hachich,* and sometimes of *tekrouri,* to the extremity of the stem of the hemp, including the leaves, the flowers, and the seed, sometimes smoked by the natives in very diminutive pipes. These smokers are mostly inhabitants of the towns and villages, and are rarely met with among the Bedouins. The Arabs call hemp *k'anal.* European hemp is styled by botanists *cannabis sativa,* Indian hemp *cannabis indica,* called *hachich* in Egypt. Mekrizy, who lived in the fifteenth century, maintains that the use of hachich was discovered by Scheikh Haider, who died in 618 Heg. (1121) ; others attribute it to Scheikh Birazian, who lived at the time of Cosroes ; and some have affirmed that it was known to the ancient Greeks.

Its employment has been repeatedly forbidden by the Mussulman sovereigns. In Algeria the French troops and colonists have only used it to become acquainted with its effects. In all parts of the regency this hemp is cultivated by the natives in gardens surrounding the towns, exclusively for the purpose of smoking, or otherwise consuming its stem. At Constantina, and in some other towns, they prepare comfits made of it, which are eaten to procure pleasant dreams. To make the madjoun, the hemp is first

little pipes exclusively used for that purpose, is the great instrument for creating kif.

The inebriation resulting from the use of these substances has generally a tranquil character : the persons under its influence have commonly brilliant eyes and a bright complexion ; sometimes a vacant laugh disturbs their features, at others a melancholy torpor settles on their face. They say that the chief object in view in using them is, because they are powerful aphrodisiacs.

The coffee-house proprietors are Turks, or Koulouglis ; and some say they are employed as spies by government, to report the conversation of Moors and Arabs, which is, however, generally vague and extravagant.

Before we leave this café, it should be added that it is built on the channel of a Roman aqueduct that descended from the hills of Mustapha, and abutted at a kind of reservoir, of which some traces were found in surveying the Jardin d'Essai. All the remains found on this spot consist of some foundations, an oval basin paved in mosaic work and cut in two by a partition wall, a medal of the Lower Empire, and some fragments of pottery.*

On the slope of the hills of Mustapha is another interesting object, consisting of the remains of a country house belonging to Dey Mustapha, and surrounded by beautiful gardens. The hills of Mustapha stretch from Bab-azoun in the direction of the Maison Carrée, and display all the richness of southern vegetation, and the remains of great luxury in Moorish architecture. The gardens of Mustapha Dey are situated above the vast infantry barracks of that name, and were a favourite resort of the proprietor, who used to keep his wives there in the fine season, and retired to that spot himself to seek repose after the fatigues of office. The Algerine people were very partial to him, and still praise his justice and kindness. These qualities probably led to his ruin, as the Janissaries wished for sterner and more uncompromising leaders. Having heard of their intention to slay him, one day that he was going from the Djeminah to the mosque of Seida,† as a last resource, he fled with his khaznadji (finance minister) to seek refuge in the sanctuary of Sid-Wali-Dada, situated at

pounded, and then mixed with honey or butter ; but the most usual way of consuming it is by smoking. The following is a common preparation of it : hemp-seed is pounded and boiled with an equal quantity of sugar and water, in the proportion of one-half to two pounds of sugar. Among the Harectas (province of Constantina) hachich-leaves are given to the horses to give them spirit on fantasia days (fête-days with sham-fights).

The curious reader who wishes to learn farther particulars relating to hachich and its ecstatic effects is referred to Ebn Beitar's Treatise on Simples, J. J. Ampère's article in the *Revue des deux Mondes*, January 1842 ; and M. Aubert Roche's experiences of its effects in the *Vocabulaire d'Histoire Naturelle* attached to General Daumas's *Grand Desert*, p. 401.

* Berbrugger, part i.

† A very pretty mosque pulled down by the French.

Ketchaoua, a little above the mosque that has been converted into a Catholic church. But the road was blocked up by mutineers clamorously demanding the head of Mustapha; and when the khaznadji entered the street, he was instantly cut to pieces. The unhappy dey followed the steps and shared the fate of his minister, being hacked to pieces by the yataghans of the Janissaries before he had time to reach the door of the marabout.

No one who visited Algiers a few years since can have forgotten seeing in the streets of that city a lame old man with a long silvery beard, whose gentle and venerable countenance attracted the beholder : he was the first native who wore the cross of the Legion of Honour over his oriental dress. This old man was the son of Mustapha Pasha; a circumstance that alone would not have secured him much consideration among his compatriots if this Koulougli had not possessed a large fortune. Out of office a man was nothing in that country; and frequent revolutions, as well as polygamy, had indefinitely multiplied the dey's children. Hence the humblest European visitor at Algiers may now have his boxes carried by the offspring of those proud pirate chiefs, once the terror of Christendom.*

Important monuments are so rare in Algeria, save in the towns, that Europeans are always wont to attribute them to an older people than the present possessors of the soil. Thus the bridge of the Haratch has been given a Roman origin, though Charles V., in his disastrous expedition of 1541, found no bridge there, and was forced to throw a flying bridge across the river. Lastly, the Arabic inscription given below removes all doubt on the subject, proving that it was built in 1149 of the hegira (A.D. 1736) by Pasha Ibrahim-ben-Ramahdan.

During a great part of the year this bridge is as useless as that over the Manzanares at Madrid ; but in rainy weather it is invaluable, maintaining the communication with the eastern tribes. The apathy of the Turks had deferred its erection for two centuries, and it was only under the last dey that it was secured by the erection of the fortress called the *Maison Carrée* by *Yahhya Arha.* The pirate government, in its usual regard for the liberties of its subjects, built it after the following plan. Every one who had to pass the bridge to Algiers was obliged on his return, if he had beasts of burden, to bring a load of sand, mortar, and bricks. Those who could not be turned to account in this fashion were forced to work at it like day-labourers; and when they asked for food after a hard day's work, they were paid by a good bastinado.

The military importance of this spot did not escape the French ; it became their out-post to the east, and remained so in 1843. Before describing the bridge, a word on its builder. Ibrahim was raised to authority on the

---

* Berbrugger, part i.

12th of Raby-el-Aouel, 1145 (23d April, 1732). This is proved by his seal; and he bore the name of Khaznadar, having been treasurer before he became pasha. He met with a good share of misfortunes in his administration, as the Spaniards retook Oran, which they had lost in 1708; and in 1147 (A.D. 1734) Algiers was ravaged by a terrible famine, corn costing three ryals (about ten francs) the saa (three-fifths of a hectolitre).* In 1740 it was visited by the plague, which came from the west, and lasted three years. In 1840 a violent storm destroyed a great many ships in the port; and in 1742 the lightning struck the Bordj Mouley Hhacan (Emperor's Fort), set fire to the powder-magazine, and blew it up with the garrison. Happily for Ibrahim, his Janissaries were not so superstitious as those of Omar Pasha, who was slain, though much beloved, because the plague, the locusts, and Lord Exmouth had come in his time. Ibrahim, like his predecessors, was almost independent of Turkey. The only war he undertook was one with Tunis, which ended in his favour by his appointing his subject Ali Bey as its governor. Ibrahim died in 1158 (A.D. 1745), and was succeeded by his khaznadji, Ibrahim Khoudja. Like all the biographies of the Algerine deys, save one or two, this notice is meagre enough, a matter perhaps not greatly to be regretted.

The position of the bridge over the Haratch is eminently unhealthy; hence it has always been garrisoned by native troops in French pay. In 1843 they consisted of native tirailleurs under the Chef de Bataillon Vergé. A native tribe, called the Aribs, has also been settled for some years in the eastern part of the Mitidja. These Arabs are natives of the province of Constantina, where being dispossessed of their own territory by more powerful neighbours, they came west, the greater part settling at Hamza; but many were granted the Rasauta district, which they peaceably enjoyed till it was given to the Polish prince *Mir.* After some discussion the mutual claims were adjusted, and the Aribs remain. Besides being shepherds, they are Makhzen, or irregular horsemen; and their chief bears the name of Ben Zekri, and professes to be descended from the great Granadan family of that name. He is said to be partial to the bottle.

An attempt was also made to establish an Arab colony near the Haratch bridge, consisting of a gathering of Arabs, Kabyles, &c. from all parts of Algeria. This rabble was christened the Beni-Ramasses by the soldiers, and has naturally failed, the Frank and Mussulman elements not admitting of an easy fusion and amalgamation.

There is now only a little village containing a hundred houses, built in a straight line, near the bridge. Its inhabitants consist exclusively of the native tirailleurs and their families. No cultivation is carried on, and the only means of subsistence of the heads of these families is derived from their pay.

---

* 8s. 4d.; three-fifths of 22·009667 gallons, *i. e.* for about 16 imperial gallons, or 2 bushels.

We here present the inscription of Ibrahim on the bridge, with its translation :

تم بناونا البل يعـ اللاهى   عن اذنـ بانيه لوجهّ اللـهى

به ابرهيم باشا بن رمضان أمر   فصا قنطرة لنا كهآترى

حعَلَ اللّه سعيه سعيا مسّكّورا   وجَزَاوّة جزاء موفورا

سنة تسعا واربعين وماية والفـ   من هاجرة منـ له العز والشّرفـ

The anxiety to make each half-line rhyme with the following has caused some words to be removed from their legitimate place by the composer. Thus, instead of فصا at the beginning of the second half of the second line, you ought to read فصار. We annex the English transcription of these lines, followed by their translation.

Tamma benaouanâ albery allâhy	an idny banyhi lioudj allahy.
Rihi Ibrahim Pasha ben Ramahdan amara	fasâra cantharatan lana kamatara.
Djaala allah sayahou sayan mashkouran	oua djezâouhou djezaan moufouran.
Sanata tesan oua arbayn oua mayet oua alf	min hadjarati min lahoualizz oua alscharf·

TRANSLATION.[*]

(The words in the inscription are supposed to be spoken by men of Algiers.)

We have finished this wonderful and brilliant structure, with the permission of him who undertook it in the sight of God.

The order came from Ibrahim Pasha, son of Ramahdan ; and the result was the bridge you see.

May God take his efforts as a work worthy of reward ; and may this reward be considerable !

The 1149th year of the hegira of him to whom glory and honour belong.

The word ' hegira' means flight, as the reader knows ; and the Mahometan era dates from Mahomet's flight from Mecca to Medina.[†] Ibrahim Pasha, of whom mention is here made, came in all probability from Turkey to Algiers, which accounts for the wording being in Ottoman and not in Algerian Arabic.

At two kilometres (1¼ miles) from Algiers, on the road to Pointe Pescade, stands a very pretty chapel, now a French cabaret. This is the marabout, or koubbah, or q'bor قبر of Sidi Yakoub. We know not on what principle the saint has made way for the cook, but certain it is that religion has here given way to the kitchen, and Bacchus has supplanted the holy Mussulman whose remains had reposed in peace on that spot for three centuries. Father Haedo calls the saint by the surname of El-hel-Desi, intended for El Andalouci, shewing that he was a refugee

---

* By M. Reinaud, Member of the Institute.   † Berbrugger, part i.

Spanish, Moor ; and he asserts that he went mad towards the end of his life. He adds, that he was one of those marabouts who were wont to take singular liberties with the fair sex of Algeria, giving a good sound beating to the poor women who flocked to kiss his hands. The sly traveller insinuates that the Algerian ladies, notwithstanding his violent habits, did not scruple to invite him and his like to visit them, hoping to get young saints by his intercession ; nor did parents or husbands oppose this, regarding the practice as a signal blessing.

The Koubbah of Sidi-Yakoub, built on the top of a schistous rock, and encompassed by fine olive-trees, is contiguous to the Hospitals of the Dey and of the Salpetrière, which have been noticed elsewhere.

Several streams issue from the rock below the koubbah, and flow over the strand. This place is named Ayoun-Beni-Menad, or the fountains of Beni-Menad.

The aged natives assert that they received their name from their builders, a tribe called the Beni-Menad, living between the western part of the Mitidja and Scherschell. These fountains and the koubbah are visited with equal fervour by Jews and Mussulmans. The same remark applies to Sidi-Ali-Zouoni ; and the Moors state that it is because these two saints shewed the Jews some favour.

The fountains being the residence of genii, a race popular with all sects, we need not be surprised at their receiving the attentions of the children of Ismael and Israel, though the latter keep to the springs nearest the Salpetrière.

There are seven fountains at Ayoun-Beni-Menad :* 1. Ain-el-Q'hha-lah, the black fountain ; 2. Ain-el-Bidha, the white fountain ; 3. Ain-el-Khadrah, the green fountain ; 4. Ain-el-Sefrah, the yellow fountain ; 5. Ain-el-Hhamra, the red fountain ; 6. Ain-loun-el-Foul, the bean-coloured fountain ; 7. Ain-Oulad-Sergou, the fountain of the children of Sergou.†

Certain sacrifices are offered up every Wednesday at these fountains, respecting which we have gathered the following curious particulars :

It is necessary to sacrifice a completely black fowl at the black fountain, a white one at the white fountain, and so forth ; and this practice

---

* عين *ayn* makes in the plural عيون *ayoun,* fountains. The names of the fountains in the Arabic character are as follows :

1. عين الكاحلا *Ain-el-Kahlâ.*
2. عين البيضاء *Ain-el-Bydâ.*
3. عين ال خضرا *Ain-el-Khadra.*
4. عين الاصغرا *Ain-el-Sfarra.*
5. عين الحمرا *Ain-el-Hamra.*
6. عين لون ال فول *Ain-loun-el-Foul.*
7. عين ولاد صرته *Ain-Oulad-Sergou.*

† The negroes in their dialect call the Christians *Oulad-Sergou.*

H

seems to be of ancient date, as Father Haedo* speaks of the green fountain as the Alame-Hader, by which he means the Ain-el-Khadrah; and he relates that in his day fowls were solemnly sacrificed there to the genii.

Sheep, goats, and bullocks are also occasionally offered up there, but rarely, owing to the poverty of the devotees. The genii are Christian, Mussulman, and Jewish; and the Mahometans relate that Mahomet, not wishing the faithful to be tempted during the trying fast of the Ramadhan by infidel genii, shuts them all up the night before the beginning of their Lent, and only releases them on the 26th of the same month,—in the same way that the police in Europe keep a good watch over well-known old criminals during holidays.

The weekly pilgrimages to, and sacrifices at these fountains are for the purpose of healing diseases : the process we shall explain presently. But the genii are not to be courted with impunity, and the health of the body is often recovered at the cost of the soul. The frequenters of these wells often become what is called *medjnoun* (possessed by genii). This disease has several developments. Some fancy themselves mendicants, and, whatever their station, go about in rags begging. Others practise what is called *djebbeb*, *i. e.* dance and leap about to the sound of a large drum, till they fall down in a kind of trance, in which they can swallow live coals, digest nails, &c. This state is evidently analogous to the μανια of the Pythoness and the trances of the dancing dervishes, and may not improbably be occasioned by the anæsthetic properties of the water or air, as in the case of the clefts at Delphi.

Sometimes the frequenters of the black fountain are seen dancing the djebbeb in black dresses on the spot. It would, however, be tedious to enumerate all the extravagances connected with this genii-worship, though the subject is interesting in a psychological and anthropological point of view.

We shall now give a brief description of the Wednesday sacrifices. Just after the gun-shot fired at dawn, when the gates open, a crowd of Moorish women, preceded by negresses with fowls, pour forth towards the Jardin du Dey. A few venerable negroes with white beards, and very fat negresses, who are the sacrificators, march in front.

Arrived at the fountains, the votaries seek their favourite spring. The old women throw grains of incense into a little stove, and toss it round the body and head of their patients, after which they bathe them in the fountain. Young girls are stripped naked behind a screen formed by the long cotton veils used by the women in the streets; and after their fumigations they soon repair the disorder of their costume, and prepare to assist at the sacrifices.

---

* Father Haedo published in 1637 a work entitled Topografia y Istoria general de Argel : Valladolid.

A negro, after drawing his knife round the neck of the fowl several times, cuts its throat. Auguries are drawn from the operation. Thus it is very unfortunate for the bird to die at once. When dead, the sacrificator dips his fingers in the blood and daubs the face of the patient. Wheat and other offerings are cast into the sea for the genii after this, and the patients depart, carrying water from the springs to complete their cure. The cure can be obtained also by proxy.

It seems strange that one of the fountains, that of the children of Sergou, should be consecrated to Christians, as that word signifies in the *guenaouya* or negro idiom. But these blacks are some of the chief actors in the scene, and many of them before conversion were Abyssinian Christians. Not that Christendom has cause to be proud of them, for they cut a most disgusting figure in the djebbeb. When thus excited, religious enthusiasm leads them not to bite the thorny leaves of the Barbary fig, or to swallow nails and hot coals, but to make a meal like that of the prophet Ezekiel,* to satisfy their depraved appetites.

Many of these curious superstitions and phenomena will remind the reader of the convulsionaries of St. Medard, of St. Vitus's dance in the dark ages, and of the Cevennes fanatics, and other epidemic disorders of the nervous system wrought by fanaticism and sympathy, and proving the uniformity in the psychological and physiological developments of man in all phases of time and space.†

* Ch. iv. ver. 12.

† Berbrugger, i. p. 68. See some other interesting particulars on native superstitions by M. Berbrugger and L. Piesse, in the Legendes Algeriennes, Paris, 1843 ; also Baron Baude's Algérie, vol. i. ; and Relation des Voyages de M. de Bréves, Ambassador of Henri Quatre in Turkey, 1628. On the djebbeb dance, see Part II. the chapter on the Negroes.

CHAPTER VII.

Interior of the Province.

CHARACTERISTICS OF ALGERIAN SCENERY—INTERIOR OF THE PROVINCE—BLIDAH —THE CHIFFA—MEDEAH—MILIANAH—THE RIVER OF SILVER—TENIET-EL-HAD—BOGHAR—THE KOUBBER ROMEAH—SCHERSCHELL—TENES—THE DARHA—ORLEANSVILLE—AUMALE—THE OASES OF THE BENI-MZAB—THE BEDOUIN TRIBES.

WHILE our European visitor is engaged in scaling the shoulders of old Atlas, or toiling along the dusty roads of Numidia, it may be some refreshment to his fevered blood to pause awhile under yon shady palms, that bend their graceful heads over the whitewashed marabout; and as he wipes the sweat from his brow, to take a survey of the broad features of African scenery.

And first, as he casts his weary limbs on the parched ground, let him mark well the fiery glories of that southern sun, which no effort of pencil or pen can conjure into the misty imaginations of patent cockneys and *badauds de Paris*. Nothing can give an idea of the sun of Africa to the absent; not even the rising of this glorious orb on the vast expanse of the ocean, nor its setting in waves of fire on the savannahs of Guiana. The sun of Africa appears gigantic and in unison with the whole aspect of nature in this terrible country. The same character of arid grandeur pervades every thing—deserts, rocks, mountains, plains; the very men partake of the nature of the lion.*

After a frugal repast of dates, and a refreshing draught of the crystal brook that laves his feet, let our new acquaintance climb that ruinous pile to the left, and gaze at the strange scene unrolled before him. His eye wanders over a vast treeless plain; and his spirit is roused by one of those mighty impulses that issue from the bowels of the earth in Africa, and to which Europe is a stranger. Large salt-lakes at his feet sparkling like diamonds, immense waves of land lost in mirage rolling away to the background, rocky arid ridges breaking the horizon on one side, a dark line in the distance seducing the imagination with Mediterranean dreams, the spectral Arab flying across the plain, and the dazzling koubbah with its venerable

---

* Madame Prus's Residence in Algeria.

plantains. As this strange solitary landscape unfolds, the spectator is filled with indescribable sadness; yet is the feeling mixed with grandeur, elevating instead of casting down the soul. The shades of ages hover over you; and these plains and mountains, the battle-field and grave of mighty nations long since gathered to their fathers, seem to retain some mysterious enchantment that inspires you. Hence the attachment felt by all who have visited it to that land of fables, prompting private or commander to escape from the monotony of the Bois de Boulogne or Elysian Fields, and to seek once more the risks, the accidents of flood and field, and those African breezes that are life to the soul.*

Meanwhile, as the traveller stands wrapt in these sweet day-dreams, let him beware of those mighty clouds that come sweeping up from the horizon, for they bring with them very unpoetical consequences. And now I fear that he is too late, and must stand the fire of African water. They say in France when it rains hard, that the devil is beating his wife, who goes and has a good cry. The devil must be very savage in Africa, for the showers in that favoured clime consist of successive sheets of water,† which have already drenched our poor friend, without throwing cold water on the characteristic ardour of a British traveller. I am glad to see that, with a red Murray's Handbook in hand, he trudges on soaked to the bone, drinking in the amenities of tropical scenery.

But the storm is past; and, forgetting his shower-bath, our honest Briton stops before the shadowy caravansary, where, seated in a family circle of social camels, his spirit holds converse with the glories of a southern night. Reclining his head on an ass couchant, he sounds the fathomless depths of that dark-blue African sky, resplendent with its millions of precious stones, till his mind wanders into the enchanted chambers of some Eastern sorcerer. The silvery light of the moon streaming over the landscape pours calm and repose over vale and mountain; whilst the abrupt ribs and ridges of the mountains, illumined at intervals by its rays, stand out like so many ghosts from the mysterious background.†

Before we make excursions through the remoter parts of this province, we shall give a broad outline of its more striking features inland. Blidah بليدة (the Bida Colonia of Ptolemy) is situated at the foot of the Little Atlas, at the entrance of a deep valley, twenty-nine miles south of Algiers. The environs of this town are rendered beautiful by the numerous orange-groves that fill the air with their delicious perfume, while fruitful corn-fields cover the sides of the adjacent mountains. At the entrance of this city you find a cemetery with peculiar sepulchral stones. Aerial minarets, cupolas, tile-covered roofs enclosed in groves of trees, and a beautiful vegetation, account for the love entertained by its inhabitants for their native place, which they used to style the second Da-

---

* Castellane, p. 364. † Ibid. p. 170.

mascus. Blidah is internally a well-built town, having regular streets much wider than those of Algiers. It is surrounded by a wall twelve feet high and one mile in circumference, with gates at each end corresponding to the cardinal points, and communicating together by a street that goes round the interior of the town. The population of Blidah, which formerly amounted to 14 or 15,000, is now reduced to 6000. The greater part of the town was destroyed by an earthquake in 1825 ; but it was rebuilt on the same site, and is now known by the name of New Blidah. The houses of the town are built like those of Algiers, and some of them looking into an inner court are surmounted by a terrace.

Blidah possesses four stone mosques, which are inferior to those of the capital ; and it contained lately many ruinous *débris* scattered about the town, occasioned by the earthquake. The country in its vicinity is well cultivated, presenting many fields of corn, potatoes, and flax, surrounded by hedges. These fields do not extend to the northward, but to the south they occupy nearly one-fourth of the slope of the mountains ; and though they contain few houses of stone, many huts of reed and wood are scattered over them. Omnibuses pass daily through Blidah, on their way from Algiers to Medeah.*

The principal passes over the chain of the Lesser Atlas, between the Mitidja and the valley of the Shellif, are : 1st, the Col de Teniah ; 2dly, the Col de Mouzaiah ; 3dly, the Portes de Fer, or Iron Gates, a cutting of the Chiffa. The Portes de Fer are situated between the peaks of Beni-Salah and that of Mouzaiah, the former 1520 metres (4985·60 feet), and the latter 1560 metres (5116·80 feet) in height. These peaks are separated by an interval of 10,000 metres (32,800 feet), forming the pass.† This is the most direct route from the valley of the Shellif to Algiers ; the pass is only 400 metres (1312 feet) above the level of the sea, and not much higher than the bed of the Shellif.

The road between Blidah and Medeah through the Lesser Atlas crosses the river Chiffa sixty-two times. The engineers have surmounted apparently invincible obstacles, and the works that they have executed are amazing. The rocks approach so near in some parts as scarcely to leave room for a man to walk erect ; and during the rainy season it was formerly impassable, being bordered on both sides for eight leagues (twenty miles) by steep mountains ; but the engineers have made a road through these defiles, confining the river and blasting the rocks. The road now rests all the way on a strong embankment confining the waters, is carried on both sides of the river alternately, and rises gently to Medeah. At all seasons it is now as good as the best English road.‡

After emerging from the pass you approach Medeah المدية which was

---

* Blofeld, p. 33.   † Baron Baude.   ‡ St. Marie, p. 18.

formerly the residence of the Bey of Tittery, under whose government it possessed a barrack for the Turkish militia. Medeah contains still a casbah and a very pretty palace, and is surrounded by a rather high stone wall one mile in circumference, in which are five gates, two of which are to the north, and the other three face the south, east, and west.

These gates were till lately weakly defended by a few loopholes, through which the besieged could fire on the assailants; whilst above the south gate there used to be two 8-pounder culverins of Spanish manufacture. The appearance of Medeah is very different from that of Algiers in the construction of the houses, all being built of stone and whitewashed with lime; but the interiors are the same, consisting of a ground-floor, a first story, and gallery supported by pillars. Medeah contains many fountains, which are, however, in general mere spouts in the walls; a pretty Moorish coffee-house; and a caravansary, where you can get a change of horses, a rare circumstance in Algeria. This town also contains, or rather contained (1843), several mosques and a public school, with a population of 6000 or 7000. Near Medeah stands a remarkable aqueduct, which has been supposed to be of Roman construction: but the minarets of the mosques are built in the same way, i.e. in stone and bricks of a peculiar composition; and the aqueduct, though ancient, is thought by some writers to be the work of the native Africans.

The environs of the town are beautiful, presenting numerous vineyards and orchards and much cultivation, displaying the agricultural industry of the possessors. The inhabitants of Medeah are much more active than is usual with the Moors and Arabs, being never unoccupied; even in the coffee-houses they knit a kind of sock for the feet, using very thick and short iron needles. Many employ themselves, moreover, in different mechanical occupations, such as those of joiners, tanners, smiths, &c.; but their principal pursuit is agriculture. Omnibuses ply to Algiers through Blidah for ten francs (8s. 4d.).*

There is every reason to think that Medeah, which stands behind the first chain of the Little Atlas, south of Algiers, is of Roman origin, as the Arab structures of the town contain several fragments of Latin inscriptions, and of pottery and other ancient materials. If the distances in the Itinerary of Antoninus are correct, it corresponds to Caput Cillani. Leo Africanus makes no mention of this town; and Marmol calls it Mehedia, which is very like its Arab name Mediyah. He describes it as an old town, built by the Romans in a great plain at the foot of a high mountain; and he asserts that it was formerly very populous, but that it was destroyed by a schismatic khalif, who subsequently built a castle there, that he called Mehedia, from his own name Madhi. Before this event the town was called Alfara. The remains of this castle, containing many Roman materials, still exist.†

* Blofeld's Algeria, p. 35.   † Berbrugger, part i. p. 59.

Two roads lead from Algiers to Medeah. The oldest, longest, and most fatiguing is over the Teniah, or Col de Mouzaiah, and descends to the Olive Wood, a narrow tongue of land separating the waters of the Chiffa from those of the tributaries of the Ouad-Djer. The other road, completed in 1842, passes through the cutting of the Chiffa, ascends the western bank of that river to reach the vale of Ouzra, and passes over Mount Nadhor, whence it reaches Medeah, running parallel to the aqueduct. This road is at times impassable in winter, and requires frequent repairs ; and like the Khyber pass in Afghanistan, it might be the grave of an invading army in the hands of a determined foe.

Medeah, standing 1100 metres (3608 feet) above the sea, has a very cold climate in winter, though the heat is excessive in summer. Abundant falls of snow occur there, obliging the inhabitants to build sloping roofs, contrary to their usual custom, which circumstance gives the scenery a European character. Olives and oranges[*] have disappeared here, to make room for pear, apple, cherry, poplar, and mulberry-trees ; yet the vine thrives notwithstanding the elevation, and the Jews make a noted white wine in the environs of this town.

Medeah is surrounded by a belt of gardens, that give the scenery an enchanting appearance. Marshal Clauzel, who succeeded M. de Bourmont in 1830, saw immediately the importance of the position, and marched on the town at the head of a French force, which took possession at once. Its gallant defence by a small garrison under Colonel Marion against a vast host of natives is a brilliant episode in the history of French Africa.[†]

Milianah is situated on the declivity of the Little Atlas, half a mile from the rich plain of the Shellif, and two leagues (five miles) from El-Herba, which stands on the site of a Roman town. Political revolutions had so materially injured the prosperity of Milianah, that we find it described by a generally accurate writer as a small village, exposed to the south and south-west, surrounded by dilapidated walls with three gates, each defended by three small towers.[‡] The fact is, that owing to the struggle between the French and Abd-el-Kader, it was almost ruined and depopulated ; especially when the latter, making Tegedempt his capital, forced many of the inhabitants of Milianah to migrate thither. The houses of this town are tiled, instead of having flat roofs covered with plaster, forming terraces, according to the custom of this country. If access were less troublesome, Milianah has several advantages to recommend it, being admirably supplied with water from the neighbouring mountain of Djebel Zeccar, one of the most considerable eminences in this part of the country. It is surrounded by many fruitful gardens and vineyards, and has a very fine view of the rich arable country of the Jendrill, Matmata, and other Arab tribes, as far as Medeah. In the spring, devotees from Algiers,

---

* St. Marie and Lamping contradict this, see pp. 134 and 138.
† Berbrugger, part i.   ‡ Blofeld, p. 73 et seqq.

Blidah, Medeah, and the neighbouring villages, used to come with great reverence to kiss the shrine of Sidi-Yousef, tutelary saint of the city. There are also several Roman remains at Milianah.*

A large tract of country has been taken from the province of Oran and added to that of Algiers, to the west of Milianah, embracing the important post of Teniet-el-Had, and the wild Aghalik of the Beni-Zoug-zoug, and of the Ouarensenis or Ouarsenis. To the south-east of Milianah you reach the Aghalik and post of Boghar, on the verge of the Sahara, which contains the two lakes called Zarhez-Chergui (east), and Zarhez-Gherbi (west).† Finally, the district surrounding Mascara embraces many new French colonies, for an account of which the reader is referred to the chapter on Colonisation.

Near Milianah you come to the Ouad-Foddah, a mountain-stream flowing through deep ravines, the scene of a daring exploit of the French army under Changarnier and Cavaignac in the year 1842.‡

The Ouad-Foddah, or river of silver, has its rise in a high rugged mountain called Wan-nash-reese, the Gueneseirs of Sanson, and the Gauser of Du Val, but properly the Ouanseris,§ eight leagues (twenty miles) to the south of Sinaab. It is commonly covered with snow, and on this account it is one of the principal landmarks of this country, being visible the whole distance from El-Callah to Medeah, towering above a number of smaller mountains. It is probably the Zalacus of Ptolemy, while Sinaab corresponds with tolerable accuracy to his Oppidoneum. After abundant rains, considerable flakes of lead, for which this mountain is famous, are brought down by the river ; and as, after being deposited on the banks, they would naturally glitter in the sun, this circumstance probably gave rise to the name of the stream, *the river of silver*. Abulfeda, with other later geographers, has been mistaken in deducing the river Shellif, instead of only one of its branches (the Ouad-Foddah), from the Wan-nash-reese, or Ouanseris mountains.

The Ouled Uxeire and the Lataff run on each side of the Foddah, and opposite its junction with the Shellif are the walled villages of Merjejah and of Beni-Reshid. In former ages the latter had a citadel, 2000 houses, and a race of warlike inhabitants, who held sway over this country as far as El-Callah and Mascara (province of Oran). But at present the castle is in ruins, the 2000 houses have dwindled into a few cottages, and the people, long subject to the Turkish government, are become equally timorous and cowardly with their neighbours, if we may believe Blofeld.

---

* Blofeld, p. 73 et seqq.

† From شرق *cherq*, 'east,' and غربي *rorbi*, 'west.' The text is the spelling in the *Tableau*. A. Gorguos, *Cours d'Arabe vulgaire*, p. 130, makes ' western' in the singular غربي *rorbi*, plural غرابة *gharaba*. General Daumas spells ' west' *J'harb*. Some Arabic sounds have no adequate expression in the Indo-Germanic languages.

‡ Blofeld, p. 73 et seqq. ; Castellane's Souvenirs, pp. 60-73.

§ Castellane, Berbrugger, and Blofeld differ in the spelling of this name : we have preferred Castellane's, as the most recent.

However, their fruits, and particularly their figs, for which they were always famous, continue to enjoy the same reputation as before, and may dispute the palm with those of the Beni-Zerouall for size and delicacy of taste. Two leagues (five miles) to the east of the Beni-Reshid, on the north banks of the Shellif, is El-Herba, with a narrow strip of plain fertile ground behind it. At this spot are several small marble pillars, of a blue colour and good workmanship ; but the capitals, which were of the Corinthian order, are much defaced : there are besides several stone coffins at the same locality.

El-Khadarah is only 13 miles in a direct line from the river Foddah, though it is much farther by the road, owing to the steep intervening mountains that give it a circuitous course. It is situated on a rising ground, on the brink of the Shellif, and is sheltered from the north winds ; while one mile to the south, Djebel-Dwee, another high mountain of a conical shape, supplies the small and beautiful plain between the ridges with a good rill of fresh water. The constant green of these plains may not improbably have given rise to the name of El-Khadarah الخضراء, or El-Chuhdary (the green), by which these ruins are known.

Seven leagues (17½ miles) east of El-Khadarah, and at a short distance from the Shellif, are the ruins of El-Herbah,* another Roman town of the same name and extent as that just now described. This appellation frequently occurs in the country, signifying *pulled down*. At this point the Shellif begins to widen through a plain as large and fertile as any in Algeria, situated at a short distance from Milianah ; and the Atlas Mountains, which from the Beni-Zerouall to El-Khadarah came down close to the river, retire at this plain to the distance of two leagues to the north of the stream.

Such is the famous district of the Ouad-Foddah, beyond which, to the south-west by west, you enter a labyrinth of defiles, fantastic cliffs, and forests, peopled by Kabyles, and known as the Aghalik of Ouarsenis, of which more anon. A little to the right of this rugged district stands the new French post of Teniet-el-Had, near a splendid forest of cedars, one of the most striking spots in Algeria.† This post is built on the plateaux called Serssous, its population amounting in December 1849 to 97 Europeans and seven natives. The last accounts state that the military defences of this advanced post are in a very forward state.‡

Almost due south of Medeah and Milianah, near the banks of the Upper Shellif, is the Aghalik and the French post of Boghar, which is a regularly fortified place with bastions and curtains, situated in nearly the same meridian as Algiers. Its population amounts to 127 persons. The chief buildings that have been erected at Boghar consist of barracks, hos-

* More correctly El-Korbah, الخربة.
† See an excellent description of this whole district in the first section of Castellane's Souvenirs, p. 40 et seqq.
‡ Tableau, p. 23.

pital, and magazines; besides which several gates have been finished, some streets have been formed, and the town is supplied with springs of water.*

The remaining portions of the province of Algiers consist of the sub-division of Orleansville on the sea-shore, west of Sidi-Ferruch, the Sahara or southern part of the province ; and the territory of Dellys on the coast, to the east of the capital. We shall describe those portions of Great Kabylia, which belong respectively to the provinces of Algiers and Constantina, under one head, in the chapter on Great Kabylia, a country deserving a special notice.

We proceed to describe a district situated along the shore of the Mediterranean, which has been taken from the province of Oran and added to that of Algiers within the last few years. This territory, which includes the towns of Scherschell, Tenes, and Orleansville, stretches from the Hadjute district near Sidi-Ferruch and Koleah to the Aghalik of Sbeah. We have previously stated, that the eastern boundary of the old beylik of Oran used to be formed by the river Massafran, after crossing which stream the traveller comes to the Koubber Romeah رومية قبر,† in Turkish *Maltapasy*, or the treasure of the sugar-loaf, supposed by some antiquarians to be the ancient family sepulchre of the kings of Mauritania, and situated on a mountainous part of the Sahel or coast range, seven miles to the east of Tefessad. A minuter description of this mysterious edifice will be found in the chapter on Archæology. Westward of Koubber Romeah are the ruins of Tefessad, supposed to be the ancient Tipasa ; and beyond this point you reach Mers-el-Amouse, or the port of Amouse, which offers a very safe refuge for shipping in westerly gales; and to the westward of this port is a considerable cape called Ras-el-Amouse, after doubling which you speedily arrive at Scherschell or Cherchell,‡ built on the site of the celebrated Jol, or Julia Cæsarea, once so renowned as the capital of Cæsarean Mauritania, of which more anon. It is recorded that Andalusian Moors, driven from Spain by the unchristian intolerance of that age, built a city on this spot in the fifteenth century, which was thrown down by an earthquake in 1738.

A strong wall forty feet high, supported with buttresses, winding for two miles through several creeks on the sea-shore, used to secure the town on the sea-side. The city, to the distance of a quarter of a mile inland

---

* P. 21. Marshal de Castellane says : "Boghar, under the same meridian as Algiers, or thereabouts, rises like an eagle's nest at the entrance of a valley leading to Medeah ; and Abd-el-Kader had lately established a cannon-foundry and important establishments there. We have converted it into an advanced post in the province of Algiers, a place of refreshment and rest for the columns operating on this side" (p. 243). He adds : "The valley we were following was green and beautiful ; . . . the nearer we approached Medeah, the more broken the ground became." Ibid.

† Tomb of the Christian woman.

‡ The name of this town, like all others in Algeria, has been variously spelled by Europeans : we have adopted that used by the Tableau de la Situation.

from this wall, lies on a plain; and after rising for the space of a mile gradually to a considerable elevation, spreads itself over a variety of hills and valleys, and loses sight of the sea.

One of the chief gates on this side is about a furlong below the top of these hills, and leads to the rugged possessions of the Beni-Menasser ; and of the two gates on the sea-shore, the western lies under the high mountain of Beni-Yifrah, and the eastern under that of Shenouah. Scherschell being thus enclosed among high mountains and narrow defiles, all communication with it on the land side may be easily cut off.

A tradition exists here that the ancient city was destroyed by an earthquake, and that the port, which was once very large and good, was ruined by the arsenal and other buildings falling into it. The cothon,* which had a communication with the western part of the port, is the best proof of this ; for when the sea is calm and the water low, as frequently occurs after strong south-east winds, you can perceive over the whole area of the harbour massive pillars and other ruins, which were probably cast there by some great natural convulsion. St. Marie, who reached Scherschell by water, states that the landing in 1845 was very bad; but the last official documents of the French Government shew the present state of the port to be greatly improved. It appears that the old Roman basin has been dug out and restored, and that it is now opened for the purposes of navigation, though it is only adapted to receive vessels of low tonnage. The jetty of Joinville, which shelters the entrance to the basin, has a development of 100 metres (328 feet), and the quays cover a surface of 17 hectares and 42 centiares (43 acres). The expense of these improvements has amounted to 388,000 fr. (15,520*l.*) As regards the defences of the town, it seems that the French have enclosed Scherschell with a new wall of masonry, including bastions, of which the plastering and the platforms were two-thirds finished in 1851. The expense has amounted to 18,600fr. (744*l.*)

A battery for four guns has been completed on the strand called Zizerin, and they have built the two intrenchments and the cart-houses which are its accessories. The parapet of the provisional battery on the *Islot de la Marine* has been partially raised ; two supporting walls have been built, one for the internal and the other for the external slopes. The provisional battery No. 3 has been completed on the slope of the port, as well as its traverse and dependent magazines. The expense of these works amounted to 4600fr. (184*l.*): and it was proposed to complete the works of the enclosure from the Gate of Tenes to the sea ; to build two permanent coast-batteries, one on Cape Zizerin, and the other on the islet of Joinville ; and to establish the batteries for the use of artillery. The estimate for these works amounts to about 79,000fr. (3160*l.*)†

---

* Κώθων, *artificial basin*, literally a goblet or drinking-vessel. See Passow's Lexicon, vol. i. p. 1381.

† Tableau de la Situation, pp. 344 and 386.

As early as 1845 the old Moorish houses of Scherschell were beginning to disappear, whilst handsome European edifices were rising in their stead. On the beach stand two little white marabouts, shaded with palm and date trees, and to the eastward you see the ruins of a picturesque Roman aqueduct; but the country is in general rather flat and covered with brushwood. The garrison musters usually somewhat strong, and is commonly composed of infantry alone.*

Not far distant from Scherschell is a rill of water which is received into a Roman basin called Shrub-oua-Krub, *i. e.* " drink and away;" as there is, or rather was, great danger of meeting robbers and assassins at this spot. To the west of Scherschell you come to Bresk and Dahmus, on the site of two Roman cities; farther on are several small islands where there is good shelter for small vessels; and beyond these you come to the large promontory of Nakkos (the Promontorium Apollonis of Ptolemy), so called from a grotto that the waves have scooped out underneath it in the shape of a bell. Approaching this cape from the coast of Spain, it presents the appearance of a wild boar's head.

Beyond Cape Nakkos is Tennis or Tenes, a town lying in a low dirty situation at a short distance from the sea. The anchorage-ground being too much exposed to north and west winds, is the frequent occasion of vessels being cast away at this spot. The Moors have a tradition that the Tenessians enjoyed formerly such a high repute for sorcery, that Pharaoh sent for the wisest of them to contend with Moses in the performance of miracles. It is certain that they are now the greatest cheats in the country, and as little deserving of trust as their roadstead.

Hammet-Ben-Yousef, a neighbouring marabout, is reported to have given the place the following character:

Arabic.	*English.*
Tenes	Tenes
Ibna ala d'ny.	Is built on filth.
Mawah shem.	The soil of it is stinking.
Madim.	The water of it is black,
Oua aoua semm.	And the air is poison.
Oua Hamet Ben Yousef	And Hamet Ben Yousef
Ma dukkul tsemm.	Would not go there.

Tenes is situated ninety miles west of Algiers; and a fine road is in process of construction, that will unite it to the inland colony of Orleansville. A new European town has been erected since 1830 at Tenes, outside the sorry old Moorish precincts, of which Hamet-Ben-Yousef gives such a deplorable account.†

The improvements that had been effected at Tenes up to December 1849 consisted of 820 metres (2689·60 feet) of principal streets and 4250 (13,940 feet) of smaller streets in a state of repair, of two squares, and of the plantation of 3000 trees. The expense of these works amounted to

---

* St. Marie, p. 193.   † Blofeld, p. 74.

A-E

960*l.* They have also built a slaughter-house and opened a cemetery, at an expense of 42,000 fr. (1680*l.*)

As regards the fortifications, they have built a battery to defend the coast; the town-wall to the east, south, and west is finished, besides barracks for 450 men; a hospital for 150 patients, and magazines for provisions, were also partially completed, together with a powder-magazine to hold 15,000 kilogrammes (33,000 lbs.).

They have also constructed a battery to sweep the anchorage, besides an arsenal and different minor edifices connected with the war-establishment.

The total expense of the various military structures amounted to 31,491*l.* 15*s.* 10*d.**

The name of Darha, which means in Arabic 'north,' is given to a mountainous district of country on the borders of the provinces of Algiers and Oran, comprised between the Shellif and the sea from Tenes to the mouth of that river, which after having formed its limit to the south, flowing in a westerly direction, turns abruptly to the north, and cuts it off in this manner on two sides. The population of this country, which is 50 leagues (125 miles) in length and 20 (50 miles) in breadth, is Kabyle.†
The soil, remarkable for its fertility, is well cultivated. It contains some magnificent orchards, and the principal branch of its commerce consists in the sale of dried figs; but the people of the Darha, being protected by their river, and seldom visited by any agents of the government, carry on another kind of industry which they find still more lucrative. Some are robbers, others are receivers of stolen goods. The latter inhabit chiefly the little Arab town of Mazouna. The subdivision of Mostaganem in the province of Oran, and that of Orleansville in the province of Algiers, are required to preserve order in the Darha.

The authority of the subdivision of Mostaganem extends over the shore-district about the mouth of the Shellif, which is in a less turbulent state. The subdivision of Orleansville, on the other hand, has to look after the most savage and vagabond tribes. The town of Tenes, situated on the sea-shore at the eastern limit of the Darha, is one of the principal points from which the surveillance is carried on. When circumstances have rendered it imperative to concentrate a large force in the Darha, the operations of troops from Mostaganem, Orleansville, and Tenes have been combined to reach and strike the enemy.‡

Orleansville, the capital of this subdivision, was founded on the Shellif in 1843. It contained in 1849 a square planted with 3000 trees; 1400 metres (4592 feet) devoted to principal, and 2400 (7872 feet) to minor

---

* Tableau de la Situation, pp. 345 and 387.

† The Kabyles are the Berber race, who are thought to be the descendants of the ancient Numidians and Libyans.

‡ Castellane, p. 121.

streets ; the population consisting of 849 persons. The military works and defences of this town are in a very forward state. In the western part along the Thigaout they have built up supporting walls to strengthen the slope. The streets of the ramparts have been completed, most of the ditches of the enclosure of the Zmala of the Spahis have been begun, as well as the main building.* (*Zmala* الزمالة, literally baggage, *i. e.* quarters, or camp.)

The enclosures of the Fort or Bordj Ain-Meran, besides the telegraphic posts of Ouled-Kosseir, of Temoulga (on the road from Orleansville to Milianah), and all those on the lines from Orleansville to Mostaganem, and from Orleansville to Tenes, have been built. The principal enclosure of the town has been finished from bastion No. 1 to bastion 7. The east and north fronts have to be raised higher, and the plantations and level- ling of the rampart streets have still to be completed. A number of public buildings have been erected within the town during the last few years, including the quarters for the officers of the garrison, cisterns, baths, windmills, &c.†

The present state of colonisation in the vicinity of this town will be described in the chapter relating to that subject.

After this cursory outline of the western subdivisions of the province of Algiers, we shall return to Medeah. That part of the Atlas between Blidah and Medeah, which reaches as far as Mount Djordjora in Great Kabylia, is inhabited by numerous clans of Kabyles. The Beni-Sala and Haleel overlook Blidah and the Mitidja ; whilst the Beni-Selim and Haleefa sometimes descend into the pasture-ground near the banks of the Bishbesh, or river of fennel, a great quantity of which grows upon its banks. Farther east live a branch of the Meyrowa, within sight of the great plains of Hamza, opposite Sour-Guzlan (Aumale) ; below them the Inshlowa and Bougaine, which overlook to the south the fertile plains of the Castoolah, noted for the feeding and breeding of cattle. Not far from the Castoolah are the Kabyles of Mount Djordjorah, of whom the Beni-Alia are the principal on the north side, and the Beni-Yala on the south.

Five leagues (12½ miles) to the south of Medeah rises the Tittery Dosh, as the Turks call a remarkable ridge of precipices, four leagues (ten miles) long, and even more rugged than the Djordjorah. On the top is a large piece of level ground with only one narrow road leading up to it, where the Ouled-Aiça (children of Jesus) have their granaries. Beyond this are the encampments of Ouled-In-anne, the principal Arabs of the district of Tittery, which lies in the neighbourhood of this mountain.

Burg-Hamza, two leagues (five miles) south of the rich plain of that name, and five (12½ miles) to the east of the Rock of Tittery, contained, before the French conquest, a garrison of one *suffrah* of Turks, consisting of a table or twenty persons, with a lieutenant called an *Oda-bashaw*, re-

* Tableau de la Situation, pp. 344-386. † Ibid.

sembling the *contubernium* of the Romans, who had ten persons in one *pavilio*, the *Decanus* corresponding to the *Oda-bashaw*. Burg-Hamza stands on the ruins of the ancient Auzia, called by the Arabs Sour, or Sour-Guzlan (the walls of the Antelope). Auzia was an ancient city, three quarters of a mile in circumference; and a great part of the walls, fortified with small turrets, still remains. Tacitus has correctly described it : "for Auzia was built upon a small piece of level ground, every where surrounded with craggy rocks and gloomy forests."*

The French have established a new and rising colony on the site of this Roman and Turkish city, to which they have given the historical name of Aumale. Placed in a central and healthy situation, at no great distance from Bugia and Setif, and united to Algiers by a good military road, Aumale is already one of the most important French settlements in the province of Algiers.

From 1847 to 1849, 21,000 fr., or 840*l*., have been expended on the improvement of the streets in the civil town, and 20,000 fr. (800*l*.) on the military town. In the former, 574 metres (1882·72 feet) of street are in a good state of repair ; and 40 metres (131·20 feet) have been recently opened. In the military town, 549 metres (1800·72 feet) are in a state of repair, 1160 metres (3804·80 feet) have been opened, and 600 trees planted. The length of the water-conduits and aqueducts is 2410 metres (7904·80 feet) ; and their daily supply of water consists of 290,000 litres (63,800 gallons).

The expenses of the military works erected between 1846 and 1848 amounted to 869,005 fr. (34,760*l*. 4*s*. 2*d*.) The fortification of the military town was finished in 1849 up to the crown work ; that of the civil town for half its development and as high as the battlements. Magazines for provisions, powder, &c., and other structures for the accommodation of the troops, of cattle, and of other supplies, have been also completed.

The population of Aumale amounted in 1849 to 557 Europeans, analysed into 463 men, 55 women, and 39 children. The proportion of different nations amongst the inhabitants presented the following results : French, 463; Spaniards, 36 ; Anglo-Maltese, 26 ; Italians, 21 ; Germans, 5; Swiss, 3; and 2 of some other origin. The natives amounted in December 1849 to 124 men, 2 women, 8 children, of Mussulman race, 6 Negroes, and 9 Jews.†

A few miles south of Sour commences ancient Gætulia, corresponding in many particulars to the modern Sahara ; and the first remarkable place in this direction is Djebel-Deera, where the river Jin-enne has its sources, a stream which, after flowing thirty miles through a sandy soil, loses itself gradually in the Chott. Most of the Gætulian or Saharian Arabs dwelling on its banks are Zaouia or Zouaia, as they call the children and de-

---

* Tacit. Annal. lib. iv.   † Tableau de la Situation, pp. 344 and 384.

pendents of their marabouts, who enjoy great privileges. The Ouled-Sidi-Aisa, the most northern of these communities, have the koubbah or sepulchre of their tutelar saint five leagues from Sour : near it, on one side, stands a large rock, on which it is reported that Sidi-Aisa used daily to offer up his devotions ; on the other side is the Ain-Kidran, or fountain of tar, miraculously bestowed upon them, according to tradition, by their progenitor. Six leagues (15 miles) farther are the Ouled-Sidi-Hadjeras, so called from another marabout. Here the Jin-enne changes its name into that of Ouad-el-Ham, that is, the river of carnage, from the number of people who have been at different times drowned in fording it. A little higher are the descendants of Sidi-Braham-Aslemmy, who spread themselves to Hirman, a dashkrah in the way to Bousaadah, at which place the palm brings forth its fruit to perfection.

Djebel Seilat lies about seven leagues (17½ miles) to the west of Sidi-Braham ; and twelve leagues (30 miles) farther in the same direction are the Theneate-el-Gannim (the sheep-cliffs). These are situated opposite the Burg-Swaary and the Tittery Dosh, at a distance of thirteen leagues (32½ miles).

A little way beyond the seven hills are the eminences and salt-pits of Zaggos ; after which are the Saary and the Zakkar, two mountains, one twelve and the other five leagues to the south of Zaggos. These, with many other rugged and mountainous districts in the Sahara, constitute what is called by Strabo the mountainous country of the Gætulians.

Six leagues to the east of the Zakkar is Fythe-el-Bothmah ; so called, perhaps, from the broad or open turpentine-trees that grow upon the spot. Seven leagues (17½ miles) from this place to the north you come to Fythe-el-Rotum, that is, the thick or shady turpentine-trees, as it is named, probably in contradistinction to Fythe-el-Bothmah. At Herba, a heap of ruins a little to the east of Fythe-el-Bothmah, are the sources of Ouad-el-Shai-er, or the barley river, a considerable stream of this part of Gætulia. Its course from Herba to the dashkrah of Boufer-joone is ten leagues (25 miles) in a N.N.E. direction. At a little distance from Boufer-joone, below a ridge of hills, there are some ancient ruins called Gahara. Besides the palm, which grows in this parallel to perfection, Boufer-joone is also celebrated for apricots, figs, and other fruit.

We are informed that to the north of Boufer-joone the Ouad-el-Shai-er acquires the name of Mailah,* from the saltness of its water ; and passing afterwards to the east of Ain-Difla, i. e. fountain of oleanders, it loses itself in the Chott. Over this fountain towers the mountain Maiherga, a celebrated haunt for serpents, leopards, and other wild animals. Six leagues (15 miles) south of Fythe-el-Bothmah are Gumra and Amoura, two dashkrahs, with their springs and fruit-trees. Beyond them, at a greater

* From ملح salt.

I

distance to the south-west, is the Ain-Maithie, and then Deminidde, which, with the dashkrahs of the Lowaate, nine leagues (22½ miles) farther to the west, are the most considerable villages of this part of Gætulia, or the Sahara of the province of Algiers. They have also in all these places large plantations of palms and other fruit-trees. Blofeld informs us that the numerous families of the Maithie, Noile, and Mellicke, with their several subdivisions and dependents, reign all over this country, from the Burg-Swaary and the river Jin-enne to the dashkrahs of the Lowaate and Ammer, who spread themselves over a mountainous district a great way to the west, the same probably with the Mons Phruraisus of the old geographers. The villages of the Beni-Mezzab are situated thirty-five leagues (87½ miles) to the south of Lowaate and Ammer,* which, having no rivulets, are supplied entirely with wells. Gardeiah, the capital, is the farthest to the west ; Berg-gan, the next considerable dashkrah, is nine leagues (22½ miles) to the east ; and Grarah, the nearest of them to Ouaregla, is similarly situated in distance and position with respect to Berg-gan. The Beni-Mzab, or Mezzab, or Mozabites, as they are sometimes styled, though they never paid any tribute to the Algerines, and though, being of the sect of the Maleki, they were not permitted to enter the mosques, yet have been from time immemorial the only persons who are employed in their slaughter-houses, and who have furnished their shambles with provisions. Blofeld describes these sons of Mezzab as being of a more swarthy complexion† than the Gætulians or Saharians to the northward of them ; and as they are separated from them by a wide desert, without even the footsteps of any living creature, if this description of them is correct, they may possibly be the most westerly branch of the Melano-Gætuli.

It is well to remark in this place, that throughout the Algerian Sahara, and even far into the heart of the Great Desert, the most usual Arabic name for small towns, especially if surrounded with walls, is *ksour* or *kessour* قَصُور, a word probably connected with the Alcazar, or palace and castle of the Moors, whose faded splendours still remain at Seville in Spain.

The numbers and organisation of the Bedouin tribes of the Sahara throughout the three provinces of Algeria being fully treated of in the chapter on the Arabs, we shall simply state on the present occasion, that their numbers in the province of Algiers amount to 290 tribes, containing 900,000 individuals, inhabiting a territory of 113,000 square kilometres

---

* Few subjects are involved in greater mystery and attended with greater difficulty than the correct nomenclature and localising of the Saharian tribes ; every traveller thinking it proper to spell Arabic names after his own fashion, and the Bedouins being as much addicted to an *alias* and an *alibi* as the most accomplished vagrants in St. Giles's.

† Castellane and Blofeld are at direct variance on this point. See pp. 26, 74 et seqq. of Blof., and 268 of Castel.

(43,505 square miles). The native authorities are, 1. Khalifas, 2. Bach-aghas, 3. Aghas or Kaids, 4. Sheicks. The 290 tribes of this province have three khalifas, five bach-aghas, and twenty aghas; and the territory is divided accordingly into khalifats and aghaliks.*

Most of the posts that had been occupied by Abd-el-Kader on the borders of the Tell and of the Sahara have been rebuilt and permanently occupied by French troops since 1844. These posts include Boghar; Teniet-el-Had, near Taza; and Laghouat, to the south of Boghar and to the east of the Djebel Amour.†

Dellys, Bugia, and most of Great Kabylia, are included by some writers in the present province of Algiers ; but we prefer giving a separate chapter to the Highlands of Algeria, as we have previously observed.

---

* Tableau de la Situation, 1849-50.   † Ibid.

CHAPTER VIII.

Excursions.

EXCURSIONS — THE ORANGE-GROVES OF BLIDAH — COLEAH, ITS DELIGHTFUL NEIGH-
BOURHOOD AND MOORISH POPULATION — THE COL DE MOUZAIA — M. LAMPING'S
EXPEDITION TO THE SOUTH — THE ATLAS — THE ARABS — THE LITTLE DESERT
— SERGEANT BLANDAN — MÈRE GASPARD — MILIANAH — EXPEDITION TO THE
OUARSENIS UNDER CHANGARNIER — THE MARCH — THE BIVOUAC — THE BLOCKADE
— TENIET-EL-HAD.

WE propose now to make divers longer excursions throughout the province of Algiers, of which the broad features must by this time be
familiar to the reader; and we shall commence by accompanying Marshal
Count de Castellane on his journey from the capital to Blidah.

"The road from Algiers to Blidah in 1842-3 went along the Bab-el-
Ouad Street, turned to the left near the tomb of Omar Pasha, and sloping
up the side of the mountain ascended as far as Tagarin (another road has
been made since). The first object at the feet of the traveller, looking
down from this elevation, was the little village of Mustapha,—its extensive
cavalry barracks, the entire bay, the Kabyle mountains, and different verdant oases, dotting the sandy margin of the sea. Proceeding further inland, this view was soon shut out, and the traveller saw nothing but the
mammelons or undulations of the Sahel covered with dwarf palms. At
length the heights of Ouad-Mandil were reached, where the eye took in
the whole of Mitidja. This plain is about five leagues (12½ miles) in
breadth, and stretches to the foot of a ridge of mountains running parallel
from east to west, or from the bay of Algiers to the upper extremity
of the level. The declivities of this ridge are covered with lentisks and
wild olives; and grey rocks tower on their summits, sprinkled with pines
and evergreen oaks.

"To the eastward, near the sea, you distinguish the Foudouk; straight
before you, in the plain, appear the shady groves of Bouffarik; to the
right, at the foot of the mountains, Blidah and its orange-woods; beyond
it the cutting of the Chiffa and the Col de Mouzaia, spots famous in
French military history; further on the Oued-Ger, the Bou-Roumi, where
much French blood has been shed; to the centre you see Oued-Laleg, the
tomb of one of the regular battalions of the Emir (Abd-el-Kader); lastly,

the lake Alloula, the valley leading to Scherschell; and to the westward, on the most distant horizon, near the territory of those famous Hadjoutes, once the terror of the whole district of Algiers, appears the Chanouan, projecting its vast ridge towards the sky, near the tomb of the Christian (Koubber-Romeah)."[*]

Another traveller (Count St. Marie) made an excursion from Algiers into the interior, passing through Lower Mustapha, Birmandreis, and Birkadem, after which they came to a plateau which commands the Sahel. Turning to the right from the Model Farm, after proceeding along a narrow road for the space of eight leagues (20 miles), they saw the Little Atlas before them. At last they came to Bouffarik, which has been previously described; and continuing their journey they reached Blidah, which looks now like a little European town, and is by far the most agreeable and healthy place in Algeria. It contains a large square, surrounded with houses having arcades as at Algiers, and adorned with plantations of plane-trees. Looking along the wide and straight streets, you see at one end of the town the Bab-el-Sets gate, and at the other that of Bab-el-Rahba. The town is walled; and one of its old mosques has been converted into a Catholic church, while the two other mosques have been turned into barracks. Some defensive works have been constructed in that part of the town called the Citadel; and the engineers enjoy elevated and healthy quarters, which were at that time occupied by Zouaves (native infantry)[†] (1845). At the other end of Blidah stands the hospital, which is large, but built of wood, and was at that time used as the quarters of the first regiment of Spahis (native cavalry).[‡] Outside the town, on a detached eminence, stands the Mimiche Fort.

Blidah is a quiet town, containing few Arabs, and peopled chiefly by Frenchmen; and during the early years of the French invasion it was taken three times, after very obstinate engagements. The Moors of Blidah were a very dissolute race, and were threatened with destruction by an old marabout named Mohamed-el-Blidah many years ago. Singularly enough, his warning was justified shortly after by the great earthquake of 1825, when nearly the whole town was ruined. Their first wish was to erect the new town at some distance; but the remembrance of its former delights acted so powerfully on them, that they rebuilt it on the same site.

St. Marie and his party left Blidah by the citadel-gate, and found their horses waiting for them outside. To the left was a gorge of the Little Atlas, and nearer at hand a water-mill on the banks of the Ouad-Kebir. Straight before them appeared the white stones of a European cemetery, and on the right the ridge of the Sahel.[§] Advancing to the right, along the walls of Blidah, they came to a perfect forest of orange and lemon trees, intersected by beautiful walks. The margins of little streams formed

---

[*] Castellane, p. 4.
[‡] Ibid.

[†] See the chapter on the French Army, Part II.
[§] Incorrectly styled *plain* by St. Marie, p. 98.

A–E*

by outlets from the river were bordered by thick bushes of the laurel-rose in full flower, and the shady trees surrounding them gave out a most delicious perfume.

This fragrant aroma is so strong that, if we may believe some writers, those who have lain down to sleep in these groves have been known to be suffocated. A large part of these forests has been cut down to clear the approaches to Blidah; and we can only form a very imperfect idea of the paradise once surrounding this second Damascus. These delicious groves are, even in their present state, superior to those of the Governor of Malta and those near Toulon, which are justly celebrated. The orange-woods near Blidah form vast gardens with little fosses between them, and enclosed on the side next the road by impenetrable hedges of aloes and Barbary figs, present beautiful contrasts in the different shades of green.

Our friends next proceeded to the westward, along the banks of the Ouad-Kebir, to gain the gigantic precipices of the Chiffa. A sloping road led them to the foot of the Atlas, having on its right a gently declining plain bounded by the entrance of the gorges. A winding streamlet crossed this valley, flowing over a bed of pebbles, with here and there bushes of laurel-roses growing on its banks. After advancing a little more in a sloping direction, they came to the banks of the Chiffa, a torrent which rises in the mountains.

The newly-constructed road through the celebrated pass of the Chiffa is broad, and here and there spaces are allotted for the erection of tents. At intervals of about six leagues from one another are little camps, each of six tents, occupied by *disciplinaires*, whose duty it is to keep the road in good repair and to guard it. Many wooden crosses are seen in the pass, erected to soldiers who have been killed by the Arabs, or in making the road. The pass is very cool, there being only a few places for the sun to penetrate, the intervening parts being constantly shaded. About half way a little wooden house has been erected, which is occupied by an engineer's guard and his family. As they advanced beyond the southern outlet of the pass, they found a great abundance of vegetation, consisting principally of oak, cork-trees, and wild olives, growing on the hills. On the opposite side of the plain they saw the camp of Nador, occupied by an Arab tribe. The water there is thought to be injurious, owing to its passing through copper ore.

On turning the hill our party found themselves surrounded by vines growing in great luxuriance; and the country displayed a highly verdant aspect, presenting here and there groves of palm, fig, and orange trees, which gave one the idea of the scenery of a vast theatre. Amidst all this rich foliage they saw before them the walls of Medeah and a few white minarets; whilst on the right an immense aqueduct of Roman construction, winding like a serpent along the plain, conveys to the town the water of the mountain springs. The arches of this aqueduct are completely lined exteriorly with creeping plants.

The gate of Medeah through which they passed into the town was an ogive arch of masonry, consisting of small pebbles embedded in cement. It opened into a tolerably large square, with young orange-trees in its centre; a fine mosque to the left, which has now become a hospital; and some ruinous houses on the right; whilst at the farther end stands a white marble fountain, backed by low moresque arcades,* beneath which some Moors and Arabs had assembled to smoke and drink coffee. On the opposite side of the square was another small gate, through which it was also necessary to pass in order to enter the town.

Storks are very common on the houses of Medeah, where they are highly venerated, and not without cause, as they destroy a large number of rats, mice, and serpents.

Medeah was at that time commanded by General Marrey, who was the proprietor of a noble tame lion named Bello. This town has also a Casbah, situated on the most elevated spot in its precincts, and contains only 600 houses. But its garrison was strong, and many soldiers were encamped without the town in tents.

An immense plain surrounds Medeah, intersected by a stream, whose course can be traced by the reeds and other plants that grow along its banks : wild boars and other game are said to be plentiful in this district. The shops in the town are narrow and few in number, nor do they communicate with the houses to which they belong ; consisting of little more than mere niches in the wall, all of which had a sort of portico projecting into the street, where the shopkeeper sits cross-legged, smoking and drinking coffee.

Half a league from the town is the old country-house of the Bey of Tittery, situated in the midst of gardens, and presenting a pretty appearance : it commands an extensive view over an open country and hills covered with brushwood, whilst to the southward appear some huts and patches of cultivation. At the distance of five miles on the plateau of Aouarah, are the vestiges of a Roman citadel, with two Roman roads leading from it, whose traces, however, are soon lost in brambles.

St. Marie and his party returned to Algiers from Medeah through the Bibans, or iron gates, and Coleah. Proceeding north-west through forests of old olives and winding paths amongst brushwood, they passed several deep ravines, which contain raging torrents in winter. Having reached the summit of a steep declivity, the pathway only gave room for one horse ; and in the evening they came to a large valley on the northern declivity of the Little Atlas, while before them was the passage of the Bibans, or iron gates. Here they stopped at a Bedouin *douar* (village of tents) for the night, where the patriarchal hospitality and manners of their hosts

* St. Marie, p. 98.

reminded the count of the early and sacred records of our race. Continuing their journey the next morning, the paths became more difficult, and they only saw a few desolate huts, but no inhabitants. At 4 P.M. they emerged from a long winding road bordered by rows of thick trees, and beheld immediately before them Coleah, situated in a deep valley surrounded by orchards and groves of palm-trees. The wall which surrounded the town was at that time almost in a ruinous condition; and a cemetery containing two marabouts is situated outside it, running in a direction from east to west. The streets of Coleah are regularly built; and in the centre of the town is a square planted with orange-trees, and having at one corner a mosque, which has been converted into a church, while facing it stands a clean and neat hotel.

Our party proceeded hence to Algiers by Pointe Pescade, which is two leagues from the capital.* The adjacent country is exceedingly picturesque, presenting on all sides pretty country houses, surrounded with blooming gardens and orchards. The mountains in the vicinity of Pointe Pescade are very steep, and the shore is covered with brushwood as far as Cape Sidi-Ferruch, where a pyramid has been erected to commemorate the landing of the French in 1830. Passing some groups of Druidical-shaped tombs, they took the main road to Algiers, and arrived about noon at the gate of Bab-el-Ouad.†

We shall next give ear to M. Lamping, who visited all these spots frequently while serving in the Foreign Legion, and was quartered in the camp at Coleah, towards the end of 1841. He states that "Coleah was at that time a true Arab town, standing on the south-east declivity of the Sahel range of mountains, in a charming little nook, and well supplied with water. It is only twelve leagues (30 miles) distant from Algiers, and three (7½ miles) from the sea; the proximity of the latter making the air extremely healthy, and the constant sea-breezes rendering the heat even then (Sept. 1841) quite tolerable. From their quarters in the camp they viewed, stretched at their feet, the vast plain of the Mitidja, bounded by the blue hills of the Lesser Atlas range. They were quartered in a fortified camp outside the town, on a small eminence which commands it.

The Mussulman inhabitants of Coleah are pure descendants of the Moors; and M. Lamping states that he had never seen the Arab so polished and attractive as at Coleah, not even at Algiers and Oran: in those towns their intercourse with the Franks having called forth all their rapacity, and spoiled the simplicity of their manners.

A corrupt Spanish, called *lingua franca*, is spoken in all the towns of

---

* St. Marie, p. 98 et seqq.

† Three forts are built on the Pointe Pescade, the west limit of the Bay of Algiers. The first at which you arrive *from* the capital consists of a semicircular battery; the second forms a large tower; and the third, to the east of the latter, consists of a rectangular battery. St. Marie (1845).

Algeria near the sea, including Coleah, which is held in great reverence by the Arabs, as it contains the vault of Abd-el-Kader, in which are deposited the bodies of several members of his family.   The French, to their credit, have spared this tomb.   The Hadjutes inhabiting the Sahel mountains, to the westward of Coleah, were the most thievish set of fellows in Africa at the time of Lamping's residence, and long kept the *banlieue* or precincts of Algiers in agitation and terror, kidnapping and violating unprotected females, and ripping up or decapitating stray soldiers and colonists.   The country is now, of course, as safe as the Boulevards or the Elysian Fields.

Coleah lies in a gorge ;* and as there is no lack of water there, the most abundant vegetation prevails on all hands, and the soldiers were delighted and astonished at the extreme richness of the scene.   The luxuriant aloe sends up its blossoms to a height of twenty feet, and a species of sedgy rush grows as high as a house of moderate size.

The chief wealth of the inhabitants consists in large herds of cattle, and in fruit-trees and gardens ; it is surrounded by the most magnificent fruit-trees as far as the eye can reach, including fig-trees and pomegranates, which were then ripe, and on which the soldiers of the Foreign Legion feasted sumptuously, though somewhat adventurously, as the figs in particular are apt to beget dysentery.

The wild laurel grows in great quantities near the town, and attains a very considerable height ; and M. Lamping adds, " I can boast of having tasted the fruit of the laurel as well as the leaf."   All the Arabs of any education or wealth used at that time to assemble in the coffee-house, to whom it supplied the place of theatres, concerts, balls, and tea-parties.

On another occasion M. Lamping paid a visit to the interior of the province, when the battalion to which he belonged was ordered to march towards Blidah across the plain at the foot of the Sahel mountains.   This ridge of the chain is low at that point ; it is highest near Algiers : but it contained most beautiful and fruitful valleys, in which were forsaken gardens and villas which once belonged to the Moors.   The heights are covered with dwarf oaks and other shrubs, which shelter great numbers of wild boars, smaller and less fierce than those of Europe.   The natives assert, that the Spaniards brought these unclean animals into the country out of spite; and as swine are an abomination to the Mahometan and may not be eaten, the breed increased rapidly.†

The sharp and broken outline of the mountains, and the dark foliage of the olives, pines, and cédars which clothe their sides, give a singularly wild and sombre character to the Atlas range.

At the time of Lamping's visit the road through the Chiffa pass

---

* The French in Algiers, &c. p. 9.   † Ibid. p. 72.

had not been made ; and his battalion had to cross the Col de Moussaia, or Mouzaia, a much more fatiguing and lengthy way : from the foot of the Col de Moussaia up to its highest point is fully seven hours' march. The Fountain de la Croix, which you meet on the Col, takes its name from a huge cross cut into the living rock, probably by the Spaniards as a pious memorial of their conquest.* A large olive-grove grows at the foot of the Col on the southern side.

Medeah is one of the oldest cities of Africa, standing on a plateau which terminates on two sides in an abrupt precipice, and is therefore easily defended. The town is surrounded by the most splendid fruit-gardens ; and a Roman aqueduct, still in good preservation, conveys water to it from a neighbouring mountain, and proves the high antiquity of the place. It is inhabited by Jews and Arabs, who seem devoted to the French. "We pitched our tents," says M. Lamping, "close to the town, beside a brook, where exquisite oranges out of a garden close by offered us some compensation for the fatigue we had undergone."

Leaving Medeah, they continued their route due south, and marched several days, their road always lying up or down hill. The heat was excessive, and their marches were at the rate of from four to six leagues (10 to 15 miles) per day. They found the valleys extensively cultivated with large crops of corn and rice, but no inhabitants ; and they only perceived a few miserable hovels of rushes and skins of beasts ; yet in former years this tract of country must have been thickly peopled, from the cemeteries of enormous extent appearing in many places. The Bedouins in this part of the country are, however, now too poor to buy tents, and hence they build the wretched hovels previously described.

They were now in the old province of Tittery, among the mountains of the second Atlas range, which at this point is not divided by any considerable rivers or valleys from the Lesser Atlas. It is, in fact, impossible to tell where the one ceases and the other begins ; all is mountain. Farther west, on the contrary, the extensive plain watered by the Shellif forms the natural division between the chains. After several days' march, the mountains, which had hitherto been covered with mere brushwood, became more wooded and romantic in appearance. They passed through immense forests of olives, firs, and junipers, the latter of which grew to a considerable height. It is remarkable, that on the very highest point of all these mountains there stands a marabout. These buildings are, at the same time, the temples and the mausoleums of the Bedouin priests. They are usually small buildings, from thirty to forty feet square, surmounted by a cupola, and commonly built of rough stone and whitewashed.†

Continuing their progress, they turned somewhat westward ; and the

---

* This conjecture is quite gratuitous on the part of M Lamping. May not the cross be a monument of early Christianity and of the North African Church?

† Pp. 49, 50.

column drew nigh to the Little Desert, or the Desert of Angad, as it is sometimes named.

One fine morning, after a two hours' march, the Lesser Desert was before them; and a most cheerless prospect did it afford. To the southward nothing was to be seen but an undulating surface of shifting sand; on the east and west alone the Atlas range was still visible. The palm grows better than any other tree in this scorching soil; but it was only from time to time that they found one, and then so stunted and withered was it that it could afford no shelter to the weary wanderer. The palm is seldom found in groups,—generally single, or in twos and threes; hence the natives call this tree the 'hermit.' To their great joy, they soon turned westward, always following the track of a caravan. The heat was burning, and they marched up to their ankles in sand; but towards evening they reached a spot recently deserted by Bedouins, with several deep cisterns of water, not good enough to drink, but sufficient for cooking, and refreshing their cattle. These cisterns are filled during the rainy season, and some water remains in them till far in the summer. The next day they turned still more to the westward; and towards evening they reached the foot of the mountains, and bivouacked beside a brook.

On their return from these burning regions of the south, the column once more crossed the steeps of the Atlas chain, where they found the air sharp and piercing, even in summer; and while they could scarcely breathe for the heat below, on the top of the ridge they buttoned up their capotes (greatcoats) to their very chins. On these airy heights they appeared to be in the land of vultures and eagles, which soared and screamed around them by hundreds.

Marching back they followed another route; and on leaving Medeah they crossed the main ridge of the Lesser Atlas to the west of the Col de Moussaia, through some defiles which took the whole day to pass. They had not, however, to climb so great an elevation as at the Col, as they followed the course of a mountain torrent which forms several considerable waterfalls, the heights on either side being covered with the finest pine and olive trees, and the whole scene being wildly beautiful.

Immediately above Milianah is the highest point* of the Lesser Atlas; and the town is built half-way up the mountain, on a plateau which falls abruptly on three sides. The town contained at that time few buildings worth looking at, except the palace of the Emir; but the French had repaired and considerably strengthened the fortifications of the place.

Dropping M. Lamping, it is our present purpose to accompany the Count de Castellane† in some of his rides in the interior of this province.

---

* This is not strictly correct, the Zakkar above Milianah being 1900 feet lower than the Djorjora belonging to the same chain. See p. 38.

†Souvenirs de la Vie Militaire en Afrique, par le Comte P. de Castellane (now Marshal), 1852.

An hour after leaving the heights of Ouled-Mandil on the Sahel, Count de Castellane entered Bouffarik. Built on an unhealthy spot, in a place where, according to an Arab adage, not even the snails can live, Bouffarik, notwithstanding its unhealthiness, which has several times devoured its population, is indebted to its central position for a certain degree of prosperity. Thanks to the works that have been begun, it is anticipated that its terrible fevers will disappear. They only passed through this embryo town, stopping a few minutes at the café of Mère Gaspard, a noted gipsy cantinière (canteen-keeper), who has erected a splendid hotel and café at Bouffarik, adorned with engravings of Horace Vernet's best pictures, presented by the artist himself.

Before you arrive at Beni-Mered, you see the column erected to Sergeant Blandan and his brave comrades. On the 11th April, 1840, the mail started from Bouffarik for Algiers, under the escort of a brigadier (non-commissioned officer) and four Chasseurs d'Afrique, who were accompanied by Sergeant Blandan and fifteen infantry soldiers who were going to rejoin their regiments. They were advancing quickly, without having seen a single Arab, when suddenly four hundred horsemen rushed upon the little band from the ravine that precedes Beni-Mered. The Arab chief rode up to the sergeant, calling out to him to surrender. A shot was his answer; and forming a square, the French soldiers faced the enemy. The bullets were bringing them down one after the other, but the survivors closed up without flinching. "Defend yourselves to the death!" exclaimed the sergeant as he was hit, " face the enemy !" and he fell at the feet of his comrades. Only five men remained of the twenty-five, covering with their bodies the dispatches that had been confided to them, when the sound of horses' hoofs, galloping up at full speed, gave them new life and hopes. Presently from a cloud of dust there darted forth a body of horse, who rushing on the Arabs sent them flying. The rescuer was Joseph de Breteuil and his chasseurs. He was ordering the horses to be watered at Bouffarik when he heard the firing. Immediately, only giving time to his troopers to seize their swords, M. de Breteuil set off full speed, followed by his chasseurs, mounted at hazard. He plunged the first into the tumult; and thanks to his rapid energy, he was able to save these martyrs to military honour. The rescuer received the same recompense as the rescued, Breteuil and the survivors being all named members of the Legion of Honour in the same ordonnance of the king.

The road to Blidah crosses the site of a wood of orange-trees cut down by General Duvivier in the name of military engineering. During two years did these trees serve to warm the troops ; and what remains of them around the town is sufficiently beautiful to make a residence at Blidah delightful.*

Having been presented to General Changarnier, they shortly left Bli-

* Castellane, pp. 5, 6.

dah, escorting him on the road to Milianah. They followed a westerly direction, hugging the mountains that rise to the south of the plain. At two leagues from Blidah they forded the Chiffa torrent, there 100 metres (328 feet) in width, with a deep and rapid current; and they soon after reached Bou-Roumi, where they halted an hour, before climbing the hills that separate the plain from the valley of the Ouad-Ger.

It was not usual to follow the valley of the Ouad-Ger when the communications between Milianah and Blidah were not open; its steep acclivities, covered with lentisks and evergreen oaks, presenting too great difficulties. The road of the columns, longer but more secure, used to pass over the ridges and abut in like manner at the marabout of Sidi-Abd-el-Kader, where they were to bivouac in the evening. Accordingly at three o'clock, after having crossed the Ouad-Ger eighteen times, they rejoined the troops that had started the night before, and their tents were pitched under the aged olive-trees that were still respected by the French axe. The next morning they crossed the valley of the Ouad-Adelia, which consisted of heavy clay, following a southerly direction. Two roads were now before them: one of which ascends towards Milianah by the slopes of the Gontas and the valley of the Shellif; whilst the other passes by the country of the Righas, and reaches the town by winding along the declivities of the Zaccar. The latter was the shorter, and was selected by them; and having reached the plateaux of the Righas, they saw before them, on the other side of an immense woody ravine, Milianah, built on the summit of a rock, surrounded by gardens and verdure.

An hour after leaving the fountain of the Trembles, they entered Milianah by its northern gate.

Zaccar signifies that which refuses or will not suffer itself to be ascended; and this name is given by the Arabs to the long rocky ridge which commands Milianah on the north side. The town is built on a plateau at the foot of the mountain, and advances like a promontory over the last slopes which continue for the space of a league as far as the valley of the Shellif. From the sides of the Zaccar, and from Milianah itself, gush forth abundant rivulets, diffusing a delightful coolness on all sides; and around the town extend those gardens so celebrated throughout Algeria: lichens, mosses of all kinds, a thousand plants with long stems, seem to encircle the white houses and tile-roofs of Milianah with a belt of verdure. At a distance the eye is deceived, and perceives nothing but a smiling scene; but if you draw near, you find nothing but whited sepulchres.

A main street designed by the French, containing a number of canteen-shops, traverses half the town, and stops short at the Arab quarter, near the minaret of a ruined mosque. The martial notes of the French clarions have long succeeded the cry of the *muezzin*, calling the faithful to prayer from its summit. This town had been a kind of advanced post down to 1841; but after that date it became, with Medeah, the basis of the French operations in Algeria. By ascending the minaret of the old mosque you easily perceive the importance of the position, for you see all the country that it commands : on one side the rolling *mammelons* or hills that separate it from Medeah ; then the valley of the Shellif running east and west ; and beyond that the rock of Ouar-Senis, commanding the Kabyle mountains, whose conquest has cost so much blood and treasure. It is a magnificent picture. The country that separates Medeah from Milianah is called the *Djendel.*

On another occasion Count de Castellane marched with his column from Medeah to Blidah, on his return from the successful expedition undertaken to put down the great insurrection of 1845. They passed through the gorge of the Chiffa, which the Count describes as one of the wonders of Africa, and one of the beauties of the world. Picture to yourself, through a precipitous cutting five leagues in length, a splendid road twenty-five feet in breadth, in some places opened through the rock by blasting, and in other places encroaching on the torrent, which has been forced to yield up part of its bed. Lichens and all manner of shrubs flourish in the crevices of the rocks ; and in some favoured places, where the mould has not been washed away, actual forests tower overhead. The river Chiffa has worn a serpentine channel through these rocks, and receives in its bosom the numerous cascades that come tinkling down the wall-like rocks. Suddenly, as you advance, the horizon expands, you issue from prison, and your dazzled eyes rest on the long range of hills that bound the Mitidja,—on the sea, which appears through the cutting of Mazafran,—and on this immense plain, so beautiful when seen at a distance.

In the course of an hour you reach Blidah. Mohamed-ben-Yousef, the traveller, whose sayings have become popular in Algeria, pronounced of Blidah, " You are styled a little town ; I call you a little rose." No description can be more exact. Blidah appears with indescribable grace among woods of orange-trees, whose perfume announces your approach while you are still far off. The French maintain that they have embellished Blidah with all the excellences of French art ; yet notwithstanding all their embellishments, it has managed to remain a very charming town, —in short, *the little rose* of Mohamed-ben-Yousef.†

The country known by the name of the Ouar-Senis extends between the valley of the Shellif to the north and the Little Desert to the south, having a length of about 15 leagues (37½ miles). It is a vast assemblage of mountains, which rise successively to the rocky ridge placed in the centre, a regular knot binding together all this labyrinth of precipices, of ravines, and of gigantic eminences. The rocky ridge in question has a length of 1500 metres (4920 feet), rises above the plateau which forms its base to the height of 600 feet, and is protected by precipitous sides. Its summit is inaccessible save by means of paths only fit for goats, and runs

* Castellane, p. 21.   Ibid. p. 248.

in an east and west direction; and at the latter end, after presenting a col or neck which answers for a pass, a rocky eminence protrudes in the shape of a dome, towering above the rest of the ridge. The reader may imagine the difficult nature of a country where narrow footpaths, on all sides commanded by eminences and woody plateaux, wind along the flank of the mountains, only leaving a free passage for one man. This dangerous ground is inhabited by wild and warlike Kabyles,* sprung from that old Berber blood which has ever offered the material of resistance to established authority. They consist of the Beni-Eyndel, the Beni-boudouan, the Beni-Rhalia, the Beni-bou-Atab, the Beni-bou-Kanous, the Beni-bou-Chaib, &c. ; tribes with republican forms, only obeying a djemaa (assembly or commission) named by the whole people ; independent, everlastingly quarrelling, yet united against the common enemy. These tribes had already encountered the French soldiers before Castellane's campaign (1843). The first occasion was at the Ouad-Foddha, of glorious memory ; later, in November 1842, they were obliged to succumb to the French columns that furrowed their territory anew ; but Abd-el-Kader excited them to arms once more in 1843. Sidi-Embarek was then in the Ouar-Senis with his regular battalions, and endeavoured to stimulate the mountaineers to revolt.

Three columns were destined to operate in that country, under the supreme command of General Changarnier. Each of them had received precise instructions, and the common rendezvous was the Medina of the Beni-bou-Deuan, a Kabyle village, or rather town, situated among those mountains. Castellane's column, under the immediate command of Changarnier, marched direct for the *cathedral*, as the soldiers had christened the rocky ridge above mentioned. The 10th of May, under a bright sun and with gay hearts, they issued from the gate of Milianah, descending the narrow path which leads in a westerly direction to the valley of the Shellif. They were accompanied by 150 horse, who were to attempt on the morrow to surprise a Kabyle village. Scarcely had they reached the valley, ere the clarions sounded a halt to give the column time to close up ; then, all in order, they started afresh. They were in a friendly and open country; and though their arms were loaded, they marched without precaution. First came the general, followed by the cavalry ; then the infantry, preceded by a company of sappers with mules carrying implements. This company was ordered to march on, without caring for the cavalry or the general. Behind them came a part of the infantry ; then the mountain guns, with their little pieces on the backs of mules with roughed shoes; the ambulance or hospital apparatus, with its red flag, followed together with a convoy of provisions ; lastly the baggage of the corps, sumpter horses, mules, or asses, under the surveillance of non-commissioned officers, and followed

* The native population is divided into two broad distinctions : 1st, the Arabs ; 2d, the Kabyles or Berbers.

by a numerous infantry that closed the march, having at the extreme rear-guard mules with *cacolets*,* in case of accidents or diseases. From time to time the general's aid-de-camps ascertained that the column advanced in good order; and at every hour the chief of the staff ordered a halt to be sounded, to give ten minutes' rest to the foot-soldiers, with their heavy load of eight days' provision. It is usual to give an hour and a half's rest about half-way, for the soldiers to make and drink their coffee-soup, *i. e.* biscuit broken up in coffee, all dipping into a common dish. Such is the common order of the march in Africa.

"We were now marching in the valley of the Shellif," says Castellane, "through magnificent corn-fields, smoking and talking, laughing and singing, or silent and thoughtful, as the mood might be; but very happily, sadness was not in fashion among us."

After a succession of halts, the column had arrived at the bivouacking place, near the stone bridge built over the Shellif by Omar Pasha; and, as usual, the portable city was pitched with an admirable dispatch.

The next morning, at the *diane*,† the band of the 58th played a gay *réveil*; and after coughing a little, and dispelling the morning fog by a draught of brandy, they resumed their march, following, as the night before, the valley of the Shellif. In the evening they stopped at the Ouad-Rouina. In the night the cavalry and two battalions started to surprise the Kabyle village of the Berkanis; but being discovered, the affair ended in a little peppering. The following evening they rejoined the bivouac; and the next morning the cavalry returned to Milianah, whilst the head of their column entered the valley of the Ouad-Rouina. A few hours later they encountered the bad roads of the Ouar-Senis. One by one, mules, soldiers, and convoy advanced in file along these narrow paths, which continually ascend, winding up the flanks of the mountains among pines. This was a hard time for the infantry; for to the right and left of the convoy some battalions were directed to protect it, cutting across the country without any track; at one time descending the ravines, at another mounting the steeps, encountering terrible fatigues, rendered necessary by war to insure the safety of all.

Though they had been in an enemy's country for two days, they had not yet met any one; everywhere they found nothing but the calm of emptiness,—a kind of desert,—when suddenly, in front of them, they beheld five or six hundred Arabs, excited and raising loud shouts, on an

---

* Portable chairs for the wounded, carried on the backs of mules, of which St. Marie (p. 22) gives the ensuing description: "*Cacolets* are a kind of pack-saddle of wood and iron, carried on the backs of mules, and supporting on each side chairs of iron, made to fold up in a small compass, so that the mule may set out with expeditions carrying provisions, and return with a load of wounded men, who must be so seated in the chair as to form a counterpoise to each other. Some of these cacolets are so ingeniously constructed as to spread out like a bed."

† The trumpet-call to rouse the troops.

eminence that commanded the narrow path. The halt was sounded. The general formed the Chasseurs d'Orleans in the vanguard ; then he started, himself at their head, to drive off the enemy. Under shelter of the figs and other trees that clothed this knoll, the Chasseurs d'Orleans escaladed it at a run, notwithstanding the fire of the Kabyles, whom they soon drove back with the bayonet. A considerable number of the natives bit the dust, and the others sustained a vigorous chase ; and the French returned with a flock which they found in the wood, some killed and some wounded,—but such is war. Meanwhile the convoy, having passed through the defile after it had crossed a ravine, had established itself near the little town of Beni-bou-Deuan. The houses of this town, which are built of wood and plaster, have a great resemblance to the hovels of the French peasants in Picardy. They are solid, defying rain and storms ; yet the soldiers had soon gutted them, for the dry wood they afforded gave out less smoke, and made better soup. Accordingly, in the space of two days, during which they awaited the other columns, not a few of them were destroyed ; and all would have been converted into fuel, had not Colonel Picouleau and his troops soon arrived.

All the Arab accounts agreed in stating that there was a gathering of the population in the direction of the Ouar-Senis. These accounts were correct ; and on the morning of the 18th of May, a few moments after crossing the Ouad-Foddha and becoming implicated in a defile, they perceived some Arab horsemen ; and on debouching on the large plateau at the base of the rocky ridge previously described, they saw the enemy.

The French arrived from the eastward, parallel to the south side of the ridge. Before them stretched away a vast plateau covered with trees, with verdure, vines, houses, and gardens. To the west the plateau terminated in a high sugar-loaf mountain, separated from the rocky ridge by a col answering the purpose of a pass. This plateau stopped short to the south, at a ravine in which there flowed a river. The ridge might be about 1500 metres (4920 feet) in length, and was surrounded by indented rocks ; the precipices of the ridge rising sharp, like walls, from the last slopes, to a considerable height. The whole mountain towered above the plateau to a height of about 600 feet. Some pines and other trees fringed the steep slopes, and stopped where the rock became vertical, but climbed higher at two opposite spots, which seemed to shew that there existed two means of access to the summit. In other respects, nothing could be more charming than this plateau ; a real oasis, which on two sides stood out in all its fresh verdure from amongst a rampart of greyish rocks, whilst towards the left the eye wandered over a line of endless *mammelons* (undulations) to the blue horizon of Tiaret. On their arrival they saw the horsemen of Sidi-Embarek ride off to the southward, and numerous Kabyles flying along the woody slopes ; but from the top of the rock a confused and muffled sound and agitation reached their ears, and some-

K

times loud cries. At intervals some Kabyles appeared on the ridge; and a singular effect was produced by scattered groups of horsemen, who, suspended on some almost inaccessible heights, stood out in bold relief against the azure sky.

The twenty-five horsemen, their only cavalry, were immediately thrown out in the direction of the col; and the Chasseurs d'Orleans, who that day formed the vanguard, throwing off their knapsacks, ran in to support the little knot of horsemen. Two other companies swept the slopes with the bayonet, while the rest of the column established its bivouac in the gardens. The attack was immediately planned. Lieutenant-Colonel Forey, of the 58th of the line, with the 6th battalion of chasseurs and some companies of his own regiment, was to attempt an escalade to the east. Two battalions of the 58th and Colonel d'Illens were to try and storm the ridge by a ravine that ran two-thirds of the way up its sides. It was about 1 P.M.; a bright sun was reflected from the arms and the rocks. The general was in the centre under some great trees, giving his orders with his usual precision and clearness. Castellane and the staff were near him, looking at this magnificent panorama, when some gun-shots and the drums beating a charge startled them on the left. These sounds gave birth to a new force, an unknown ardour in the soul. A few seconds later, the company of chasseurs whom they had seen exchanging shots with the Kabyles in a fir-wood, and trying to avoid the masses of rock rolled down upon them by the enemy, passed on to rejoin its battalion, Captain Soumain at its head, all bruised by an ox that had been cast down upon him. The firing became sharper to the east; and a horseman soon rode up to announce the capture of the smala of Sidi-Embarek by the Duke of Aumale.

At this moment Castellane moved to the east, near the Chasseurs d'Orleans. Arrived at the foot of the rock with a part of the battalion, Lieutenant-Colonel Forey, an old chasseur officer, ordered the men to unsling their rifles. "We have to escalade the rocks, my lads, with spirit: remember you are the Chasseurs d'Orleans." Immediately the charge sounded, and on they dashed over the roots and rocks and broken ground, climbing and leaping like apes, mastering all obstacles, despising the balls that fell direct among them, and the rocks rolled down on their heads by the Kabyles. Thus climbing up on all-fours, they reached a point beyond which to advance was impossible. Every Kabyle who peeped over the ridge was reached by their bullets, but many of their hands were crushed by the rocks sent down from above. It was a curious sight; a scene of the middle ages; you might have taken it for the assault of one of those ancient castles built on the brink of some precipice.

As soon as the general came up, he ordered a retreat, to save the lives of the brave chasseurs. A Kabyle prisoner pointed out two narrow paths by which the enemy had reached the summit, which they considered impregnable, the tracks being so bad that the cattle had to be drawn up by ropes. But as there was no water, a blockade of three days was sure to enforce a surrender.

The 58th had bravely tried to mount, but had been balked by a rocky ravine with a slight loss, including Colonel d'Illens.

The column, divided into two corps, guarded the north and south-east slopes; while the reserve and convoy were established in the midst of gardens, where the pomegranate-trees, interlacing their red flowers with the large vines that ran from tree to tree, afforded an agreeable shelter to the weary troops. At night the bivouac fires sparkled like so many stars along the slopes of the mountain; and an enormous flame, no doubt some signal, shone forth at the east end of the rock; whilst overhead towered the limpid vault of heaven, into whose depths the eye loved to wander. A large fire of olive-wood gave a pleasant warmth to the staff-officers, who passed the evening in smoking and chatting; while Captain Carayon-Latour, one of the best trumpet-players in France, woke up the magnificent echoes of the mountains with his hunting airs.

The blockade continued till the 28th, when the thirst on the ridge reached an extreme that forced the chiefs to demand *aman* (terms) from the general about twelve o'clock. It was a wild sight to behold the flocks rush like an avalanche down the dizzy steeps to the river; while from the rock whole tribes of men poured down like a torrent, amidst shouts, tumult, and dust. Sheep, goats, oxen, women, and children, altogether ran down to the water, while the men, with fierce countenances, suffered in sullen silence. The soldiers made a glorious supper that night on Kabyle sheep.

Thus all the population of the southern part of the Ouar-Senis was subdued at a blow; but the northern tribes had still to be brought into subjection. Accordingly on the 24th they started with ten thousand head of cattle for Teniet-el-Had, a new post established at the watershed three leagues from the plateaux of the Serssous. Two days afterwards they passed through the magnificent cedar-forest, from which you get a sight of Teniet-el-Had. The variety of views and of the scenery, its extent of nearly five leagues, and the splendid size of the trees, make this forest one of the most curious spots in Africa; yet it is not safe to venture there alone, as on all sides there may be seen traces in the shape of a hand-grenade, which indicate the presence of lions. Colonel Korte, of the 1st Chasseurs d'Afrique, was then the commandant superior of Teniet-el-Had; a man of estimable character, of a daring heart, and a perfect horseman. In July 1842, under Changarnier, he made a gallant razzia on Ain-Tesemsil, a plateau of the Serssous. With 200 chasseurs, supported by zouaves, he made a dash at a post of flying Arabs guarded by 1500 horsemen. The least hesitation would have been destruction; but he knew his men and his foe, and he cut off the retreat of the fugitives, throwing them back on the French column. There was much firing, and many chasseurs bit the dust; but Korte brought into camp two thousand camels, eighty thousand head of cattle, an immense booty, and a great number of prisoners.

While this razzia, justly celebrated throughout the province of Algiers, was related to them, the staff and column reached the new post. Teniet-el-Had (the col or pass of Sunday), thus named from an Arab market that is held there on that day, had only been occupied by the French since May (1843). No building had at that time been erected, and a simple earthen ditch protected the soldiers who were encamped under the great tents of the administration; but the climate was healthy, and the *morale* of the troops excellent, hence there were but few sick. The column found provisions prepared for them there by the foresight of the general, and after a short stay they departed once more for the mountains of Ouar-Senis. But the lesson they had received had quelled the insurrection of the mountaineers; they received the submission of numerous tribes, and were forced to return to Milianah on the 7th of June through lack of provisions.[*]

* Castellane's Souvenirs, p. 59.

GATE OF ORAN.

CHAPTER IX.

Province of Oran.   The Coast.

OUTLINE OF THE COAST — MOSTAGANEM — ARZEU — ORAN — NEMOURS — ORAN —
MERS-EL-KEBIR — THE GULF OF ARZEU — ANTIQUITIES — ST. MARIE — ORIGIN
OF MOSTAGANEM.

THE next province that we shall describe and analyse is that of Oran,
following the series adopted by the *Tableau* and M. Berbrugger.
This land of the west, the cradle and home of Abd-el-Kader, has been the
nursery of the boldest spirits and the theatre of the most daring exploits
in Algeria.

The province contains 102,000 square kilometres (39,270 square
miles) ; and 275 tribes, including 600,000 souls, besides 35,246 Europeans
and 21,630 natives in the towns : total 656,876.

We shall, as usual, first give a broad survey of the province, beginning
with the sea-shore. Following the coast to the west of Tenes we come to the
Darha district, part of which belongs to the subdivision of Mostaganem, in
the province of Oran; and after passing the Djebel Minis, or mountain of salt,
and the Zour-el-Hammam, we come to the mouth of the river Shellif, the

largest and most celebrated stream in Algeria.* It flows during the greater part of its course in the province of Algeria, and has been already noticed in detail. A short distance to the west of the Shellif we come to Mostaganem, so called, according to Blofeld, from the sweetness of the mutton fed in its neighbourhood. This town is built in the form of an amphitheatre, with a free prospect of the sea ; but in every other direction it is enclosed by a circuit of hills which overhang it. The inhabitants have a tradition, confirmed by some vacant spaces, that the present town is composed of several contiguous villages. In the middle of it, and near one of these vacancies, are the remains of an old Moorish castle, erected, as it appears from its construction, before the invention of fire-arms. The north-west corner of the town, which overlooks part of the port and the ditch, is surrounded by a strong wall of hewn stone, where there is another castle built in a more regular manner. But Mostaganem being too much exposed to every troop of Arabs who can take possession of the hills behind it, its chief strength lies in a citadel situated upon one of these eminences, which has a full command of the city and of the surrounding country. The population in 1843 consisted of 2500 persons, exclusive of the French garrison. Passing through a fine country. sheltered by a chain of hills which bounds it to the south and south-east, the traveller comes to Mazagran, a small mud-walled town situated on the western declivity of a chain of hills, within a furlong of the sea. This is the place where, a few years ago, it was stated by the French Government that 123 French soldiers had successfully resisted 7000 Arabs for three days. After calling forth the powers of poetry and painting, Colonel Lievre's exploit turned out to be a fabrication, if we may believe Mr. Blofeld.†

A short distance beyond this place is the river Sigg. The Habrah, another considerable stream, falls into the former, whose mouth is called El-Mockdah, the Ford, which, save in the rainy season, is entirely occupied by the sand, leaving the passage without water. Not far hence, under some steep rocky cliffs, are two small ports, one of which opens towards Mostaganem, the other towards the port of Arzeu, five miles beyond.

Arzeu, called by the Moors the port of Beni-Zeian, from the Kabyles living near it, was formerly a large community. The land many miles behind it presents a rich landscape ; but towards the sea rises a range of steep rocks, forming a breakwater to the country. The water used now by the people of Arzeu is brackish, being drawn from spots much lower than the sea ; but the whole city was once built on cisterns, which

---

* Lieutenant De France, who was taken captive by the Arabs at Arzeu in 1837, describes the Shellif as the principal river of the country, rising in the mountains south of Miliana, running east and west, and falling into the sea near Cape Ivi, between the Cape of Tenes and the Gulf of Arzeu : *The French in Algiers,* translated by Lady Duff Gordon, p. 124.

† Berbrugger records it as a fact, part i.

still remain ; and numerous ruins of aqueducts, temples (one in particular in very good preservation), and other large buildings which lie scattered along the coast, prove that formerly a very considerable city existed on this spot. Leaving Arzeu we come to Cape Ferrat, remarkable for a high rock which stands out to sea. At a short distance from this cape is Oran.

Oran is an important fortified city, about a mile and a half in circumference, built upon the declivity and near the foot of a high mountain. It is naturally a place of great strength, and has been made much stronger by art ; yet it is commanded by the neighbouring hills. Oran was taken by the Spaniards in 1509, retaken by the Algerians in 1708, and taken once more by the Spaniards in 1732, who left it finally in 1792, having adorned it with several beautiful churches and other edifices in the Roman style during their occupation.

With a fair wind, the passage hence to Carthagena in Spain takes fifteen hours.

The country surrounding Oran presents a variety of pleasing prospects and cool retreats, numerous plantations of olives, picturesque rocky precipices, and rills of water trickling or rushing down them. Five miles beyond Oran is Mers-el-Kebir, the Portus Magnus of the Romans, so called by Pliny from its great size. This harbour is formed by a neck of land which advances almost a furlong into the bay, sheltering it from the north and north-east winds. Two leagues to the west is Cape Falcon, beyond which are the isles of Ha-beeba ; and farther on is Figalo, not far from the Sinan, the last of the brooks which fall into the Ouad-el-Mailah, "the salt river," whose sources are situated at the southern confines of the plain of Zeidoure, through which the stream glides in a variety of beautiful windings. It may not improbably be near this river, which might occasionally be swollen by the rains, that the elder Barbarossa, after flying from Tlemsen, scattered about his treasure when he was pursued by the victorious Spaniards, his last though ineffectual effort to retard the pursuit of his enemies. The Ouad-el-Mailah, a little after its union with the Sinan, discharges itself into the Harshgoun.

To the west of the latter are several ancient ruins called Tackumbreet, where the city of Siga or Sigeum, once the metropolis of Syphax and other Mauritanian kings, was situated. Opposite Tackumbreet is a small island, the Acra (Ακρα) of the ancient geographers, forming the port of Harshgoun,* where ships of the greatest burden may lie in safety. Tackumbreet is on the western banks, near the mouth of the Tafna, the ancient Siga, whose volume is formed by the Isser (Assanus), the Barbata, and other tributaries.†

A short distance beyond the Tafna stands Djama-Ghazouat, which has been named Nemours by the French, and constitutes their last post towards

---

* Rashgoun, according to the Tableau.   † Blofeld, p. 83.

the frontiers of Morocco on the sea-board. This little town contained in December 1847, 503 Europeans; in December 1848, 429; and in December 1849, 405. The number of natives in 1849 was 42. In December 1848, 950 metres (3116 feet) of water-conduits, and 250 (820 feet) of sewerage, had been opened, for 13,431 fr. (537*l.* 5*s.*); and 2492 fr. 46 ct. (99*l.* 14*s.* 2½*d.*) had been devoted to the improvement of the fortifications. A *débarcadère*, or landing-place, has been also built (1847), 44 metres (144·32 feet) in length, at an expense of 10,421 fr. Analysing the fortifications, we find that the town wall, or *enceinte*, begun in 1845, has been finished; that the curtain 6-7 has been organised together with the interior communications; that a cavalier has been built on the *terre-plein* of bastion 3; and that a mule's road has been opened from the town to the heights of Touënt. The expense of the works has amounted to 32,720 fr. (1308*l.* 16*s.* 8*d.*), and it is calculated that their completion will cost 200,000 fr. (8000*l.*).*

As Nemours may almost be reckoned a new French colony, seeing the small infusion of natives in its population, we shall revert to the state of its agricultural and commercial industry in the chapter on Colonisation.

Six leagues to the west of the Tafna is Cape Hone, the foreland protruding from the ridge of the Trara mountains, and corresponding to the Great Promontory of Ptolemy. This cape nearly coincides with the 1° 40′ W. long. of Greenwich. A short distance to the west of Cape Hone is the river Twunt, which with the Trara mountains has been commonly regarded as the western limit of the province of Oran and of Algeria generally.

Before we pass to the survey of the interior, we shall linger a little longer about the coast, and dwell more minutely on its individual features, beginning with its capital.

The voyage from Algiers to Oran is usually performed in thirty hours, touching at Scherschell, Tenes, Mostaganem, and Arzeu. Let us now accompany Baron Baude and his disciple St. Marie to Mers-el-Kebir, which they represent as a better port of refuge than Gibraltar, where the sea is sometimes tremendous, the action of the winds terrible, and the anchorage bad.† In December 1825, fifteen ships were cast away on its shores; but nothing of this kind is found at Mers-el-Kebir, where the sea is not dangerous, and the anchorage might easily be made unassailable. The possession of the fort of Mers-el-Kebir used to be regarded as the key of Africa. It is of considerable extent, and the fire of its batteries sweeps the bay, at the farther end of which is the city of Oran, unapproachable by large ships on account of reefs and shallows. Mers-el-Kebir is a good refuge for vessels in storms, situated at the eastern entrance of the channel between Spain and Africa; and the currents of the shore, together with the westerly winds which prevail during two-thirds of the year, drive into

---

* Tableau, &c. p. 48.   † Baron Baude, vol. ii. c. x. p. 119.

the bay vessels coming out of the Straits of Gibraltar, and check the course of those seeking to enter the Atlantic.* Hence this would be a good place to intercept the communication between the Mediterranean and the Atlantic; and the correspondence of the Peninsula and Paris through Perpignan might pass directly through Carthagena and the richest part of Spain.† The most prominent part of the fort is surmounted with a lighthouse. A broad well-made road leads hence to Oran; and when arrived half-way, St. Marie descended to a cavern forming two grottoes, in the centre of each of which is a basin of water three feet below the ground-level. The water is brown, cool, and brackish, the temperature 40° of the centigrade thermometer, and is said to be useful in curing cutaneous diseases.‡

The position of Oran is delightful, forming an amphitheatre along two banks of a shady ravine, commanded by the solid and lofty walls of the Casbah. The appearance of the place shews its former importance; and the inhabitants are not in the miserable condition of the other towns of Algeria. The men look strong and vigorous; and naturally remembering that the Spaniards, who once held the place for a long time, thought it best at length to withdraw, probably anticipate the same result with the French. Hence they are animated by an innate feeling of pride and independence, which nothing can subdue.

"Having crossed the ravine," proceeds St. Marie, "we entered a broad well-paved street, planted with old trees, and leading by a gently winding declivity to the highest point of the city. Here we came to the gates of a barracked camp, adapted for infantry and cavalry, and at that time (1845) occupied by the Foreign Legion."

When St. Marie reached this point, he entered on an extensive plain beyond the city, where 6700 Arab Gooms (irregulars) of the division of Oran were being reviewed by General Thierry. The French possessions in the province of Oran covered in 1845, according to St. Marie, a superficies of 200 square leagues, the produce of which does not supply even the city.§ Five leagues (12½ miles) to the north of the city is a barren plain, called by the Spaniards Telamina, the soil of which is mixed with salt; and to the south, masses of ruins shew the sites of Roman settlements, probably abandoned on account of insalubrity. There are some plantations of cotton and madder in this direction; and prior to the French occupation in 1832, the country about Oran presented a flourishing aspect, but in 1845 there was nothing but ruins.

The city of Oran is built on two long plateaux, having a deep ravine between them, containing a river which turns several mills and supplies

---

* Baron Baude, c. x. p. 119; St. Marie, p. 167, *et seqq.*
† Baron Baude, *ubi supra.*   ‡ St. Marie, p. 167.
§ The Count must have made another grievous mistake here.

the city with water. It was given up to the dey in 1791 by the Spaniards, after an earthquake had destroyed every thing except the fort. Mount Rammra rises 500 metres (1640 feet) above the sea, and commands the city to the west, being surmounted by a fort called the Bastion of Santa Cruz. At the outlet of the road from Mers-el-Kebir stands Fort St. Gregory ; and to the south, on the sea-shore, is the Fort of Moune Point. The western part of the city is terminated inland by the old Casbah, which is used as barracks for infantry, the fortifications being in ruins. In the opposite part of the city, on an eminence overlooking the sea, rise the fine ramparts of the new Casbah : begun by the Spaniards, it was finished by the bey,* who made it his residence. At the south end of this part of the city stands the fort of St. Andrew.

In 1845 the houses at Oran were all in the Morisco style, with flat terraced roofs; the streets were broad and straight; and it was remarkable for the beauty of its chief mosque, ornamented with exquisite open-work sculpture. The ravine between the two parts of the city is chiefly occupied by gardens and orchards, in which the pale green of the banana contrasts finely with the rich tints of the citron and pomegranate trees. But European houses are already beginning to be built in this valley, so that these blooming gardens will doubtless disappear by degrees.

We will next take a survey of Mers-el-Kebir and Oran, in the company of our respected friend M. Berbrugger.†

Travellers seeking to reach Oran by water are commonly forced to land a little to the westward, as merchant-vessels seldom, and ships of war hardly ever, anchor before the town. The usual landing-place is Mers-el-Kebir, or the great port, which you reach passing to the westward of Oran, and leaving to the left the fort of Mouna, or rather Mona ; the name being probably of Spanish origin, and bestowed on the place on account of its being frequented by monkeys. Above Mona rises Fort San Gregorio, itself commanded by that of Santa Cruz, which placed on the culminating point of the mountain was held to be impregnable. The little rocky summit on which it is built forms, with the extremity of a neighbouring ridge, a very remarkable embrasure, answering the purpose of a landmark to seamen at some distance out at sea.

After passing Point Mona you enter the roads of Mers-el-Kebir, the best shelter on the coast of Algeria, and the only spot where great ships can hibernate. This bay is encompassed by very high land, save at its extremity, where a decided sinking of the hills creates the embrasure previously noticed. Violent squalls are apt to sally forth from this gully, even in summer ; and the Spaniards used to call these gusts of wind *polvorista* (dust-bearing).

* St. Marie says ' dey ;' which is clearly an error, as the dey always resided at Algiers, leaving Oran and Constantina to the tender mercies of his beys ; p. 175.

† Algérie, part ii.

The usual anchoring place is near the fort, a fine and solid Spanish structure, built by the convicts of the *presidios* (garrison). It is partly cut out of the rock on which its foundations stand, has an oblong form, and occupies almost the whole of the little peninsula that forms the northern point of the bay, and whose neck is closed by a bastion covered by a little *demi-lune*.

This part of the fortification is formidable; the ditches, which are entirely dug in the rock, having a mean depth of at least 40 feet. Broad platforms of paving-stones solidly cemented together can receive numerous pieces of ordnance, which would be protected on the sea and land sides by traverses in masonry of unusual strength and great frequency. When M. de Bourmont ordered the evacuation of Oran in 1830, they blew up part of the sea-batteries; but the mischief has since been repaired. The fort of Mers-el-Kebir, when the French took possession in 1830, had 44 guns, 24 to 36 pounders: they were of Spanish origin.

At the east end of the fort stands the pharos, a little tower painted white, whose summit only rises 28 metres (91·84 feet) above the sea. It was for a long time only furnished with a tin lantern and a long candle: it was provided in 1843 with a fixed light, raised 26 metres (85·28 feet), and discernible a league off.

There are two ways of reaching Oran, one by sea and the other by land. Though the distance by water is only three miles, the passage is often retarded, especially during east and north-east winds. The road by land follows the maritime declivity of the mountains that form the bay.

The new road, that has supplanted the old footpath, doubles Point Mona under Fort St. Gregory, and passes over a vast grotto hollowed in the mountain, into which the sea enters by an artificial opening, and where vessels are sheltered by artificial means that are placed there in stormy weather. In the same locality are great excavations, that answer the purpose of warehouses, that have been made in the rock itself, which is easy to work. Yet these vast underground passages, which have been too much extended, do not possess all the solidity that is desirable, and since the year 1831 they have been injured by several landslips.

Immediately after having passed the point of Mona, Oran is before you. This town is situated at the bottom of a vast inlet to the west of Cape Ferrat, between two strands of sand, and on the two ridges of a ravine (Ras-el-Ain, the source of a stream), in which flows an abundant stream. That part which stands on the left bank is badly built, and ruined in many places in consequence of the earthquake of 1790; this is the old town, which was inhabited by the Spaniards. On the right bank is the new town, crowned by the new castle or Casbah.

The position of Oran is highly picturesque; and when the traveller descries from the deck of his vessel the two groups of white houses (the old and new towns), bisected by a ravine dotted with very pretty gardens, in the form of an amphitheatre, cut by tongues of land, whence a number of streams come gushing down, setting several mills in motion by the way, the eye dwells with pleasure on the charming features of the scene.

But as soon as you land, and crossing the beach, you enter the quarter of the Marine, which precedes the two others, the illusion ceases, and you experience a feeling of disappointment, not uncommon on entering African towns. After having crossed this quarter, you reach, at length, the gate of the town itself, for the Marine quarter is only a kind of European appendix to Oran. The first thing that meets your eye at the gate is the pretty minaret, giving one of the most favourable specimens of Moorish architecture in the town, these being but very few in number. (See Cut, p. 149.)

Standing in front of the gates, you have the old town to the right. When seen close at hand, the whole deformity of this mass of crumbled buildings is exposed to view; their ugliness being increased by the loss of the usual coat of plaster that gives some degree of decoration to the commonest structure in Africa. To the left of the elegant minaret of the great mosque you see the Casbah. This castle extends above a lofty and solid rampart raised by the Spaniards, the only modern people whose massive erections call to mind the time-defying structures of the Romans.

The great artery of the town is called the Rue Philippe, adorned with sturdy and luxuriant trees, which give it at first the appearance of one of the boulevards of Paris. But on a more minute inspection of the houses bordering this avenue, and of the population circulating in the artery under this canopy of verdure, the traveller soon discovers that he is in Africa. The low houses surmounted with terraces, the white walls, and especially the men, of lofty stature, with bronzed features set off in a characteristic relief by the capuchon or hood of their black bournous or cloaks, some of them pacing along with Mussulman gravity, whilst others, gathered up near the shop of some Moor, preserve such an immovable attitude that you might take them to be the signs of the shops;— all these features stamp a special character on the locality, and very quickly dissipate all idea of analogy to European cities.

The Rue Philippe abuts in the square, and thence it continues, under the name of Rue Napoleon, to the south gate. This artery traverses the whole town, and constitutes the principal feature of the place; for the whole trade and circulation of Oran is centered there ; and if the pedestrian ventures into the side streets, he finds the bustle diminish in proportion as he leaves the main street, and at their upper extremity he encounters only ruins and solitude.

The reader must not imagine that the Colosseum of Hassan Pasha, represented in the adjoining cut, is an old Roman ruin, as its name, applied by Europeans, would seem to imply. The building, now converted into barracks, was built by Hassan, the last bey of the western province, to accommodate his harem. Its general character is bold and elegant,

COLOSEUM AT ORAN.

p. 157.

and it is a matter of regret that it has not been employed for a more congenial purpose.

We have already described a Moorish house; and as this only differs from others in being larger, we shall not enumerate its compartments; and we shall take a future opportunity to speak of an interesting phase of Eastern life, we mean the institution of harems.

The following description of the entrance to the port of Oran is from the graceful pencil of Marshal Castellane :.

" Entering the bay of Mers-el-Kebir at dawn, the traveller is greeted with a magical spectacle. First appear the houses of Mers-el-Kebir, clinging to the walls of the old Spanish fortress; next, the dismantled towers of St. Michael and the line of mountains, which for the space of one league borders the bay, separating the port from the town of Oran; lastly, the Fort of St. Gregory, proudly perched half-way up to the right, at the foot of Santa Cruz, an eagle's nest built at the summit of an arid ridge, commanding the town and the country. Beneath the fire of the batteries of St. Gregory, the houses of the town wind along the sides of the hill, and stop at the walls of the Chateau Neuf, a vast structure raised, facing St. Gregory, by the soldiers of Philip V. To the east, along the line of cliffs that frown upon the ocean, the eye discovers the mosque, which has been converted into the quarters of the Chasseurs d'Afrique, and was built by their labour ten years ago; farther on, along the shore facing Mers-el-Kebir, rise the naked slopes of the mountain of the Lions, and, in the horizon, the rocks of the Iron Cape (Cap de Fer). Not a shrub is to be seen on the whole of these hills and mountains, though a little verdure may be perceived at the entrance of the ravine of Oran, which is almost concealed by the angle of the mountain of Santa Cruz. A neat village, with its white houses, peeps out of the middle of gardens at the foot of the mountain of the Lions, on the sea-shore; and a slight haze often contributes in softening the harsh features and outline of the land, from which the breeze wafts a sweet perfume over the sea.

" The distance from Mers-el-Kebir to Oran is a drive of an hour and a half; and during the first years of the French occupation you were obliged to follow a narrow and steep footpath, which led by the fort of St. Gregory, and ascended 400 feet above the houses of Oran. Whenever your horse or your mule stumbled, you ran the risk of being thrown down into the sea. All these dangers are now removed. The soldiers of the garrison of Oran laid down their muskets on returning from an expedition, and took up the spade, which they wielded so efficiently under the directions of engineer officers, that they cut in the side of the mountain a wide and convenient road, in which various descriptions of vehicles may now run with ease and expedition between the town and the port."*

* Castel'ane, p. 295.

We are informed by Baron Baude that the Spaniards invaded and captured Oran in 1505, under Cardinal Ximenes; and that they lost it in 1708, at the time of the troubles occasioned by the war of the succession. Their administration was clever, and they managed to subdue the Arabs in a radius of 15 to 20 leagues (38 to 50 miles), destroying the ports of Hone and Haresgot; whilst Tlemsen and Mostaganem paid them tribute down to 1551, and the tribes of the Habra, of Canastel, of Agobel, and of the Beni Amer, made common cause with them.[*]

On the 15th of June, 1730, the Count of Montemar, with 27,000 men, landed again at Oran ; and they retained it with a nominal garrison of 3000 men, often reduced to half that number, till the earthquake of 1791. General Damrémont took the place in 1830, and the French were well received by the natives ; but the Bey Hassan, califa of Sidi-Ahmed, a chief of Tunis, to whom the province was yielded by General Clauzel, alienated their minds, in consequence of which the French government would not acknowledge his acts, and he resigned. Since then the province has been under a French governor, who is almost independent and absolute.[†] Lamoricière filled the post many years with credit.

The surface of Oran, within the walls, is 75 hectares (187 acres) ; and the population shews that there are 331 individuals per hectare ($2\frac{1}{2}$ acres). At this rate, if it had the density of that of

Sedan	. .	181	per hectare ($2\cdot47$ acres), it would reach	. .	13,575
Metz	. .	241	,,	,, . .	18,075
Paris	. .	264	,,	,, . .	19,800
Bayonne	.	415	,,	,, . .	31,875
Toulon	. .	524	,,	,, . .	39,300

In 1839 the French formed one-seventh of the total population; whilst the Spaniards composed one-fourth, the Jews one-half, and the Mussulmans, who before the French occupation were the dominant body, are losing daily their relative importance.

In 1833

The Europeans amounted to	.	.	.	.	1042
Mussulmans	,,	.	.	.	440
Israelites	,,	.	.	.	2372
Total	.	.	.	.	.	3854

In 1839

The Europeans amounted to	.	.	.	.	4837
Mussulmans	,,	.	.	.	1003
Israelites	,,	.	.	.	3364
Total	.	.	.	.	.	9204

In December 1847 the European population amounted to 15,191, in Dec.

---

* Baron Baude, vol. ii. p. 137.
† Ibid. c. x. p. 115. Rozet, Voyage, vol iii.   ‡ Ibid.

1848 to 15,324, and in Dec. 1849 to 17,281; the native population at the latter date numbering 7564. Hence the total population of Oran in Dec. 1849 was 24,888.*

In the twelfth century, when the coasts of the kingdom of Tlemsen and those of Andalusia were united under the sceptre of the caliphs, there were found at Oran vast bazaars and flourishing manufactures, and its port was full of Spanish ships.† In 1373 the Pisans formed great establishments in these seas by a treaty of commerce, whose precision and equity could hardly be surpassed by the diplomatists of the present day.‡

Oran stands in 35° 45′ 57″ N. lat., and 2° 40′ 52″ W. long. of Paris; 66 leagues (165 miles) west from Algiers. The harbour has from four to six fathoms of water, and is defended from the north-west by the point of Fort Mouna. The landing is situated between Fort Mouna and the town. Fort Mers-el-Kebir advances like a mole into the sea; and the best anchorage is found in this place, as it is the most sheltered part of the bay. The port of Mers-el-Kebir is about five miles by sea from Oran, and the intermediary intercourse is carried on by boats called *alléges* (lighters), owing to which circumstance it requires sometimes fourteen days to unload a vessel.§

According to Lieutenant Garnier, of the French navy, six line-of-battle ships, six frigates, and fifty smaller craft, can anchor at Mers-el-Kebir, if some of them employ four anchors.‖ The trade of Oran is still considerable, consisting of grain, cattle, leather, &c.; and there are also manufactures of burnouses at this town. The street of St. Philippe joins the two parts of the city, which is built on very diversified grounds, and possesses six gates. Shortly after the French took Oran, redoubts and blockhouses were constructed around it, and the garrison was raised to 4350 men. In 1837, a military colony of Spahis (native cavalry) was established at Messerguin, near Oran; and the forts of St. André and St. Philippe have been re-established by the French.¶

The land surrounding Oran consists chiefly of pastures, but to the east some arable land occurs.

We shall now lay before the reader the latest improvements effected at Oran, consulting the French official documents for our facts.

The *basin of refuge* of Oran, undertaken in 1849, intended to contain 4 hectares (10 acres), is destined for the reception of ships bound to Oran, but which are almost always obliged to moor at Mers-el-Kebir. The docks

* Tableau de la Situation, pp. 96, 113.
† Marmol's Africa, b. v. Edrisi, p. 230.
‡ This painfully interesting specimen of unhappy Italy's palmier days under the sun of liberty exists in MSS. in the archives of Pisa. " Mantissa veterum diplomatum populi Pisani a nobili viro Navaretti recollectorum quæ apud equitem J. Schippisium diligenter asservantur."
§ Rozet, vol. iii. p. 274.   ‖ Tableau de la Situation, 1839.
¶ Baron Baude, vol. ii.

had been established in 1850. Thirty metres (98·40 feet) of jetty, and 150 metres (492 feet) of quays, built in 1847-48, occasioned an expense of 388,000 fr. (15,520*l.*) ; and the quays and dry dock at Mers-el-Kebir cost 248,000 fr. (9960*l.*) From 1833 to 1849, 3000 metres (9840 feet) of principal streets, and 1100 metres (3608 feet) of branch streets, have been opened, costing 280,000 fr. (11,200*l.*); eight squares have been cleared and planted with 150 trees, besides a promenade planted with trees.

The aqueduct of Ras-el-Ain has been made (1841-42) 3100 metres (10,168 feet) in length, supplying 4,500,000 litres (99,000 imperial gaions) daily, and costing 70,000 fr. (2800*l.*) ; and the aqueduct of the Ravin Blanc, 1300 metres (4264 feet) in length, supplying 350,000 litres (77,000 gallons) daily, and costing 25,000 fr. (1000*l.*), was finished in 1845. Three water-conduits have also been built,—one at Oran, the second on the road to, and the third at Mers-el-Kebir, —at a cost of 244,000 fr. (9,760*l.*) : the second is 900, the third 5000 metres long (16,400 feet). 410 metres (1344·8 feet) of sewers have been opened in the ravine of Ras-el-Ain (1844-48), at an expense of 114,000 fr. (4560*l.*) ; and 700 metres of other sewers (2296 feet) in the streets of Oran were finished between 1837-39, for 30,000 fr. (1200*l.*) Oran possesses a palace of justice, built in 1837 for 10,500 fr. (420*l.*) ; and a civil prison, built in 1841-42 for 13,000 fr. (520*l.*); a school-house, costing 37,000 fr. (1480*l.*) ; and two churches, costing 149,997 fr. (5999*l.* 18*s.* 1*d.*), of which that of St. Louis was finished in 1850, holding 1200 worshippers ; two cemeteries, established in 1841-43, cost 19,000 fr. (760*l.*); and a douane in 1845, 181,157 fr. 53 cents. (7246*l.* 6*s.* 3*d.*) A hospice des femmes was erected in 1847-48 for 7300 fr. (292*l.*); and a caravanserai, afterwards turned into a hospital, was built at the same date for 163,270 fr. 56 cents. (6530*l.* 17*s.* 1*d.*).

As regards the fortifications of Oran, between 1832 and 1849 the defences of the coast cost 417,510 fr. 23 cents. (16,700*l.* 8*s.* 6*d.*) ; and the land-defences cost 1,083,000 fr. (43,320*l.*) The chief works consist in repairing and improving the town-wall and the detached forts ; in repairing the sea-face of the fort of Mers-el-Kebir; in beginning the coast batteries, save that of the Spanish jetty now in progress; in making cavalry and artillery quarters, barracks at the Chateau Neuf, the old Casbah and the Colisée, magazines, hospitals, &c.*

We have now completed our survey of Oran, and shall take a ride along the coast to Mostaganem, in company with Count St. Marie and the former excellent Bishop of Algiers, M. Dupuch.

"After breakfast they mounted their horses, the bishop wearing a violet-coloured robe with a gold cross on his bosom, and a three-cornered hat with two gold tassels. Over his robe he had thrown a white burnouse,

---

* Tableau, p. 387. (1849-50).

which was merely fixed round his neck; but the two vicars who accompanied him were in black; and they had two men besides, as escort and guides. They took the road to Arzeu, ten leagues (25 miles) from Oran, crossing a plain intersected by difficult ravines; the soil presenting a mixture of clay and sand, whose fertility was obvious from the healthiness and vigour of the vegetation, growing in patches. They observed some thistles and other plants six feet high; but the country looks uncultivated and desolate, and some fine olive-trees which they passed still bore traces of bivouac-fires.

Advancing, they passed through a village whose Arab name is Kerguenta, containing the ruins of a monument called the Medersa, constructed by the first bey who occupied Oran after the Spaniards retired. Within the building was a small mosque, containing two beautiful tombs of white marble. This mosque was surrounded by pillars, and surmounted by a dome, open at top; in the centre was a large palm, which reared its stately head above the ruins, and overshadowed them with its massive foliage. After passing through the village they saw the ruins of an aqueduct, almost hid beneath thick acanthus-plants, with water issuing from the midst of the ruins. They came soon again into the plain, where all vegetation, save dwarf palms, became more and more rare as they advanced; and at 4 P.M. they arrived at Arzeu, where they were obliged to go to a miserable hotel by the sea. St. Marie and Baron Baude, as usual, agree in pronouncing the little port much better sheltered than that of Mers-el-Kebir, and the surrounding locality has been prepared by nature for commerce and shipping. The water is unfortunately only deep enough for third-class vessels; and the indolence of the natives has left an evidence of the great quantity of corn once exported, at the time when the Spaniards forbade the natives to traffic in the port of Oran. The vessels which came to Arzeu for cargoes of grain threw their ballast into the sea, which has left an accumulation that obstructs the anchorage nearest the coast.*

The town is commanded by a fort guarded by veterans; and a little islet situated in front of the port serves as a jetty, at the end of which a large lantern used to be fixed up (1845) as a lighthouse. Very extensive ruins, and numerous Roman medals scattered about the plain at Arzeu, shew it to have been the site of an important city, and have occasioned some archæological discussion. The Spaniards built at this place vast magazines for barley, wheat, and salt, besides a quay of freestone; but after the abandonment of the province, it fell once more into the possession of the Arabs, who have suffered the buildings to decay, and ruined the port.†

* Captain Despointes, in his survey of the bay, states precisely the same fact.

† St. Marie treads on the heels of Baron Baude in his description of Arzeu. Captain Despointes' survey of the bay (1833-4), published in the Appendix of Baron Baude's Algérie, contains the same expressions as those employed by the Count. Thus the Captain

The country around is rich in salt-mines; the salt they yield being better than that obtained in Spain and Portugal, and only requiring that kind of labour for which the Arabs are adapted, namely, that of collecting and transporting. Arzeu was once the port of the kingdom of Tlemsen, which comprised all the valleys of the Shellif.

Next day they left for Mostaganem, which is about fifteen miles distant. It is wonderful that in this undisturbed district the French had not constructed good lines of road in 1845, communicating between Mostaganem, Mazagran, Arzeu, and Oran. Yet the works would be easy and highly advantageous to those towns. The Arabs in this part of the country are industrious, and the women of Mostaganem make the most approved haicks and burnouses. A Spanish merchant, M. Canapa, has established a house of business at Mostaganem, which appeared likely to answer.

The importance of the port of Arzeu, the largest on the coast of Algeria, induces us to extract a description of its hydrography by Captain Despointes, who was stationed there in the corvette Alcyone from May 1833 to March 1834 :

"Between Cape Ferrat and Cape Yvi you see a great inlet, to which the name of Gulf of Arzeu has been given. Almost all along the shore which forms this coast you find anchorage, in general open and offering little security in winter ; one alone appearing to me to unite all that constitutes an excellent shelter ; it is that which is named Arzeu.

During the winter that the Alcyone passed in these roads, it was observed that in strong gales, those blowing from the sea or north and north-east did not enter much into the bay ; only the swell became very high, and gave a rise of almost five feet, so much the more inconvenient because the broken sea occasioned by this swell often lays the ship on its broadside. The bottom, consisting of white sand mixed with plants, only diminishes insensibly in depth, which renders the holding ground excellent.

Save in storms, the prevailing winds come from the eastward, passing by the north to the west ; those that blow most violently come from the north-west and west. The sea is almost calm during the prevalence of land-winds ; and, however strong the land or sea winds may have been, during a six months' stay the Alcyone was always able to communicate with the shore, and a merchant-ship would never have been obliged to interrupt its loading.

A stone quay was formerly carried out at Arzeu for a considerable

---

says : " The numerous ruins, &c. on shore prove that formerly a considerable city occupied this spot. . . . Some Roman medals found at a slight depth, &c. . . . The Spaniards had built at Arzeu vast warehouses, sheltered by their solidity from the attacks of the Arabs. These warehouses were destined to house wheat, barley, and salt. It appears proved that the Spaniards not only carried on in that country the corn-trade, but that of feathers, of carpets, &c. ; and caravans even came there." It is almost too delightful to find witnesses agree so closely. St. Marie, p. 186; Baude, App.

distance seaward, and must have allowed ships to come in themselves and take up their cargoes. The old warehouses are still in good condition ; but the quay would require many repairs.

To give the port its ancient depth, and permit even large ships to anchor further in, sheltered from all winds, several dredging machines, &c. would be required, and very great care on the part of the officer commanding the station.

The Roman road that led to Mascara abutted near the port.

Continuing to follow the coast, at the distance of four miles, and almost S.S.E. of the point of Arzeu, on the height you see an Arab village, improperly called the village of Arzeu. The neighbouring country is very well cultivated, and shews a good vegetation; and the village contains many Roman remains. Ships wishing to take in their cargoes to the village must come and anchor at a cable's length from the coast, with a depth of seven fathoms ; and their communication with the shore would be often interrupted by the swell.

From this point to the bay of the Macta, which takes its name from the river that falls into it, you reckon three miles from west to east, and some degrees south. You may cast anchor all along this coast in sixteen fathoms ; still, though the bottom is good, consisting of mud and sand, it would not be prudent to trust it except in the fine season.

At the east of the point that forms the cove of the Macta, the anchorage is better, from the nature of the bottom, which is soft mud. Large vessels cannot enter inside the point ; they anchor in nine or ten fathoms, and are exposed to the N.W. and N.N.W. winds, which sweep the coast and give rise to a very heavy swell. Boats and small craft can find easy shelter in some species of basins, the works of man, and which probably served formerly as receptacles for the galleys. It would be very easy to fortify the cape, which forms almost an island near the dry land ; but the river which is found at the bottom of this cove is barred at its mouth.

The whole of this part of the gulf presents charming views.

Behind the bar, and for a mile up the river, you find four metres (13·12 feet) of water ; and it would be easy to make this river accessible for barges of thirty or forty tons.* The whole shore of this bay is also scattered with vestiges of Roman edifices, including a very perfect temple.

Leaving the Macta, you proceed along the east coast; and after having taken cognisance of the village of Mazagran, inhabited by Arabs, and only defended by low walls, you see Mostaganem, a rather considerable town, surrounded with walls and provided with a casbah. The least bad an-

* See the Reconnoissance hydrographique de M. Garnier, lieutenant de vaisseau.

A–F*

chorage (for none are good) off this place is at six cable-lengths from the
shore, in twelve or fourteen fathom, muddy bottom. You open the citadel
then, to the east, 40° south. In these moorings you are exposed to the
N.N.W. winds, circling round to the west, which reign rather frequently
on this coast during the winter. Save in this locality, the bottom is
every where scattered with rocks, rendering chain-cables indispensable.
The N.N.E. and E. winds that come down in squalls from the mountains need not occasion any anxiety; they can, however, be felt in the
bay, and enable ships to make out to sea. In winter it is especially
necessary to guard against the N.N.W. and N. winds; and it is prudent
to set sail when the swell rises from this side, and the weather seems
uncertain. When once the breeze has begun, it soon freshens, and you
would be overtaken by bad weather at your moorings. The communication with the land is bad enough, on account of the almost continual swell
that exists on this shore; and the Moorish boats that come from Oran
to seek for vegetables and poultry, and other slight goods, are forced to
draw up on land; accordingly, they often come and anchor at Arzeu, to
wait for the wind permitting them to make this manœuvre."

M. Despointes did not examine the coast of the bay beyond Mostaganem; but M. Jules Tessier, commissary of the king in that town, has
signalised at three miles to the eastward, a creek surrounded by rocks,
where, according to him, you might, with an expense of 100 fr. (4*l.*), procure an excellent shelter for small trading vessels.*

M. Lamping, who was quartered at Mostaganem in 1841 with the
Foreign Legion, states that it contained at that time from 4000 to 5000
inhabitants, consisting of Arabs, Spaniards, and Jews, besides the French
regiment in garrison there. The town must have been formerly much
larger, as is shewn by the number of ruins scattered without the walls;
but with the exception of a few mosques, there is no building of any importance. The former citadel, called the Casbah, was then in ruins, and only
garrisoned by some fifty or sixty pairs of storks, who have founded a
colony on the extensive walls.

Almost as much Spanish is spoken there as French or Arabic; nearly
all the natives speaking a corrupt Spanish, a kind of *lingua franca*. The
younger generation, however, *i. e.* boys from ten to fourteen, converse in
French with tolerable fluency, but somewhat marred by their deep guttural
tone. The ease with which the settled Arabs and Bedouins continue to imitate whatever they have but once seen or heard is very remarkable. The
district south of Mostaganem may be called the home of the Bedouins,—if,
indeed, these wanderers have a home. There the richest and most powerful tribes fix their tents, sow and reap their corn, and feed their flocks;

* See Baron Baude, vol. iii. Appendix.

purposes for which the country is well adapted. The large plains between Mostaganem, Mascara, and Oran, and the fertile valleys of the Shellif* and Mina, afford these nomades excellent pastures for their numerous herds, and an unlimited room for their horses and camels. During the whole winter, and till the month of June, which is their harvest-time, the Bedouins camp in these places ; but when the heat has burnt up whatever pasture was left, they retreat into the valleys and defiles of the Atlas, where food of some sort, though scanty, is still to be found for their flocks and herds.

In October, a few days are sufficient, after the parching heat of summer, to call into existence, as it were by magic, the most luxuriant vegetation : the richest verdure has sprung up beneath the withered grass, the leaves of the trees have lost their sickly, yellow hue, the buds have begun to burst, and the birds to sing their vernal songs; in short, this is the African spring. The burst of vegetation was the strongest in the vale of Matamor, which divides the fort of that name from the town, and which is watered by a stream. Every inch of ground there is turned to the profit of man. Magnificent fruit-trees, pomegranates, figs, and oranges, and the most various vegetables, cover the ground; and Spaniards, Arabs, Jews, and French are diligently employed in cultivating the fruitful soil.†

Baron Baude supplies us with the following important statistical and general facts relating to Mostaganem. At the end of 1839 the population of Mostaganem consisted of 1428 Mussulmans, 406 Jews, and 282 Christians. The Mussulmans are very industrious, and their women work hard, manufacturing haicks, burnouses, and all sorts of clothing. The markets of the town are greatly frequented, especially since the merchants Puggimondo, Bigarelas, and Canapa, a Jew from Gibraltar, have established themselves there to export grain to Spain. The Koulouglis,‡ who are those amongst the natives that are most inclined to make common cause with the French, were the particular objects of Abd-el-Kader's animadversion. Those of Mazouna and El Callah having shewn their inclination to place themselves under the protection of the French, the Emir caused their dwellings to be sacked, and part of the inhabitants to be carried off to people his new town of Tagadempt. The Koulouglis formed the chief

---

* Lieutenant de France describes his visit to the plain of the Shellif in these terms : " On the 23d August, at five A.M., we again left Kaala, and marched northwards ; and after a march of seven hours, we encamped on the very edge of the plain of Mostaganem, near the river Shellif. Our camp stood in a grove of ilexes and gum-trees, on the top of a mountain commanding the plain ; just such a spot as was selected by knights of old to build their castles on, for the better convenience of robbing travellers, &c. . . . I am too poor a hand at my pen to attempt a description of the beautiful and fertile plain which lay at our feet, covered with crops of various kinds, fruit-trees, herds, flocks, and tents." P. 120.

† The Foreign Legion, p. 87.

‡ Offspring of Turkish Janissaries and Moorish women.

part of the Mussulman battalion kept by the French at Fort Matamor in 1841. Many Arab tribes find an opening for their produce at Mostaganem. First come the Achems and the Aribs: the latter living near the Shellif, have flocks yielding fine wool, oxen and horses of a large size; the Achems, whose territory is adjacent on the south to that of Mostaganem, have likewise an extensive cultivated district.

Turning to our friend M. de Castellane, we find that " an Arab tale relates that two children were once playing during the Rhamadan (fast) on the banks of a stream that flowed on to the sea after running a league. In the midst of their play, the youngest, gathering a reed, carried it to his mouth ; and after giving it to his companion, said, ' Muce-kranem' (suck the sweet piece of cane)."* Hammud-el-Alid, the powerful chief of the tribe of Mehal, was at that moment debouching on the hills, and heard the words of the children. Wishing to found a town on that spot, Hammud had been puzzled as to what name he should give it ; the two children freed him from his difficulty, for he called his city (in A.D. 1300) by this name, according to the legend. However widely spread this legend may be, the warrior has left more durable traces of his doings. The fort of Mehal still exists ; and the works executed by the care of his three daughters have made his memory dear to all the inhabitants : for they owe aqueducts to the beautiful Seffouana ; their gardens to Melloula the graceful ; whilst Mansoura, a woman of great piety, drew down the blessing of Heaven on the town by building a mosque that became her tomb. It is no doubt to her prayers that Mostaganem owes its prosperity, which it always enjoyed, even under the Christians. A ravine watered by a stream separates the town from a little hill called Matamore ; the numerous *silos* (underground granaries) that the Turks had dug in it, enclosed by a wall with loopholes, having given it its name.

The principal military establishments occupy the crest of this hill, whence you discover a magnificent view. At your feet the town, its houses, its gardens ; in front the sea, with its mighty surges, incessantly moved by the west winds ; on the right, at a league's distance, high mountains ; towards the left the eye follows the woody slopes of the hills that fringe the sea in the vast bay of the Macta, that rise up to the point of the Cap de Fer, and shoot up the naked ribs of their grey rocks to the blue sky ; whilst at a distance, in the mist, the eye distinguishes the mountain of the Lions (Oran). The horizon is immense, yet the eye discovers, without difficulty, all its details ; but if the air is humid, if no air agitates it, as often happens when *dirty* weather is at hand, by a singular optical effect, distances become nearer ; and it would appear that a few strokes of the oar would suffice to bring you to the harbour of Arzeu, which you

---

* Compare this account of the origin of the name with that given by Blofeld, p. 150, i. e. ' sweet mutton,' et puis revenons à nos moutons.

perceive, with its white houses, on the opposite shore, at a league from the cape. Four thousand natives, colonists from all countries, and a numerous garrison, live together in a friendly way at Mostaganem, passing every day without cares or grief. The Mussulman says, " it was written ;" the Christian, " never mind;" and the result is the same, for they know that the French commandant watches over all.*

* Souvenirs, p. 347. The *Tableau* gives Mostaganem 1300 metres (4264 feet) of street, and 1700 of water-conduits, supplying 700,000 litres (154,000 gallons) per day : the town is surrounded with an embattled wall, 10 to 13 feet high, 10,640 feet in circumference, and flanked with towers; the coast-battery has 5 guns, and the powder-magazine contains 55,000 lbs. Pp. 344, 354, 387.

CHAPTER X.

Province of Oran. Interior.

OUTLINE — TLEMSEN — MASCARA — TAGADEMPT — MAZOUNAH — A TOUR THROUGH THE PROVINCE — ST. DENIS — MASCARA — SIDI BEL ABBESS — TLEMSEN — NEMOURS — THE FAR SOUTH — TIARET.

L ET us now proceed with a broad survey of the interior of the province of Oran, after which we will analyse it more minutely.

Returning eastward from the river Twunt, and five leagues (12½ miles) south of the mouth of the Tafna, is Tlemsen (according to Arabic pronunciation, Telemsen or Tlemsan), almost surrounded with trees, and situated upon a rising ground, beneath a range of rocky precipices, the *Sa-rhatain* of Edrisi. These are part of the Middle Atlas chain; and upon their first ridge (for there is a much higher one to the south) is a large strip of level ground, from which a great number of fountains gush forth. These, after uniting gradually into small brooks, and turning some mills, fall in a variety of cascades as they approach Tlemsen, which they supply with an abundance of water.

In the western part of the city there is a large square basin, of Moorish construction, 200 yards long and 100 broad. The inhabitants have a tradition that formerly the kings of Tlemsen entertained themselves upon this water, whilst their subjects were here taught the art of rowing and navigation. But the water of the Sacratain, as Leo informs us, being easily turned from its ordinary course, this basin may have been employed as a reservoir in case of siege, being used at all other times to supply the beautiful gardens and plantations situated beneath it. Edrisi notices a structure of this kind, into which the fountain of Om-Yahia discharged itself. Most of the walls of Tlemsen have been moulded in frames; a method of building, according to Pliny, used by the Africans and Spaniards in his time. The mortar is composed of sand, lime, and gravel, well tempered and mixed, and as solid as stone. The several stages and removes of these frames are still observable, some of which are 100 yards long and two in height and thickness, by which the immense quantity used at one time may be seen. About 1670, Hassan, dey of Algiers, laid most of this city in ruins to punish its rebellious character, and only about one-sixth of old Tlemsen now remains. When entire it might be about four miles in circuit. Tlemsen contains many vestiges of ancient times ; and its houses are like the others in the province, low and mean in appearance, forming a great contrast to the ruins. It contains a fort capable of holding 5000 soldiers, with walls 40 feet high, circular in shape, like most Moorish forts in the inside. The population of Tlemsen was reduced in 1843 to about 20,000 souls, of which 1000 were Israelites. A few years ago a cannon-foundry was established there by Abd-el-Kader.

At the distance of half a mile from the present city is an immense enclosure, with walls surrounding it, and the remains of a half tower, about 60 feet high by 20 square. Half a league farther on towards the Tafna exists half of a similar tower ; and the Arab legend relates that these two halves once formed one, but that agreeing in a separation, the latter one fine morning walked away from its better half ; but some say that being built by an Arab and a Jew, they quarrelled about their claims, and the Jew's half took wing one night and perched on its present site.

On the banks of the Isser, the east branch of the Tafna, are the baths of Sidi-Ebly ; and after you have passed them commence the rich plains of Zeidoure, which extend through a beautiful interchange of hills and valleys to the banks of the Ouad-el-Mailah, for a distance of thirty miles. About the centre of them is the Shurph-el-Graab, or "pinnacle of the ravens," a high pointed precipice, with a branch of the Sinan running by it. The Ouled-Halfa and Zeir are the principal Arab tribes in this neighbourhood.

Six leagues south of the Sinan is Djebel Karkar, a high range of rocky mountains bending to the south ; and beyond them are the mountains of the Beni-Smeal, with the Arab tribe of Harars living a short distance south of them in the Sahara. Beyond these again, and at the distance of five days' journey to the S.S.W., are the villages of Figig, renowned for their plantations of palm-trees, and whence the western parts of the province are supplied with figs. Beyond the river Mailah, as far as Oran, is the Shilka, as they call a very extensive plain of sandy, saltish ground, which is dry in summer, but covered with water in winter. The Ammeers have their encampments in this neighbourhood ; a tribe which, from its intercourse with the Spaniards when the latter held Oran, have adopted some of their manners, To the south of the Shilka are the mountains of Souf-el-Tell and Taffarowy, which form a part of the Atlas chain ; the extensive ruins of Arbaal lie on one side of them, and those of Tessailah on the other. The latter, which, from their name, may be the remains of the ancient Astacitus, are situated on some of the most fertile plains of the country, cultivated by the Ouled-Ali, the enemies of the Ouled-Zeir and Halfa.

Crossing afterwards, almost in the same parallel, the rivers Makerra and Hamaite, both of which fall into the Sig, we come to Mascara, a collection of mud-walled houses, built in the midst of extensive plains, ten leagues (25 miles) from Mostaganem, with a small fort to defend it from the neighbouring Arabs. The Hachems, who are the Bedouins of this part of the country, are called *jowaite*, or gentlemen ; before the French conquest they were exempt from taxes, and served only as volunteers when required by the government of Algiers. Mascara is built on some table-land, between two small hills, commanding a view of the immense plain running north-east and south-west for several leagues. Its population in 1843 consisted of about 15,000 persons, including 500 or 600 Jews.* Five leagues north-east of Mascara is El-Callah, the largest market of this country for carpets and burnouses, and which, though much larger than Mascara, is a dirty, ill-built town, without drains, pavement, or causeways, being built, as the name implies, upon an eminence among other mountains. Several villages are scattered around it, all of them profitably engaged in the same sort of manufacture. El-Callah possesses a citadel in which the Turks kept a garrison ; and from the large stone and marble fragments found there, it may have been a Roman city, perhaps the Gitlui or Apfar of Ptolemy. Some leagues farther is the river Mina,† which falls into the Shellif at El-Had, near the plain of El-Mildegah, where the Swidde have their principal place of abode. El-Had may mean a mountain, by way of eminence, such as those of the Benizer-ouall deserve to be called, forming a ridge which runs here parallel with the Shellif. This part of the Atlas is famous for the quantity and delicacy of its figs, resembling those that the elder Cato praised when he threw them down in the senate, saying, " The country where this fine fruit grows is only three days' voyage from Rome ;" and history adds, that from that day he never concluded a speech without introducing the words, " mihi quoque videtur Carthaginem delendam esse." Sidi-Abid, a noted sanctuary, is situated four leagues farther, near the influx of the Arheu into the Shellif. On the opposite bank of the latter stream is Mazounah or Mezounah, a dirty, mud-built village, that contains no traces of the fine temples mentioned by Dapper and Marmol. It is, however, as remarkable for its woollen manufactures as Mascara and El-Callah, and it stands in a beautiful situation, under the side of the Little Atlas. The Ouled-Solyma are the neighbouring Bedouins.

Almost under the same meridian as Mazounah, and at the distance of

* Blofeld, p. 83 et seq. Mascara has 3,960 metres (13,018 feet) of street, and two squares. Tableau, pp. 345, 354.

† Lieutenant de France describes the country traversed by the Oued-Mina in the following terms : " Soon after midday we saw the village of El-Bordj, but we made a *détour* to avoid it, as it was market-day. Towards night, after travelling over various hills, many rocks, and much brushwood, through a savage and uncultivated country, we reached a little village at a few leagues from the falls of the Oued-Mina. The situation of this village, at the foot of a mountain, near several streams, is delicious ; rhododendrons, poplars, almond, fig, peach, and apricot trees, cover the whole plain ; and the gardens are kept fresh and green by a plentiful supply of water." P. 165.

eighteen leagues (45 miles), is Tagadempt, consisting of the extensive ruins of one of the oldest cities in Africa, which was governed by the ancestors of Abd-el-Kader, who tried to restore it, and made it for some time the capital of his dominions. In 1841 it contained 5000 inhabitants, including 200 or 300 Jews; one straight street, thirty feet wide, built in the European style; with two cafés, and also a manufacture for guns, which was able to make eight per day. On the advance of the French he destroyed the town, and forced all the inhabitants to desert it; and lions are now the principal inhabitants in this vicinity.

Returning to the Shellif, four leagues (10 miles) from Sidi-Abid, is Memunturroy, a large, old, square tower, being probably a Roman monument, and so called by the Ouled-Spahi, who live near it. Five miles farther from the banks of the Shellif are the ruins of Memon and Sinaab, two contiguous cities, the latter about three miles in circumference, and much the larger of the two; but nothing now remains of either of them, save some large fragments of wall and some large cisterns.

The most important French post in this part of the province is Tiaret, a little south of the Ouanseris district, in the province of Algiers.

To fill up our picture of this province, we have still to notice a few remarkable features in its eastern part, and especially the great plain of Mina. Starting from Touiza,* the valley widens to the last hills which sink down gradually at the distance of two leagues into the great plain of the Mina. This plain takes its name from a river which has its source on the high plateaux of the Serssous, crosses the country of the Sdamas, borders the Flittas district, and debouches at the south-west of this great plain; flowing in an almost straight line for the space of three leagues and a half ($7\frac{1}{2}$ miles), till it reaches the mountain of Bel-Assel. Then taking an oblique direction, it follows for three leagues ($7\frac{1}{2}$ miles) this new course, till it falls into the Shellif, which comes from the opposite quarter, i. e. the east; and the united waters fall into the sea at the distance of fifteen leagues (38 miles) from the confluence. Not a tree or any kind of shelter is to be found in this immense plain; here and there are scattered a few bushes of wild Barbary figs (jujubiers), slight undulations in the soil, and a salt lake. This dismal stretch of land has a framework of naked and misty hills; several parts of the plain deeply channelled by the rains are impracticable in winter. The Mina itself flows in a chasm twenty-five feet deep, that has been hollowed out by the winter floods. The fertility of this part of the plain, which is called the Lower Mina, is proverbial. The

* " Scarcely have you left the plain of Mina," says Castellane, " before you enter the valley of Touiza. This valley precedes the mountains of the Flittas, parallel to the sea and to the east, forming a large basin amongst these mountains, covered with lentisks, with here and there clearances sown with corn. To the south, and facing Touiza, is the defile of Tifour; to the west, two leagues off, opens the passage of Zamora; to the east, in this natural basin, winds a road leading to the Oued Melah, in the direction of Guerboussa. This road abuts at the khamis or magazine-post of Beni Ouragh." P. 226.

soil, formed in part of alluvial earth, can be partially irrigated, thanks to the embankments that the Turks erected at Relizann, and which the French have restored. Some day this African Bœotia will be covered with fine cultivation ; but in 1845 it resounded with the dropping shots of the Arabs.*

" At length," says Castellane, " in December 1846, the order was given to prepare for departure; but it was not for a very perilous expedition. The general treated us something like children to whom you give a little plaything to engage their attention ; he was going to take us a peaceful trip through the districts, where we were only to meet friendly Arabs congregated to salute the head of the province. Our little troop had soon finished all preparations for departure. By an invitation from the general, a companion joined us in the shape of a M. de Laussat, concessionary of the fine property of Akbeil, ten leagues from Oran. We all loved his merry yet serious mind, and his benevolence full of delicacy ; we therefore shook him cordially by the hand, when, punctuality itself, he arrived at 8 A.M. in the court of the Chateau Neuf. He was mounted on a bay horse, the only one that could be procured in haste ; but its transparent skin, and its thinness savouring of famine, caused the poor beast to be christened *Apocalypse* from the outset, amidst shouts of laughter. Notwithstanding the bad weather, the reader will perceive that blue devils were not our portion when we took the road to Mascara.

" At the moment of our departure, a violent west wind was sweeping the clouds before it; and so soon as we had cleared the first league, nothing met our eyes in the long distance but naked land, extending from Fort Sainte-Croix, and the arid ridges which terminate to the west of Miserghin, as far as the great salt lake, which we left to the right, and to the mountains of Tessalh, rearing themselves up in front of us in a line parallel with the sea. All was bare and leafless, for from the Basin of Oran the olive-forest of Muley-Ismael cannot be seen. To the eastward, near the sea, we saw mountains, hills, and these large stretches of country, —every where desolation. Still as we advanced, the tents of the tribe of Douairs seemed to thicken; and we shortly entered the fertile plain of Melata, where the Arab husbandmen were tracing shallow furrows with a plough similar to that which we see in the drawings of the first ages of Rome."

Proceeding they found an auberge (or country inn), built of boards, and a *petit verre d'eau-de-vie* (glass of brandy) to dispel the damp, on the desert banks of the Tlelat, where the industrious Martin, Lamoricière's well-known maitre d'hotel, was able to get up a kind of cross between French and Arab cookery. " Whilst we were breakfasting, the rain was resolved to share in the banquet, and we were obliged to mount our horses, the hoods of our *cabans* (light woollen greatcoats) drawn over our

---

* Souvenirs, p. 219.

faces to ward off those sheets of water that fall in all their glory in Africa. Happily the road crossed the forest of Muley-Ismael, and the stony ground resisted the hoofs of our steeds, joyful at having at last quitted the slimy and muddy soil of the Melata plain. In time of war crossing the wood is dangerous, and many engagements have taken place there. A little to the right we passed the mound where Colonel Oudinot, of the 2d Chasseurs, was killed in 1835 in a brilliant charge at the head of his regiment. Near the water-trough which General Lamoricière had established in the middle of the wood, in order to quench the thirst of the columns on their way, an old wild olive-tree is pointed out, covered with little bits of cloth, and piled round with stones. It is the tree under which the Cherif of Morocco, Muley-Ismael, stopped, when, 140 years ago, at the head of a numerous cavalry (of which the Douairs and Abids formed a part), he advanced to attempt the conquest of the country. This forest has taken its name from his defeat. Every woman whose husband is at the wars, faithful to the popular belief, throws a stone in passing at the foot of the olive, and attaches to it a bit of her clothing to preserve him from evil. At three o'clock P.M. we crossed the wooden bridge, and the drummer of the station saluted the entrance of the general into the village of Sig, composed of wooden huts and one stone house. As to the other buildings, they were either half finished or on paper ; and those of the colonists whom the fever had not driven to the hospital, passed their time in disputing. The previous year, when they built the enclosure of the village, all believed in its rapid prosperity. This part of the plain was healthy, the land proverbially fertile, the cannon resounded through the valley, the Arab horsemen were galloping full-tilt along the channels made for irrigation, discharging their muskets to salute the arrival of water in the plain. In fact, it was a great day ; for, under the skilful direction of the captain of engineers, M. Chapelain, the old Turkish dam had just been restored. Nothing could be more beautiful than this piece of masonry, 100 feet wide, raised with large blocks of stone, almost all taken from Roman remains, which covered the ground within a radius of 4000 metres (13,120 feet). Stopped between two rocks by this dam, the waters spread over the two banks by two principal channels, carrying into all the fields abundance and fertility. When, standing on the little bridge of the sluices, you turn towards the plain, whilst at your feet you hear the redundant waters leaping over the barrier, and rolling as they roar into the ancient bed, your eyes discover an immense horizon, a verdant and fertile plain, hills lost in the mist ; and on your right, eight leagues from the Sig, the marshes of the Macta and a series of sand-hills spreading out like the meshes of a net.

The general wished to ascertain the causes which prevented the development of a village placed in the best conditions for prosperity ; he, therefore caused it to be announced that after five P.M. he would receive all the colonists who wished to speak with him. This interview and its results will be described in the chapter on Colonisation. Suffice it to say here, that owing to the active measures adopted by the governor, a few months later, any one crossing the Sig would no longer have recognised St. Denis, so greatly was that village transformed.

" A little beyond St. Denis," proceeds Castellane, " you enter the gorges of the mountains which separate the valley of the Sig and the Habra from Mascara and the valley of Eghris. The night was dark when we crossed these defiles to arrive at the bridge of Ouad-el-Hammam (the river of the bath), where we were to bivouac. The next morning we had to start forthwith : we left behind us the little redoubt, where, in the revolt of 1845, a canteen-keeper, an old non-commissioned officer of some regiment, having been shut up in the blockhouse with two stout companions, held his post against the Kabyles, and was relieved by a detachment going to Mascara. The rain began again to pour down with still greater violence as we left the road usually followed by the *prolonges* or baggage-waggons, and we climbed the cross-road at the risk of falling into the ravines ; but at length we cleared the famous ascent christened by the soldiers *crève-cœur* (break heart), and we soon after met General Rénaud, who came to meet General Lamoricière, with a great number of officers, of Arab chiefs, and with the commandant of the city, M. Bastoul, who was regarded as the Solomon of the place. We had reached Mascara. The history of Mascara is connected with the most glorious recollections of the people of the province of Oran. In 1804 Bou Kedach, the dey of Algiers, confided the command of the west to one of his favourites, a young man twenty-four years of age, named Bou-Chelagrham (the father of the mustachio). Ambitious, active, and intelligent, Bou-Chelagrham had sworn to avenge the death of his predecessor, the Bey of Chaban, killed by the Christians of Oran ; but before he turned his arms against the infidel, he wished to reduce the whole province under his authority.

Until then the town of Mazouna, situated in the Dahra, between the Shellif and the sea, had been the residence of the beys ; but being too distant from the centre of the province, they had seen a great number of tribes escape from their authority. The first act of the new bey was to quit Mazouna, and to transport the seat of the Turkish power to the other side of the first chain of mountains, to a spot called the country of the Querth, from the name of a Berber tribe which inhabited it. This position, which permitted the cavalry of Bou-Chelagrham to flank the tribes of the plains, of the Mina, of the Illil, of the Habra and the Sig, placed them equally within reach of the southern tribes, which up to that time had dared to defy the orders of the beys. The Turkish chiefs posted at Mascara had, moreover, an easy communication with Tlemsen by the lofty table-land near Sidi-Bel-Abbes.

The town of Mascara (Ma-askeur, the mother of soldiers) was built

MASCARA.

p. 174.

upon the last slopes of the chain commanding the fertile plain of Eghris. This place became the residence of the beys up to the time when they drove the Christians from Oran ; it soon prospered, and contained a numerous but not a very moral population, if we may believe the traveller Mohammed-Ben-Yousef, who says : "I had conducted the rascals to the walls of Mascara ; they found shelter in the houses of that town." Its inhabitants might be sad scoundrels, but it is quite certain that their military position was excellent. Accordingly at all times Mascara was considered by all military men as the key of the country ; and when General Bugeaud, having formed a strong column at Mostaganem, was uncertain whether he should march upon Tegedempt, the new post founded by Abd-el-Kader on the borders of the Tell, or upon Mascara, to establish his forces there, as General Lamoricière had advised him ; General Mustapha-ben-Ismael, being asked his opinion, gave this answer : "At the time of the insurrection of Ben-Sheriff (1810), there was a great council of greybeards of Turks and of Arabs. They discussed what it was best to do,—to go to Mascara, or to make war on the tribes by *razzia*. The men who were cunning in council, and all who were firm in their stirrups, were unanimously of opinion that they should go to Mascara. I have not the presumption to think that I know more than they, and that which they then said, I say now : 'Go to Mascara, and remain there.'" The army, nevertheless, marched for Tegedempt ; but they were soon obliged to return to the advice of old Mustapha and General Lamoricière. Established in this town during the winter of 1841-42, without provisions and without resources, General Lamoricière was commissioned to undertake, and successfully concluded a campaign, which secured the peace of the province, and struck the hardest blow at the power of the emir (Abd-el-Kader); whilst General Changarnier, the mountaineer, as old Bugeaud called him, by his daring energy forced the populations of the province of Algiers to sue for quarter.

Twice ruined, Mascara has now only a few Arab inhabitants ; on the other hand, its European population is numerous; and houses, barracks, and sundry military establishments have been erected on all sides, giving the place the appearance of a French town. Built upon two hills separated by a stream, whose waters turn a mill, surrounded by gardens and orchards, containing olives, figs, and other fruit-trees,—this ancient capital of the emir commands the fertile plain of the Eghris, the territory of the Hachems, which extends at its feet ten miles in breadth and twenty-five in length. Here and there large orchards of fig-trees break the monotony of this plain, the eye rests on the long ranges of hills, and to the westward on the lofty mountains which appear on the distant horizon, where their summits seem always floating above the mist. The Arab traveller Mohammed ben-Yousef has said : "If thou shouldst chance to meet a proud, dirty, and fat man, make sure he is an inhabitant of Mascara." " See if the saying of Mohammed-ben-Yousef is not true," added Caddour Myloud, the Douair officer, pointing out to us with his finger the first Arab whom we met at the gate of Mascara; and he began to laugh with that silent laugh which the habit of ambush-fighting gives a man. We were compelled to join in the opinion of Caddour Myloud, for in the midst of that motley crowd which pressed forward to salute the general, the native of Mascara could be easily recognised. Yet, Heaven knows there was a goodly show of Arabs and Kabyles with patched haicks. As for the Europeans, each man had the costume of his own country, of the north or south, of Spain as well as of Italy; there were specimens of all lands ; and at the moment when our horses could hardly make way through the crowd, our travelling companion M. de Laussat, who was at my side, suddenly heard himself called by his name and addressed in the purest *patois* of the Pyrenees. Astonished, he turned his head ; it was a Bearnais (native of Bearn) who had spoken to him, a man with a bold and manly face, quite delighted to have met *Monsieur* there. As soon as he had recognised his countryman, a stroke of the spurs obliged Apocalypse to cross the road, and the hand of M. de Laussat squeezed with emotion that of the native of his paternal village. Merry and contented, this Bearnais had a pretty government grant among the gardens of Mascara ; all went well with him, and he made M. de Laussat promise to come to his house and taste the wine of his own vintage. The halting place was in the square or *place*, situate in the centre of the town, near a large and carefully preserved mulberry-tree. Scarcely was he dismounted, when the general began to hold a full court for the expedition of business, whilst the band of the regiment played its flourishes ; for it was Friday, and on that day the *twelve women* of Mascara dress themselves in all their finery, under pretext of going to hear the music, and coquet with their looks with those of the garrison who are off duty, and who, when their service is ended, come to walk away their ennui, smoke their cigars, and take their glass of comfort at Vives, an illustrious confectioner. Vives, who had arrived with the first column that occupied the town, and at first could only boast of a canvas tent, had afterwards a wood hut ; at last, a stall in the street ; and his fortune progressed on a par with the town.*

We spent two days at Mascara; then, all affairs being finished, and the Bearnais wine having been tasted by M. de Laussat, we set out for Mostaganem ; but, instead of striking off in a line to the right, by the road which follows the ravine of the Beni-Chougran, we took the route of the

* Lieutenant de France says, their " camp was pitched at the foot of the mountain which bounds the plain of Mascara on the north, and a little stream, whose banks were covered with oleanders, ran through the midst of it. Mascara stands in the centre of a mountain gorge, on a steep and precipitous hill ; the white and cheerful-looking houses are surrounded by a perfect grove of fig-trees, and a few graceful poplars and slender minarets rise like lances amongst them." P. 144. Its walls are completed, and its powder-magazine contains 66,000 lbs. Tableau, p. 387.

prolonges (baggage-train), and marched to the west, in order to visit El-Bordj (the fort), whose outer wall had been erected by the soldiers. We were to breakfast and bivouac there, at the foot of the mountain, by the fountain whose waters are lost in the plain of the Habra. Whilst chatting, we arrived on the little table-land of El-Bordj, where we were to receive the hospitality of Caddour-ben-Murphi. The great tents of the bivouac, all of white canvas, were pitched at the gate of the enclosure, which caused this spot to be named the Fort (El-Bordj). A detachment of soldiers of the garrison of Mascara were occupied at this moment in raising the wall, and building in the interior (at the expense of the Arabs) stone houses for the agha and his horsemen. The general was enchanted with these works, which he justly regarded as very important: for the Arab will not be actually reduced under our sway till the day when, through all the country, the stone fixing him to the soil, he will not be held, as now, to the earth merely by the stake of his tent. He encouraged by his praises those brave soldiers who, as soon as peace is restored, dropping the musket, shoulder the pickaxe, and give their sweat, as an instant before they would have shed their blood, for the grandeur of France. It was past noon before the general had finished looking at every thing; and after having been on horseback since five in the morning, our stomachs cried hunger. Our pleasure was great, therefore, when we found ourselves seated with legs across on the carpets of the great tents, and saw the large dishes of cous-coussou,* the ragouts with piment, and roast mutton, marching in on the heads of the Arabs.† Advancing farther, the west wind had brought up clouds, and the clouds, after their confounded fashion, the rain in large drops, which soon made our horses slip in the muddy declivities of the mountain; very fortunately, rain and wind ceased an hour before we arrived at the fountain, where we passed the night. The next day, at an early hour, the country sparkled under a beautiful sun, and we traversed the fields, which were adorned with their first verdure; saluted by the sharp cries of the women of the Douairs, uttered, according to the custom of the Arabs, to do honour to the chief of the province. The spectacle which surrounded us was truly singular. Animated by the ride, every one looked brilliant and joyous. On all sides was heard the sound of arms and spurs, all the noises which are the precursors of combat; one might have said, indeed, that we were preparing to run to danger, whilst we had only one hour's march before meeting General Pelissier, commanding the subdivision of Mostaganem, who awaited us at *the three marabouts* with the 4th Chasseurs-à-cheval: bronze faces, with long moustachios; tall men, proudly seated on their little horses. This regiment was worthy of the cavalry whose name alone carried terror into the enemy's ranks. Colonel Dupuch then commanded that valiant troop, whose flourishes

---

* A kind of porridge and soup combined.   † Castellane, pp. 341-2.

M

animated the march as we crossed the valley of the gardens which precede Mostaganem. This valley, covered with fruit-trees and figs, is sheltered from the sea-winds by the hills along the coast : it is the usual promenade of the inhabitants of the town of Mostaganem.*

After this trip Lamoricière and his staff returned to Oran, where they made a short stay, before undertaking another promenade *pacifique* to Tlemsen, &c., which Castellane describes in the following terms : "After the departure of the Mareschal and the deputies, nothing more detained the General de Lamoricière at Oran. He gave orders, therefore, to prepare to depart. We were going to traverse the west of the province, as we had a short time before run through the circles of Mascara and Mostaganem. The following day at twelve, after having been accompanied on our journey by a companion of joyous temper, a beautiful sun which made the moistened grass sparkle, so that it seemed just sprung as by enchantment from the earth through the early rains, we arrived at the Roman ruins of Agkbiel. These ruins, which extend to the south of the hills of Tessalah, belonged to M. de St. Maur, who came to receive us at the limit of his domains, followed by two harriers, his only subjects. The impression which you retain of these places is very singular. If the traveller climbs the highest ruin and allows his eye to wander over the immense plain, he is seized with one of those sensations which issue in Africa from the very bowels of the earth, and which the scenery of France has never begotten. Before him, at his feet, the great salt lakes, whose crystallisations shine like diamonds in the sun ; to the right are the undulating lines of the earth, which unite with the mirage of the air, and seem to float and disappear in the mist ; on the left you behold verdant and woody hills, whose semicircle closes at Miserghin, to shoot up again in a rocky ridge, and whose slope gradually rising, attains the summit of Santa Cruz,—a rocky bluff on which the Spaniards chose to found a fortress, whence the eye wanders over all the country. More distant, blending with the blue sky, the spectator discovers a dark line ;—it is the sea, whose waves bathe the shores of Provence; but on the right, the wild aspect of the Mountain of the Lions reminds him that he is very far from France. At some distance from the Roman ruins, our neighbours of Bel-Abbes, the Goums of this post, were waiting for us. As the rain continued to fall in torrents, so soon as the ground permitted us, we set off at a round trot, and at five o'clock our horses were fastened to the cord in the camp formed by two battalions of the Foreign Legion, which was bivouacking near Bel-Abbes. Situated behind the first chain of mountains, eighteen leagues (45 miles) to the south, upon the meridian of Oran, the post of Bel-Abbes commanded the flanks, and assured the security of the plain of the Melata, presenting to our columns a prompt means of drawing supplies

* Castellane's Souvenirs, p. 346.

when they had to carry on operations at the extreme edge of the Tell and
Serssous.   Founded in 1843, under the name of Biscuitville, by General
Bedeau, the establishment of Bel-Abbes belonged to the series of magazine-
posts which every twenty leagues—*i. e.* every three marches of the infantry,
and every two marches of the cavalry—were raised upon two parallel lines
running from the sea-shore to the interior, throughout the whole extent of
the province of Oran.   When the war took a decided turn," continues
Castellane, " we owed a great part of our success to two different causes,—
the creation of magazine-posts, and that of the Arab bureaux, or offices.
The magazine-posts indeed multiplied our forces, by approximating resources; and the Arab bureaux, by securing a proper employment of them.*
The following day we took the route to Tlemsen, under the escort of two
fine squadrons of African chasseurs; for since the Beni-Hamer had been
led to Morocco by the Emir in 1845, the year of the great revolt, all the
country from Bel-Abbes to the Isser was empty and delivered up to high-
waymen.   The sole inhabitants now of these fertile hills were some lions,
whose traces we often saw in the shape of large footprints majestically
engraven on the earth, some hyenas, and wild boars.

We disturbed their repose by giving them a vigorous chase; and this
did very well as regards the wild boars and hyenas, but the lion was generally respected.   This chase is not without danger; not on account of the
boar,—with a little skill and coolness one can always avoid the strokes of
his tusks,—but these cursed Arabs who accompanied us, without troubling
themselves as to whether we were in front of them, did not cease firing,
at the risk of missing the beast and sending the ball through us.   It was
far from Bel-Abbes to the Isser, where we were to bivouac; and it was
quite dark when the little column arrived at the bank of the river : without moon or stars, we did not know where to set foot; and it was necessary to find out the ford, for the river is rapid and wide in this spot.
The first who attempted the passage tumbled over, a second was not
more fortunate, but a third gained the opposite side.   Then lighting some
branches of the wild jujube-tree, torn from amongst the neighbouring
bushes, we stuck these torches on the top of our sabres, and the whole
troop passed without difficulty.   At daybreak the trumpets of the chasseurs sounded the réveil.   The air was sharp and animating ; a few clouds
were floating over the blue sky and the tops of the mountains, forming to
the east and south a kind of horse-shoe, that marked out the basin in
which Tlemsen is built.   The Mansourah and its admirable waters, which
spread fertility through the environs of the town, was in front of us ; on
our left, a little behind, we perceived the hills of Eddis, where, about the
end of December in the year 1841, the solemn interview was held which
decided the subjection of the greatest part of the country.†

* Castellane, p. 367.   † Ibid. p. 369.

This country of Tlemsen is not, however, easy to govern; at all times
it has been the theatre of great struggles ; and many centuries ago, Si-
Mohamed-el-Medjeboud (mouth of gold) said, "Tlemsen is the stony
ground in which the hook of the reaper breaks. How many times have
women, children, and old men been abandoned in its walls !" The history of this town is only a long description of war, since that famous siege
of Tlemsen, in 1286, by Abi-Said, brother of Abou-Yakoub, the Sultan of
Fez,—who during seven years kept the Beni-Zian in a state of siege, and
caused a tower to be constructed within his camp, the ruins of which still
exist,—to a blockade which the Commandant Cavaignac sustained behind
the walls in 1837, with the volunteer battalion (*bataillon franc*). We
arrived at the bridge which had been thrown over the Safsaf by the
Turks, and before us extended the large olive-trees which shaded the
entire country, and spread themselves out like a green carpet at the foot
of the town. Nothing could be more beautiful, more graceful, or more
charming than this city, whose white houses rested, on one side, against
the slopes of a rocky mountain, which poured forth in majestic cascades
its spouting waters, irrigating at their feet a rich enclosure of fragrant
gardens ; whilst in the distance, hills succeeded hills, and mountains were
piled beyond mountains, blending with the blue line of the sky."*

M. Berbrugger gives the following description of General Clauzel's
march to Tlemsen in 1836 :

"It was on the 8th of January, 1836, that the French army left Oran,
under the command of Marshal Clauzel, and took the road to Tlemsen.
There was an urgent necessity for this expedition, as the French auxiliary
chief, Mustafa-ben-Ismail, and the garrison of Turks and Koulouglis whom
he commanded, had just experienced a somewhat serious check ; they were
closely besieged by Abd-el-Kader in the citadel named Mechouar, and provisions as well as ammunition were on the point of failing them. Now,
after having encouraged them to resist the Emir with energy, it was out
of the question to desert them in misfortune.

The first day's march took the army by Meserguin to the Ouad-Bridia,
on the northern shore of the great Sebkhah (salt lake), which at that time
contained, instead of water, a kind of yellow mud or deposit. On the
second day they halted for the night on the banks of the Ouad-el-Malahh,
which is also called Rio-Salado, the Spanish translation of the Arabic appellation. By the way they discovered two emissaries of Mustafa-ben-
Ismail in the brushwood. This chief announced that Abd-el-Kader was
in Tlemsen, and that he was arranging to carry off the inhabitants the
moment that the French appeared. He added that their arrival was
anxiously expected by the Koulouglis.

On the third day they encamped in a pretty circular valley formed by

* Castellane, p. 375.

the Ouad Senan and another small river. They observed at this spot the ruins of a fortress built with blackish stones of a volcanic appearance, and forming the remains of the citadel called Qasr*-ebn-Senan in Nubian

MEETING OF MARSHAL CLAUZEL AND MUSTAPHA.

geography. Edrisi even asserts that there was a considerable town on this spot in his time.

The following day, at Ain-el-Bridje (the fountain of the little fort), they arrived at the remains of some Roman structure, situated near a fountain, where a stone belonging to an ancient sepulchre has been discovered. This spot appears to have been the site of a kind of fortress; but the expedition only brought to light a single tumular inscription void of interest.

On the 12th of January they had reached the Ouad-Amiguera, and were only separated by five leagues from the end of the expedition. They shortly learnt from Mustapha that Abd-el-Kader had departed, taking away 2000 inhabitants; and on the 13th the expedition approached Tlemsen in two columns; the main body, under Marshal Clauzel, advancing along the high-road to the town, reached Ouzidan, a truly delightful spot, whose beauty was increased by its contrast to the barren country that they had just traversed. The marshal was soon after met by Mustapha-ben-Ismail,

* Qasr or Ksour, the same word as Alcazar.

and after a short interview entered Tlemsen with his army, amidst the salutes and cheers of the Koulouglis."

The antiquities of Tlemsen are described in another chapter, and we shall here simply state that it is a very ancient Moorish town, built near the site of a Roman city. We cannot thread the mazy web of Moorish dynasties that have held sway in Tlemsen, which was once the capital of a great kingdom. Omitting many details in its history, we proceed to observe, that the dynasty of the Beni-Zian falling into disgrace through the abuse of despotism, saw its vast empire dismembered. Mostaganem, Mazagran, Tunis, and many other towns, had chosen individual sovereigns when the Spaniards conquered Oran ; and the Turks, masters of Algiers, strove to extend their authority westward. The dissensions that arose in the family of Beni-Zian favoured the general tendency to dismemberment that manifested itself in the kingdom of Tlemsen. The usurpation of Bou-Hamou, who seized the reins of government to the detriment of his nephew Abou-Zian, increased the confusion. Baba-Aroudj, or Barbarossa, the lucky Turkish corsair, who had just founded an empire at Algiers, was then engaged in reducing Tunis ; and learning the events at Tlemsen, he resolved to profit by them. He advanced with his army as the supporter of Abou-Zian, and the gates were opened without a blow being struck, on his promising on the Koran to restore the legitimate sovereign, Abou-Zian, whom, however, he at once strangled, exterminating all the other members of the family on whom he could lay his hands. The Spaniards were annoyed at his neighbourhood, and sent an expedition from Oran to dispossess him, under Don Martin de Argote. Barbarossa, shut up in the Mechouar, was soon reduced to great straits for want of provisions, which induced him to attempt a flight by an underground passage ; and though he scattered gold and silver on his path to delay the pursuit of the Spaniards, he was overtaken on the banks of the Ouad-el-Malahh, or Rio-Salado. After a desperate fight, Garcia de Tineo, a Spanish officer, killed Baba-Aroudj, and cut off his head; which being sent to the governor of Oran, was forwarded to the monastery of St. Jerome at Cordova. To this trophy was added his vest of red velvet embroidered with gold, which the monks used as a priest's vestment (chape).

Bou-Hamou was replaced on the throne by the Spaniards ; but Khair-eddin, brother of Barbarossa, soon re-established the Turkish power by becoming the patron of Messaoud, who disputed the throne of Tlemsen with his brother Moussa-abd-Allah, both being sons of Bou-Hamou. At length, under the rule of Salah-Rais, pasha of Algiers, the Turks became complete masters of Tlemsen, driving away Mouley-Hhaçan, the last prince of the Beni-Zian dynasty, under the pretext of his holding relations with the Spaniards at Oran.

Henceforth the annals of Tlemsen became blended with those of Algiers, the last event of note in its history being its siege and partial destruction by Pasha Baba-Hhaçan, in 1081 of the Hegira (A.D. 1676).*

The territory of the town of Tlemsen, backed by the mountain of Tyerm, is contracted between the river Ouad Safsaf,—which lower down, before falling into the Isser, is called the Sikak,—and the Ouad-Hermaya, one of the tributaries of the Tafna. Numerous brooks of fresh water, some of which are employed as water-power for mills, irrigate this fertile soil, whose powerful vegetation presents in a small compass the trees of Europe and Africa combined. To the west and north the outskirts of the town are decorated by a complete forest of magnificent olive-trees, regularly planted, and yielding a considerable return.

The old enclosure of Tlemsen, which has a development of five thousand metres (16,400 feet), consists of walls composed of a mortar of sand, lime, and small stones that have been cast into moulds. This structure, remarkable for solidity, has suffered much less from the ravages of time than more recent edifices raised in the same place. The modern enclosure, scarcely a third of the ancient, is an earth-wall (*en pisé*) flanked with towers. It is often broken, is without ditch, and surmounted with terraces on the east and south sides, having on the former side an angle with a demi-lune before it.

The interior of the city exceeds even most Arab towns in the irregularity of its thoroughfares. It contains such complication, and is such an inextricable web of confusion, that the stranger once involved in its labyrinths can scarcely find out his starting-place. As a compensation, it used to enjoy the luxury (in hot-climates) of streets covered with trellis-work, but civilisation has of course banished them; and the houses, which consist of one story only, are not whitewashed outside, as at Algiers, which gives the town externally a dull appearance.

The mosques of Tlemsen are numerous, but of little importance, save the Great Mosque, whose minaret is not deficient in elegance, but unhappily intestinal wars between the Koulouglis and Arabs have much injured it.

The most remarkable monument of Tlemsen is the *Mechouar*, a citadel situated south of the town, which it touches, but which it only imperfectly commands. This fort, which has no ditches, contains a hundred houses and a mosque. The garrison maintained there by the Turks used sometimes to amount to 3000 men, from which the size of the mechouar may be inferred. The French have built handsome barracks in it.†

Outside the town, at the distance of about one mile to the west, you meet a vast enclosure of earthen walls (*en pisé*) called *Mansourah*. It is stated that a town used to stand there, though not a vestige of a house re-

---

* Berbrugger, part ii.
† Castellane, p. 377 ; Tableau, p. 387.

mains.   The minaret of a destroyed mosque is the only ruin on the spot;
and this monument, which is built in rather a bold style, is ornamented
with arabesques in very good taste.

MOSQUE, ETC. AT MANSOU RAH.

The outskirts of Tlemsen are tolerably well cultivated, and present
several villages of considerable size: including Ouzidan, near the interior
bridge of the Ouad-Safsaf; El-Abhad, better known by the name of Sidi-
Bou-Medin or Medina, a marabout who is interred there in a splendid
koubbah, which has been sadly injured since the French occupation; Ain-
el-Hhadjar, at six kilometres ($3\frac{3}{4}$ miles) to the north-west of Tlemsen;
Ain-el-Hhout, at four kilometres (2·4 miles) to the north; El-Hannaya-
Tralemt and Melitia. You find, moreover, some genuine villages of
Troglodytes, whose inhabitants are called Rharanizah (people of caverns),
at Qalaâly, Chelebi, &c.   It is supposed that their dwellings are excavations made at a remote period in quarrying.*

From Tlemsen Lamoricière and Castellane journeyed west to the post
of Lela-Marghnia, on the frontier of Morocco,† by a road much in-

* Berbrugger, part ii.   † Souvenirs, p. 379.

p. 184.

TOMB OF SIDI-BOU-MEDINA.

fested by lions. They halted the first day by some hot springs, in one of the strangest situations imaginable. Around them was a dark stony ground, red sandstone soil, with sombre olives clothing the hills. Suddenly, at the turn of the road, a magician's wand conjures up a fairy-scene, a garden of Armida. Enormous palm-trees shoot up, bound together by the creepers of vines and parasitical plants; and under this dome of verdure the boiling waters bathe the foot of the gigantic trees. The scene exceeds the wildest dream of Oriental poet. It seems like the enchanted shades where a mysterious genius makes his abode; and it has its wild legend.*

In the evening they reached the French post of Lela-Marghnia, a quarter of a league from the frontier, and separated by a plain of six leagues (15 miles) from the Morocco town of Ouchda. This immense plain is watered by the Oued-Isly, and is the scene of the great victory gained by Marshal Bugeaud over the hordes of Morocco in 1844. After crossing the scene of the battle, Castellane's party reached, at two hundred paces from Djema, the funeral column raised to the fallen French, under the shade of large carob-trees, in the midst of a meadow; and five minutes after they entered Djema. This mazazine is built on the sea-shore, at the mouth of a little river, between two steep cliffs, where you perceive the ruins of villages formerly the nests of pirates. Barracks in planks, a loop-holed wall, large magazines, some cabarets; on the shore some fishermen's barks, and small craft belonging to the French navy; and in the roads some transport-brigs, or at times a war-steamer; and, amidst all this, busy soldiers, cantinières, and tradesmen;—such was Djema, or Nemours, in 1846.

It is a dull place of residence, the chase and study being the only resources of the officers. Their mess-room and café was a hut of deal planks, and their fare blue wine; instead of the elegant saloon of the Frères Provençaux, its gilded panels, mirrors, and nectar. But then they were jolly companions every one.

The next morning nothing detained them at Djema-Ghazaout,—or Bugtown, as it was then christened. This sobriquet will explain their anxiety to leave it. The road back to Oran passed through Nedroma, a cool and shady town surrounded by good solid walls, with rich and industrious inhabitants, where, according to the report of evil tongues, it is said that money is so beloved, that nobody inquires about its source.†

Leaving Nedroma, they began to ascend the Kabyle mountains; and they found on the road a population furious at being obliged to submit, but paying their dues without daring to say a word, for the sight of a regiment that escorted the governor made them as gentle as lambs. Passing on, they chased a hare under a brilliant sun, after regaining the plain, and before crossing the col that brought them to Ain-Temouchen,

* See Part II., Chapter on the Arabs.
† Is not this the character of places nearer home?

on the road from Tlemsen to Oran. In the heavy soil of Sidour, a kind of dismal swamp of the province of Oran, the rain pouring down in torrents, their horses kept floundering and sliding, and the officers indulged in a cross-fire of oaths.

At Ain-Temouchen, in the revolt of 1845, the post had but very little ammunition. Colonel Walther Esterhazy attempted a relief with 500 Arab goums; and ordering the march, a caid made some observation and refused to obey. The colonel blew out the caid's brains, and two others shared the same fate. This act of energy overawed the wavering, and the post was relieved. This spot, called Chabat-el-Lhame (the flesh-defile), is also noted for the heroism of 1000 Spaniards, who fell almost to a man, facing the enemy, overwhelmed by numbers.* Twenty alone escaped to Oran, which was reached the same evening by Lamoricière and the staff.

We shall next accompany Castellane in an expedition to the oases, unveiling some of the wonders of that region of the sun. While serving in the west, a servant of the Rhomsi† rejoined Castellane's party, bringing a letter from the commandant of their little column, giving them order to return as soon as possible, because their squadrons were leaving for Saïda.

" We took in much haste the direction of our bivouac ; and we learned on arriving that we were destined to form a part of the column of General Renaud, which was to leave on the 1st of April for a long excursion in the oases of the south. This was, to our minds, a rare piece of good fortune; and when, a few days after, the column with its long convoy quitted Saïda, we were all delighted to penetrate at length into those regions, of which so many strange things are related. A train of mules carried a supply of water, as whole days would pass without our finding any ; two thousand camels belonging to the Hamians and the Harars were loaded with provisions, and extended in single file, descending the slight elevations, and mounting the little hills, to the monotonous songs of their conductors. The hares fled by hundreds before these new *rabatteurs* (men who beat up the game in battues); and the camel-drivers, frightening them by their cries, and throwing their knotted sticks at them, soon got the best of it, and those which escaped them fell under the teeth of our greyhounds. At night our bivouac resembled a vast market. The Arabs carried the game they had got in the day from fire to fire. Upon the table-land of the Serssous, political economy might for once have justified one of her axioms; for it was with much difficulty, whilst offering a hare in one hand and holding out the other, saying, 'Donar soldi,' that the Arabs succeeded in getting rid of their merchandise, so terrible had been the massacre of the morning. Two days after, we bivouacked upon the border of the Chotts. These immense salt lakes, dried up in summer, are only passable in April in a very

---

* Castellane, p. 392.
† A tribe of Arabs in the province of Oran.

few places. The day after, at the réveil, every one was ready; alas! we had been awakened long before by the lowing of the camels, which their conductors were loading in order to be in time. These cries are one of the punishments of an expedition in the south, on the other side of the Chotts. We were going to seek the Bled-el-Rhela (the country of void); but at the dawn, before we planted our foot upon the other side, the long file of camels appeared to assume the most grotesque forms in its narrow passage. One seemed to have only an immense head, others swelled out like sails, many appeared to send out flames and to float in the air; again, several walked with their legs uppermost, and in per- petual motion. This was one of the singular effects of mirage, so com- mon in the Chotts, and which are considered fabulous by those who have not seen them. Our guide was an Arab *de proie*—a man of the Hamians; a freebooter of the high lands, an adventurer, with the hooked nose of a vulture, eyes black and liquid, with a thin, bronzed, calm, and impassible physiognomy, a true type of a Saharian. He now directed us to the wells, where, under the branches which covered them, we found an abun- dance of pure water. At our departure the branches which protected them were religiously replaced; for a well in the Sahara is a sacred spot, which demands the care and protection of every traveller. Our march then con- tinued in this country of void, whose wastes have not the grandeur of other wastes. They oppress the heart instead of elevating it. It seems as if a heavy curse lies on all around; and we advanced into these naked plains, seeing right and left, and as far as the horizon, arid mountains without vegetation, offering nothing on which the eye could rest. In fact, that part of the Sahara we were then crossing was of sad celebrity, and it is never more than a passage for the nomadic inhabitants of these countries. A small body of western Hamians (Garabas), not subjected to France, were with their flocks at about twenty leagues (50 miles) from us. The general heard this from his scouts; and as for several days we bivouacked only in the hollows, and during the day the mirage prevented the dust raised by our column from being seen, we were certain that we were unobserved. Therefore, at 3 o'clock P.M., a picked body of six hundred infantry, with the cavalry and the general, left in the hope of effecting a bold stroke. The rest of the infantry and the convoy directed their march towards the wells of Nama, where we were to meet them the next day.

"The heat was overpowering; but these men, inured to fatigues, feared neither the burning sun nor the chilling rain. At six in the morning the column halted; the Arab scouts returned, announcing that the camels of the Hamians were grazing at a distance of three hours' march from us. This was an evident sign of their security. The infantry had already marched fifteen hours; and from the spot on which we halted to the wells of Nama, we were four hours' further march. If the attack proved a failure, that would make nearly thirty hours of active duty. The general dared not send off the cavalry alone, and to the extreme regret of the Arabs, who reckoned on the booty, the order was given to take the direction of Nama. At one o'clock, after having crossed these sandy downs under a burning sun, without having found a drop of water since the day before to refresh our parched lips, we arrived at the place of bivouac with only five men on the cacolets (or sick-list), and that caused by accident. The cavalry had gone on before ; and when on the summit of one of these sandy undulations, our squadron perceived an immense sheet of water, whose banks were reflected in the transparent waves as in a Swiss lake, there was a universal shout of pleasure, and we hastened to unbridle our horses to water them; but as we advanced we saw the water recede before us at a distance of about six feet, so that we soon discovered our error. We were again the dupes of a mirage. However, we found water in the sand-hills at seventy paces on our right. It was necessary to draw the water from the wells, in order to pour it into the troughs which surround them. The next day the baggage and the rest of the column had rejoined us a few hours' before, when a most frightful hurricane swept over us. In ten minutes the whole sky became a curtain of clouds ; the thermometer fell of a sudden, and whirlwinds of snow succeeded the most overpowering heat. Happily we were all together, otherwise it would have been all over with us. At three paces off we could not see each other ; and for fear of straying, we were obliged to gather, to the sound of the trumpet, the broom which covered the downs,—this being the only aliment we could find to feed our fires. The day after the ground was covered with snow. Imagine what the sufferings of that night and the two following days were, for this darkness continued during that period. At the first returning rays of the sun, the sands of the stony ground in the plain absorbed the melted snow. The air, however, continued icy cold; but we were advancing south, approaching the mountains, of which we soon reached the highest passes.

" We met occasionally a pistachio-tree of meagre foliage, or the violet-flowering broom, growing amongst the limestone rocks and the reddish soil. Our men, marching in open column, descended a steep slope in the direction of Chellala. The same sullen, desolate, melancholy aspect pervaded the whole district ; and our horses trod on nothing but the *alpha*, a sort of little round rush, or those small shrubs whose salt-flavoured leaves are so much liked by the camels. When the eyes have been unrefreshed for many days with the sight of verdure, it is scarcely possible to imagine the delight with which pure running water, foliage, large leaves, and trees whose shade shelters him from the sun, is welcomed by the weary traveller. For several days the heat of the sun had been insupportable ; therefore, when we arrived at the oasis of Chellala, our previous sufferings were enough to make us find its sickly fig-trees and scattered palm-trees very delicious.

" The general received the homage and the tribute of the town (if an assemblage of mud-built houses deserves the name), whose narrow miry streets displayed a wretched sickly population. There, as every where else, the grasping Jew has taken up his habitation, and meddles in all the transactions of the place. This was the first ksour or town that we had passed in our journey; but our stay was only short, as our destination was now towards Bou-Semroun, an oasis situated more to the south, whose inhabitants had refused to pay tribute.

" A rather broad and sandy valley has to be crossed in order to reach Bou-Semroun. On either side rise arid mountains; and parallel with them towers aloft a rocky eminence, in the form of an inverted shell, leaving a space between the foot of the mountain and the base of the rock. A minaret gives notice of the proximity of the town, which is only hidden from view by a small eminence. From the summit of this sand-hill, its gardens of palm-trees, enclosed in a narrow ravine of two leagues (five miles) in length, appeared like a stream of verdure between two banks of sand. The inhabitants had fled; but gun-barrels glittered on the minaret, from whence several fanatics, wishing to die in the sacred cause, fired upon the infantry sent to occupy the ksour.* The column bivouacked south of the town, between it and a marabout of elegant architecture. Who could have constructed it in this remote country? Without doubt, some Christian prisoner. The Greek crosses introduced in the ornaments made us assume this. The ksour resembles a citadel, surrounded by a broad ditch and good mud walls, having but two outlets. Bou-Semroun could defy mere plunderers; and in these narrow alleys, and in these houses of two stories, the merchandise, the corn, and the riches of the nomadic tribes, are in safety. Happily the unsubdued inhabitants had not thought of defending themselves; otherwise it would have been necessary both to sap and mine, to have taken their fortress. Open doors enabled us to enter their houses, amongst which several overlooking the ravine had a certain degree of elegance about them; they were no doubt the dwellings of the chiefs. Our bivouac, with its movable houses, had been established close to the gardens. After descending the arid dry declivity, the scene suddenly changed into one of freshness, of calm, and of repose, cooled by the abundant waters of a pure limpid stream. Here every field is surrounded by an earthen wall (*en pisé*), very solidly constructed; and a wooden lock protects the barley and grass, the pomegranate and the fig trees of the inhabitants of the ksour. Enormous tufts of palm-trees shoot towards the sky, their lofty crowns meeting above. It was a magnificent park in which to repose after our fatigue. The gardens supplied us with fresh vegetables, green barley for our horses, besides the cane of the palm-tree, which each foot-soldier cut as a remembrance of this expedition to the

---

* The name given to the Saharian towns and villages.

south. To our great joy, we remained in this lovely spot for a whole week; and we occupied ourselves during the halt in inventing fresh pleasures and amusements.

" Inaction was a fatigue, and motion a necessity to us. Thus one evening, to the sound of the trumpet, as on a village-green, a grand steeple-chase was announced for the next day in the gardens of Bou-Semroun. The general, as umpire and mayor of the place, was invited, according to ancient custom, to preside at the fête. Every body flocked to it; the exquisites on horseback, and the humble trooper with cane in hand. A *cantinière*, nominated Queen of Beauty, was to present the winner with a beautiful pair of pistols offered by General Renaud. The stake was worthy of the peril; for never did the Croix de Berny offer, in its most prosperous days, greater difficulties : 2400 metres there and back; walls, gates, impediments of every kind; tufts of palm-trees, of which we had to keep out of the way; and to sum up our difficulties, after one stone wall came another of earth. The horse had to jump, at a height of three feet, through an opening just large enough to admit his body; whilst the rider had to throw his legs on to the neck of the horse in order to escape injury. Such was our race-course.

" All things came off according to rule. A member of the Jockey Club, a *real* member, gave us the starting word in English, and the galloping avalanche cleared gates and impediments. But, alas! there was more than one fall; and I assure you that it is no joke, when on the point of reaching the goal, to find yourself under your horse's hind-legs, with every chance of having your jaw smashed at his slightest movement, were it not that the poor beast itself is half dead; then to see the hoofs of all the other horses levelled at your head, as they drop close to it, before they can clear the unexpected obstacle which you present to them. The sensation of all this is singular and rapid, and has at least the charm of surprisal. Without broken bones, or even a scratch, we were all well, and each of us laughed at his mischances to keep up the general hilarity.

" Thus the time flew rapidly away. Without care, without disquiet, without sickness, the column was in condition to have supported the severest fatigues. The onions of Egypt were regretted by the Hebrews in the desert. Our soldiers may be pardoned, therefore, for having sighed more than once at the remembrance of the small tender onions of Bou-Semroun, when the time came for returning northwards; our course being first to the east, then towards the south, in order to reach l'Abiot-Sidi-Chirq, a celebrated marabout village in that country. The descent in the road was very steep. At length, after passing the last defile, an immense extent of horizon opened before us. On our right, high crests of mountains formed a half horse-shoe; and the chain extended eastward on our left. At the foot of the mountain, sand-hills crossed and recrossed each other like a network; and these yellow sandy billows mingled with the distant outline of the horizon. In front of us a flinty plain, two leagues in length, separated us from the four villages of the Ouled-Sidi-Chirq, encompassed by their fresh shady gardens.

" The depression which this desolate country had wrought in our minds disappeared on seeing these vast distances, leaving in its place an inexpressible sentiment of elevation and grandeur. A mosque, held in veneration by the faithful, occupied the centre of these villages. The chiefs of this important tribe, whose religious influence extends over all the Sahara, and even over a portion of the Tell, came to meet the general, to offer their homage and the expected tax. It was the 30th of April, and for a month there had been no news of France. More than 120 leagues (300 miles) of sandy desert separated us from the coast ; and here, at the gates of these mysterious countries, we were going to celebrate the *fête du roi* (king's birthday).* The evening before, the small howitzers that we use in the mountains announced the fête to the people of the south ; and the morrow each soldier exercised his skill to obtain the prizes offered by the general. Horse-racing, racing in sacks, sheep-shooting, games of all kinds as in a village fête, took place, accompanied by gay sallies and laughter. Each man forgot his fatigue, and scarcely thought of the distance which separated him from his family and from France. Two little negroes, with some ostriches and haiiks (Arab cloaks), presents to the general, reminded us, however, that we touched upon unknown lands ; as well as the rumbling of thunder, which is heard every day at the hour of prayer (three o'clock). (By a singular phenomenon, every day in summer, towards this hour, gusts of wind and a storm arise in Abiot, and continue about two hours.) These distant peals seem echoes of those far-off lands of which so many wonders are told.

" Indeed it seems that this mountain-chain, whose base forbids the farther extension of this vast sea of sand, is a barrier placed by the hand of God to stay the northman if he attempt to penetrate into these unknown regions. From the summit of these arid peaks, broken only here and there by narrow passes, the traveller can contemplate these solitudes and these sands, to which the voice of the Lord has said, as to the waves of the ocean, ' So far shalt thou go, and no farther.' But if the Christian must for a time abstain from travelling over them, the Arab, under the protection of his Moslem creed, knows not these obstacles ; and every year, attracted by the allurement of gain, numerous caravans furrow the desert, following the same routes of which Herodotus gives an itinerary.

" The generally impassible Arab experiences that sensation of uneasiness which every man feels before embarking on a long sea-voyage, when on the point of hazarding an expedition in the desert : in fact, these long journeys are much the same thing, since the same organisation and dis-

---

* This was in the reign of Louis Philippe.

cipline as on board a vessel are requisite to overcome similar dangers. Here, as at sea, when the passage is more than usually dangerous from robbers, one caravan awaits another to double its strength, and then they emerge together from the sheltering oasis, and advance without fear. The respect which is paid to the adventurous traveller on his return is a proof of the fatigues and dangers he incurs."*

Castellane gives the following description of the district of the Flittas, between the plain of Mina and the Ouanseris : " One day that we had set off hunting after the natives, very early in the morning, we had penetrated into a frightful ravine extending to the west of the watershed as far as the Mina. The road that we were following was two feet in width, and advanced along the steep slopes of a hill, abutting at the bottom of a ravine, whose left side it had previously followed. Evergreen oaks, lentisks, and other shrubs covered this dangerous ground. In the centre of the basin, the waters had worn a wide ditch through the rich mould, forming a ravine in a ravine. During the winter the unbridled waters rush furiously, forcing a passage, dragging trees along with them, and boring underground passages to arrive the quicker at this great central artery, 50 feet wide and 30 deep." But in summer and its five months' drought these caverns are accessible, serving as catacombs to conceal the persons and property of the rebellious Flittas, who were smoked out by the French, disgorging a torrent of men, women, children, and goats.

The reader will here remember the terrible tragedy of 1845, when Colonel (now General) Pelissier suffocated 1600 Arabs, or Kabyles, in a cave in the Darha, not far hence. Advancing with Castellane farther inland by the territory of the Kerraich, Temda, and the Ouad-Teguiguess, we approach Tiaret.

" The country changes completely on approaching Tiaret. Woods of evergreen oaks, some cedars, large prairies, and springs of water, take the place of the grey and naked shadows of the hills. A troop of gazelles fled before our horses, sometimes bounding through the trees, at others stopping as if to provoke us, but quickly vanishing if they perceived that they were seriously pursued. Occasionally the sun shone out from the clouds, . . . throwing its pale light (it was winter) on a part of the wood, whilst the long mountain of Tiaret prolonged the shadow of its wall-like precipices. At length we reached the pass of Guertoufa ; and then there opened before us, at the height of 200 feet, the crevice through which we had to penetrate. To reach it you have to traverse a stone avalanche or slip, and to climb the side of the mountain by a zigzag path. Eagles were majestically sailing over our heads. Nothing was heard but the ringing of our horses' hoofs, or of our sabres against the rocks. Amidst these obstacles, the soul is roused, and the sublimity of the view fills it with noble

* Castellane, p. 367.

thoughts. Then after we had reached the summit, what an imposing and magnificent sight met our eyes ! At our feet was unrolled, immense and luminous, the cascade of rocks that we had just passed, over which the bayonets of our infantry were glancing and flashing ; beyond these were woods, verdure, and-meadows ; further still an endless succession of hills, undulating like a sea to the horizon. At the extreme limit of the Guer-toufa, rose, lighted up by the sun and amidst bluish vapours, the lofty mountains of Bel-Assel. A little to the right, the two peaks of Tegui-guess stood out, after the fashion of a promontory ; and this sea of moun-tains was prolonged for 20 leagues (50 miles), till it met the foot of the Ouar-senis, whose long solitary ridge commands the country in a radius of 60 leagues (150 miles). Its form, resembling a fluted obelisk, gives it the appearance of an ancient cathedral topped by a majestic dome. The scenery breathed a sublimity and calm that carried back the thoughts to primitive times.

The defile, which extends 500 metres (1640 feet), brings you to Tiaret. This post, built of fine masonry, on the limit of the Tell and of the Little Desert, is renowned for the sweetness of its water. The Tell, the foster-mother of Africa, produces corn, just as the Serssous nourishes numberless flocks. It seems as though God wished to establish a barrier between these two countries, whereof one is the slave of the other, being mutually separated by a rampart of mountains. The mountains of Tiaret are the highest of all this chain, and can only be crossed by three passes. From Tiaret you discover a part of the Serssous. Beneath your eye stretches a plain of little rocky hillocks, and between each hillock or mamelon gushes forth a spring ; and, thanks to the kindly waters, a thick and sub-stantial growth of grass shoots up, nourishing immense flocks of sheep.*

According to the *Tableau de la Situation* for 1850, the European civil population of Tiaret amounted in December 1847 to 85, in December 1848 to 65, and in December 1849 to 81 persons. The natives at the latter date amounted in all to 63 individuals.

In connexion with its military works, it appears that from 1845 to 1849, the expenses amounted to 311,074 fr., appropriated chiefly to the construction of two barracks, a hospital, and a cattle-fold or stockade.

* Castellane, p. 238.

CHAPTER XI.

Probince of Constantina. Coast.

THE COAST — DJIDJELLI — COLLO — PHILIPPEVILLE — BONA — THE PORT — THE
TOWN — THE BUILDINGS — THE POPULATION — SANITARY CONDITION — MOUNT
EDOUGH — TRIP TO LA CALLE — AN ARAB TRIBE — LA CALLE — BASTION DE
FRANCE.

THE existing province of Constantina, or the eastern province of Algeria,
has a surface of 175,900 kilometres (67,721·5 square miles), containing a population of 1,300,000 inhabitants, and 580 tribes.

This province lies between the meridians of the Djidjelli and Zaine ;
but the old beylik used to extend westward to the Booberak, thus embracing the whole of that remarkable district known by the name of Great
Kabylia, to which we shall devote a special notice. According to the
older division, this province was nearly equal to the other two in extent,
being upwards of 230 miles long and more than 100 broad. The sea-coast
all the way from the Booberak nearly to Bona is mountainous, whence
it obtained from Abulfeda the name of El-Adwah, or the lofty.

The present eastern limit of the province on the sea-shore is half-way
between Dellys and Bugia, in the aghalik of Sebaou. From hence to the
kaidat of Ferdjiounah, you follow the coast of Great Kabylia, passing the
town of Bugia. As we shall devote a special notice to that portion of Great
Kabylia which is comprised in the province of Constantina, and also to
the portion now embraced in the province of Algiers, we shall simply
state on the present occasion that it presents a wild, mountainous region,
watered by several rivers, of which the principal is the Summam, falling
into the sea near Bugia ; while the highest summits belong to the great
range of the Djordjora, chiefly in the province of Algiers, overhanging
the Mitidja plain, and peopled by the Kabyles of the Zouaona tribe, who
are thought to be descendants of the Vandals.

Passing to the east of Great Kabylia, we shall first follow the coast-
line of this province, which brings us to Djidjelli (Igilgilis), situated near
the eastern extremity of the Bay of Bugia.

M. Lamping, who was quartered some time at Djidjelli, or Dschidgeli,
in 1841, with the Foreign Legion, has given the following description of
the place : " Dschidgeli has only a small roadstead, and is built on a rock

rising out of the sea. It belongs to the province of Constantina, and lies between Budschia and Philippeville : it is inhabited by Turks and Arabs, who formerly drove a thriving trade in piracy. Although the town looks like a mere heap of stones, it is said still to contain much hidden treasure. Notwithstanding that Dschidgeli lies nearly under the same latitude as Algiers, its climate is far hotter and more unhealthy; and the oppressive heat has a very remarkable effect upon all new-comers, whose strength deserts them from day to day, so that men who were previously as strong as lions creep about with yellow, pale faces, and with voices as small as those of children."

Fort Duquesne stands upon the sea, and defends the south-east side of the town. This fort is built upon a rock rising so abruptly from the sea, that a few half bastions towards the land are sufficient for its defence.*

The latest official accounts state, that from 1844 to 1849 the sum of 20,000 fr. (800l.) has been expended on improving the streets of Djidjelli. From 1843 to 1848 a channel was made to bring water to the town, whose depth is 20 centimetres (7·80 inches) ; that of the siphons is 95 millimetres (3·795 inches). The chateau d'eau, or reservoir, can receive 15,000 cubic metres (17,640 cubic yards). The expense of this work was 107,100 fr. (4284l.); and 640 metres (2099 feet) of sewerage have been made, at an expense of 2500 fr. (100l.) They have also built a civil prison at Djidjelli, from 1843 to 49, for 6500 fr. (260l.); besides a school, erected also between 1843 and 49, at a cost of 3500 fr. (140l.)

Independently of these structures, a church and a mosque have been built at Djidjelli between 1843 and 49; the first of which cost 3000 francs, and the last 7300 francs (292l.).

A market-house and slaughter-house were built at this town between 1843 and 1849, for 21,300 francs (852l.); and it has been provided during the same time with a cemetery, for 7100 francs (284l.).

The greater part of the military works at Djidjelli have been completed very recently; amongst others, the new Porte Constantine lately finished.

The new wall enclosing the town has been continued; and steps have been adopted to defend its approaches on the sea side. A permanent battery for nine cannon has been built in front of the hospital; and a provisional battery for seven guns has been established at Fort Duquesne. The total expenditure on these works has amounted to 83,000 francs (3320l.).

The remaining works that were required in 1850 were, the completion of the town wall (*enceinte*), and the construction of the permanent batteries on the coast.†

Baron Baude, who visited Djidjelli in 1849, states that it stands 13 leagues (32½ miles) east of Bugia; that the port is defended to the westward by the peninsula on which the town stands; and a chain of rocks

---

* The Foreign Legion, p. 22.   † Tableau de la Situation.

breaks the sea in front of the harbour. It is one of the best maritime
stations on the coast; standing on a headland, instead of at the extremity of a bay, and having moreover a good harbour. The building
slips of Djidjelli were once in high repute; and the town contained in
1839 about 200 sailors.

Djidjelli, or Gigel, was once a bishopric during the sway of the North
African Church; and Roman roads led hence to Bugia, Setif, Constantina,
and Hippo Regius. The town was encompassed for many years after the
French occupation by numerous blockhouses, placed in a semicircle on
the surrounding heights. Fort Duquesne stands on the sea-shore, defends
the S.E. side of the town, and rises abruptly from the water.

St. Marie* gives the following account of Djidjelli, which he visited in
1845 : " Left Bugia at 11 at night, and next morning at sunrise we were
off Gigelly. The port is defended on the west by a peninsula stretching
towards the north, on which the fort is built. Towards the offing it is
imperfectly defended by a chain of rocky islets, between which the sea
rushes with great violence in strong weather. This chain, which joins the
end of the peninsula, and runs east parallel with the coast, is more than
200 metres long (656 feet).

" Ancient Igilgilis was intersected by some Roman roads leading to
Bugia, Setif, Constantina, and Hippo. The French, Genoese, Venetians,
and Flemings had commercial houses at this place, which traded in leather
and wax. " On the 22d July, 1664, the Duke of Beaufort took possession
of it; and in a small fort commanding the town, which still exists, he left
400 men, who, dispersing afterwards, were massacred by the Arabs. A few
Maltese now carry on the coral fishery, and the French garrison is of no
importance. Leo Africanus gave Djidjelli or Gigel 600 hearths or fires at
the beginning of the 16th century ; and Aroudj (Barbarossa) took the name
of Sultan of Gigel in 1574; but in 1725 Peyssonel only found 60 houses
there."†

The Ouad-el-Kebir, or great river (Ampsaga), falls into the sea 10
leagues (25 miles) east of Djidjelli; beyond it are the Sebba-Rous, or seven
capes, where the Sinus Numidicus of the Romans may be supposed to
begin, and where also the river Zhoora has its influx. The Ouled Attyah
and the Beni Friquanah, two principal clans of the Sebba-Rous, use the
water of this river; and, unlike other Kabyles, they live in caves scooped
out of the rock, or found ready-made. When a ship comes near the shore,
they run in crowds to the coast, and pray to God to give it up into their
hands; reminding one of the Cornish clergyman, who, hearing of a wreck
in church, desired his congregation to give a fair start, that he might have
a chance.

---

* Every particular recorded by Baron Baude is naturally chronicled by Count St. Marie
au pied de la lettre, including the wax and leather exports, and the Duke of Beaufort.
† St. Marie, p. 201. Baron Baude, vol. i. p. 155.

Baron Baude informs us that the rocks of the Sebba-Rous consist of limestone; and that the crests are fringed with pines and carobs, and sprinkled with a few patches of cultivation.*

The Ras-el-Kebir is formed of basaltic prisms of pale green, which are found as far as Collo. You may discover the same formation, through a glass from a ship, on the top of the lofty peak of Coudia, where it answers the purposes of a landmark to point out the anchorage. Turning a rock, you now discover the masts of some *sandals*;† then the end of a quay, and a kind of warehouse built of rough unhewn stones; some fine trees, planted without symmetry; a mosque; and behind, on the slope, some houses of a miserable appearance, covered with hollow tiles. This is all that you see of Collo; yet it fills the whole of the little space intervening between the extremities of two hills washed by the sea, and behind which is a pretty plain. It was reported to have 2000 inhabitants in 1840; but it did not appear to Baron Baude to contain so many.‡

Leo Africanus calls it the most opulent and the safest place on the coast. This may not be strictly correct; but its vicinity possesses forests of oak, where the Algerian navy obtained its timber; and specimens of copper ore have been found there. The inhabitants manufacture a coarse stuff, and carry on a coasting trade with Algiers and Tunis. The anchorage before Collo is excellent, and quite sheltered from N.W. winds. Frigates can anchor at 500 metres (1640 feet) from the coast; and near the land you find 5 metres (16·40 feet) of water. At 3 leagues (7½ miles) to the south of Collo, the small lake which, according to tradition, confirmed by numerous ruins, formed the old port, has retained all its depth. It is only separated from the sea by a tongue of land of 100 metres (328 feet) in breadth; and the Oued-Zeamah is navigable 3 leagues up the country, and has its influx here.

Collo, sometimes written Cull, or in Latin Cullu, stands in a picturesque situation under the most eastern of the seven capes; but, like Igil-gillis or Jigel, its present condition is very poor; and it contains but few antiquities. Its harbour is small, though larger than that of Jigel; and the neighbouring waters and coast are said to contain many beds of coral; but the wild tribes of the vicinity have hitherto, in a great measure, neutralised this advantage.

Baron Baude, who touched at Collo on his passage from Algiers to Bona, was visited by many natives in boats, bringing fowls, apes, &c., and what they called little tigers. Many of the men have blue eyes, clear skin, and light hair. The same features are found among the Spahi Kabyles of Youssouf at Bona; and they must be the descendants of the Vandals. Collo is a very likely place for these children of the north to have retired to when Gelimer fled to Mount Edough (Pappua Mons). Procopius says,§

---

* Baron Baude, vol. i. p. 159.   † Native boats.   ‡ Baron Baude, vol. i. p. 159.

§ Belisarius Gelimerum persequens usque munitam venit civitatem, juxta mare sitam,

"Belisarius following Gelimer, came to a fortified city situated near the sea, which they call Hippo the Royal. He heard there that Gelimer had fled into the Pappuan Mount, and that it would not be easy to capture him there. This mountain, which is situated at the extreme limit of Numidia, is steep and difficult of access; being surrounded on all sides by very lofty precipices, where the barbarous Mamusii are friends and military allies of Gelimer." According to the opinion of Baron Baude, Collo stands on the site of the ancient Cullu. The anchorage he represents as good and sheltered, so that frigates can moor at the distance of 500 metres (1640 feet) from the shore; and 30 fathoms are found in many places close in shore. The neighbourhood of Collo has many natural advantages; forests of oak clothe the land in its vicinity, which is also said to possess copper veins.*

Advancing eastward, at the distance of 8 leagues (20 miles) you reach Stora, at the bottom of a cove formed by abrupt mountains. It was completely deserted in 1840, when visited by Baron Baude; but it contains more vestiges of antiquity than Philippeville. It stands on the site of Rusicada; and some paces from the sea are the ruins of some reservoirs, fed by a neighbouring source: the waves also bathe the foot of some old walls of rough stones and brick, which may not improbably have contained a fort for troops; but the hills surrounding it are too steep to have allowed of a large establishment. To the east the slope is wooded, and capable of culture; but the vale of the Oued-el-Kebir is very open, and turns in the direction of Cirta. Ancient Rusicada stood on a height that commands its mouth, and the ground on that spot is covered with its ruins. At an equal distance from Cirta and Hippo, it was united to both by a Roman road; and the country seems very easy to cut through by turnpike or rail roads. The anchorage of Stora is only preferable to that of Collo for small craft; it could not conveniently hold more than two corvettes: and, according to Baron Baude, it is not well adapted for a port.

The *Tableau de la Situation* observes, that the port of Stora, at a short distance from the new colony, is safer than the port of Philippeville.†

Stora is chiefly remarkable as the port of Philippeville, which we must now proceed to notice.

---

quam Hipponem regiam vocant. Ibi Gelimerum audivit in Pappuam montem confugisse, nec facilem a Romanis captu esse. Hic enim mons in Numidiæ finibus extremis, valde quidem abruptus, adituque difficilis, petris undique altissimis communitus, in quo Mamusii barbari habitantes Gelimeri amici ac bello socii.—*De Bell. Vand.* c. i.

* The Company of La Calle had an agent at Collo, and procured there honey, grain, a little cotton, oil, and 300 or 400 metrical quintals (88,000 lbs. avoirdupois) of wax at the fixed price of 180 francs, besides 30,000 raw hides. Those of oxen and milch cows were assessed at 4 francs 50 centimes, and at 2 francs 80 centimes. These relations, long interrupted, were renewed in 1816 ; and in 1820 the people of Collo drove out the Turkish garrison, and pronounced themselves independent ; but they soon recalled it, in order to recover the French trade which they had lost thereby.—*Baron Baude*, vol. i. p. 162.

† Tableau, 1839.

Madame Prus, one of the latest visitors to Algeria, gives the following description of this French colony, which was founded in 1838 on the site of Rusicada, and at the distance of half a league (1¼ mile) from the gulf of that name :

" The town of Philippeville, which was built by the French on the site of ancient Rusicada, has the appearance of a fine provincial town thinly inhabited. The walls which surround it defend it from the attacks of the Kabyles, who, notwithstanding this, succeeded in setting the town on fire a few years ago. Speedy measures, however, were adopted, and the flames were prevented from spreading ; but from that time the Bedouins of the country were forbidden to remain in the town after sunset.

" At every step vestiges of the old Roman city meet the eye, but it is impossible to obtain any account of them. The town is peopled almost exclusively by emigrants from Provence, Marseilles, and Corsica, as is the case with all the principal towns of Algeria. It presents a sombre aspect, as many of the houses are shut up ; and the number of bills for lodging visible in every window are a sufficient proof of the depopulation of the city. The hospital and barracks, however, are fine large buildings.

" The general impression conveyed by Philippeville is, that there exists a necessity of filling much ground, without that of accommodating many inhabitants."*

The line of coast that we have been now describing was passed in 1845 by Count St. Marie, who has left the following account of it. After his description of Djidjelli, he proceeds : " We soon arrived on the north of Mers-el-Zeitoun, with Cape Bougaroni on the east. The mountains, whose bases are washed by the sea, are like those to the west of Bugia, wild and rugged, but without the picturesque. They have a grandeur of effect, owing to their stupendous masses ; but though verdant, they do not present any of those pleasing spots on which the eye of the traveller loves to dwell. These shores are said to abound in coral ; but, unfortunately, the ferocity of the neigbouring tribes does not permit the fishers to approach the coast.

" Beyond Cape Bougaroni the coast becomes deeply indented ; and it is indebted to this configuration for its appellation of Djebel Saba Rous (the mountain of seven heads).

" This mountain, which is of calcareous formation, is crowned with pines and carob-trees, the brightness and freshness of the verdure denoting the vicinity of springs. Near each of these sources appear a cluster of huts, as rude as the *mapals* of the Numidians, and almost buried amongst the trees.†

" The Ras-el-Kebir consists of pale-grey basaltic prisms, which reappear beyond Philippeville. At length, on turning a rock, we entered

---

* Madame Prus, Residence in Algeria, 1850.

† Sallust. Jug. cap. 18. Baron Baude, vol. i. p. 159.

a pleasant little hollow, at the end of which was situated the European
town of Philippeville, distant two days' journey from Constantina, to which
it answers the purpose of a port, serving as the chief point of communication with that interesting part of the province. All the houses of Philippeville are new and well-built; and the European inhabitants seem happy
in having established themselves on a fertile soil surrrounded by a good
air, with plenty of sweet water."* The count proceeds to make the same
remarks, almost verbatim, as Baron Baude, on the Ouad-Zeamah, the vale
of the Oued-el-Kebir, and the woody and fertile slopes around the town.

It appears from the latest official documents, that the French have
undertaken, since 1847, civil improvements at Philippeville at an expense
of about 231,000 fr. (9240*l.*); including 1023 metres (3454·4 feet) of new
streets on a large scale, and 5305 (16,400·40 feet) on a small scale, making
a total of 6328 metres (20,755·84 feet). The Rue Nationale, uniting the
Grande Place and the landing-place at Philippeville with the road to Constantina, and comprising in itself alone a length of 1023 metres of streetage on a large scale, has a paved road, six metres (19·68 feet) in width,
for a distance of 511 metres (1676·8 feet) on the slope towards the sea.
The slope towards the gate of Constantina had been macadamised, in
1850, for a length of 512 metres 80 centimetres (1681·98 feet), with a
width of 6 metres.

Fountains and Drains.†—It appears that the Roman cisterns have
been restored, consisting of eight great basins, which had to be emptied.
The walls, which were in a dilapidated state, have been renewed. The
conduit between the cisterns and the walls has been restored, and 3752
metres (12,306·56 feet) have been cleared for a channel to bring the waters
of the Beni-Melek to them. Another plan is in agitation for bringing the
waters of the Filfila to them ; the expense of this undertaking being estimated at 500,000 fr. (20,000*l.*). As regards drainage, sewers have been completed in the Rue Nationale ; that part of it between the sea and the Rue
du Cirque being on a large scale, as well as that of the Rue des Citernes.
These two drains answer the purpose of main-sewers. The expense of
the sewers in the Rue Nationale was 38,395 fr. 25 c. (1535*l.* 16*s.* 9½*d.*)

Branch drains have been made in the Rues de Stora, Vallée, Marie-
Amélie, Joinville, Nemours, &c., costing 133,854 fr. 58 c., and executed
between 1842 and 1848. The new church of Philippeville was not completed in 1850, though it had then cost 154,643 fr. (6185*l.* 15*s.*)

Philippeville being a sub-prefecture of the province, it has been found
necessary to erect a building for that purpose, which was completed in
1847, at a cost of 2097 fr. 77 c. (83*l.* 18*s.* 4*d.*) Among other recent
civil works completed or in course of erection at this new colonial city,
we may specify a police-station and a cemetery; a douane has also been
partially built, at an expense of 111,000 fr. (4440*l.*)

* St. Marie.   † Tableau de la Situation, p. 356.

Under the head of military erections, we find that the arsenal of Philippeville has cost 110,108 fr., and that the wall and ramparts have been repaired; magazines have been built at the expense of 308,000 fr. (12,320*l.*), a hospital at 551,000 fr. ; and the general fortifications for the defence of the place have cost 1,915,118 fr. (76,604*l.* 15*s.*): consisting chiefly of three provisional batteries to command the anchorage of Stora and Philippeville, and to be superseded by permanent ones ; of the arsenal, containing a well in its precincts ; barracks, especially that of the Numides ; the city wall (*mur d'enceinte*), quarters of cavalry, and the hotel of the commandant, &c.

As regards the population of Philippeville, the European inhabitants amounted in 1847 to 5499, analysed as follows : French, 3354; Maltese, 1088 ; Spaniards, 223 ; Italian, 625 ; German, 82; Swiss, 46; divers, 81. Men, 2885; women, 1496; children, 1118.

In 1848 (Dec. 31) it amounted to 4501:—French, 2756 ; Maltese, 1320 ; Spaniards, 162; Italian, 138; German, 15; Swiss, 26; divers, 84. Men, 2260; women, 990; children, 1251.

In Dec. 1849 it amounted to 6653 :—French, 2142; Maltese, 2408; Spaniards, 120; Italian, 1426; German, 330; Swiss, 13; divers, 175. Men, 2796 ; women, 1749, children, 2108.*

The statistics of births and deaths at Philippeville from 1840 to 1850 present the following figures : In 1840 the births amounted to 16 ; in 1845 to 149; in the first six months of 1850 to 97; the maximum being in 1848, 262. The deaths were, in 1839, 1 ; in 1845, 244; in 1849, 657.†

We shall simply enumerate the colonial villages surrounding Philippeville, as they will be minutely analysed in another place.

Vallée, Damremont, and St. Antoine, are the oldest of these centres of population in the territory annexed to Philippeville. Several more recent colonial establishments have been formed on the road to Bona and Constantina. Of these, more in another place. The port of Stora, now one of the chief stations of intercourse with France, and the fine road uniting Philippeville to Constantina, Biskara, and the Sahara, must shortly make it a place of considerable commercial and general importance.‡

Philippeville was founded by Marshal Vallée in 1838, on the bay of Stora, and has, according to the *Tableau*, a good sheltered harbour. The citadel at Philippeville is called the Fort de France, and the fort to the west is Fort Royal. At the opposite extremity is Fort d'Orleans ; and eastward, on a height in the plain, is Fort Vallée. Detached forts have also been erected on the heights surrounding the valley in which Philippeville is situated. The land surrounding this rising town is rich and good; and its distance from Constantina is twenty-two leagues (55 miles), the

* Tableau de la Situation, 1850.
† For further particulars see the statistical tables.
‡ Near Philippeville, St. Marie saw some fine plantations of tobacco.

road passing first through an open plain, and then through the defiles of the Little Atlas. The population of Philippeville amounted in 1839 to 290 French and 221 foreigners, 97 women, and 108 children ; total, 716.*

East of Philippeville is the small port of Gavetta ; and after doubling Ras Hadeed and proceeding four leagues (10 miles), you come to the eastern boundary of the Sinus Numidicus and an island called Tackeesh, with a village of the same name on the opposite continent. Proceeding eastward, you pass Cape Hamrah (or red), the *Hippi Promontorium ;* and after doubling this you reach the Fort Génois, beyond which you arrive at Bona.† But we must dwell a little longer on its approaches. Numidia was more favoured than any other part of Africa by the Romans ; and the best part of Numidia is the plain enclosed between the slopes of the Atlas and the outliers that detach themselves from it to form to the east Cape Rose, to the west the abrupt shore of Stora. The sea bathes it to the north by the two indentures called Gulf of Bona and Gulf of Numidia or Stora. Mount Edough, whose long and narrow mass rises like a rampart, separates this plain from the sea, running between the two gulfs for fifteen leagues (37 miles); and passing behind the mountain, you proceed in a straight line from Bona to Stora by a road parallel to it. The Seybouse falls into the sea at the gates of Bona, and the Mafrag at five leagues (12½ miles) to the east ; both are navigable from their mouth to the entrance of the valleys of the Atlas.

This plain, by which the French possessions touch the regency of Tunis and approach the islands of Sicily and Sardinia, being 180 leagues (450 miles) in extent, is better adapted than any other part of Africa for colonisation, but has been less resorted to than any other part of Algeria hitherto, chiefly owing to the want of drainage in the surrounding marshes, and the want of a good port at Bona.‡

What is called the port of Bona is only a shallow anchorage with bad holding-ground, weakly defended from the sea by the point of the Lion, and lower down by that of the Stork (*Cigogne*), which advances 60 metres (196·80 feet) into the sea. The anchorage consists of a bed of sand stretched over the rock, stirred up and moved in bad weather by the surf, and offering no resistance to anchors. A year seldom passes without shipwreck in the bay of Bona ; and on the 25th January, 1835, fourteen vessels, including one brig of war, perished there ; eighteen days after, six other ships experienced the same fate, being the last vessels left in the roadstead. But to the north of this dangerous station, a high coast, which ends in the Cap de Garde, runs for two leagues (5 miles) in a northerly direction, and presents in its indentures the anchorages of Caroubiers and of Fort Génois.

The first is at the distance of two miles, the other three from the town.

* Tableau de la Situation for 1839, and Baron Baude.   † Blofeld, p. 43.
‡ Baron Baude, vol. ii. p. 1.

BONA.

When Bona was more frequented, marine assurances only applied, in case of accidents, to ships anchored in those two harbours, from the 15th of May to the 15th September; and during the remaining eight months of the year, they were only given to vessels mooring under the Fort Génois. Since the year 1835, the largest ships of the French navy, such as the Jupiter, the Suffren, and the Montebello, remain all the winter in the anchorage of Fort Génois, whence, however, there was no road to Bona in 1841. In the time of the Romans, the quiet and deep waters of the Seybouse gave them a good port; but for thirteen centuries the alluvial deposits have gained on the sea, and the regular bottom of the river is behind a bar, alternately open or shut, according to the predominance of the fluvial current and the winds in the high sea.

According to Baron Baude, the only place near the town fit for a port is the creek bordered by rocks, before the Stork Fort and the Point of the Lazaretto; the sea at this spot being deep, the approach easy, and the accumulation of sand impossible.*

Having brought the reader to the gates of Bona, we shall enter the town in the society of some select friends, prefacing a broad outline of its most prominent features.

Bona is the Frank name of this city, and is thought to be a corrup-

* Baron Baude, vol. ii. p. 10.

tion of the Latin Hippo Regius, a Roman town situated at the distance of one mile on the Seybouse, and from whose materials it was originally built by the Saracens. The Arabs call Bona Blaid-el-Aneb, or Anaba, عَنَابَة the city of jujubes; or simply Anaba, from the quantity of those fruit-trees growing near it; and Leo Africanus informs us that Blaid-el-Aneb was built out of the ruins of Hippona.*

St. Marie says that the town stands on a flat space of ground, forming a pentagon of fourteen hectares (35 acres) in extent, surrounded with wretched walls. He adds that its population is not numerous, and that it has no trade.†

Baron Baude assigned it a population of 5338 Europeans in 1840, adding, that a bad wall shuts in this population in a pentagon of fourteen hectares (35 acres), and separates it from the sea. In 1850, Madame Prus gave it a population of 12,000, of whom 4000 were French, chiefly Provençaux (natives of Provence). ‡

St. Marie, who landed at Bona in 1845, informs us that at the Quay they observed in the harbour many barks about to start for the coral fishery. On landing, they saw before them a great Morisco Gate, like that of Medeah (in 1845); and on one side was a pretty broad street, which, after some turning, led to a square surrounded by houses in the European style, as at Algiers.

Bona is situated low down on the south side of the coast.§ On a summit, only remarkable for a rapid ascent, is the Casbah, whose guns command the anchorage of the Cassarins. Open on all sides, the surrounding ground offers no shelter for the advance of an enemy, who would find it impossible to mask himself by entrenchments, because, throughout nearly the whole line of approach, the pickaxe, at the first stroke, comes in contact with the solid rock. The Casbah, built by Peter de Navarre, is inferior to nothing in modern art. The trifling trade of Bona is, according to St. Marie, in the hands of the Jews. Madame Prus, who had the advantage of a longer and more recent residence at Bona (1850), gives us the following particulars: "The Rue Constantine, which is a kind of suburb to the town, is composed both of Arab and French houses. The Rue Damremont and the Place d'Armes are built entirely in the French style; but the roofs are surrounded with terraces, where linen tents are pitched, under which the inhabitants spend a great part of their time, breathing the cool evening air.

"A beautiful church was commenced three or four years ago on the outskirts of Bona, near the Porte Damrémont; but the want of funds, the great obstacle to all undertakings of this nature, prevented its further progress in 1850. The wits of the town compare the building of this edifice

---

* Blofeld, p. 43. † Page 209. ‡ Mad. Prus, pp. 36-38.
‡ Bona stands in 36° 52′ N. lat. and 7° 45′ E. long. of Greenwich.

to that of the triumphal arch at Paris near the Barrière de l'Etoile : it would be unfortunate, were the same result to take place in both cases.

" I must not forget to add, that Bona is more backward in civilisation than any other town occupied by the French in this country. Its finest habitations, like that of Karesi, offer a strange mixture of modern luxury and ancient barbarism. Bona, the distant, the uncivilised Bona, presented but few attractions, either to the philosopher or the traveller. The interest of all has ever been centered in Algiers : thither were dispatched the first specimens of Parisian commerce ; there the first French settlers formed their establishment ; and, thanks to twenty-one years of civil and military occupation, this chief of Algerian cities has lost much of its primitive character."

The 4000 French inhabitants of Bona flocked in after the army, and the Maltese had even forestalled them. The latter have a monopoly of provisions and of household goods; and the French stand no chance in competing with them, as they are very sharp in business. The Maltese lend out on interest, the ordinary rate of usury being 10 per cent, whilst in urgent cases it is raised to 25, 30, and even 40 per cent. They lay claim to the office of street-porter; are a very strong, laborious, and industrious race, sleeping on the floor of their warehouses, without taking off their clothes; they are, moreover, possessed of great muscular strength, four of them being able to carry easily a great cask of wine, suspended by ropes to the tops of wooden poles, which they place on their shoulders. Their treatment of the Arabs, like that of the colonists generally, is very bad and insulting ; a circumstance resulting, in a great measure, from the inefficient state of the police, and the charities of Christendom.

Madame Prus was particularly struck with the singular appearance of a Maltese wedding, which presents a greater likeness to our notions of a funeral. On the bridal procession the women go together, their heads covered with black aprons, including the bride, and followed by the men, dressed in a uniform costume, not unlike that of English sailors. The catalogue of Maltese charms is crowned by Madame Prus pronouncing them perfidious, cunning, and superstitious, with all the vices of Italians, and without any of their virtues.* It is truly gratifying to find that these excellent and honest people are classed by the French official documents as *Anglais*. Verily, the Union Jack covers a multitude of sins.

We propose to enter somewhat minutely into the statistics of Bona, and the causes of its unhealthiness, which has become proverbial. In 1841, Baron Baude remarked that at that time there were fewer women in proportion to men in Bona than in any other town in Algeria. The French scarcely composed one-third of the number, being much less numerous than the Maltese, who had not, however, brought their women with them. The proportion of women to men was, in 1839,

* Residence, &c. 1850, pp. 36-38.

	Men.	Women.	Children.	Total.
Europeans.	1975	552	158	3095
Mussulmans	699	679	582	1966
Israelites	140	80	63	283
Total	2812	1311	803	5338

The Europeans were analysed thus : French, 1114 ; Maltese, 1206 ; Italians, 115 ; Spaniards, 550 ; German, &c. 110. The proportion of women to men in this number is 28 per cent. There die in France 1 in 39·5, at Bona 1 in 13·2, though the colonists are not old people. The following table contains the statistics of births and deaths from 1833 to 1838 :

Years.	Births.	Deaths.	Years.	Births.	Deaths.
1833	16	71	1836	73	140
1834	57	108	1837	70	170
1835	56	136	1838	110	209

So much for the Baron's statistics.* The French official documents from 1847 to 1849 present us with the following tables :

1847. French, 2387 ; Anglo-Maltese, 2268 ; Spaniards, 144 ; Italians, 1344 ; Germans, 352 ; Swiss, 18 ; divers, 122. Men, 2998 ; women, 1659 ; children, 1978. Total, 6635.

1848. French, 3152; Anglo-Maltese, 1541 ; Spaniards, 195; Italians, 1510; Germans, 152 ; Swiss, 49 ; divers, 113. Men, 3194 ; women, 1981 ; children, 1537. Total, 6712.

1849. French, 3229 ; Anglo-Maltese, 1047; Spaniards, 230 ; Italians, 489 ; Germans, 101; Swiss, 72; divers, 82. Men, 2802; women, 1387; children, 1061. Total, 5250.†

The unhealthiness of Bona is proved by the hospital returns. The garrison has seldom exceeded 4500 men ; and, independently of numerous invalids in the regimental infirmaries, the hospital contains habitually one-tenth, and occasionally one-third, of the troops. The hospital returns from 1833 to 1839 are as follow :

Years.	Mean.	Dec. 31st.	Deceased.
1832	312	452	459
1833	442	331	1526
1834	534	762	466
1835	383	391	376
1836	344	765	369
1838	450	717	651
1839	445	804	640 ‡

The garrison on the 8th of April, 1832, consisted at first of 100 men of the 4th regiment of the line, and rose till October of the same year to 3500 men. As every soldier enters about twice every year, the above

* Algérie, vol. ii. p. 33.   † Tableau de la Situation, 1850, pp. 94-96.
‡ Baron Baude, vol. ii. p. 10.

hospital returns do not really give a sufficient list of sufferers. It is somewhat remarkable that the district of Bona was once very healthy ; but the cause of the present infectious fevers that prevail there is well known. The space between old Hippo and the modern town was in remote times a cove of the gulf. Earth brought down by the Seybouse, and driven up by the sea, has converted this cove into a plain, many parts of which are hardly on a level with the sea. The sand-hills raised by the wind along the coast have made a number of reservoirs in the low places, into which the waters of Mount Edough, of the vale of Kharezas, and sometimes of the surf, flow ; these waters not being able to flow off again, stagnate and give birth to miasmas under the action of a hot sun. There are four chief depôts of marshes near the town, the most remote being about 1500 metres distant (4920 feet).

M. Carette in 1833-4 estimated the amount of land to be drained at 15·27 hectares (38·17 acres), and the amount of matter required to fill them up 100,000 cubic metres (109,200 cubic yards).

St. Marie describes the pestilential exhalations as occasioned by the Herbeyra marsh, adjoining the Constantina gate.

Two thousand metres (6560 feet) from Bona, to the right of the mouth of the Seybouse, is a piece of low and wet ground called the cuve (pit or hollow) Herbeyra, whose exhalations reach the town. It consists of about 70 hectares (175 acres), and might be easily filled up with the sand of the neighbouring sand-hills, or planted over. The marshy meadows of Vale Kharezas ought also to be drained.*

Another cause of the unhealthiness of Bona may be traced to the filthy state of many of its streets ; dirt having accumulated in them since time immemorial, and having raised the soil considerably in some places. Many of the houses, according to St. Marie, are buried 2 metres (6·56 feet) in ruins.†

An additional cause of insalubrity is presented by the scarcity of water, the aqueducts having been destroyed in 1832, when Bona was taken by the French, who have only lately attempted to remedy the evil. Baron Baude states, that it used to have seven fountains carefully kept up under the Turks; but in 1841 there was only one ; and every household had to go half a quarter of a league from the ramparts every day to get theirs. The fact is, that in 1832 Achmet-Bey wished to destroy Bona, and cut all the water-conduits.‡

In 1834 Baron Baude states, that the 14 hectares (35 acres) embraced within the walls contained 674 houses, of which 288 belonged to the authorities (the Domaine), 266 had been appropriated to barracks, and 22 to the civil service.§

The latest particulars respecting the sanitary condition of Bona, and

* Baron Baude, vol. ii. p. 21.　　　　† St. Marie.
‡ Baron Baude, vol. ii. p. 18.　　　　§ Ibid.

the statistics of its public and private buildings, which are now presented to the reader, are obtained from official documents.

From 1834 to 1848, 1233 metres (4044·24 feet) of principal streets were opened, of which 608 metres (1994·24 feet) were paved, and 625 metres (2050 feet) macadamised ; 3540 metres (11,611 feet) of smaller streets have been paved, and may be considered in a good state of repair. Seven squares have been macadamised and planted with 338 trees. These works have cost 306,000 francs (12,240*l.*).

Several important works have been undertaken to supply Bona with water ; *e. g.* 7750 metres (25,420 feet) of water-conduits, and 5670 metres (18,597·60 feet) of sewerage, have been opened from 1833 to 1847. Sixty-four *regards*, or watermen, look after the works. The great conduit which brings the water from Mount Edough to Bona forms two siphons ; the first of which has a length of 3157 metres (10,354·96 feet), and the second of 793 metres (2601·04 feet). · Out of the 5676 metres (18,617·28 feet) of drainage, 350 metres (1148 feet) are vaulted, and can be entered and examined from within. The expenses incurred in forming these conduits and drains have amounted to 597,500 francs (23,900*l.*).

A civil prison and tribunal of justice have been erected at Bona since 1845, costing 25,000 francs (1000*l.*) ; and a school-house was built in 1845-6, at an expense of 27,000 francs (1080*l.*).

The church of Bona, which had already cost 180,000 francs (7200*l.*) in 1850, was not at that time completed.

The house of the sub-prefecture of Bona, begun in 1846, was not finished in 1849 for want of funds, having cost 79,000 francs (3160*l.*). A market-place was built there in 1846-47, costing 1200 francs (48*l.*) ; and a cemetery at the same date, estimated at 24,000 francs (960*l.*).

The Douane of Bona, built in 1844, cost 109,000 francs (4360*l.*) ; and a caravanserai, afterwards converted into a native market, was built there in 1843, at a cost of 70,000 francs (2800*l.*).

The military works erected at Bona by the French government within the last few years have cost 1,991,800 francs (79,672*l.*), from 1832 to 1849 inclusive. A battery for ten pieces of ordnance has been built at the Fort Cigogne, with magazines, &c.; and a battery for twelve pieces has been begun on the rock of the Lion. The town wall and the Casbah have been improved ; barracks for 1100 men and 360 horses have been established ; a workshop for 300 convicts has been formed, and a residence for the commandant built, besides a powder-magazine at the Casbah for 30,000 kilogrammes (66,000 lbs.).* M. Berbrugger gives the following graphic description of Bona :.

" Before reaching the anchorage of Cazerain, you pass the Cap de Garde, or Cap Rouge. The last name is the literal translation of Ras-el-Hamrah, the appellation applied to it by the natives: the ancients called it *Hippi*

* Tableau de la Situation, 1850, pp. 313-381. (Travaux Publics.)

promontorium. Roman marble-quarries, whence they obtained the materials for the construction of Hippo Regius and of Aphrodisium, are still
seen there, together with fresh traces of their quarrying labours ; and a
great number of fossil shells, some of large size, are found incrusted in the
schistous rock of the cape.

A little beyond the cape, towards the town, stands forth the Fort
Génois; a building of a shining white colour, built on a rock lashed by the
waves. When Bona was subject to the kings of Tunis, they granted the
monopoly of the coral fishery, reaching from the mouth of the Seybouse
to Cape Rosa, to the Genoese. The fishermen, whose industry was interrupted by pirates, obtained permission to build a fort on a rock in the
bay; and though opposed by the inhabitants, they succeeded; hence the
Fort Génois.

After the anchorage of Fort Génois comes that of the Caroubiers; then,
underneath the Casbah, the Ras-el-H'mâma (Cape of Pigeons), whose extreme point has, when seen afar off, a considerable resemblance to a lion
couchant,—hence the Europeans have christened it the Lion's Rock. Beyond this point you come to the anchorage of Cazerain, whence you
behold the town of Bona, and the ruins of Hippo Regius, or more properly the woody hillock where they lie hid under the olive, jujube, and
Barbary fig-trees.

To the west the eye rests with admiration on the imposing mass of the
Djebel Edough. From its summit, which rises above the clouds, two
strongly-marked ridges descend to the sea, where, spreading out, they
form the Cape of Garde and that of the Pigeons. Mount Edough is an
infallible barometer to the good people of Bona; and when, on a winter's
day, the clouds are seen trooping up and shrouding its grey sides with a
misty belt, you may reckon that you will soon be soaked by one of those
deluges of rain characteristic of Africa. This mountain, whose access is
very difficult, is inhabited by Kabyles. The Romans, who called it Pappua, found it very difficult to penetrate into its recesses, when they wished
to pursue certain contumacious native princes who had fled thither. When
Belisarius recovered Africa from the Vandals in 532, Gelimer, who could
find no security in the towns which had been dismantled by Genseric,
sought refuge in the Pappua. The people of that mountain, who had remained in primitive barbarism, though so near the splendid growth of
Roman civilisation, are reported to have viewed with wonder the effemi-
nate character of the fugitive Vandals. So completely had they been
enervated by the abuse of the luxuries that success had showered into
their laps, that they had sunk far beneath even Roman degeneracy and
corruption.

At the foot of Mount Edough, and a little in front of the point of rocks
overhanging the Stork's Fort (*de la Cigogne*), is situated the Coral Fishers'
Bay (*des Corailleurs*), which on fête-days is encumbered with the coral-

o

fishers' boats, whose crews come to the chapel built on that spot to thank God for a successful season, or to supplicate His favour if they have failed. The next object is the Lazaretto, and then the Stork's Fort; and after passing the latter spot, you face Bona, whose appearance is not at all imposing. It stands on the site of Aphrodisium, so called from Aphrodite, or Venus, to whom the inhabitants had raised a handsome temple, and whom they had chosen as their patroness. Bona is built of the remains of that little city, and of those of Hippo Regius. The natives call it

بلد ال عنّابة Blad-el-Anabe (the town of jujubes), or more commonly Anabah, on account of the number of jujube-trees that grew around it formerly, but which were cut down by the French soon after their occupation, in order to free the approaches of the town, and to remove all shelter for the Kabyles, who used to practise at picking off the French soldiers with their long guns.

Bona was founded soon after the destruction of Hippo by the Arabs (A.D. 687). Its inhabitants lived always independent till the time of the Turks, regarding the kings placed over them rather as patrons than as sovereigns; and when the latter sought to tighten the reins, they threatened to surrender to the Christians. The latter took possession of it under Charles the Fifth, at the time of his expedition to Tunis, Alvar Gomez Zagal being left there with 1000 foot and 25 horse. He kept the town and plundered the country with this weak force; but on Zagal's death, the emperor ordered the town to be abandoned, and the fortifications to be razed. The Turks took it afterwards from the kings of Tunis, who were too weak to hold it; but it changed masters several times subsequently.

Beyond Bona you perceive a river, the Boudjema (the *Armua* of the ancients), which, after having watered the vale of Kharesas, passes under the bridge of Hippo Regius, and reaches the bay by trickling through the sands that obstruct its mouth. A little beyond is the Seybouse, a rather broad and deep river, once the port of the Romans; but its entrance is now barred by a shifting sand-bank. The brig *Rusé*, which was wrecked there in January 1835, altered the bar considerably; and this fact may lead the way to clearing the mouth of the river eventually, by suggesting some mode of deepening the channel.

Between these two streams rises a green hillock, terminating on the sea-side the chain of hills that limit the vale of Kharesas on the S.E. On this spot once stood Hippo Regius, a very important city, of which more anon.

On the left bank of the Seybouse begins a vast plain extending beyond the Mafrag (the Rubricatus). The splendid pastures that it presented afforded to Ahmed Bey of this province the means of paying the annual tribute to the Dey of Algiers, and of pocketing a balance of 100,000 fr.

(4000*l.*) Being once in difficulties from immediate want of money, the plain of Bona alone gave him 500,000 francs (20,000*l.*) in the space of a few days.*

"We entered Bona," says Madame Prus, "by the Porte Constantine. It was most curious to see the Arab market, which was held outside the gates, but within the fortifications. Imagine a number of white figures, of the same colour as the walls which surround them, moving busily to and fro among the stores of provisions laid out for sale. These are the Arabs of the district, wrapped in their white burnouses, or sheepskins. Their wares consist of different kinds of fruits, which grow abundantly in this country, curdled milk in earthen vessels, butter, &c. A certain degree of courage is necessary to penetrate through this crowd and gather in one's stock of provisions, as the want of cleanliness, both in the articles of food and in the persons of those that sell them, is most revolting. They use no ablutions, except those prescribed by the Koran, which are limited to the hands and feet. Their clothes actually swarm with vermin, and a visit to the Arab market can never be made without disastrous consequences; but the inconvenience being unavoidable, the best way is to bear it with stoical firmness, and to overcome the disgust which the scene described causes to our more refined feelings.

"Nothing can be more picturesque than the view from the market-place. The old Arab town, half concealed by its high embattled walls, is built in the form of an amphitheatre; and about a hundred steps farther you see the terraces belonging to the more modern parts of the city. The military hospital, and the minaret with its pointed roof, are imposing edifices, situated in the Place d'Armes. On the left is the lofty chain of Mount Edough, at the base of which is a lovely valley; and on the right the blue waters of the Mediterranean. On the sea-shore are to be seen the tents of the Bedouin salt-merchants, with their camels lying on the sand, and their small lean horses picketed to the ground. In the distance the Twin mountains appear in bold relief against the blue sky; on these were erected the sumptuous edifices of ancient Hippona, several imposing remains of which are still to be seen. Reservoirs of enormous size, and the beautiful ruins of the Church of Peace, of which St. Augustin was the first bishop, attest the grandeur of the ancient city.

"The higher classes of the Moors, though they do not conform to all our customs, have adopted many of them from the mere impulse of imitation. They speak French; which, indeed, they find indispensable, from their frequent contact with French society. But this accomplishment is practised by the men only. A Moorish lady has never been known to accompany her husband on any visit, or to make the least change in traditional usages."†

"Here we are," observes M. Berbrugger, "in the interior of Bona, in

* Berbrugger, part iii.   † Madame Prus.

the Place Rovigo, where the principal streets meet, leading to the gates of the Marine, Constantine, Zikhan, and of the Casbah. To the left is the house of Youssouf, and at the end you see the taper minaret of the mosque. To the right you see a tree whose trunk is surrounded with boards, on which are commonly pasted up the proclamations, notices, and other official publications. This *place* is also the seat of a kind of permanent fair, which took a remarkable development after the return of the second Constantina expedition. The arms, the carpets, and even the dresses of the conquered were exposed there for sale. The Jews and Maltese, who had followed the army with views somewhat foreign to glory, let the French soldiers reap the laurels ; and, after gathering in a more lucrative and less honourable harvest, they came back from Constantina with the fruits of their industry, which they displayed at Bona. There was a complete fever at the time for what were called the *souvenirs de Constantine.*

" The interior of Bona is like that of most towns in Algeria. Seen from a distance, almost all appear pretty ; but when you enter them, it is soon discovered how remote the reality is from the appearance. But in Bona the streets appeared, even in 1843, less narrow and obscure than those of Algiers, which proceeds merely from the circumstance that the houses are not so high. Save this difference, the nature of the dwellings is about the same. In the business-streets appear little shops without any communication with the building to which they belong, and which seem so many niches raised four feet from the ground. Every where else in the Arab streets you see only completely bare walls, in which you find nothing but some openings through which a child's head would pass with difficulty ; within, a court surrounded by a gallery supported on columns ; two or three long and narrow chambers opening into this gallery, and only receiving the daylight through it ; above is the terrace, an almost universal appendage at Bona.

" The other side of the square is built in the European style, like the Rue de Rivoli at Paris, but in much more modest proportions. This kind of building is onerous to the landlords, but it is very agreeable to pedestrians, who find under the galleries a refuge from carriages, horses, and other cattle, and a shelter against the sun and the rain. In 1843 this square was almost the only part of the town where French architecture had appeared, all the other parts remaining in their ancient state, save some demolitions rendered necessary to clear the roads for the French wagon-train, which gave a ruinous aspect to many streets at that time."[*]

The chief mosque of Bona contains some splendid Corinthian capitals and beautiful fluted columns, which appear to be the relics of some Roman temple. They may possibly have belonged to the famous Basilica of

[*] Berbrugger, part iii. p. 8.

INTERIOR OF THE GREAT MOSQUE AT BONA.

p. 212.

Peace, existing at Hippo in St. Augustin's time; or they may perhaps be the remains of the temple of Venus at Aphrodisium. The first Christians built their churches with the materials of Pagan temples; afterwards came the Arabs, who built their mosques with the ruins of the two preceding worships. The same stones and marbles have been devoted by succeeding races to form the House of God; as immortal as religious ideas, they have only changed in form and arrangement.

The great mosque of Bona resembles all buildings of this kind. Its three parallel galleries call to mind the nave and collateral aisles of our churches. At the bottom is a niche turned to Gobla, or Mecca : there stands the priest (*imam*), or the person commissioned to direct public prayer. A modern staircase terminated by a platform (*mombeur*) is seen to the left; it is a kind of pulpit, which the imam mounts every Friday, before mid-day prayers (*el-eulem*), to preach to the people. The ground is covered with mats, with carpets on the top of them, where the slipperless worshippers kneel. When the crowd is great, those who fear unpleasant exchanges take their slippers with them, instead of leaving them at the door.

Lamps with several jets hang from vaults by iron chains; but they are never lighted except during the Ramahdan, when the exterior of the mosque is also illuminated. An elegant minaret (*smâ*) shoots up over the mosque, and is crowned by a gallery whence the *mouedden* calls the faithful to prayers five times a day. Further particulars respecting Mussulman worship will be found in another place.*

With the reader's kind permission, we shall now take a stroll to Mount Edough; and we doubt not that he will gladly exchange the miasmas of the marshes for the fresh mountain-breezes.

Mount Edough offers limestone all along this coast, a formation of which it is deprived for 80 leagues (200 miles) from Bugia to Cape Roux. The eastern slope of Mount Edough is uninhabited; and at the foot of the mountain are the remains of the immense plantations of olives formed in the 17th century by Mustapha de Cordenas, a rich Moorish refugee from Spain, which Peysonnel found in all their vigour in 1725. On the other side of the cape there has been established from time immemorial the poor and inoffensive tribe of Ali, which was visited by Baron Baude. He found the solitary slopes of Mount Edough carpeted on that side up to the highest summits, with that humble arborial vegetation which in Africa issues from the struggle between the vegetative force of the soil and the devastating teeth of the cattle. Some fruit-trees, vines, little fields of maize and corn, and sheds built of unhewn stones, are signs of the tendency existing in the Ouled-Ali to plant and build on a larger scale, if they possessed the means; but the tribe is so limited, that it only constitutes a small family.

* Berbrugger, part iii. p. 7.

A–H

The only reproach that St. Augustin could make to the Kabyles of Mount Edough, during his residence of thirty-six years at Bona, was, that they spoke Punic, and did not understand Latin, in which the word of God was preached.

Four leagues and a half south-west of Bona (11¼ miles), the Lake Efzara occupies 10 square leagues (62½ square miles) at the foot of Mount Edough. The valley of Kharezas opens in a direct line from Bona to Lake Efzara, between the foot of Mount Edough and the hills of Belelida ; and, according to Desfontaines, whenever the waters of the Lake Efzara are swollen by winter rains, they flow by this valley into the Mediterranean.

Bona is represented by Baron Baude as a better military station than the ancient Hippo, from which it is separated by an interval of 1000 metres (3820 feet). The north-west angle of the plain, which extends on the left of the Seybouse, is closed between the head of the Edough and the sea by a hillock of 108 metres (354·24 feet) in height, separated from the mountains by a narrow valley. Bona is situated at the bottom and on the south side of this hill ; and the summit, which is reached by steep slopes, is crowned by the Casbah, whose cannon, as previously stated, sweep the anchorage of Cassarins.

There are 70 square leagues (437½ square miles) of plain between the Efzara lake and the river Mafrag. This surface is divided into two almost equal parts by the river Seybouse. The eastern part is a rectangle, and touches the walls of Bona by its north-west angle ; the sea and two navigable rivers defend three sides of it, and on the fourth the Atlas is not practicable. The cultivation of these 110,000 hectares (275,000 acres) of land can be always safely carried on. The river Mafrag, the west limit of this plain, crosses it at five leagues (12½ miles) from Bona, about parallel to the Seybouse, is as broad and as deep, and the navigable part of its course appears to extend as far as that of the Seybouse, amongst the branches of the Atlas range. Like the latter stream, the Mafrag is also barred up with sand at its mouth most of the year.

The ruins of Hippo Regius, which will be circumstantially described in another place,* are situated 1000 metres (3280 feet) from Bona, near the mouth of the Seybouse ; and it appears that the city was grouped at the foot of two mamelons, one 80 metres (262·40 feet), the other 38 metres (124·64 feet) in height, and called in Arabic Bounah and Gharf-el-Antram.†

The agricultural and other colonies surrounding Bona, and on the road to Philippeville, of which a doleful account has been given by Madame Prus, whilst the *Tableau de la Situation* speaks of them in favourable terms, will be fully described in the chapter on Colonisation. It will suffice here to mention the name and situation of the most important, which are Penthièvre, Mondovi, and Barral.

* Chapter on Archæology, Part II.
† These particulars are obtained from Baron Baude, vol. ii. c. 8.

Mr. Dawson Borrer, who sailed along the line of coast from Algiers to Bona in 1846, gives the following description of the scenery that it presented. "The night-wind blew cold from the snow-clad heights of the Djorjora, as, bidding adieu to Bugia, onward we glided, cleaving the glittering waters of the gulf. Gigantic rocks jutting forth from the rugged shores increased in grandeur as the shadows of darkness fell upon them. Here and there, among the wild recesses of the mountain heights, the glimmering of Kabyle watchfires might be seen. Then, as we turned Cape Cavallo, a new scene burst upon us. A vast tract of mountain-side presented one glowing sheet of flame. Towering heights were clothed with fire; chased by the reflection, the silver rays of the pale moon no longer danced upon the rippling surface around us. Thus does the Kabyle clear a space upon his brushwood-clad mountains, that he may cast in his grain, the sowing season being at hand. After touching at Djidjelli, where the French inhabitants are annually decimated by the malaria, and at Philippeville, we turned Ras-el-Hamrah (the Hippi Promontorium of the ancients) about 5 P.M., the third day of our voyage, and soon dropped anchor in the bay of Bona. I like Bona. Its wood-clad heights overlooking the wide blue sea, and its rich plains watered by the Seybouse, the Boojeemah, and the Ruisseau d'Or, please me. So do also its gardens, in the fertile soil of which luxuriate flowers and vegetables of all sorts, skilfully irrigated by that most industrious class of colonists, the refuse of Spain and Malta, who, never idle, cultivate the land by day, and rob and cut throats by night.

"Again, how interesting are the moss-clad ruins of ancient Hippona, shadowed by groves of olives, jujubes, and carobs! The wind sighs through those now-deserted courts, from which the venerable St. Augustin so nobly combated the ruinous march of Roman luxury, and those various heresies which then tore the Christian church in Africa. And was it not within those walls that, borne down by the evils which assailed the empire and the church, he died?—Vandal shouts ringing in his ears, as, in pursuit of the unhappy Boniface, they filled the courts of Hippona with their Arian hordes." *

Having surveyed the sea-board of this province from Djidjelli to Bona and Hippo, we shall finish our description of the coast-line before we analyse the inland parts.

The Seybouse and the Mafrag, the principal rivers between Bona and Tabarca, seem to be the Ubus and Rubricatus of the ancients. Beyond Cape Rose, five leagues from the Mafrag, is the *Bastion*, where there is a small creek, and the ruins of a fort that gave rise to the name. The factory of the French African Company had formerly their settlement at this place; but the unwholesomeness of the situation, occasioned by the neighbouring ponds and marshes, obliged them to remove to La Calle,

* Dawson Borrer, p. 326.

another inlet, three leagues (7½ miles) more to the east. About two miles to the E. of La Calle is the little river Zaine (Tusca), which has served for centuries as the limit of the two regencies of Algeria and Tunis.

There is a little island at the mouth of the Zaine, which continued many years in the possession of a noble Genoese family, from the time of Andrea Doria, to whom the Tunisians gave it, with the consent of the Sublime Porte, as ransom for a prince taken prisoner by Doria. This place was defended by a good castle, and protected the coral fishery in those seas; but in 1740 the Bey of Tunis took it by treachery from the Genoese, put some of them to the sword, and the remaining three or four hundred were made prisoners.*

Having taken this rapid survey of the ground, we shall accompany Baron Baude in a trip that he made to La Calle from Bona, in the company of M. Prosper de Chasseloup, going by the foot of the Atlas, and returning along the coast. They went with letters of recommendation to the Sheikh of the Merdes, Sidi-Mahmoud, and returned to Merdes on the 23d of September.

From Draan, a station in the plain 12½ miles from Bona, to the douar of the Merdes, you travel seven leagues (17½ miles) all in the plain. To the south lies a high mountain, and to the north you leave the isolated hills of Sidi-Denden and of Kennader. The soil consists of a clay mixed with sand, of which the fecundity is attested by the vigour and perfection of the thistles and other large plants that cover it, which sometimes rise higher than a man even on horseback: there was, however, no cultivation, and they only saw a dozen scattered trees on the road. They had crossed, by the fords of Sidi-Denden and Sidi-Abdelaziz, the Seybouse and the Mafrag, which, on issuing from the mountains, offer a narrow bed, which is, however, as navigable as that of the Saône at Lyons. They were now in the true Arab country; and ascending, for half an hour, the right bank of the Mafrag, they reached the top of a mountain, whence their Arab escort dashed on to the douar, and Sidi-Mahmoud came forth to meet them, pressing his hand to his heart in the oriental fashion. Two poor sources of water, and some plantations of maize and tobacco, are the chief merit of the valley. At a short distance are the remains of a Roman dwelling; and from the neighbouring summits the eye roves E.S.E. along the extensive valley by which the Mafrag descends from the Atlas. At this point the Arabs call it the Ouad Merdes, from the name of the tribes on its banks: that of Mafrag, which is given to it lower down, comes from the bar of sand raised by the wind at its mouth; for the rivers and brooks of Barbary frequently change name as they pass from one tribe to another. The rock of the mountain at this spot consists of red sandstone, which is a very extensive formation in this vicinity; and no other rock is seen from Draan to La Calle, and from La Calle to Bona. These

* Blofeld, p. 43 et seqq.

first degrees or steps of the Atlas have the same character as the Boudjareah, near Algiers. The rich clay of the plains ascends to the foot of the rocks, a luxuriant verdure clothes their sides; and the reason why the trees do not grow higher is, notwithstanding Sallust's opinion, the destruction of the inhabitants. Such is the Baron's view of the case, to which we do not pledge ourselves.

At midnight they started with Sheikh Hafsi, of the tribe of the Beni Urdjin, with sixty horsemen, by a fine moonlight. During the first hour they marched over a heavy land; then they came to marshy ground, dried up by the sun and covered with great reeds. This district is called Ras-Mafrag, or the head of the Mafrag, and is the parent of unhealthy miasmas. After passing it you come to a strong and stiff soil, watered by many brooks, where they halted for three hours. The neighbouring woods, which had been ignited by the Arabs, presented the appearance of a vast conflagration. At the morning dawn all the Arabs of the party threw themselves prostrate in prayer, presenting a striking and patriarchal scene; and after performing their devotions, they proceeded, and at 8 A.M. they were received with semitic and scriptural hospitality by the principal douar of the Ouled-Djeb, at which spot the plain ends, woody mountains enclosing its rich pastures.

There are no palms, agaves, or cactuses in this neighbourhood, which give an African character to the country around Algiers. The forest by which the douar stood clothed the two faces of a mountain, whose foot was bathed by the lake El-Malah. Red sandstone here and there pierces the sand of which the soil is composed, which, however, is often very moist. Cork-trees (*chêne liège*) are almost the only timber in these woods; but no use is made of them, though they might be turned to such a useful account. The Arabs burn down large tracts of these celebrated forests, which extend northwards to the sea, westward to Cape Rose, and eastward to the frontier of Tunis, embracing a surface of no less than 20,000 hectares (50,000 acres), and all forming what are called *propriétés domaniaux*, or government property. Baron Baude represents in strong language the folly of neglecting such a valuable possession, not only for its intrinsic value, but also because if the country were stripped, it would speedily become a desert.

When our party had arrived at the end of lake El-Calah, it was not far to La Calle; and they found the country delightful, though rather marshy. They observed that the Arabs in that vicinity had learnt many expressions of the Provençal dialect from the old French mariners and merchants who were wont to frequent La Calle. Strange that these primeval cork-forests should witness the marriage of the *gaie science* and the Prophet's sacred tongue, and that troubadours and marabouts should shake hands on the ruins of Carthage!

On the 25th they were on horseback at daybreak, and soon arrived on the banks of the lake El-Garah, where the scenery reminded them of that of Scotland ; and two hours after starting, they beheld La Calle at their feet. This town, which was burnt to the ground on the 27th of June 1827, was taken by M. Albert Bertier on the 22d July 1836 with only fifty zouaves (native troops).

The port* is commanded to the south by the post of the windmill. On the *plage du fond* (or beach) at the bottom of the harbour, besides a well and an excellent spring, are the ruins of a lazaretto and a mosque. The town is built on the peninsula of rocks, three hectares in extent ($7\frac{1}{2}$ acres), which encloses the port ; and over the land-gate is inscribed the date of 1677. The only remains left of the old buildings consist of a few vaulted warehouses and walls ; and the rock is, unfortunately, a sandstone (*gres*) of a loose texture, in which the sea makes inroads, undermining the pavement. The French establishment at La Calle is of the same date (1520) as the first occupation of Bona and Constantina by the Turks. Francis I., Henry II., and Charles IX., were allies of the Turks and Selim II. against the house of Austria, the old hereditary foe of France. Between 1560 and 1604 was the period of the greatest concessions and power enjoyed by the French at La Calle ; and from the time of Francis I. to Louis XIV., the whole secret of this anomalous coalition between the very Christian king and the head of Islam may be traced to the great contest between France and Austria.

After the decline of the latter power, the concessions came to be only commercial ; and in 1694 was concluded the *traité de Pierre Hely*, which was the basis of the French relations with Algeria down to the conquest. Its chief enactments were repeated almost verbatim in the conventions of 1714, 1731, 1768, and 1790. Pierre Hely and his company enjoyed the monopoly of the whole trade of La Calle, Cap Negre, Bona, Bastion de France and its dependencies, for 34,000 gold roubies (105,000 fr.; 4200*l.*). In 1719 it passed to the French India Company; in 1741 to the African Company and to Marseilles, with a capital of 1,200,000 livres (48,000*l.*). Six per cent was always obtained by the shareholders ; and from 1772 to 1777 each shareholder received 300,000 fr. (12,000*l.*) every year as his dividend. The African Company was in a particularly flourishing state under Director Martin ; but it was suppressed, with all other monopolies, in 1794.

The garrison used only to consist of fifty veterans with a captain. We have already referred to the prevalence of Provençal among the tribes surrounding La Calle, whose population in 1794 amounted to 600 persons, amongst whom, no women being allowed, it is reported that most deplorable immorality prevailed.†

---

* Baude, vol. i. p. 182.

† See Abbé Poiret's Lettres écrites de l'ancienne Numidie pendant les années 1785-6, 2 vols. Paris, 1789.

A mean exportation of 90,000 hectolitres (247,610 bushels) of wheat used to be annually effected by the Company. The price of the local load (153 kilogrammes; 5·06 bushels) varied from 7 fr. 50 c. (6s. 3d.) to 15 fr. (12s. 6d.) per load; thus making 5 fr. 51 c. (4s. 7d.) per hectolitre, or 3 bushels 40 lbs.

The Company used also to export considerable quantities of barley, maize, and beans.

At the present time there are fine plantations of tobacco around lake El-Hout, and among the Ouled-Djeb, the Djeballah, and the Seybas; and M. Pasquier, director of the administration of tobacco, pronounced the specimens that he saw equal to those of America, which fetch 100 fr. (4l.) per metrical quintal (220 lbs. avoirdupois, or 4¼d. per lb.). The cork-forests in the vicinity of La Calle might also be the source of a flourishing trade, and provide abundance of wood for the army.

The lofty hills that border the coast of La Calle are covered with shrubs, and above the post stands a very fine group of mulberry-trees. A beautiful panorama is unfolded to view on the top of these hills; the land gently dips to the lake El-Garah to the southward, and to the east to lake El-Hout, whose waters bathe the feet of the green slopes. Rich valleys extend between woody hills, whose varied summits project in one place into the azure sky, and in other places stand out from the dark sides of the Djebel Koumir.*

The lakes, whose Arabic appellations are mercilessly disfigured by the French, have also long enjoyed European sobriquets, applied by the Provençal traders to La Calle. Thus, the Guilta-el-Malah was the *Etang*, or pond, of the Bastion; the Guilta-el-Garah, the *Etang de Beaumarchand*; and the Guilta-el-Hout, *l'Etang de Tonegue*. "The plain near the latter," says Baron Baude, "when we visited the douar of Moussa, was the *Plaine de Terraillane*." The territory of La Calle is shut in by three lakes, two of which, those of Tonegue and of the Bastion, flow into the sea, the third of which almost shuts in the space between the other two. The Etang de Beaumarchand is at the distance of 1000 metres (3280 feet) from that of the Bastion, and at 2000 metres (6560 feet) from that of Tonegue; and you might thus enclose by 3000 metres (9840 feet) of ditch an extent of three or four square leagues (25 square miles), embracing some very good land. The centre pond is thought to be the cause of the fevers that prevail there from June to September, which are, moreover, aggravated by the frequency of the southerly winds. This pond is shallow, and might be drained.

All property belongs to the crown at La Calle, and the carcasses of the houses that cover its surface are gratuitously granted for five years if the tenants make them habitable; but when buyers congregate and capital pours in there, they will be sold.

* Baude, vol. i. p. 205.

Leaving La Calle, the baron's party proceeded along the heights to Bona; and after advancing two hours, they saw, on the banks of a cove of white sand to the right, the ruins of the old Bastion de France. All the rivalries of empires and the passions of human nature have contended on this little spot, which is now nothing but a ruinous tower.

There is a lake to the south of the bastion; and a channel, 600 or 800 metres (2624 feet) long, leading to it, is almost dry at times, the waters of the lake being frequently very low. The variations in the level amount to two metres (6·56 feet). This lake used at one time to be a port for coral boats and for the bastion; its depth at the lowest water-mark is two or three metres (9·84 feet); it penetrates two leagues (5 miles) inland, and it contains an area of about 2500 hectares (62£0 acres). Its navigation would be useful, from the nature of the woods and land surrounding it; and it is said to be well supplied with fish.

The cork-forests continue to Cape Rose; but they are interrupted by the delicious valley of Djeballah, whose soil, defended from the south and sea winds by the elevation of the surrounding hills, consists of a rich and light loam, made still richer by irrigation. There is a good landing-place in the stream that waters it, and this little harbour is called by the Italian coral-fishers Porto delle Cannelle—in French, Port Canier (Reed Harbour). It is a port of refuge for small ships from westerly gales.

Almost all the land of the Djeballah is divided into cultivated fields, and produces an abundance of tobacco, maize, wheat, &c.; and the oxen and horses show in their forms the goodness of the vegetation. The tribe of the Djeballah has about fifty tents, almost all of which are scattered, and not congregated into douars, these Arabs not being nomadic, but settled. A range of cliffs is detached from Cape Rose, and runs gently down to the mouth of the Mafrag. The land slopes back from it to the south; and the water from this range flows back into the Mafrag in an opposite direction from the sea. A large sandy zone, furnished with shrubs, extends along the gulf of Bona; but as soon as you come to the first village of the Seybas, you find all the fertility of the plain. Here and there appear fine fields of tobacco and corn; and the tribe of the Seybas, though ravaged by the plague, has 100 tents, and is one of the richest in the plain.

" Towards half-past four in the morning," proceeds Baron Baude, " we left our friends the Seybas, and at 7 A.M. we entered the large gully that forms the mouth of the Mafrag in the sea through the sand-hills : it was completely choked by a bar about twenty-eight metres (91·84 feet) in thickness, composed of sands heaped up by the waves of the sea;—we passed over dry-shod. To our left the river was at least 200 metres (656 feet) wide, and seemed very deep ; in its rise it forces the bar, and nothing is more variable than its entrance. The sand-hills that border the sea to the right and left of the mouth of the Mafrag are of pure sand; but by the effect of filtering, the bottom of the soil is almost moist there, consequently they are covered with the richest verdure ; they are crowded with the olive, the carob, and the cork-tree, whilst the vine entwines them in its festoons. The douar of Sheikh Hafsi was at a short distance from the Mafrag, on the banks of the lake Beida ; but notwithstanding his hospitable entreaties for us to remain, we went on. The territory comprised between the Mafrag and the Seybouse is occupied, under our protection, by the tribe of the Beni-Urdjesi, whom General Uzer wisely established there, when it fled the persecutions of Ahmed, bey of Constantina. It touches the gate of Bona, and has become rich by trade. From the Mafrag to the Seybouse you follow the whole of a valley which runs between two parallel lines composed of sand-hills formed by the sea; at high water the two rivers sometimes communicate through it. An excellent ferry-boat has taken the place of the floating isle of rushes on which the Arabs used to cross the Seybouse."

From La Calle to Bona is a march of thirteen hours and a half. They met no lions on the road, though these quadrupeds are reported to be common there.*

Having now taken the traveller along the coast of the province, we shall give first a broad outline, and secondly an analysis of the interior.

* Baron Baude, vol. i. p. 213. The *Tableau* gives La Calle a population of 400 inhabitants in 1849, and states that the town-walls have been improved, and a battery established to defend the port. *Tableau*, 1850, pp. 96, 113, 345.

CHAPTER XII.

Province of Constantina. Interior.

INTERIOR OF THE PROVINCE—BROAD OUTLINE—ANALYSIS—BARON BAUDE—NATURAL FEATURES OF NUMIDIA—ST. MARIE—CONSTANTINA—MADAME PRUS—BORRER — GUELMA — GERARD, THE LION-KING — CONSTANTINA — BETNA—AOURESS—EL-GANTRA—BISKRA— THE OASES.

THE whole of this province, between its old limits, the rivers Booberak and Zhoore, from the sea-coast to the parallels of Setif and Constantina, is mostly a continued chain of very high mountains. Near the above parallels it is diversified with a beautiful variety of hills and plains, with a greater or less adaptation for cultivation, till it ends up the Sahara in a long range of mountains, probably the Buzara of the ancients. The district of Zaab is immediately under these mountains; and beyond Zaab, at a great distance in the Sahara, is Ouadreay, another collection of villages. This part of the east province, including the parallel of Zaab, answers to Mauritania Sitifensis; or the first Mauritania, as it was called in the middle ages.

The mountainous region between the rivers Zhoore and Seybouse is of no great extent, seldom reaching more than 6 leagues (15 miles) within the continent. From the Seybouse to the Zaine, except near Tabarca, where it begins again to be very mountainous, the country is mostly plain, though sometimes diversified by hills and forests. The same variations are found below Tuckush, along the encampments of the Hareishah, Grarah, and other Bedouins, as far as Constantina, where may be occasionally seen a small species of red deer not met with in other parts of the colony. Beyond this parallel is a range of high mountains, the Thambes of Ptolemy, extending as far as Tabarca, behind which you find pasture and arable land, ending in the Sahara, as Mauritania Sitifensis did before in a ridge of mountains,—the Mampsarus, probably, of the ancients.

Part of the Africa Proper of Mela and Ptolemy, the Numidia Massylorum, the Metagonitis terra of the classical authors, was comprehended in this part of the province.

Leaving that portion of the province which belongs to Great Kabylia*

---

* Consisting of the two great basins of the Ouad-Summam and the Ouad-Adjeb, or Bousellam ; the first draining the high lands above and around Aumale, and the latter coming down from Setif, and joining the former a little above Bugia, where they both fall into the sea.

for another occasion, we proceed to remark that Mount Atlas, throughout the province of Algiers (formerly Tittery), as far as Mount Jurjurah, runs parallel to the sea; but, after passing that point, diverges to the S.E. In the same direction rise the lofty mountains of Ouan-nougla and J'aite; succeeded afterwards, but in a direction more parallel with the sea, by those of Oulad-Selim, Mustewah, Aouress, and Tipasa, which run into the Regency of Tunis. Three or four leagues south of Mount J'aite is Messeilah, the frontier town of the province to the west. It is built on the southern skirts of the plains of El-Huthna, 9 leagues (22½ miles) to the S.S.W. of Sidi-Embarak-Es-mati, and 16 leagues (40 miles) S.W. of Setif. Messeilah is a dirty place, like all villages in this country; the houses being built with reeds daubed with mud, or tiles baked in the sun. The air is too cold for dates in this spot, and other places on the skirts of the Sah'ra; and the gardens surrounding it only contain peaches, apricots, and the fruits of North Africa. Messeilah means a situation, like that of this town, on the banks of a running stream. At the same distance on the other side, *i.e.* north of the Djebel J'aite, commences the plain of Medjana, shaded to the northward by the Dra-el-Hammar, and to the west by the mountains of Ouan-nougla. These plains are large and fertile; but numerous pools of foul water, as the name denotes, filled in the rainy season, and stagnating in the spring, give birth to agues and fevers, &c. Several heaps of ruins are scattered about, of which the Turks have built a fort. The country presents nothing remarkable till, passing by the village of Zammora, *i.e.* of olive-trees, we come to Setif (Sitipha or Sitifi), the ancient metropolis of this part of Mauritania, which made a brave resistance to the invading Saracens. This city may have been perhaps a league in circumference, and was built on rising ground facing the south; but it scarcely contains a fragment of Roman remains, the few structures that are now seen being the work of later inhabitants. The fountains, which continue to flow very plentifully near the centre of the town, are equally convenient and delightful.* The town contains four good streets, and is well fortified.

Setif is situated to the west of Constantina, and at the distance of about 20 leagues (50 miles) south from Bugia, and contained in 1849, 646 Europeans and 436 natives.

The ancient Sitifis colonia, after being the capital of a fine province during the Roman sway, presented in 1839 nothing but a heap of ruins, near which the Arabs held a market every Sunday. This town is situated on an immense table-land, whereof the elevation above the level of the sea is represented to be 1400 metres (4592 feet); accordingly it is exposed to severe cold, and snow is seen there during almost six months of the year; the wind, moreover, sweeps over this high land with extreme violence, driving vast clouds of dust before it.

* Blofeld, p. 43 et seqq.

Setif is perhaps the healthiest spot occupied by the French in the whole of Algeria; and it is supplied with excellent water.

The distance from Constantina to Setif is about 30 leagues (75 miles), and is traversed by two roads. The shortest passes through the territory of the Abd-el-Nour, presenting a rich country entirely stripped of trees, and without the vestige of a town or camp; the only traces of human structures consisting of a great number of ruined Roman monuments, which offer, however, little interest. The other road passes by Milah, Ma-Allad, and Djemilah. Both roads are impassable for carriages.

The plains and rich pastures of Cassir-Attyre lie a little to the south of Setif, and are cultivated by the Raigah, a clan of Arabs famous for breeding cattle, especially horses, which are considered the best in the country. Near the Raigah are the Ammers, a powerful tribe. Eight leagues (20 miles) S.E. of Setif are the ruins of Taiggah and Zainah, situated half a league from each other, in a fruitful champaign country, under Djebel-Mustewah, the principal abode of the Ouled Abdenore, a very numerous and powerful clan. Taiggah and Zainah are rarely mentioned apart, but from their contiguity are conjointly called Tagou-Zainah. A small brook runs between them; and at Zainah, among other ruins, is a triumphal arch, supported by two large Corinthian pillars. Five leagues to the east of Tagou-Zainah, on the northern skirts of the Djebel-Aouress, is situated the sepulchral monument of Medrashem, or Mail-Cashan, which is similar to, though not larger than, the Koubber Romeah,* and has a cornice supported by pillars like the Tuscan order. The district near this spot is named Ain-yac-coute, probably from the Ain-yac-coute, or diamond (*i.e.* transparent) fountain, situated near the centre of it. Fragments of Roman highways and other ruins are scattered all over it; among which the principal are those of Om-oley, and Sinaab, a league or more to the west of Medrashem, on the road to Zainah. Tattubt, bordering on the Ain-yac-coute to the N.E., is about four leagues (10 miles) from Om-oley and Sinaab, and about eight leagues (20 miles) to the S.S.W. of Constantina. It has been formerly a considerable city; but at present is almost entirely covered by earth and rubbish. Tattubt seems to be the same place as the Tadutti of the Itinerary; and lying between Lambese and Gemellæ, as the ancients called Tezzonte and Jimmeilah, may justly lay claim to this situation. Ten leagues (25 miles) to the south of Tagou-Zainah, and 12 leagues (30 miles) from Medrashem, are the remains of ancient Thubana, now Tubnah, situated in a fine plain near Bareekah and Boomazooze. Seven leagues (17½ miles) S.S.W. of Tubnah, and 16 (40 miles) S.E. of Messeilah, is the village of Em-dhou-Khal, surrounded by mountains; and at this spot you meet with the first plantation of date-trees; but the fruit does not ripen so well as in the Zaab district, which is at no great distance from this spot. The Shott is a large valley or plain,

* See Chapter on Archæology, Part II.

which runs, with few interruptions, between two chains of mountains, from the neighbourhood of Em-dhou-Khal to the west of the meridian of Messeilah. The word commonly means the sea-shore, or the banks of some lake, &c.; but the meaning in this case is, the borders or area of such a plain as, according to the season, will be covered with salt or water. Several parts consist of a light oozy soil; and after inundations, its quick-sands are very dangerous.*

Crossing the Bou-ma-zooze, opposite Tubnah is a large mountain of very good freestone. It is called Muckat-el-Hadjar, *i.e.* the quarry; and the Arabs have a tradition that the stones employed in building Setif (and doubtless other neighbouring cities) were brought from this place. Four leagues to the north of this quarry is Boo-muggar, a fruitful little district, with some traces of ancient buildings. Between it and Ras-el-Aiounne is the village of Nic-Kowse, or Ben-Couse as it is called by the Turks. The inhabitants are chiefly Zaouia (or members of a religious college and confraternity); and the village is situated in a valley, with a circle of mountains at a moderate distance from it. A rivulet runs by the village to the west; but being impregnated with nitrous particles, which are numerous in this neighbourhood, the water is seldom used for drink. Nic-Kowse contains vestiges of an ancient city; and the inhabitants pretend to show the tombs of the Seven Sleepers, asserting that they were Mahometans, and that they slept at this place, and not at Mount Ochlon, near Ephesus, from A.D. 258 to 408.

The powerful clans of Lakhader, Coussoure, and Hirkawse inhabit the mountainous district to the east of Tubnah and Nic-Kowse, as far as Djebel Aouress. The latter is the Mons Aurasius of the middle ages, and the Mons Audus of Ptolemy: it does not consist of one mountain merely, but forms a large knot of lofty eminences, with several beautiful valleys and glens between them. Both the higher and lower parts of Djebel Aouress are very fertile, and form the garden of the province. This group or knot of mountains is reckoned to be about 120 miles in circuit. The northern part is possessed by numerous clans, such as the Bou-zenah, Lashash, Maifah, and Bou-aref; and the district is so fortified by nature, and defended by so brave a people, that the Turks could never subdue it. A high pointed rock, on which their dashkrah is situated, is probably the Petra Geminiani, or the Tumar of Procopius. Numerous ruins are scattered over the hills and valleys of this district, including the remains of Lambese or Lambasa. The Kabyles of these mountains of Aouress are quite different to their neighbours in appearance, their complexion not being dark, but fair, and their hair of a deep yellow. Though Mahometans, and speaking the Berber tongue, yet their physical characteristics make it probable, that if they are not of the tribe mentioned by Procopius, yet they must be a remnant of the Vandals, who, though dis-

---

* Blofeld, p. 55.

P

possessed at the time of their strongholds, and dispersed among African families, may have collected together afterwards. Between Djebel-Aouress and Constantina is the high mountain of Ziganeah, at the foot of which is Physgeah, formerly a city of the Romans, where there is a plentiful fountain and reservoir, according to the name, the water being formerly conducted by aqueduct to Constantina.

This city, the capital of the province, was called Cirta (Sittianorum) previous to the time of Constantine the Great. It is situated beyond the Little Atlas, 48 miles from the sea ; and in history it appears as one of the principal cities of Numidia, which is proved by the extent of its ruins. Its position was, and is, very strong, the greater part of the city standing on a high peninsular promontory, inaccessible on all sides except the S.W., where it joins the continent.* This promontory is a mile in circumference, somewhat inclined to the south; but to the north it ends in a perpendicular precipice of 600 feet ; hence it commands a beautiful view over valleys, mountains, and rivers, the prospect being bounded to the eastward by an adjacent ridge of rocks much higher than the city. But to the S.E. the country is more open, with a distant view of the mountains of Sidi-Rougeese and Ziganeah. In these directions the peninsular promontory is separated from the continent by a deep narrow valley, with perpendicular cliffs on both sides, where the Rummel, or Ampsaga, conveys its stream. On the most elevated point of the city, at the Naugh, is the Casbah, an old edifice now used as French barracks, and commanding Constantina. Below it are corn-mills, turned by the Ouad-Rummel ; and there are many gardens on the banks of this river, in the part called El-Hamma. The streets of Constantina are paved, but narrow and winding ; whilst almost all of them are steep, the houses being generally two stories high, the most beautiful being built of Roman remains. The street of the Jews is remarkably singular, overhung with vines richly laden with fruit, and very shady; and at one end is a minaret with a glittering crescent. A pleasing calm prevails, not found in European cities. The appearance of the buildings, the gravity of the customs, the imposing step, the faces of the Moors and Arabs in the silent shops, compose a pleasant scene.† The ancient palace of the Bey is a remarkable monument. Ahmed Bey, before the French conquest, had employed in its decoration the columns and materials, &c. of the finest buildings in the province. Hence Constantina is rich in antiquities. The chief gate of the four in this town is on the neck of land facing S.W., about half a furlong broad. All this spot down to the river, with a strip of plain ground parallel with the deep valley already described, is covered with ruins. Ancient Cirta stretched as far as this ; but modern Constantina is not so large, but con-

---

* Blofeld, p. 59. Dr. Shaw.

† Blofeld, p. 59. E. Carette, Exploration Scientifique. Recherches sur la Géographie et le Commerce, &c. E. Carette, p. 243. Madame Prus, p. 159.

fined to the peninsular promontory. The gate to the S.W., and that facing the S.E., are both splendid monuments of Roman architecture.*

Constantina has thirteen principal mosques, besides a great many inferior places of worship. The inhabitants are industrious, many of them being tradesmen and artisans. Saddlery and shoemaking give occupation to very many persons; but the principal riches of the country arise from the cultivation surrounding the city. Horned cattle and sheep are numerous; and from the wool of the latter the natives fabricate coarse cloth, which meets with a quick sale. The women spin and weave capital burnouses. The climate of the country, and city in particular, is very healthy, but cold, though the plains near it are generally very hot.† Ruined in 311, in the wars of Maxentius against Alexander, a Pannonian peasant who had assumed the purple in Africa, it was re-established under Constantine, who gave it his name. Its population, which consisted before the French conquest (1837) of Moors, Turks, Kabyles, and Jews, is reported by the natives to have amounted to 40,000. The Kabyles formed one-half, the Moors a quarter, the Jews and Turks the remainder.‡

Constantina, fortified as it is by nature, and by the works which are in process of erection on its precipitous front, would defy the most powerful force.

Below the bridge the Rummel turns north, and runs nearly half a mile through underground passages, with openings for the natives to get at the water. Were it not for this outlet, the river would form a vast lake, and lay a great part of the neighbouring country under water. A quarter of a mile to the east of Sidi-Meemon, the Rummel falls from its subterranean channel in a large waterfall; and the highest part of the city lies above it, whence, till lately, criminals were cast into the river. A little beyond the falls is Kabat-bir-a-haal, a neat transparent fountain full of land-tortoises. These animals, which are devoured by the natives, are thought to be demons; a mythos containing, as usual, an ingredient of truth, since their flesh is the occasion of fevers and other maladies.

Five leagues north of Constantina is the city of Meelah (Milevum, Mileu), built among beautiful mountains and valleys. The surrounding gardens are full of fountains, one of which has a Roman basin; and this place chiefly supplies Constantina with herbs and fruits, the pomegranates of Meelah being, in particular, very large and fine, and held in high repute. Leo and Marmol speak of the excellence of its apples, and assert that the city Mileu took its name from them.

Proceeding eastward from Constantina, you pass by Alligah and Announah, containing ruins, and arrive at Hammam Meskoutin, i.e. the hot or enchanted baths, situated on low grounds, and surrounded by high mountains. It consists of several very hot fountains, which afterwards flow into the Zenati; and not far from them are other springs which are

---

* See Archæology, Part II.   † Blofeld, p. 59.
‡ Ibid. Pananti gives it 100,000 inhabitants: Aventure, vol. ii. p. 11.

intensely cold,—an instance of the sharp contrasts so dear to nature. A few ruined houses stand near the springs. All this part of the country, from Constantina to the Zenati, consists of fruitful hills and valleys, mixed with some beautiful plantations of olives and forests. The district of Boukawan is eastward of the Hammam Meskoutin, on the north of the river Seybouse; and on the other side of the district of Mounah are the possessions of the Beni-Salah, a warlike clan, with the ruins of Ghelma, or Kalma as the Turks pronounce it. A modern town has arisen on this spot, out of the ruins of the Calama of antiquity; and it promises, under French protection and patronage, to match, and even outstrip, its ancient prosperity. Behind Mounah is Tiffesh (Theveste or Thebæ), the only city in the district of Hen-neishah, and a place which has retained its ancient name, though the walls have been destroyed by the Arabs. It stands in a fine plain containing a brook, and is nineteen leagues E.S.E. of Constantina. Near Tiffesh is the country of the Hen-neishah, not only a powerful and warlike, but a graceful and pleasing tribe. This district is the most fertile and extensive of Numidia, comprised between the rivers Hameese and Myskianah, the latter the most southern, the former the most northern branch of the Me-jer-dah; almost every acre of the territory is watered by a brook ; and there are but few of these without a city on them or near their banks, though most of them are now in ruins. To the south of Hen-neishah, near the banks of the Melagge, is Tipsa (Tebessa, Tipata), now a frontier city, standing in a fine situation, not far from mountains, and containing an ancient gate and some part of the old walls. This was formerly a place of importance, and a large underground quarry is situated in the mountains near it. The river Melagge, a little to the north of Tipasa, is a continuation of the Myskianah, and has its sources at Ain-Thyllah, the western confines of Hen-neishah. A little farther the Melagge, flowing to the N.E., takes the name of Serrat, and forms the east boundary of Algeria. This stream, when joined at a little distance by the Sugerass coming from Millah by Hameese, and Tiffesh to the west, assumes the name of Megerda (Bagradas). Near the western banks of the Serrat, ten leagues from Tiffesh, is Collah, Gellah, or Gellah-ad-Snaan, a good-sized village built on a high pointed mountain, with only one narrow road leading up to it. This place, which could only be reduced by famine or taken by surprise, was formerly a convenient sanctuary for criminals from Tunis and Algeria.*

That part of this province which belongs to the Sahara contains, exclusively of the distant city of Ouerghela or Ouaregla, and village of Engousah, the two considerable districts of Zaab and Ouadreay, with their numerous ksours and villages. These places are commonly a collection of dirty huts, built entirely of mud walls, with rafters of palm-timber ; and all their inhabitants are employed in the cultivation of the date.

* Blofeld, p. 69.

The district of Zaab (the Zebe or Zabe of the ancients), once a part of Mauritania Sitifensis, and also of Gætulia, is a narrow tract of land lying immediately under the mountains of the Greater Atlas, and displays a chain of villages, with few intermissions, from the meridian of Messeilah to that of Constantina. Dousan, Toodah, Sidi-Occ'ba, Biscara, and Oumil-hennah receive their rivulets from the Tell; but the fountains and brooks that contribute to the others rise within the Sahara, or else ooze from the southern skirts of Mount Atlas. The Dued-Adje-dee, or Djedi (that is, the river of the kid), receives these streams; and after running to the south-ward looses itself in the Melrir, an extensive tract of the Sahara, of the same saline and absorbent quality as the Shott above described. This river is probably the Garrar or Jirad of Abulfeda. There are no other great streams on this side of the Niger, and it may possibly be the Gir of Ptolemy, though placed by him much more east or south-east, among the Garamantes.

Biscara, the capital of Zaab, was once the residence of a Turkish governor and garrison, and contains a small castle built by the Bey of Constantina. Its chief strength consisted in six small cannon. Surrounded by a brick wall, this city has much trade in slaves, &c. and other produc-tions of Nigritia. Many of its inhabitants migrate to Algiers, where they work as porters, &c., and form a corporation. The village of Sidi-Occ'ba is famous for the tomb of the Arab general of that name, who is its tute-lary saint. The tower of Sidi-Occ'ba is reported to tremble when you call out, *Sizza bil ras Sidi-Occ'ba* ('Shake for the head of Sidi-Occ'ba'). This wonderful tradition may, like others, be founded on fact, resulting from one of the mysterious miracula of gravity and acoustics. Nor would it be the first stumbling-block to shake the faith of the sceptic;—a tower at Rheims exhibits somewhat of a similar phenomenon when you ring one of its bells. Roman remains are scattered throughout the district, with traces of the care they bestowed on the channels of irrigation.

The eating of dog's flesh is said still to be a common practice in Zaab, as among the Carthaginians and the Guanches of the Canary islands, which thence received their name. It is also well attested that there are human puppies in the Sahara, where they present the same phenomena and cha-racteristics as in the Elysées and Regent Street, with a slight difference in their exoteric development.

Ouad-reay is another collection of villages like those of Zaab, twenty-five in number, running in a north-east and south-west direction, their capital, Tuggurt, standing on a plain without a river. There are no foun-tains in this country, but they obtain water by digging 600 or 1200 feet, at which depth they invariably reach it; the ground being perforated by innumerable subterranean streams called Bahar-taht-el-Erd (the under-ground sea), a phenomenon noticed by Dr. Shaw. They dig through several layers of sand and gravel till they reach a flaky stone like slate, known always to lie above the Bahar. This layer is easily broken through, and the water rushes up so quickly that the man who digs through it is sometimes drowned.

Thirty miles south-west of Tuggurt is Engousah, the only village of several in this situation which existed in the time of Leo. After Engousah, at five leagues distance to the west, is the noted and populous city of Ouaregla, the most remote community of any size and importance this side the Niger. These several cities and villages, together with those of Figig and of Beni-Mezzab, are justly compared by the ancients to so many verdant spots in a great expanse of desert, and belong probably to the country of the Melano-Gætuli.

After describing Gætulia, Ptolemy reckons the nations to the southward, among which the Melano-Gætuli and the Garamantes were the principal. These nations certainly extended behind the greatest part of Algeria, Tunis, and Tripoli ; or from the meridian of Siga, near Tlemsen, to the Cyrenaica, 35 deg. more to the east. And as, inclusive of the Bedouins, there are no nations in this direction besides the Figigians, the Beni-Mezzab, the inhabitants of Ouad-reay and Ouaregla to the west, and those of Gaddeniz, Fezzan, and Oujelah to the east, it is probable that the Melano-Gætuli must have been the predecessors of these western Libyans, as the others to the east were, for the same reason, the successors of the Garamantes.[*]

The country of the Beni-Mezzab is very fertile ; and besides a considerable commerce with Gadamis, Bornou, Timbuctou, and the whole of Soudan, it disposes of the produce which it draws from those countries to the inhabitants of Tunis and Tripoli. In short, it has to a considerable extent a monopoly of the *roulage* or carrier-trade of north-west Africa.[†]

Proceeding to analyse the ways and by-ways of this province, we shall join several parties of travellers, some of whom are old friends. And first we shall follow the expeditionary column under Marshal Clauzel, that marched from Bona to Constantina in the autumn of 1836. Baron Baude[‡] and M. Berbrugger,[§] who both accompanied the column, have left a minute diary and description of their adventures.

Marshal Clauzel, who was then governor of Algiers, commanded the expedition, which was accompanied by the Duke of Nemours, who had with him General Edward Colbert, Colonel Boyer, and Lieutenant-Colonel Chabanne. The marshal was himself escorted by nine aides-de-camp ; Colonel Duverger was chief of the staff, and had eleven officers under his orders. Colonels Tournemine and Lemercier commanded the artillery and engineers; and each *état major* or staff reckoned six officers. The administration was confided to M. Malain d'Arc, military intendant of the army of Africa ; and the chief surgeon was Dr. Guyon.

The army, consisting of 8766 men (7410 French and 1356 Turks, &c.)

* Blofeld, p. 73.   † Ibid.   ‡ Baude's Algérie, vol. ii. ch. ix. p. 44.
§ L'Algérie, historique, etc. part iii.

started on the 13th of November from Bona, when the marshal, with the main body, reached at 7 P.M. the right bank of the stream of Bouinfra. On the 14th they ascended towards the ancient Ascurus, and the marshal stopped on the banks of the rivulet of Nechmeya, two leagues (five miles) from the bivouac of the previous night. On the 15th they started early, the weather being fine, and reached the Seybouse. From the camp of Draan to the Bouinfra the country is very broken; mountains clothed with

BIVOUAC ON THE BANKS OF THE NECHMEYA.

shrubs, isolated from the Atlas chain, and only to be compared to truncated volcanic peaks, rise like islands from the middle of the plain. For part of the journey the soil is meagre and light; but it became excellent again on approaching the Bouinfra. After passing this stream, you enter, not to leave it again till beyond Constantina, a country of jura limestone formation : you have before you a branch of the Atlas, that encloses on the north side the valley of the Seybouse ; a narrow hill is detached perpendicularly from it, and advances like a spur into the plain. The road begins by following its back, and passes near the ruins of Ascurus, traversing a historical country scattered with ruins. About the Bouinfra the ground is well wooded, and crossed by several limpid streams. From the Col you descend along a pretty valley to the thermal waters of Hammam-Berda, which are probably the Aquæ Tibilitanæ of the Itinerary of Antoninus.

They flow into a basin of masonry, and are abundant, clear, insipid, and in-
odorous ; their temperature is that of ordinary baths, *i. e.* from 25 to 30°.*
The site is agreeable, the soil fertile ; and the vigour of the rose-laurels
announces that the streams, whose courses are marked by festoons of
flowers and foliage, are rarely dry. The Roman establishment on this
spot must have been considerable, but the foundations alone remain. The
attention devoted by the ancients and the Orientals to multiply baths
depends on hygienic causes, which cannot be neglected with impunity.
The vale of Hammam-Berda debouches into that of the Seybouse, opposite
Guelma : the river has at this place a width of 60 metres (196 80 feet),
and its current is very rapid ; its left bank is covered with marshes.
Guelma, or rather the heap of ruins which bears that name, is on the
other side of the Seybouse, 1500 metres (4920 feet) from the river, on the
even but rather steep slope of a hill. The ancient enclosure of Guelma
(Calama) contains a space of from seven to eight hectares (20 acres).

On the 16th the troops ascended the vale of the Seybouse, finding but
little cultivated ground on the road, but numerous flocks of sheep within
reach. At 2 P.M. they halted at Mjez-Amar, at the foot of the Ras-el-
Akba, where the Seybouse receives the Oued-Cherff, which takes its source
15 leagues to the south-west, not far from the ruins of ancient Tigisis. It
makes a curve to the north to turn the Ras-el-Akba, by the deep cutting
at the entrance of which are the famous thermal springs of Hammam-
Meskoutin. The little plain in which it debouches is raised from 20 to
30 metres (98·40 feet) above its bed, the banks being rocky and almost
vertical. The road followed this day was the scene of Jugurtha's triumphs
over the Roman Aulus, of which more anon. On the 17th of November,
crossing the Seybouse, they began to climb the Ras-el-Akba. The Arabs
relate wonderful stories of the altitude and marvels of this mighty natural
pile, which may be compared to the Col de Tarare in France, save that
the forms of the rocks at the Ras are much sharper, and the Col is commanded on two sides by lofty rocks. The mind is almost filled with a
feeling of oppression and discouragement at the aspect of this country.
You see, as far as the eye can reach, mountains swelling up in gigantic
masses, between which you can perceive no way to steer ; all around is
naked ; and in this immense horizon you seek in vain for a tree or a little
brushwood. Halting at the foot of the Ras-el-Akba on the 18th, some of
the party drew nigh to the ruins of Announah, which are still considerable,
and are situated in a singular inaccessible position half-way up the cliffs.
On the 19th the column, after having crossed, marching westward, two
offshoots from the Ras el-Akba, came about 10 A.M. to the banks of the
Seybouse, not far from the marabout of Sidi-Tamtam. The Seybouse is
here called the Oued-Zenati, from the name of the tribe whose territory it
crosses : it has only a small stream of water ; hence the great volume of

* Reaumur.

water that the army crossed the day before must have come from the valley of Alliga.

The Ras-el-Akba forms a kind of promontory, round which the Seybouse doubles. The distance from Mjez-Amar to Sidi-Tamtam is 22 kilometres (13·66 miles) by the mountain, and 36 kilometres (22·49 miles) by the banks of the river. Following the gorges of the Hammam Meskoutin, you meet, at 20 kilometres (12·42 miles) from Mjez-Amar, the vale of Alliga, which takes the direction of Constantina, and where you find the traces of the Roman road from Sicca Veneris (Keff) in Tunis to Cirta.

PASSING THE SEYBOUSE.

By this road the distance to Constantina is only 46 kilometres (28·58 miles) ; while continuing to ascend the valley of the Seybouse, in order to descend that of the Bou-Merzoug, you make a circuit of 74 kilometres (45·98 miles).

On the 20th of November the army marched from 8 A.M. to 5 P.M., a cold wintry wind sweeping across their path. On quitting the basin of the Seybouse, it enters a rich well-cultivated plateau, on which are many douars. The column turned to the south of a group of rugged mountains, and descended by the vale of the Oued-Berda into that of Bou-Merzoug, which throws itself into the Rummel above Constantina. At last they arrived at the clayey table-land of Soumah, whilst the winter's sun shone on a group of white houses 3 leagues (7⅔ miles) N.N.W., half masked by the plateau of Mansourah. This was Constantina. The army halted around a Roman monument, of which a further description will be given elsewhere.

On the 21st the army reached with difficulty the banks of the Bou-Merzoug, a torrent which, swollen by the recent rains, rolled its furious waves over the slippery rock in its channel. The column met with less delay in traversing the lesser tributaries, and about 2 to 3 P.M. they arrived together on the plateau of Mansourah, when they beheld the whole of Constantina, from which they were only separated by the deep ravine, at the bottom of which rages and roars the Rummel.

The depth of the channel of the Rummel beneath the highest part of Constantina is 100 metres (328 feet) ; and the towers of Notre Dame at Paris, if you seek an object of comparison, are only 66 metres (216·48 feet) high. The river traces a cincture of 1500 metres (4920 feet) at the foot of the town ; it has a fall of 75 metres (246 feet), and the precipices on all hands are vertical. The frame is worthy of the picture. " Mountains covered with snow surrounded us," says Baron Baude, " on all sides, whilst the damp clay was the only bed of the soldiers. The plateau of Mansourah alone is formed of alternate beds of rock and of marl." Without dwelling on the hardships and sufferings of this brave army, or criticising the errors of the government or commanders, who have been respectively blamed for exposing it with insufficient means at a most inclement season,—it will suffice to say, that the attempts at storming failed; and provisions also failing on the 24th, the army began its retreat, after destroying its tents, baggage, &c.

The distress of the column on the retreat was very great, and many veterans of the Russian campaign (1812) declared that its horrors and sufferings were exceeded ; yet all was borne with heroism and a *British* patience by the French troops. The retreat is also remarkable for a display of coolness by which Changarnier made himself conspicuous for the first time. The circumstance was as follows. On the 24th the French army marched slowly, amidst the continual fire of the Arabs of Achmet Bey ; it held them in check by its tirailleurs, and the foe fled as soon as the French soldiers faced about. However, half-way to the monument of Soumah, the battalion of the 2d closing the march, the enemy, reckoning on their superiority of numbers that the victory would be secure, decided to charge.

Commandant Changarnier rallied his men, running in to form square, and awaited the enemy at twenty-five paces. "They are 6000, and we are 250," he said to the soldiers : "you see very well that there is nothing to fear !" The volley, directed with the steadiness of parade, dispersed the Arabs in two minutes. There were thirty-four killed or wounded in the square ; but it stood firm, and saluted with the cry of " Long live the King !" (*Vive le Roi*) the flight of the enemy, and some tirailleurs detached in pursuit killed the dismounted men.* This warm reception prevented

* Baron Baude, vol. ii. c. 9.

a repetition of any attacks on the part of the enemy during the retreat, in the course of which the troops ransacked the silos, or corn-holes, of the natives most successfully. Cæsar has given an exact description of these silos, which must have been identically the same then.*

On the 25th and 26th of November the army continued its retreat, and on the 27th arrived at the broken plain of Sidi-Tamtam, which stretches on the left bank of the Seybouse, whilst on the right bank the first slopes of the Ras-el-Akba embrace in their concavity the bend that the river describes in this place. The French army drew up on the mountain at 7 A.M., and beheld the same spectacle that Cæsar recorded 1881 years before, when 30 Gaulish horsemen, on his retreat to Ruspina, drove back into the walls of Adrumetum 2000 Moors who pursued them. "We were on the slope," says Baron Baude, "as on the steps of a theatre ; the 3d Chasseurs d'Afrique alone remained in the plain, drawn up in line perpendicularly to the river, and separated from the Arabs of Achmet by the bivouac we had just left. Suddenly a savage cry arose, and the Arabs rushed like famished jackals on the abandoned camp. Like sheep before the dogs, the Arabs ran away amidst the laughter of the spectators, scattered by the charge of Captain Morris." (Compare Cæsar *De Bello Afr.* c. 6.)†

From Constantina to the Ras-el-Akba the country is very fertile, but very melancholy in its character, though picturesque. The soil consists of a bed of tenacious clay, without any mixture of pebbles : it is well fitted for the cultivation of corn, almost every where grassed over, and pierced at intervals by banks of limestone. "For 20 leagues (50 miles) we only saw one little copse of half an acre at some distance from the road, and one shrub on the plateau of Oued-Berda. At the gates of Constantina alone some vegetation reappears, without the soil having in appearance changed nature. The thickness of the turf; the beauty of the corn, of the barley, and of the beans found in the Arab silos ; the excellency of the chopped straw for the horses,—announce a very great productive energy in the soil. A numerous population existed here under the Romans, and you meet with ruins every where : not of rustic structure, like those of Hippo ; masonry is every where employed, and there must have been plenty of wood in the country at that time for the use of such cities."‡

The wooded vales of Mjez-Amar and of Calama appeared the more beautiful from their contrast to the naked declivities of the Ras-el-Akba. On the 28th of November the staff passed the Seybouse to go to Guelma. The surrounding country is rich, graceful, and woody, like the left bank.

* " Est in Africa consuetudo incolarum, ut in agris et in omnibus fere villis sub terra specus, condendi frumenti gratia, clam habeant, atque id propter bella maxime hostiumque subitum adventum præparent. Qua de re Cæsar certior per indicem factus," &c.— *De Bello Africano,* c. 65.

† " Accidit res incredibilis, ut equites minus xxx Galli Maurorum equitum duo millia loco pellerent, urgerentque in oppidum." ‡ Baron Baude, ibid.

We shall here take leave of our brave column, which lost many men in hospital at Bona from the hardships incurred in the expedition, who were, however, amply avenged next year, 1837, on the fall of Constantina.*

As regards the plain of Bona, Baron Baude remarks further, that much is said of its fertility, and that it is the only point on which all testimonies are agreed. " I ran over it in many directions ; and notwithstanding some thin and marshy ground, I know in no department in France a similar extent of land so good. The soil is a mixture of sand, clay, and marl ; the banks of clay are almost every where adapted for making bricks, and in many places for pottery ; they preserve the freshness of the ground, preventing it from absorbing too quickly the rain ; and most probably, by sinking wells, you could obtain good water there, such a stratification giving good hopes for the success of Artesian wells.

" A great advantage is also found in a vast bed of hydraulic limestone 4 leagues (10 miles) from the Seybouse, and 9 from Bona, on the road to Guelma."

So much for Baron Baude. We shall now accompany our old friend Count St. Marie, who travelled from Draan to Constantina, in 1845, with two squadrons of the Chasseurs d'Afrique. "The French post of Draan is five leagues (12½ miles) east of Bona, and stands on a height which rises with gentle acclivities like an island in the midst of an immense plain, on which nothing is to be seen but thistles parched by the sun. After leaving the camp of Draan, we found the country before us scattered with little hillocks, as if they were detached masses from the chain of the Atlas." Every object on the road is noticed almost in the same words as those of Baron Baude ; yet it were, perhaps, uncharitable to hint at plagiarism, the country being bald, and its characteristics few. The volcanic hills; the light soil, the rich banks of the Bouinfra, the detached spur of the Atlas, the ruins of Ascurus, the springs of Hammam-Berda, the foliage of the laurel-rose, Guelma and its ruins, the Seybouse and its breadth and velocity, which they broke by making some of the cavalry stand higher up the stream,—all these features are chronicled almost in Baude's words, as likewise the dreary view from the Ras-el-Akba, the ruins of Announa, and the unfortunate Cornelia, who only *vixit annos* XIX. We shall spare the reader a verbatim repetition of the Baron's description, merely adding, that precisely the same features are noticed by St. Marie as those previously noticed by the Baron, and in the same words ; and after passing Sidi-Tamtam, Bou-Merzoug, and Soumah,† we are delivered at the gates of Constantina. We have purposely confined ourselves to the tenderest criticism on the Count ; but it must be admitted that it is somewhat unfortunate that he shews such a close affinity to the Baron. Arrived at Constantina, however, we may safely trust him, as the Baron never even

* Baron Baude, ibid.   † St. Marie, pp. 232-237.

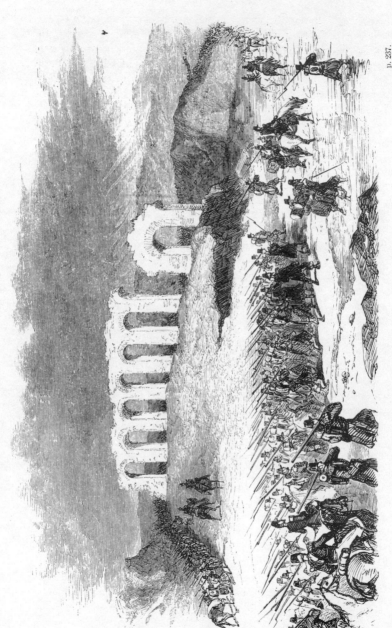

p. 237.

THE FRENCH ARMY PASSING THE RUMMEL.

peeped over the walls ; and we can check any of the Count's disposition to build castles in the air by posterior authorities whom he could not copy.

" Constantina," says the Count, " is encircled by the river Rummel, and commanded by the heights of Mansourah and of Sidi-Mecid. The last is the Jewish burial-place, and its summit is 350 metres (1148 feet) above the city. On the south-west the heights of Condiat-Atz, fronted by a little hill covered with Mussulman tombs, also commands the approaches to the city. The table-land on which the city is built overlooks extensive and fertile plains. The Ouad-Rummel leaves the city at Sidi-Rachet, where it forms a cascade falling into a great ravine, which extends along the south-east and north-east sides. At the northern extremity of the city stands the Casbah. Here the Rummel forms a new cascade, the Tortoise-fall, and then leaves Constantina, continuing its course to the north. At the point of El-Kant'ra, the river for a little distance takes a subterranean course; and after flowing once more a short distance over ground, it again disappears. In this manner it is lost sight of four times, being concealed beneath a natural bridge of from 50 to 100 metres (164 feet to 328 feet) in width. The three gates, Bab-el-Jedid, Bab-el-Ouad, and Bab-el-Ghabia, are united by an ancient wall 30 feet high ; but there are no moats. Outside the Bab-el-Ouad (Water-gate) you find a little suburb inhabited by artisans. At this place are also situated the leather, wax, and wool markets. A mosque, in good preservation, stands next to the old building once used as stables of the Bey, and capable at that time of containing 800 horses. The walls, however, are not very solid, and have no proper foundation. The bridge of El-Kant'ra is broad, rests on three tiers of arches, the lowest of which is Roman, and crosses the river at the great chasm intervening between the city and the mountain. At the highest part of the city rises the Casbah, which contains now nothing but barracks, and is only mounted with a few guns. Lower down are some corn-mills, set in motion by the Rummel. Gardens and orchards line the banks of the river on the north side of the city, in the quarter called El-Gemma. The form of the city of Constantina is compared by the Arabs to that of a burnouse spread out at full width, the Casbah representing the hood."[*]

The city contains three squares, to which the French have given a look of regularity, by pulling down many old buildings, and planting trees; thus converting them into pleasant promenades. The palace of the late Bey Ahmed is remarkable for the fine columns of marble that adorn its front.

" The chief branches of industry at Constantina are the manufacture of saddles, boots, shoes, and a sort of gaiters worn by some of the Arabs. There are also some forges, in which iron brought from Tunis is wrought into agricultural implements, bridle-bits, spurs, and horse-shoes. The

[*] St. Marie, p. 237.

burnouses and haicks made by the people of Constantina are the best in Algeria."*

Madame Prus soon arrived at Guelma, which town is situated on the summit of the mountain Serdj-el-Aonda, and must have been a place of considerable extent and importance in the time of the Romans. This town abounds in antiquities; for the examination and correct valuation of which, a scientific commission has just been appointed by the government.

All the towns of Algeria resemble each other. The houses are square, whitewashed with chalk, surmounted with a terrace; and the walls deprived of all ornament in the shape of windows, and only provided with small apertures to admit air.

Guelma must have been very difficult of access in times of siege; defended on one side by the steep mountain on which it is situated, and on the other by a lake formed by continual showers on the marshy plain, it afforded but little chance of success to a besieging force. The Seybouse, in its frequent inundations, overflows the road, which prevents all communication; thus, during six months out of twelve, it is necessary to use horses, mules, or camels to go the short distance between Guelma and the neighbouring towns. Guelma was the ancient Calama of the Romans (it is often spelled Ghelma).†

Passing to Constantina, Madame Prus remarks that several houses in the town have tiled roofs instead of terraces; the interior arrangement is generally the same as at Bona, and the other towns of Algeria. The aspect of the town is gloomy, and the streets are narrow and dirty in the extreme, although the greater number of them have been recently paved.‡

Vaulting into the saddle, we shall next accompany Mr. Borrer in his tour from Bona through Constantina to Biscara.

"Oct. 19th, 1847, at 4 A.M. we left Bona by the Porte de Constantine, with guns slung at our backs, pistols in our holsters, and muffled in a thick burnouse, which did not, however, keep out the cold. With a spahi for escort, we galloped over in three hours to the camp of Drean, the first military post on the road; having traversed an open naked country, which was, however, well adapted for the growth of corn. You pass an ancient Roman bridge over the Bou-jeema, or a branch of that river about two hours from Bona on this route. Drean is a mere collection of barracks formed of planks, with an earthen rampart, and a slight foss surrounding

---

* St. Marie, p. 237.   † Madame Prus, p. 159.

‡ Madame Prus (p. 192) describes Constantina as situated on a high mountain, level at the top, and surrounded on three sides by the Ouad-Rummel, a deep ravine with precipitous banks. To the south this plain joins the hills on the left of the Rummel by the isthmus of Condiat Atz, while at its north-east angle a gigantic bridge is thrown over the ravine, consisting of two rows of arches, one above the other. This bridge was constructed by the Romans, and restored by the Spanish engineers, and serves as a means of communication between the town and the table-land of Mansourah, over which lay the route of the French army before it arrived at the fortifications of Constantina.

it, situated on a naked mamelon, and containing at that time a troop of spahis. Leaving Drean, they soon reached the next military post, *Nechmaya*, a stone building and some huts of planks serving as stables, canteen, &c. Soon after leaving Nechmaya, we traversed a district covered with brushwood, wild hills and massive isolated rocks varying the scenery. The road from Bona thus far is good enough; but here in winter the traffic is sometimes quite stopped by the swelling of the streams. Soon after we came to a fountain of clear water, both warm and mineral, and containing the remains of Roman brickwork, which shew that they were not neglected in antiquity. This spot is called by the Arabs Hammam-Berda. Oleanders and other shrubs luxuriate upon the margin of this ancient spring; numerous mocking-birds find covert in their branches, and fill the air with sweet music. There is fine pasture-land around Hammam-Berda; the soil being irrigated, besides the bath-stream, by another brook, whose rapid and plentiful current turns the mill of some French speculator. Farther down, before it joins the Seybouse, this stream waters many beds of water-melons belonging to the gardens of Arab and French colonists. A wooden bridge has been erected over the ford of the Seybouse, at which we now arrived, close to Guelma. This town (once Calama) is situated on the S.E. bank of the Seybouse, about a quarter of a mile above the bridge, and on the slope of a hill of gentle inclination.* As the traveller slowly wends his way up the winding road to this French post, and beholds strewn around him vast blocks of fine stone, skilfully squared by the Roman chisel, mingled with fragments of marble columns, he can but meditate on the instability of human power, and how the might of nations is entombed by time. Draw back the veil of ages, and the double-peaked summit of the wooded Maouna overlooks a noble city; her towers, her temples, and her palaces radiant with Parian marbles. Forty thousand inhabitants within her walls bow to learning, art, and luxury; beyond her gate extends a richly-cultivated plain, its teeming slopes watered by the winding Armua, which, leaving its tribute here, then hastens on its rapid course to refresh the delightful gardens of the royal Hippo; ships freighted with Oriental luxuries are borne upon its bosom, before it mingles its waters with the sea. *Tempora mutantur!* The Mare's Saddle, as the Arabs name Mount Maouna, which forms the background of Guelma, now overlooks some few score French houses, garrisoned by about 800 men, chiefly spahis and the Foreign Legion."

The amount of the civil population of Guelma in December 1849 was 1399 souls.†

It contains one street of considerable length; the houses built chiefly

---

* The reader will have noticed that two gentlemen, Baron Baude and Mr. Borrer, describe Guelma as seated on a slope, while Madame Prus says that it stands on the top of a hill. We leave it to the gentle reader's gallantry to determine if the opinion of one lady should outweigh that of two gentlemen.

† Tableau de la Situation, p. 96.

of massive Roman remains turned up on the spot, and therefore not quite so likely to be tumbled down by the first slight earthquake or violent rains, as most of the pasteboard settlements of the colonists. Building was actively going on there (1847), notwithstanding the financial crisis of the colony, money fetching 20 and even 30 per cent at Guelma in 1846. The original plan drawn for this French post is based on a population of 7000 souls. In 1842 there were 92 inhabitants; in 1843, 108; and at the end of 1844, 317.* There is one French regulation much to be admired, that of allotting gardens to their soldiers. At Guelma a fine piece of ground has been cultivated by them, and those industriously inclined spend their leisure-hours there. There is also at Guelma a *pépinière* to supply the colonists with seeds, trees, and other vegetable products: "it was shown me," says Borrer, "with much boasting of its beauty and promising nature; yet rank weeds and little else luxuriated in it. The circle of Guelma, as far as I have visited it, is, I should say (barring its distance from a sea-port, viz. 26 leagues, and that not to be traversed by merchandise in the rainy season), a promising point of colonisation; for its plains and its valleys are rich, and watered by numerous streams. The chief of these are the Seybouse, the Oued-Scherf, the Ouad-Bou-Hamdon, and the Oued-Zenati. Tobacco, mulberries, and corn would undoubtedly flourish here together; and it is reported that cotton and indigo have succeeded in the *pépinière*,—which one would have scarcely expected at such an elevation. There is as yet very little land brought into cultivation in the vicinity of the town; but the soil is good; and to the west of the town is a delightful little valley, with a fine stream, where there are gardens sufficiently extensive to supply their owners with vegetables. The climate is much the same as that of Constantina, though not quite so cold in winter. Dysentery and intermittent fevers are the prevailing diseases, as in most colonies; but every thing considered, it is a healthy position. Guelma is rather a desirable spot for the sportsman. Hares, red-legged partridges, quails, and at the present moment (autumn) the little African quail, remarkable for having only three toes, abound there. As for lions, the lion-king, as the Arabs have christened that renowned Nimrod, Spahi *Gerard*, has made them rather shy in the neighbourhood. He is incorrectly reported

---

* The last *Tableau* or Blue-book, gives the following account of the statistics of Guelma:—Population in December 1847, 736; in December 1848, 1102; in December 1849, 1399. As regards streetage, 810 metres of large, and 2853 of small streets, have been opened from 1845-9 for 18,100 fr. ; and 400 metres of drains have been opened for 35,700 fr. A school has been built for 16,200 fr., and also a church and presbytery, costing 69,875 fr. A cemetery was also opened in 1847, at an expense of 5000 fr.

Under the head of fortifications, it appears that they have finished the curtains of the *reduit* between towers 1 and 18, 3 and 6, and the tower 18. They had likewise finished in 1850 the little bastions of the city-wall, and a part of the ditch. Barracks have also been completed to hold 400 infantry and 170 horse, besides a hospital and magazines. The total expenditure from 1843 to 1849 was 231,510 fr. 4 cents. (9260*l.* 8*s.* 8*d.*) Tableau, 1850, § Travaux publics.

to have been decorated for his courageous feats in lion-slaying. No less incorrect was the late report of his death. It is a curious fact enough, however, that he owes his life to a lion; and thus it was. He was one of the unfortunate battalion which was a short time back leaving Guelma for Tebessa, a French post lately established on the confines of Tunis; and who, deceived by the apparent friendship of an Arab sheikh, fell a prey to Numidian treachery, every father's son of them being most barbarously massacred, save Gerard. The spirit of Nimrod watched over our spahi. A lordly lion, crossing the route of the battalion, a short time before it fell into the hands of the Philistines, was fired at and grievously wounded by Gerard, who, dismounting, swore by his beard that he would have the skin of the beast. Plunging into the thicket, he followed the lion all that and the next day, when he at length reached the king of beasts, and slew him. The chase over, our hero turned back to the route of the battalion; but he wandered many days and found it not. During this time his comrades were all killed, and he was thought to be among the dead. But one fine morning he marched into the auberge at Guelma usually frequented by him, with a fine lion's skin, and asked for breakfast from the landlord, who, petrified, thought he saw a ghost. But he ate so well, that they soon found, to their joy, it was Gerard himself in the flesh. Even hostile tribes often apply to him to slay lions; and so great is the license he has gained, that his superior officers allow him to absent himself à *discrétion* when thus summoned to the chase. The darkest nights are those chosen by him for his sport, the glare of the lion's eyes then offering the surest mark.*

Borrer spent three days at Guelma,† and walked one day with his gun some distance up the banks of the Seybouse to the west of the city, where they became wild and craggy. The Arabs of this part of Algeria struck him as far superior in aspect and manners to those of the province of Algiers. Many are handsome, with fine oval countenances, large black eyes, small aquiline noses, and snowy teeth. The women wear large silver rings in their ears, of great weight, and as large as anklets; besides numerous other ornaments, such as little looking-glasses, and especially a wooden hand attached to the breast; the fingers representing the number 5, to which they attach a special virtue.

The following day he left Guelma for Constantina, with an escort of two spahis, and arrived in the evening on the borders of a wide plain called Beni-Simsen. Here he found the Zenatia, a powerful tribe who drink the waters of the Oued-Zenati. "I had a letter for the caid of this tribe. It was the hour when the flocks and herds were wending their way to the

---

* On Gerard, see *Leaves from a Lady's Diary of her Travels in Barbary,* vol. i. p. 271 (1850).

† Captain Kennedy, who passed through Guelma in 1845, describes the garrison as forming almost the whole population of the place. *Algeria and Tunis,* vol. ii. p. 221. The only hotel at Guelma in 1848 was a *guinguette* (pot-house), called Hôtel des Voyageurs. *Lady's Diary,* vol. i. p. 268.

douar, there to seek, within the circle of tents, shelter during the night-season from wild beasts and robbers. The caid with the elders of the tribe were sitting in a ring upon the ground, withdrawn a slight space from the douar, holding council. Immediately upon our approach he arose, and proceeding to the douar, after the usual salute, *salaam alikum,* showed us into a good tent spread with matting." A large carpet in the centre, used as the seat of honour, was assigned to Borrer. A sheep was then driven before him and slain, and an enormous dish of cous-coussou was brought in about midnight. The mutton being rather tough, his spahis kindly tore off the most fat and delicate morsels with their fingers, and stuffed them into his mouth. This douar consisted of about 900 tents, and the caid said that he could at any time lead forth 2000 horsemen equipped for battle. The chief soon retired; but the noisy conference of the spahis and Arabs, and the rushing forth of the dogs to chase away hyenas and jackals, effectually chased away all slumber from Borrer, who lay on his saddle till 3 A.M. Mounting in the dark, and amidst much rain, they forded the Oued-Zenati, and proceeded across the vast and naked plains of the Beni-Simsen. In the act of crossing the Oued-Zenati, Borrer's saddle turned round, and he found himself on his head in the mid-stream, with one of his feet hanging in the stirrup. He released himself; but his steed galloped off in the dark, and trouble enough they had to capture him. A good ducking, and his gun-barrels full of water, were the fruits of this un-expected evolution. "Day had long appeared, but still we were upon this eternal plain, and the rain fell in torrents. At last we arrived at the base of a vast range of mountains, of which Djebel Bahbara and Djebel Bougareb are, I believe, the most lofty, the latter being about 1300 metres (4208 feet) in elevation. A furious wind assailed us in the gorges of the mountains, howling among the savage rocks, and at times almost sweeping us from our horses ; added to which, the rain had begun, and galled us so severely, that we were several times compelled to halt and turn our backs to it.

"It was now 4 P.M., and not an Arab tent had we seen ; not a morsel of any thing except hail had entered our mouths since 8 o'clock the evening before ; moreover, one of my spahis' horses broke down with fatigue." They had passed numerous remains of ancient, apparently Roman, struc-tures in the plain. About three leagues (7½ miles) from Constantina they met an Arab, from whom Borrer begged a handful of dates, and de-voured them with gusto, though hard and full of worms. They saw to their left the remains of a Roman post on a height named Soumah, and forming a shoulder of the Djebel-Oued-Msetas, the elevation of which is about 1183 metres (3880·24 feet). About 8 o'clock they descended into the valley watered by the Bou-Marzeg, which has its confluence with the Oued-Rummel (the Ampsaga of the ancients) about a mile south of Constan-tina. Soon after, they reached the bridge of El-Gantra, suspended over the fearful chasm in the rock, 700 feet in depth, which forms a natural moat to this strangely perched city of Constantina.*

"I am just returned," writes Mr. Borrer, "from breakfast with General Bedeau, the commander-in-chief of this province. He appeared to me a man of profound understanding, united with great fluency in expressing his ideas. He is, indeed, generally admitted to be one of the most able men the French have in Africa. General Bedeau and Marshal Bugeaud ran their course together; the former having distinguished himself as lieutenant, under the command of the latter, at the siege of Saragossa (1809).

" The palace in which the general resided was built by Achmed, the last bey of this city. It is a spacious and handsome specimen of Moorish

* The following are the latest official statistics of Constantina :—European population in December 1847, 2013 ; December 1848, 2590 ; December 1849, 2050. Native population in December 1849, 20,944 ; analysed into 16,835 Mussulmans, 673 negroes, and 3436 Jews.

In 1848, according to the *Lady's Diary*, there were 20,882 natives, and 1919 Europeans. The town, according to the same authority, is divided into two parts, one native, and the other European. No carts pass along the narrow streets ; camels, asses, &c. taking their place. 350,000 fr. are annually spent in providing the town with water (vol. i. p. 222). The Hôtel de l'Europe is a dismal Moorish house.

As regards the civil works, in 1849 none of the larger streets of Constantina had been opened in accordance with the plans and surveys made and levels taken, save 205 metres (672·40 feet). The old pavement of the bridge of El-Kantara has been replaced by a pavement of sandstone (*en gres*). They have planted 150 feet with trees, and cleared 800 metres (2624 feet) in the Place de la Brêche from rubbish. In the 640 metres (2099·20 feet) of small streetage they have replaced the old limestone pavement by a causeway with side-gutters, all built of sandstone. All rubbish has been cleared away, and they have levelled the Places du Commerce, du Palais, and the Carrefour d'Orleans. The expenses for streetage from 1843 to 1849 amounted to 50,949 fr. 80 cents. (2038*l.*) In connexion with waterworks, the conduit from Midi-Mabrouck to the cisterns was finished in 1848, consisting of two tunnels,—one of 202 metres (662·56 feet) in the rock, the other of 764 metres (2505·92 feet) of masonry,—a principal conduit of 2447 metres (7826·16 feet), and secondary conduits of 1467 metres (4811·76 feet). A siphon of a large size receives the waters on the plateau of Mansourah, and brings them to the Casbah : expense, 438,000 fr. Other waterworks were in contemplation. Four fountains have been also established ; 800 metres (2624 feet) of new sewerage, and 1200 (3936 feet) of old opened. The fountains are those of El-Kantara, of the Pont d'Aumale, of La Pépinière, and of Sidi-Mabrouk. A canal is also being built from the Rummel, called *canal de dérivation*, 2 metres (6·56 feet) broad, and 1 metre 4 centimetres deep. When finished, it will be 6751 metres (22,143·38 feet) long, and cost 120,000 fr. (4800*l.*) The other hydraulic works have cost 536,100 fr. (21,440*l.*) They have installed provisionally a *tribunal de première instance*, built a school for 18,000 fr., and opened a church and presbytery, by appropriating an old mosque, for 19,570 fr. The fortifications from 1838 to 1846 have cost 2,739,520 fr (105,588*l.* 16*s.*) : consisting of the Port Vallée, and that part of the front belonging to it ; the post of the Casbah ; and the greater part of the curtains 2 and 3, and half of bastion 4. The infantry barracks can hold 2000, the cavalry, called Bardo, 920 men. A hospital on the east side can hold 650 sick. Large bomb-proof magazines have been built near the Port Vallé, and under the barracks of the Casbah ; and two powder-magazines, holding 60,000 kilogrammes. An arsenal has also been constructed at a cost of 142,492 fr. *Tableau*, pp. 345, 356-7, and 388-9.

Captain Kennedy, in vol. ii. chap. xii. of his *Algeria and Tunis*, gives a good description of Constantina in 1845. He pronounces the town to be an assemblage of densely crowded houses, with tiled roofs. The Place Royale was in 1848 a heap of rubbish, and the Place Nemours consisted of miserable Moorish houses. Two streets run from it, Caraman Street and Ronand Street. *Diary of a Lady*, &c. vol. i. p. 227.

A-I

architecture, with its marble pavements and cloistered courts, surrounded by extensive galleries, on which, as usual, the doors of the various apartments open. The walls of the vestibule of this building are ornamented, or rather disfigured, with Arabic frescoes, delineating, with great contempt for perspective, Istamboul, Algiers, and numerous other seats of Islamism, together with sea-fights and other designs."

Mr. Borrer left Constantina about 7 A.M. on the 28th of October, with an escort of two spahis, *en route* for Biscara *viâ* Betna. Their first halt took place at 10 A.M. at *Ain-el-Bey*, a source of sweet water; and you find there steps and brickwork apparently of Turkish construction, and near at hand are the remains of a Roman fort. About 4 P.M. they reached the great plain of Ismoul or Bayla, rich in pasture, and containing innumerable Arab herds. The country traversed this day before coming to the plain was undulating, totally unwooded, but containing much corn ; and the Arabs were busy ploughing it, as it was the season of the first rains, when they sow wheat and beans. They are rather odd agriculturists, beginning by running the plough round a given space, leaving a furrow ; they then cast in the seed upon the rough space thus marked out, and then plough it ; and mark out and plough several successive spots in the same manner. Their ploughs are very light, have only one handle, and only serve to scratch over the soil. The Arabs are too indolent to pull up the stumps they meet, and run the plough round them. When one spot is much overgrown, they go to another, as there are no landmarks or fences. The vast plain of Ismoul is enclosed on the east, west, and south by lofty limestone mountains, the summits of which are broken, presenting forms often bizarre enough. Successive convulsions of nature have upheaved the rocks, the rains and winds have removed the *débris*, and thus these natural turrets stand alone on lofty pedestals. " Opposite our encampment to the east rose the vast Guerioun, of about 1700 metres (5576 feet) in height, presenting on this side one unbroken precipice. A great marsh between us and the mountain was swarming with wild fowl, especially Egyptian geese, which are beautiful birds, and whose representations are found sculptured on the banks of the Nile ; swans, spoonbills, &c. also abounded."

The unhealthy spot where they were had been chosen for a settlement, though very subject to malaria, probably because it presents fine pasture-ground, plenty of water, and no dwarf palms. Hence this locality is better adapted for poor colonists than the arid spots often chosen in the province of Algiers, which are frequently at a distance from water, and produce nothing but cat-weed. Much rain falling in the night, converted the plain into a vast sheet of mud and water, full of deep holes, dangerous to pass. Drawing near a wretched douar, Borrer fell into a *silo* full of water ; and in the douar his gun went off accidentally, singeing a greybeard, and hitting a tent full of Arab women. Happily the natives, though astonished, thought it a British *fantasia*.* Soon after they came to the foot of a

* Arab welcome, when they dash up full tilt, and fire their guns under your horse's belly.

mountain called Bec de l'Aigle, in Arabic *Djebel Nefensser.* A large cemetery met with here shows the unhealthiness of the spot. A neighbouring mountain is called Djebel Harlouf, the hog's mountain : and when Borrer asked some questions relating to it, an Arab, thinking that he called him a hog, threw a cannon-ball at him ; but Borrer fortunately rode him down, avoiding the blow.

After passing two salt lakes, divided by a muddy isthmus, and crossing another great plain, where they drank at a noted spring, *Ain-yac-coute* (the diamond fount), they were preparing to bivouac on the plain in the wet, when they saw the glimmer of Arab fires, and reached a wretched douar of seven tents. Little sleep did Borrer get that night, as herds of goats, which formed part of the family establishment, amused themselves in dancing minuets upon his person. Near this douar is the fine monument called Medrashem by the Arabs, and by the French the tomb of Syphax, of which a description is given elsewhere. An hour beyond Medrashem they came to a rich plain and numerous tents of the Haractas, amounting to 300, divided into douars containing from 10 to 20 each. The inquisitive, simple, and waggish sheikh of the tribe asked Borrer how long the French sultan would live, believing Europeans to be omniscient. He also gravely clipped the beards of the elders with Borrer's scissors, and unceremoniously brushed them with his tooth-brush. At half-past four they came to Betna, having traversed a hilly country, here and there slightly wooded, chiefly with stunted junipers. On the way he noticed many Roman vestiges, particularly of the great *via* from Constantina to Betna, and probably leading from Lambese to Sitifis.

The ruins of Lambese are very extensive, and above two leagues (5 miles) east of Betna, in a nook at the north bar of the Djebel Aouress. They will be noticed at a future place.

There were 2000 troops at Betna* in 1846, partly cantoned in tents, and partly in small barracks. The camp formed a square, enclosed by a slight foss, or ditch, and an earthen rampart. A camp had been proposed in 1843 by the expeditionary column under the Duc d'Aumale, near the base of Djebel Soulthan to the west ; but the fierce attacks of the mountaineers made them recede farther into the plain.

The Kabyles of Djebel Aouress are a peculiar race, very fair, and more like Germans than Arabs ; they speak the Showiah dialect of the Berber, and are warlike and industrious. Those who are subdued pay more regu-

---

* The *Tableau* spells Betna " Batna," and gives it in 1847, 268 European inhabitants ; in 1848, 385 ; and in 1849, 340. 400 metres of streetage were in good repair in 1849, and 7500 feet of trees had been planted in the promenade along the Pépinière ; 1800 metres (5904 feet) of large street, and 4500 (14,760 feet) of small *voirie*, were opened between 1845 and 1849. Expense of all these works, 12,500 fr. (500*l.*) A fountain, *lavoir*, &c. have been established in the part of the town south-east of the military quarter, expense 15,100 fr. (604*l.*) Drains and a water-conduit had to be made in 1850. The town wall, and that of the military quarter, has been raised to a mean height of 2 metres 30 centimetres. Two infantry barracks are completed for 1248 men, and officers' quarters, magazines, and stables : expense, 760,000 fr. (30,400*l.*)

larly the taxes than the Arabs of the plains ; for the latter being vagabonds, can cut and run when they list, whilst the dashkrahs of the Kabyles are stationary.   At present those within the circuit of Betna are quiet.

These Kabyles are living tokens of the Vandal hordes that came from Spain in the fifth century.   A forest of fine cedars is found on the Djebel Soulthan ; and Djebel Aouress abounds in walnuts, Spanish chestnuts, and other of the more hardy fruit-trees, on this its north aspect ; whilst its southern valleys produce grapes, oranges, lemons, peaches, apricots, &c. in great abundance, and of very good quality.*

Borrer left Betna November 5th, after waiting two or three days for the sheikh El-Arib, who was on his way from Constantina to his winter residence in the neighbourhood of the Zaab of Biskra.   The principality of this noble Arab embraces three khalifats, extending over a portion of the Djebel Aouress, as well as a vast tract of the Sahara.   He is said to be the most powerful sheikh in alliance with the French, and his riches are reported to be enormous.   Three days after leaving Betna, Borrer passed his *smala* (household), consisting of numerous women and children, 1000 camels, and many fine falcons and greyhounds.   He could not wait for the sheikh at Betna any longer, owing to the severe cold ; and at 11 A.M. he commenced his journey, escorted by three spahis and six Arab goums (irregulars). At 5 P.M. they found a little douar, after crossing an uninteresting naked country.   " We were now on the plain of Merdjet-el-Ksour, or plain of the castle ; so called because of the ruins lying on it bearing the name of Ktar-el-Louz, the fountain of the almond-tree.   This plain is ten leagues (25 miles) from Betna, and five (12½ miles) from El-Gantra ; and the douar where they stopped is called Ben-Juraba ; whilst a little river, a quarter of a mile west of it, is named El-Ksour."   Next morning they crossed the remainder of the plain, and entered among some rugged and sterile mountains.   Here and there strata of limestone were exposed to view ; and hills, whose profile was polished by the action of the weather, lay in their course.   This district was watered by numerous mountain streams, the chief of which was the Ouad-Fdala.   The waters of the rivers this side of Betna flow towards the desert, the country forming an inclined plane to the south ; whilst on the other side of Betna they flow northwards to the sea.   The Sahara is said to be on a level with the Mediterranean.

The approach to the oasis of El-Gantra is striking.   After toiling for two days across wide-spread plains, entirely without trees, and succeeded by rugged mountains more arid than the plains themselves, suddenly the traveller comes to the base of a tremendous wall of rocks, rising several thousand feet into the air, and seeming to bar all progress.   Presently a file of camels, with dates from Zaab or Tuggurt, comes forth ; and turning a sharp mountain cape, the wanderer beholds a narrow breach of perhaps forty feet in width, through which rushes a mountain torrent.   This was the Calceus Herculis† of the ancients, where the athletic demigod was

* Borrer, p. 369.   † Kick of Hercules.

reported to have kicked a gap in the mountain. It is, in fact, the gate of the Sahara, or, as the Arabs call it, the *mouth* of the Sahara. Through this *gate* the tribes of the east Sahara pass and repass to and from Constantina with their long trains of camels laden with dates, haicks, and other produce of the desert and of its inhabitants. The precipice on the right-hand of the gorge, as you come from Betna, is the abrupt west end of the Djebel Aouress ; that on the left is the east face of Djebel Metlili. A beautiful Roman bridge of one arch spans the Oued-el-Gantra, and the road passes between beetling cliffs after crossing the bridge. Emerging from it, suddenly rich groves of palm-trees, pomegranate, fig, and apricot trees, meet the astonished gaze, and the murmuring Oued rushes into this terrestrial paradise. The town of El-Gantra is shut in by mud walls, with chopped straw and palm-leaves mixed in it, the whole being baked in the sun. Watch-towers are found at equal distances on the walls, strengthened by rafters of palm-timber, and built of the same materials as the wall. The houses are all built in the same way, roofs of palm-trunks being laid lengthways, the interstices filled up with mud, and overlaid with long palm-branches. The whole oasis inside the walls is divided into innumerable small square enclosures, each of which is further confined within its own mud wall. The only way to get into these gardens is through a hole in the surrounding wall of each. A door is attached to these holes, made of palm-branches, small palm-trunks, or a rough slab of stone, by pushing which aside, and almost on hands and knees, you obtain entrance.*

The Ouled-Zaid inhabit this truly African town, of which tribe Sidi-Mokaran was the old caid. The roof of his vestibule was supported on square pillars of mud ; a raised platform at one end was covered with carpets and mats, whilst the horses occupied the other end. Borrer and escort were regaled with dates, cous-coussou, chickens peppered with chilis, and pancakes swimming in honey. The old sheikh tore up the meat and fed his guests with his greasy paws. The caid† was very tall, thin, and pale, with a silvery beard and gentle manners. At night, a snoring chorus of the old caid and his escort drove out poor Borrer into the night air, where he was richly rewarded by a lovely moonlight scene in the oasis.‡

Dates, capsicums, and chilis constitute the riches of this plain, which is the most northerly point at which dates arrive at perfection. Passing through rocky hills and rugged basins to the south, Borrer halted in two hours at a hot spring, thirty feet square and from two to four feet deep, surrounded by a marsh. Tradition says that El-Hammam was carved by Hercules, and Roman steps may be discovered at the north-east corner of the water. An hour hence they passed the base of a high mountain, called the Salt mountain, consisting entirely of rock-salt ; and at 11 A.M. they

---

* Borrer, p. 369 et seqq.
† We presume the same individual in this case enjoyed the dignity of sheikh and caid.
‡ Borrer, ut supra.

came into the plain of El-Outaia. Here is another city on the banks of the Ouad-el-Gantra, now called Ouad-Outaia. You find corn, a little pasture, and gazelles in the plain. A great many tents are pitched in the plain, close to the town ; about an hour after passing which, to the southeast, they observed two lofty monticules and a Roman station. At the southern limit of the plain they saw the wide-spread Sahara, and in the foreground the palm-bearing Zaab of Biskra.*

The Sahara was graphically compared by the ancients to a tiger-skin, the oases answering to the dark spots.

Biskra is overlooked to the north and east by the range of the Djebel Aouress and Djebel Nemenchia. To the north-west are those of Djebel-bou-Ghezal, Djebel Matraf, and Djebel Silga, the southern face of which is often white with the snow blown by the desert winds. The northern face is that of dark limestone rocks ; but beyond Biskra, and to the southwest, the eye roves over a vast unbroken expanse.

A few salt streams water the oasis of Zaab. The mud walls of Biskra are overshadowed by fine palms; and the French citadel (1845) was made entirely of a great number of palm-trunks, and of cedar-wood from Mount Aouress. The fort is built on a mound in the centre of the oasis ; and they are talking of building a new one on a rocky mound in the N.E. part of the oasis, the present position being too much enbosomed in palm-groves. In 1844 every officer of the garrison was massacred by the treachery of the Ouled-Nail, who were admitted into the citadel. The garrison was thought by Borrer to be too small, as Biskra is at the distance of four days' forced march from Betna, and eight from Constantina. The citadel had only three or four little guns, and would fall an easy prey to an enemy. At that time there were many encampments of nomadic Arabs around Biskra, there being a great exchange trade between the Sahara and the Tell through Biskra, which obtains from the latter (Tell) grain, cheeses, wool, figs, horses, asses, arms, &c.

The Ouad-Biskra, in the eastern part of the oasis, is a turbid and salt stream ; but the French were engaged in making an artesian well. The Bahar-taht-el-erd (underground rivers) is a common phenomenon here.

The inhabitants of this oasis still eat dog's flesh, but only in cases of fever as a remedy. There are forty oases around Biskra, which contains 3000 native inhabitants, and 115,000 palm-trees in its precincts.† The caid was in 1846 a handsome man; and the governor, M. St. Germain,

---

* Borrer, p. 369 et seqq.

† The European civil population of Biskra amounted in 1847 to 132 ; in December 1848, to 89 ; and in December 1849, to 98. 2000 trees have been planted around the new post, and divers works of levelling have been effected. The outer wall of the Fort St. Germain has been raised to its proper height on three faces ; the fourth face had reached the battlements. One small bastion is quite finished, and the other three are raised to 3 metres (9·84 feet) in height. Barracks for 400 men were being built, with subordinate buildings connected with the military department. *Tableau*, p. 389, &c.

gave a soirée to him and the beauty and fashion of Biskra, while Borrer was there. There is the minaret of an ancient mosque just outside the S.W. walls of the citadel, and some Roman columns stand near it; but there are no other ancient remains. Some hot springs, known to the Romans, exist near Biskra, which produce the finest petrifactions. The officers' gardens are N.W. of the mosque, and at the foot of the citadel, consisting of four or five acres, enclosed by a mud wall, and containing palms, chilis, capsicums, millet, and water-melons.*

The climate of Biskra is very hot. The best dates come from Oued-Se-ref, to the S.E. of Biskra, and are called by the Arabs *de-glet-en-nour* dates. The harvest begins at the end of October. In planting palms, the young tree is put into deep holes with manure, as much sand being cleared as possible.

Sidi Occ'ba is an oasis eight leagues (20 miles) S.E. of Biskra, taking its name from the famous Arab general, contemporary with the Prophet, who built Kairoan and worked miracles. This oasis is renowned because of his tomb, and on account of a tower which trembles visibly if you shout " Tizza-bil-ras-Sidi-Ok'ba.". Borrer was unable to visit it, on account of the revolt of some tribes; but he went to Tolga, an oasis twelve leagues (30 miles) S.W. of Biskra, visiting *en route* the oases of Bouchayroun, Lichâna, Za'dch'a, and Farfar, all very similar to Biskra, and containing little mud-built towns. The caid of Tolga was a noble-looking man of forty, mounted on a fine black mare, who gave Borrer a grand enter-tainment of dates, pilau, fricaseed chicken, stewed cucumber, cakes in honey, and a grand dish of cous-coussou. All these dishes were made very hot with chilis and capsicums. Afterwards came coffee and pipes. He here beheld the largest scorpion that he had ever seen, adventurously killed by an Arab with his bare foot.

Tolga, which was almost laid in ruins by Abd-el-Kader in 1844, is an oasis comprising three mud-built towns and extensive date-gardens.†

It is with regret that we here take leave of Mr. Borrer for the present, while we proceed to give a description of the most recent condition of some points that we have not visited in the interior of this province and of that of Algiers, constituting the wild mountainous region known by the name of Kabylia, or Great Kabylia, which, from the remarkable features of the territory, and the singular character of its population, the αὐτόχθονοι of Algeria, has appeared to us deserving of a separate notice. The antiquities and colonies of the province of Constantina will be de-scribed in special chapters.‡

---

* Borrer, pp. 331-370.　　　† Ibid. p. 355 et seqq. to the end of the chapter.

‡ The reader will find many additional particulars relating to Bona in Captain Ken-nedy's *Algeria and Tunis*, vol. i. chap. xiv., and vol. ii. chap. ii.; and also in the *Lady's Diary*, vol. i. p. 244 et seqq., and vol. ii. pp. 1 to 38. Both authorities agree in praising its theatre, and in condemning its port and its Maltese population. The Lion d'Or was a good inn in 1845, near the Grand Square; and the *Lady's Diary* pronounces Bona the pleasantest town in all Algeria.

CHAPTER XIII.

Great Kabylia.

AUTHORITIES — BROAD OUTLINE — THE DIFFERENT KABYLIAS — GREAT KABYLIA — ETYMOLOGY — HISTORY — ANALYSIS OF ITS TOPOGRAPHY — BUGIA — ITS ROAD-STEAD — ITS TRIBES — EXPEDITION OF MARSHAL BUGEAUD — THE ZAOUIAS OF SIDI-BEN-ALI-CHERIF — KUELAA — DELLYS.

THE following description of this singular region of Algeria is derived from three principal sources: 1st. The *Exploration Scientifique*, by Captain E. Carette; 2d. *La Grande Kabylie*, by General Daumas and Captain Fabar; 3d. Dawson Borrer's *Campaign in the Kabylie.**

Algeria has, like France, its north and south poles, its langue d'oc and its langue d'oil, its industrial genius and its poetic genius: in a word, Kabylia, the focus and home of workmen; and its Sahara, the nursery of speculators and adventurers.

All the mountaineers of Algeria come under the appellation of Kabails, Kabyles, or Djebalis: the former term being hypothetically derived from the Arabic *kabail*, a tribe; and the latter proceeding more certainly from the Arabic word *djebel*, a mountain. But Kabylia *par excellence,*— Kabylia properly so called, as M. Carette styles it; or Great Kabylia, as it is named by Colonel Daumas,—is that large mountainous district which forms a stern barrier between the provinces of Algiers and Constantina, and that frowns to the eastward over the Mitidja plain; being, in fact, a ramification of the Little Atlas, which, after running parallel with the sea-coast throughout Algeria, inclines about thirty leagues (75 miles) S.E. of Algiers, more to the S.S.E., throwing out at the same point a series of exceedingly lofty mountains, the most elevated of which is the ridge of the Jurjura, or Djorjora (the Mons Ferratus of the ancients), which gives its name to the greater part of the mountainous district above referred to. The northern extremity of this almost inaccessible region, laved by the Mediterranean, presents, according to Borrer, a sea-face of about sixty or seventy leagues (150 or 175 miles), commencing seventeen leagues (42½ miles) east of Algiers. Its depth inland is from twenty to forty leagues (50 to 100

---

* Captain Kennedy, vol. i. chap. xiii. gives a generally correct account of Kabylia, interspersed with occasional errors.

miles) ; and its breadth extends from the eastern limit of the Mitidja to Philippeville. Its limits are, however, in reality very undefined ; and a great part of the territory was, even in 1848, independent, though the most exposed tribes have for some years been nominally subject to the French. Its population is considerable, and it is estimated that it can muster 80,000 fighting men.[*]

The surface of Great Kabylia, according to M. Carette, embraces 7800 square kilometres (3003 square miles), with a population of 370,000 ; which would give 47 inhabitants to each square kilometre (122 per square mile), and 5·24 acres to every inhabitant. In France the proportion of the population to the territory is 60·288 per square kilometre, ·60288 individuals per hectare, or ·244 per acre, or about 1·65 hectares, making about four acres to every inhabitant. Therefore the proportion of the population of Great Kabylia is four-fifths of that of France ; or, taking the population of France as unity, it stands as 0·77942.

The specific population of Great Kabylia is four and a half times greater than that of the rest of Algeria, which only contains, at a mean estimate, 7·67 inhabitants per square kilometre (247 acres).

We have already stated that the population of Great Kabylia amounts to 370,000 persons; and the number of villages being 1533, each village has a mean of 245 inhabitants, and a maximum of 3000. The whole surface of Great Kabylia being 780,000 hectares (1,926,600 acres), and the number of villages 1533, each centre of population occupies a mean space of 500 hectares (or 1235 acres).[†]

Great Kabylia is distinguished from the other parts of Algeria by three special features : 1st, the exercise of professional arts ; 2d, the taste for, and custom of work ; 3d, the stability of the dwellings.

Kabylia properly so called occupies, according to M. Carette, on the sea-shore an extent of 146 kilometres (90·71 miles), comprised between the mouth of the Ouad-Nessa to the west, and that of the Oued-Aguerioun to the east ; the former stream flowing near Dellys, the latter towards the extremity of the Gulf of Bugia.

---

[*] These remarks are from Dawson Borrer's *Campaign in the Kabylie*, p. 1 et seqq.

[†] The foregoing calculations are mainly derived from E. Carette's *Kabylie proprement dite*, vol. i. l. ii. p. 113, in the *Exploration scientifique*. We have found it necessary, however, to rectify a serious error of that author, or his printer, by which he estimates the population of Kabylia at 4⅔ persons per hectare, and that of France at 6 individuals per hectare. Now, as he gives the surface of Kabylia at 780,000 hectares, and its population at 370,000; as, moreover, 780,000 hectares make 7800 square kilometres, and he gives Kabylia 47 souls, and France 60, per square kilometre,—it is evident that M. Carette or his printer has made the proportion per hectare ten times too great. To verify our conclusion, and accustom the reader to decimal calculation, we give the comparative French and English measures of surfaces again :

100 square kilometres = 1 square myriametre = 38·5 square miles.
100 hectares = 1 square kilometre   247 acres.
10,000 square metres = 1 hectare = 2·47 acres.

A–I*

On the land side it is circumscribed by various groups of tribes ; and the approximative surface of the whole region is nearly 800,000 hectares (about 2,000,000 acres) ; that of the island of Corsica being 980,510 (2,451,275 acres).

The general idea which has been held respecting the continent of Africa, and the false inferences drawn from partial information, have long given currency to serious errors as regards Algeria, which is considered as a country of plains and marshes ; while the accidents and the dryness of the soil, on the contrary, are its characteristic features. The shore of Algeria is almost always mountainous. Between the frontier of Morocco and the Tafna exists the chain of the Traras ; and Oran, like Algiers, has its undulating Sahel.

From the mouth of the Shellif as far as that of the Mazafran, that is, for a length of sixty leagues (150 miles), with a depth of from ten to twelve leagues (30 miles), rises and branches out the chain of Dahra. That of the Little Atlas is connected with it by the Zaccar, and shuts in the semicircle of the Mitidja. Having reached this point, the mountain-system rises to a greater elevation, widens, becomes more complicated in its character, and decorates the whole extent of the coast as far as the neighbourhood of Bona. This is not all : we must reckon, moreover, in the interior, the Ouarenseris, which faces the Dahra, commands it in elevation, and exceeds it in extent ; besides other great masses parallel with the preceding ones, and which separate the Tell from the Sahara in the same way that they have cut it off from the Mediterranean. Such are the Djebel Amour, the Aouress, &c., of which we have already treated.

These mountainous regions embrace nearly the half of the Algerian territory, and nearly all of them are inhabited by Kabyles, a race, or mixture of races, quite distinct from the Arabs. The various Kabylias have no political tie between them : each of them constitutes merely a sort of nominal federation, in which exist so many independent unities—of weak or powerful, religious or warlike tribes, subdivided in their form into fractions and villages, all equally free. Although they present a striking analogy in manners, origin, and history, the proper analysis of facts requires that they should be considered separately. All these Kabylias constitute so many detached pages ; such as those of the Traras, of the Ouarenseris, of the Dahra, of the Little Atlas, of the Jurjura, and many others. It is with the latter alone that we are at present concerned, the Kabylia of the Jurjura, which by many writers has been emphatically styled *the Kabylia,* and which we shall call, on account of its relative importance, *Great Kabylia.**

This region embraces all the surface of the vast square comprised between Dellys, Aumale, Setif, and Bugia. These limits may be brought under the foregoing distinct heads ; and though they are fictitious limits,

* La Grande Kabylie, General Daumas, p. 3.

inasmuch as they do not result from geographical configuration, they are rational limits in a political and historical point of view.

The Kabylia, which is about to occupy us, has engaged the popular attention in France more than any other. Many causes have contributed to this effect. Its extent, riches, and population; its proximity to Algiers, which has naturally become the source of some commercial relations ; its ancient renown for independence ; and its inaccessibility, owing to the great mountains that cover it,—have combined to fix the public attention on this important region : and during some years there has been much uncertainty about what policy should be followed with regard to it. Important events have lately settled this question, at the same time that they have thrown much light on all its phases.

The learned are not agreed upon the etymology of the word *Kabyle.* Some assign a Phœnician origin to it. Baal is a generic name of Syrian divinities, and כ in the Hebrew language serves to unite the two terms of a comparison (*K-Baal,* כ בל‎, as the worshippers of Baal). In support of this hypothesis, which would also determine the primary cradle of the Kabyles, the partisans of this derivation cite analogies of proper names, such as Philistines and Flittas (Kabyles), or Flissas ; Mohabites and Beni-Mezzab, or Mozabites ; besides some others. But Colonel Daumas rejects this etymology, because it is not supported by the writers of antiquity. In Herodotus we find the name Kabal applied to some of the Cyrenaic tribes, but we find it nowhere else among the classical authors ; and no trace of it exists amongst the numerous authors of the Roman epoch, historians or geographers, who have left so many documents concerning the two Mauritanias.

It was only after the invasion of the Arabs that these mountaineers began to be called Kabyles ; hence the origin of the name is more probably Arabic, and ought to be derived from one of the three following roots :

> Kuebila = Tribe.
> Kabel = He has accepted.
> Kobel = Before.*

The first would result from the national organisation of these highlanders in clans. The second, from their conversion to Islam. Compulsion, here as elsewhere, would have enforced at least an exoteric profession of the new creed ; and they would bow to the crescent to escape taxation or the sword. They would accept the Koran. The third derivation is not less plausible. In calling these mountaineers *Before,* they would have published a fact in harmony with all tradition, history, and experience ; *i. e.* that the αὐτόχθονοι are invariably driven to the mountains, the last

* Q'byla, *tribe ; plu.* q'bâyl قبيلة, قبايل. 2. Q'bal, *he has accepted* قبل. 3. Q'bel, *before* قبل.

strongholds of independence, by the succeeding tides of invasion. Amongst the Kabyles, the mixture of the German blood left by the conquest of the Vandals is still betrayed by physical traits ; and etymologists endeavour to add to this some additional evidence derived from the approximations of names, such as Suevi and Zouaouas, Huns and Ouled-Aouan.*

We shall lay no great stress on these apparent linguistic affinities, which are subject to much uncertainty.†

For the history and language of Great Kabylia we refer the reader to the chapters on those subjects. It will not, however, be inapposite to make a few remarks on the names of Gouraya and Jorjora. Above the town of Bugia, the *chef lieu* of Kabylia, rises a vast mountain mass called Mount Gouraya, and inhabited by a Kabyle tribe, the Beni-Labeos, that is undoubtedly of Vandal origin. The term *Gora* in the Sclavonic language signifies mountain ; and there can be little doubt respecting the derivation of this name. Gourgoura appears also to be the Berber name of the culminating peak of Kabylia, which has been altered by the Arabs into Jorjora. There can also be little doubt respecting the origin of this term, as it is pure Russian for the mountain of mountains (*Gorgora*), and is evidently a northern importation. The Prince of Mir, a Polish refugee who, as before stated, occupied the Rassautah, a villa near Algiers, in 1841, informed Baron Baude that a considerable number of Sclavonic words occur in the Kabyle tongue, or rather special dialects thereof.‡

We have described Great Kabylia as a vast square, whereof the corners extend to Aumale, Dellys, Bugia, and Setif. The sides of this square are formed, by more or less broken lines, as follows :.

West face. Between Aumale and Dellys, the new road from Algiers ; the Oued-ben-Ahmoud as far as its confluence with the Isser, at the bridge of Ben-Hini ; the Isser as far as Bordj-Menaïel; the Oued-Sebaou, from the Bordj of the same name to its mouth.

North face. From Dellys to Bugia, the strand of the sea.

East face. From Bugia to Setif, nearly a straight line.

South face. From Setif to Aumale, the road of the Bibans, followed in 1838 by the column coming from Constantina ; and afterwards the Oued-Lekal, after leaving Kaf-Radjala.

The country within these limits covers a surface of about 500 square leagues.§ Colonel Daumas gives it 250,000 inhabitants, disseminated in the proportion of 500 persons per square league. ‖ This fact does not

---

* Ouled signifies child, descendant.

† La Grande Kabylie, General Daumas, p. 6.   ‡ Baron Baude, pp. 131 and 69.

§ 500 square leagues would give about 3279 square miles ; somewhat more than the estimate of M. Carette.

‖ 1 square league = 6 square miles, at 2¼ miles to the league. This gives 85¼ persons per square mile, an estimate differing from that of M. Carette by one-third.

correspond with the appearance of the valleys of the Summam, the Sebaou, and the Adjeb, which are as populous as most French departments ;
but we must bear in mind the solitary character and barrenness of the
numerous rocky ridges.

It would exceed our purpose to enter into all the details of the physical and political geography of Great Kabylia. The reader would be
wearied by a minute enumeration of names and localities that could leave
no definite impression on his mind. On the other hand, a bold outline
of the broad features of this curious land and people cannot be unacceptable to the intelligent reader.

There exists a strong analogy between the moral and material physiognomy of the country. The territory exhibits a number of little valleys separated by the chief and presiding chains, and constituting real
arteries in which the principal vitality of the country circulates. On
examining these primary basins more closely, a number of secondary
valleys are discovered opening into them, their sides being formed by
elbows of the principal ridge, and carrying off its waters. These little
rivers in their turn receive torrents, and these torrents are fed by rivulets
or waterfalls ; thus you ascend by a chain of perpendicular systems, from
the basins to valleys, from valleys to dells, from dells to ravines ; and each
of these geographical elements has its proper name and details, and would
admit a particular description. But to simplify the features of this region
and make them comprehensible, we shall confine ourselves to the three
great valleys : that of the Oued-Adjed, which is, however, properly only
a branch stream ; and the two principal basins of the Sebaou and of the
Summam, having their issue in the sea.

The first of these water-courses descends from the vicinity of Setif,
where it bears the name of Bou-Sellam ; and meeting Mount Guergour,
it pierces a narrow passage through rocky masses. But this cutting is
almost every where inaccessible ; consequently the road from Setif to
Bugia can only reach the course of the river lower down. The latter continues traversing a broken country as far as the Summam, running along
the side of mountains of a middling height, but irregular, chaotic, and
impracticable. This broken ground is nevertheless covered with good
vegetable mould, and conceals many mines in its bosom.

The chain of the Djorjora, which is the highest ridge in the country,
determines the existence and the form of the two other almost concentric
basins, which are those of the Summam and of the Sebaou. The chain in
question runs parallel to the shore comprised between Bugia and Dellys.
Its rocky pinnacles rise more than 2000 metres (6560 feet) above the level
of the sea. Save in the case of some naked ridges, pathless hollows, and
accidental rents, the soil is generally covered with a thick bed of vegetable
mould, a rich and productive soil : wanting neither wood nor water, it
seldom presents insurmountable obstacles, and in every respect is much

better adapted for travelling and intercourse than any of the other Kabylias.

The watershed becomes naturally a geographical and political frontier, between the northern waters flowing into the Mediterranean, and the southern slopes, whence the eye descries an endless succession of mountains and valleys, and embraces as it were a sea of solid waves. Not only do the basins of the Summam and of the Sebaou describe on opposite sides of the Jurjura two concentric rings, but their very topography presents, moreover, a symmetrical contrast : their slopes follow an opposite development ; the Sebaou flows from the east to the west, and falls into the sea after having encircled Dellys ; whilst, farther on, the Summam descends in an inverse direction from the west to the east, but similarly encircles Bugia before it empties itself into the sea.*

The principal town of this remarkable region is Bugia. Let the reader imagine a narrow and rocky beach on the sea-shore, then a very steep declivity about twenty metres (65·60 feet) in height ; afterwards a gentle slope, forming a kind of plateau, which runs up to the precipitous sides of the Gouraya ; and towering above all, that mountain itself, spread out like a curtain behind the town, raising its indented crest to about 700 metres (2150 feet) above the level of the sea.

Such is the situation of Bugia. One essential feature fixes the attention at this spot ; namely, the ravine of Sidi-Touati, which divides the town in two, and carries off the waters from the Gouraya, below the Gate of the Marine, almost down to the landing-place. Seen from the sea, this cutting leaves to the right the hill and quarter of Bridja, one of the extreme points of which closes in the anchorage of the town, and commands it by the guns of Fort Abd-el-Kader, built on its sides. To the left of the ravine you see the hill and quarter of Moussa, commanding the opposite declivity, and embracing two forts in a respectable condition for defence :—first, the Casbah, almost at the edge of the shelving beach ; and Moussa, facing the mountain.† Historical associa on, as well as the romantic position of this town, perched upon the rocks at the foot of Mount Gouraya, the base of which is laved by the waters of the noble bay, renders it interesting to the wanderer. The population of this frontier town of Kabylia, which figured before the French invasion in 1833 at several thousands, is now diminished to about 500 Europeans and to a very few natives, and is almost wholly composed of vendors of such necessaries of life as are required by the garrison, by which they are attracted, and from which they gain their subsistence.‡ Bugia thus ranks third in population, compared to the other points occupied by the French on the coast of Algiers ; Bona and Philippeville containing a superior population,

---

* For this excellent sketch of the physical geography of Great Kabylia, I am indebted to Colonel Daumas and Captain Fabar. See *La Grande Kabylie,* chap. iv. p. 133.

† La Grande Kabylie, pp. 84-5.　　　　　　　　‡ Dawson Borrer, p. 161.

and Djidjelli, Dellys, and La Calle an inferior." This statement of Mr. Borrer cannot include the province of Oran.

The distance from Bugia to Algiers by sea is thirty-five leagues (87½ miles);* and it is situated thirty leagues (75 miles), rather N.W., from Constantina; twenty leagues (50 miles) from Setif (Sitifis); and fifty (125 miles) from Bona (Hippona), the ancient episcopacy of the venerable St. Augustine.

St. Marie thus describes the approach to Bugia by water: "After doubling Bouac Point, we came in sight of the Monkey Valley and of the Marine Garden, the verdure of the latter presenting a fine contrast to the gloomy rocks surrounding it. Then passing Fort Abd-el-Kader, after having nearly doubled the great jetty formed by the Gouraya, we descried Bugia, situated on some rapid declivities fronting the south. Notwithstanding the forts, and the large extent of the ground it covers, Bugia is only a mass of huts, not a town; and its streets are, in point of fact, nothing but rough footpaths, running, without any order, between rows of irregularly-built houses. The ruined *débarcadère*, or landing-place, had been complained of by Baron Baude in 1841, and was still a national disgrace to the French in 1845."† Two thousand men then occupied a barracked camp on a point suited for the defence of the place, but deficient in water, the stream that used to supply the town being lost among ruins. The French might recover this, if they had the intelligence and zeal of the Romans and old Arabs. From the camp to the summit of the Gouraya there is a road opened, under the direction of General Duvivier, in the rear of the great walls. This road extends to the length of 4000 metres (13,120 feet), over a calcareous rock, covered by a stratum of argillaceous earth.

The lentisk, the mastic, the vine, and the wild olive, grow here luxuriantly, and would flourish vigorously if the cattle were prevented from ranging among them. The summit of the Gouraya is 682 metres (2236·96 feet) above the sea; but St. Marie is mistaken in stating that on the northern side the elevation is 700 metres (2296 feet), and on the southern 2000 (6560 feet). The effect of this prodigious mountain-pile is quite magical.

The marabout of Sidi-Bosgri, on the top of the Gouraya, was thought as efficacious a pilgrimage for the infirm as that to Mecca; but being taken, after a hard fight, in 1833, by the French, a fort has been built on its site, which commands the mountain. Colonel Larochette has improved its defences by making a path from the fort, following the crest of the Gouraya, and descending to the plain, passing by the precipice of the Dent. This road is so constructed, that you can always see the move-

---

* Borrer, p. 161. Mr. Borrer says in another place, that Bugia is 45 leagues to the east of Algiers,—which must mean by land.

† St. Marie, chap. vi. pp. 197-200.

ments of your assailants and mask your own, whatever they may be. Still, when you wish to go along it, even at present, it is necessary to have an escort of about thirty tirailleurs to clear the borders. This road leads down to the blockhouse of Doriac.

Five advanced posts complete the defence on the land side.* St. Marie states that the marabout of Sidi-Bosgri was heroically defended by the Kabyles in 1833; and that the blockhouse was nobly defended, at a later date, by ten Frenchmen for three days against a host of Kabyles. The walls riddled with shot attest the heat of the combat, in which the French, with the chivalry for which they were once famous, refused to fire on a sheikh's widow, who urged on the assailing Kabyles with the greatest energy. The cattle and the soldiers of the garrison did not venture for many years beyond the five advanced posts before alluded to, for fear of being captured or slaughtered by the Kabyles. The cattle, when sent out to graze, used to be accompanied by dogs to beat about the bushes, as in a hunt, and drive off the Kabyles.† Baron Baude, who appears to be copied by Count St. Marie, gives the following description of the country beyond the Gouraya : " In the midst of the chaos at your feet, as you stand on the top of that lofty pile, a deep hollow opens, which becomes bifurcated at the distance of three leagues (7½ miles) from Bugia. This is the vale of Soumah, and beyond it lie the beautiful plains of Zamoura and Setif." The dingles in this neighbourhood show traces of cultivation; but the villages of Dharmassar and Sumnia had been burnt at the time of St. Marie's visit.‡ The bottom of the cistern, which forms the plain of Bugia, may contain about 6000 hectares (15,000 acres); but it is only cultivated on the right bank of the Summam.

The Gouraya towers to the east and north of the town, is connected in the interior with Mount Tondja, and being prolonged into the sea, gives birth to Cape Carbon. To the southward, a pretty bay entered the land to receive the waters of the Ouad-Summam.§

On the sides of Djebel Gouraya was once situated the famous koubba, or domed tomb, of the fair Kabyle saint, Lella-Gouraya, which is now replaced by a French fort commanding Bugia. Upon the right is the

---

* Baron Baude, vol. i. p. 133. In November 1833, the year of the conquest of Bugia, four blockhouses had been constructed : i.e. those of Bou-Ali, covering the plateau of Moussa ; and those of Salem, Rouman, and Khalifa, situated on the western plateaux. At the same time Colonel Lemercier was also laying the foundation of a very fine work, in erecting Fort Gouraya. In the beginning of 1834, Commandant Duvivier built another outwork, the *blockhaus de la plaine;* and in 1836-7 were erected the Fort Lemercier and the towers of Doriac and Salomon. La Grande Kabylie, pp. 93-96 and 125. We learn from the *Tableau* that the defences and the landing have been improved, and a lighthouse erected at Bugia.

† St. Marie, p. 200.   ‡ Ibid. p. 201. Baron Baude, vol. i. p. 133.

§ Col. Daumas, Grande Kabylie, p. 93. The bay of Bugia is described by E. Carette as a large bight or indenture, comprised between Cape Carbon to the westward, and Cape Cavallo to the east. La Kabylie proprement dite.

Ouad Messaoud or Summam, forming here the east boundary of the plain ; the opposite shores being covered with massive groves of olive-trees, and overlooked by wild mountains clothed with wood, and held by the fierce Beni-Bou-Messaoud, who, with the *Mezaya*, an equally warlike tribe, long kept the people of Bugia cooped up in their walls, rendering it, even down to the visit of Dawson Borrer (1847), a mere military post held by the French.

Two entrenched camps have been made near Bugia, one higher and the other lower, constructed on the Gouraya range ; and a road has been made from the camps to the summit, 4000 metres (13,120 feet) in length, at an inclination of one-tenth. The lower camp, which is 120 metres (393·60 feet) above the sea, is calculated to contain 2000 men. There is every probability that Bugia, under an enlightened government, would recover much of its ancient political and commercial importance,* its position being central and convenient, and the district of Great Kabylia containing the most industrious race in Algeria. According to the observations followed in the *Cabinet Atlas*, Bugia is situated in 36° 49′ N. lat., and in 5° 28′ E. long. of Greenwich.† General Daumas, in the map accompanying his work on Great Kabylia, places it in 36° 45′ N. lat., and in 2° 46′ east of Paris.‡

The country surrounding Bugia is very fertile. The river Bou-Messaoud is here of great depth and of considerable width, with a muddy bed ; and in winter its channel is much subject to overflow, through the operation of the mountain torrents. The Summam closes its career flowing through an agreeable plain of moderate extent, surrounded on all sides of the horizon by a framework of picturesque mountains.

Bugia, suspended amongst rocks that seem ready to swallow it up, and the waves that eat away their base, only communicates with the smiling valley, descried from its walls, by a somewhat narrow tongue of land. Hence the mountaineers form its nearest and most formidable neighbours, owing to the nature of the locality and other accidental circumstances. It so happens, moreover, that the tribe of the Mzaias, which is in possession of those heights, is reported to be one of the most warlike, poor, and savage of all. Its territory is carefully cultivated, but the spots of good mould are not sufficiently abundant to support the inhabitants. Accordingly a certain number go forth to work elsewhere; and those who remain are never backward in any thievish or warlike enterprise. They can muster 800 foot-soldiers. The plain belongs to two tribes—the Beni-Bou-Messaoud and the Beni-Menioun; which can each of them bring from 500 to 600 firelocks into the field, with a small body of horsemen. Their district

---

* Great quantities of wax used to be exported from Bugia ; whence came the French name for wax-candle, *bougie*. Kennedy, vol. i. p. 261.

† Universal Gazetteer, in the Royal Cabinet Atlas, p. 20.

‡ See the Chart, p. 488, of La Grande Kabylie.

is more thriving; for instance, they can boast of fine flocks, of corn, flax, a great many bee-hives, olive-trees, and some tolerably flourishing villages.

Still, neither of these three tribes is so powerful as those more in the centre of Great Kabylia.*

The roads of Bugia are the best in Algeria. They are, it is true, somewhat exposed to squalls and to a heavy swell; but these evils are remedied by their excellent anchoring-ground. To seaward of a space of about 60 hectares (150 acres) situated before the town, and suited for merchant-ships, the anchorage of Sidi-Yahia can receive, from Pointe de Bouac to Fort Abd-el-Kader, four line-of-battle ships, six frigates, and a considerable number of smaller craft. The Turks were in the habit of putting up their fleet in Bugia roads in the winter. Recent travellers agree that the famous inlet at Cape Carbon, into which, according to ancient geographers, ships could enter under full sail, would now scarcely admit a boat.†

Behind Bugia rises Mount Gouraya, 670 metres in height,‡ whose rocks consist of limestone, and are covered to the top with argillaceous earth, whose fecundity counteracts the usual effects of exposure to the south. The lentisks, carobs, vines, and wild olives which clothe its sides and summit, only require protection from the cattle, to supply the base of the mountain with abundant sources, by attracting and retaining the rain. The great rents of the Simplon, St. Gothard, and Splugen offer nothing comparable to this prodigious up-heaving of mountains. The view from the Righi is more extensive, but less imposing, than that of the Atlas from the Gouraya, which reminds one of the imperfect work of the Titans, described in Virgil:

> Ter sunt conati imponere Pelio Ossam
> Scilicet, atque Ossæ frondosum involvere
> Olympum.   *Georg.* lib. i.

Approaching Bugia by water from the south-east, the rocky mass of the Gouraya seems detached from the shore; and the deep gorge intervening between it and the mainland indicates at once the position of the city of Bugia, and the course of the Roman road which led from Rusguniæ and Rusucurrum, and descended to Saldæ (Bugia) on the south reverse of the mountain.

The Arab and Mussulman population generally appears to have almost entirely deserted Bugia; and the European population, which at one period since the conquest amounted to 740 persons, scarcely numbered 100 in 1841. It has been, in fact, merely a military hospital; and all travellers agree in condemning the folly of the French government in not improving

---

* La Grande Kabylie, p. 94.   † Baron Baude, vol. i. p. 139.
‡ Borrer, p. 161.

the port, which affords such fine natural advantages. There are many channels for commerce in the neighbourhood of Bugia : to the south-west, the valley of the Adouse ascends, following the base of the Djorjora to the plain of Hamza, whence you descend towards Algiers; to the south, the Adjelly pierces in a direct line the chain of the Atlas; and its valley opens at 20 leagues (50 miles) from the sea on the fertile plains of Medjana. It cannot be expected that the French will derive any benefit from the conquest of Bugia, till by force of arms or arts they can prevail on the fierce highlanders, by whom they are encircled, to allay the bitterness of hostility with which they regard the invading Christians. As for any colonist who may be tempted by visions of hecatomboian cattle reared upon the fertile shores of the river Bou-Messaoud, his lot will be but an unhappy one in the present state of affairs at this point; for Bugia is, in fact, a mere military post, the very sentinels upon the walls being ever and anon hailed by the whistle of a Kabyle bullet. A certain Scherif Mohammed, who has annoyed the French considerably from time to time, lives at present in the neighbourhood, encouraging the spirit of revolt; but from the checks he has lately received, he is now compelled to content himself by sending out occasional marauding parties; keeping up a kind of guerilla warfare, which holds in a state of harass and alarm both the garrison of the town, and the few allied Arabs in the neighbourhood.

" A night or two before my arrival at Bugia," writes Mr. Borrer in 1847, " a band of this mountain-chief's foragers were outwitted by an ambuscade of indigenous cavaliers in the French service, and sadly mauled. In fact, there is a continual sparring going on between these sturdy sons of the Mons Ferratus (Gouraya) and the present tenants of Bugia. No sooner are the French flocks, or those of the allied Arabs, led forth to revel in the fat pastures of the Oued-Messaoud, than hungry eyes gloat upon them from the thicket-clad heights around, and a sudden swoop carries off shepherds and sheep. If, on the other hand, the hostile mountaineers are tempted to descend with their own herds, the same fate awaits them; so that a system of aggression and retaliation keeps both parties in a delightful state of *qui vive*."

We shall now give the reader a peep into the wilds and recesses of this Alpine region, ere we pass on to consider its ethnology.

Our old friend Mr. Dawson Borrer accompanied the French expedition under Marshal Bugeaud in the spring of 1847, which penetrated into the heart of Great Kabylia and subdued all parts of it, except its most retired and rugged fastnesses. We shall present the reader with an outline of his progress, to break the monotony of dry details.

After leaving Algiers they marched to Arba in the Mitidja, a district which we have already described. The column, consisting of eleven battalions, two squadrons, and two sections of mountain guns, advanced thence to the foot of the Little Atlas, which they reached about half an our after quitting Arba. The slopes of the mountains are there clothed with brushwood, chiefly lentisk, stunted bellotas, and myrtle, intermingled with the bright-flowered coronilla and the dwarf gum-cistus. A road has been cut along the face of the Djebel Moussa, leading to a newly-established French post named Aumale (the Sour-Guzlan of the Arabs, and the Auzia of the ancients), which lies about four days' march S.E. of Algiers.

The mountains they were now traversing are intersected by very deep and beautiful valleys, up the steep slopes of which were clustered numerous *gourbies*, or huts forming villages, or *dashkrahs* as the mountaineers name them. These huts are constructed of rough stones or masses of turf, the interstices filled up with mud and cattle-dung. The roofs are thatched with coarse straw or reeds and branches of trees. The extreme lowness of these dwellings is remarkable, the walls of few being more than three feet in height, so that the branches covering the roofs often touch the ground at the eaves. One large apartment alone is found in each hut, a portion of which is enjoyed by the family, and the rest by their live-stock. It is only in the centre that you can in general stand upright, immediately under the ridge of the roof. In the neighbourhood of these villages the land is well cultivated, and crops of remarkably fine bearded wheat were at that season (May) shooting up from the ground.*

Without accompanying the column all the way in its victorious course down the valley of the Summam, whence, after subduing most of the tribes by violence or terror, and after forming its junction with General Bedeau's column from Setif, it marched on to Bugia, having subdued the greater portion of the lowlands of Great Kabylia,—we shall dwell on some of the most striking features of the region.

Marshal Bugeaud encamped with his troops on the 15th at Sidi-Moussa, on the banks of the Summam. On the opposite bank the rich but strong country of the Beni-Abbas rose in the form of an amphitheatre. Their numerous villages, clustering together, are perched on a series of steep summits, the most inaccessible and populous being Azrou, which was stormed, sacked, and burnt by the French. This example struck such terror into the neighbouring tribes, that most of them submitted, especially the confederation surrounding the zaouia of Sidi-ben-Ali-Cherif, forming a little theocratic state. This sacred college and kind of monastery is situated near Chellala, on the opposite or left bank of the Summam, and is the nursery of numerous tolbas (*savants*) and of wonderful legends. It contains three venerated tombs: those of Sidi-Mohammed-ben-Ali-Cherif the founder, of Sidi-Said, and of a famous marabout Milah. The family of Sidi-Said holds the chief authority, and all his descendants are reputed to have been blessed with one male child and

* Campaign, &c., by Dawson Borrer, p. 29 et seqq. Compare chap. xiv. p. 282.

heir. But these unlucky chiefs, like the Abyssinian olive-branches, are bound never to leave the territory of the confederation. One daring fellow who peeped over was struck blind, like our peeping Tom. Near the founder's tomb are two colossal walnut-trees, which must not be touched without the permission of the tolbas, or before the fatah has been said over them. A sly taleb venturing to pocket a nut, a leech falling from heaven bit out his eye. This zaouia, pre-eminent for strict morals, is served by the villages of Chellala and Ighit-ou-Mered, whose inhabitants are forbidden to have any education, that they may not aspire to become masters instead of servants.* This zaouia possesses vast property, and is supported by ready donations.†

Leaving the column, we shall proceed to analyse the unsubdued district of the Zouaouas, the singular town of Kuelaa, and finally Dellys.

The country of the Zouaouas‡ embraces the highest and most arid part of the mountains. Their soil is poor and affords little grain, the tribe preferring to cultivate vegetables, flax, and tobacco. Fruit is not wanting, including carobs, olives, figs, pomegranates, apricots, apples, &c. Sweet acorns are very plentiful, and eaten in cous-coussou by the Zouaouas. They have much game, including hares, partridges, quails, pigeons, &c. Lions are rare, but panthers are more common; and to destroy them they often employ a kind of infernal machine, with a piece of meat near it as a bait. §

The Zouaoua mountains also contain a host of hyenas, wild boars, jackals, &c., and especially vast numbers of apes;|| but the produce of the country would be quite insufficient for its inhabitants, if they were not a highly industrious race.¶

Most of the towns of Algeria seem built under the impression of fear; and Kuelaa is a veritable miracle on the score of unassailableness: the only exposed approach is Bouni, on the side of Medjana. A natural phenomenon indicates clearly the separation of the Arab and Kabyle territories at this spot. Near the village of Djedida a colossal gate opens between the rocks, separating two countries of strikingly opposite characters. To the south is the rich Medjana plain with its golden harvests. To the north an abrupt and rugged ground and sterile soil, yet, as you advance, improving and displaying picturesque mountain beauties. Passing mighty rocks and a splendid cataract, you reach the plateau of Bouni, separated from Kuelaa by three leagues (7½ miles) of broken territory, whose difficulty

* How like this to the superior wisdom of some enlightened classes nearer home !
† La Grande Kabylie.
‡ The name of the Zouaouas is frequently extended to all the Kabyle tribes inhabiting the ridge of the Jurjura, between Dellys and Bugia.
§ La Grande Kabylie. || See the Fauna.
¶ See following chapter.

exceeds the fabulous, the path being for the most part along a ridge like Mahomet's razor, with fearful precipices on both sides, and only at times one metre in width. At length you reach the plateau of six kilometres (4½ miles), only united to earth by this narrow ridge, standing on wall-like precipices, and commanding a vast vat-like basin. This sport of nature holds four villages, composing the town of Kuelaa. Ruins at the north-east point, called Bordj-el-feteun, point out the civil dissensions of its brilliant rulers the Mekhranis, one of whom built the Casbah, now in ruins, and brought four vast cannon of European origin to Kuelaa. The people are now governed by a natural Djema, and can raise 700 firelocks. They belong to the soff of the Beni-Abbas. (See chap. xiv.)

The aspect of Kuelaa is smiling. The houses, built in the Moorish style, are often white-washed, always tiled. The great mosque commands the town, and has a graceful appearance, the porch being decorated with poplars.

Unhappily the town has no water. Seven basins have been dug in the rock by an alley separating the quarters of Ben-Daoud and Ouled-Aissa (the son of David and the children of Jesus), but the water only trickles there in drops. In winter they have plenty of rain, but in the droughts they have to resort to the Oued-Beni-Hamadouche, winding at the bottom of the ravine, half a league (1¼ miles) off by the steepest roads. The banks of this river present a little cultivation, but the people would starve were it not for their great industry. Men and women work hard, making immense quantities of woollen garments, and many of them migrating to the towns of Algeria and Barbary. The women are noted for their beauty and toilette; and the strong position of Kuelaa has made it for ages a kind of sanctuary for person and property in this anarchical country.*

Turning to Dellys (in Arabic *Teddel*), the west limit of Great Kabylia on the seaboard, we find that this town stands on the supposed site, and is built of the remains of, Rusucurrum, one league from the mouth of the river Booberak, and forty-five miles east from Algiers. St. Marie, who passed Dellys in 1845, on his voyage from Algiers to Bona, describes it as the first well-inhabited place on the coast within a distance of twenty leagues from Algiers. The surrounding hills show careful cultivation ; and a succession of delightful gardens indicates amongst the inhabitants a certain love of order and repose, not to be met with in other parts of Africa.†

* La Grande Kabylie.

† Baron Baude, vol. i. p. 127. St. Marie, p. 197. Diary of a Lady's Travels in Barbary, vol. i. p. 155. Nicholas de Nicolai, who was at Dellys in 1551, remarks: " C'est une ville habitée d'un peuple fort récréatif et plaisant, dont presque tous s'adonnent au jeu de la harpe et du luth." He gave it 2000 fires ; and Gramaye agrees in his statement.

It appears from the latest official documents,* that a landing-slip, fifty-five metres (180·40 feet) in length, and built of masonry, was constructed at the port of Dellys in 1847-8, costing 46,611 fr. 29 cents. (1860*l*. 9*s*. 5*d*.) The Place Nationale was partially cleared of rubbish in 1850 ; and 1790 metres (5871·20 feet) of principal streets, and 1470 metres (4785·20 feet) of branch streets, were opened from 1844 to 1849. The springs within the walls supply daily 43,200 litres (9504 gallons) of water ; whilst the ain, or conduit, of Mezel-el-Foukani, finished between 1844 and 1849, at an expense of 7260 fr. (290*l*. 8*s*. 2*d*.), has a length of 225 metres (738 feet), and yields a daily supply of 21,600 litres (4752 gallons). The conduit of Ain-Bouabada, called Sidi-Souzou, was finished in 1849, at a cost of 15,400 fr. (616*l*.), having a length of 500 metres (1640 feet), and yielding a daily supply of 28,800 litres (6336 gallons). The latter conduit has been brought in as far as to the fountains within the walls.

A building connected with the maritime service, and called *direction du port*, answering to our harbour-master's office, was built in 1844-6, costing 6469 fr. (258*l*. 15*s*. 10*d*.) ; as well as a *bureau Arabe*, built at the same date, at an expense of 16,717 fr. (668*l*. 14*s*. 2*d*.)

The precincts of Dellys are occupied by a certain number of petty tribes, who in a great measure identify their interests with those of the town, forming a distinct confederation from the other Kabyles. Its principal members are the Beni-Slyems and the Beni-Thour, and they can raise 1400 muskets. Dellys numbers about 1339 inhabitants, of whom 308 are Europeans.†

After leaving Dellys, as you proceed eastward along the coast of Kabylia towards Bugia, you pass the port of Zuffoone, commonly called Mers-el-Fahm (the port of charcoal) ; and doubling Cape Ash-oune-mon-Kar, where stood the ancient Vabar, the next remarkable place you come to is Mettsecoub (the perforated rock). The Spaniards have a tradition that Raymond Lully, in his mission to Africa, was in the habit of retiring to this cave for meditation. Not far hence is Bugia.‡ St. Marie, who also sailed along this coast, speaks thus of its appearance : "Leaving behind us Cape Sigli, we saw at sunrise the islet of the Pisans, a wild rock, on which innumerable sea-birds alight.§ This part of the coast is rocky and mountainous, and their forms indicate a calcareous soil. Here and there thin black spaces mark the spots where the Kabyles have burned the dwarf-palms and other wild vegetation, to clear the uncultivated ground for sowing."‖

Having completed our survey of the topography of Algeria, we proceed

---

* See the Tableau (1850), p. 344.
† Diary, vol. i. p. 155.   ‡ Blofeld, p. 43.
§ Query : might it not contain a deposit of guano ?
‖ We shall revisit this interesting region in a future chapter.

in the following chapters to analyse the physical characteristics, manners, customs, and laws, the arts and sciences, of the different strata of humanity that have been deposited on this shore by the tide of time.*

* Baude, vol. i. p 127. St. Marie, chap. vi. p. 197. For a description of the topography, &c. of Algeria in the earlier years of the French occupation, see *Nouvelles Annales des Voyages*, Dec. 1833 ; *Aperçu historique et statistique sur la Régence d'Alger, &c.* par Sidi-Hamadan Ben Othman Khoja ; *A Review of Rozet's Voyage*, par Laurenaudière ; *Appel en faveur d'Alger et de l'Afrique du Nord ;* and the works of Poiret, Hoest, Norberg, Bruns, Langier de Tassy, Renaudor, &c. Many additional details relating to the topography of Algeria in 1845 and 1848 will be found in Captain Kennedy's *Algeria and Tunis*, and in the *Diary of a Lady's Travels in Barbary*. Our limits prevent us from dwelling any longer on this branch of the subject ; but we especially commend to the reader's attention chaps. i. ii. and xii. of the first volume of Captain Kennedy, and sections 1, 2, 3, 4, 5, 11, 12, 13, and 14 of the *Diary*, on the city of Algiers.

PART II.

STATISTICS AND HISTORY,

POLITICAL, SOCIAL, AND NATURAL.

CHAPTER XIV.

𝔗𝔥𝔢 𝔎𝔞𝔟𝔶𝔩𝔢𝔰.

NATIVE POPULATION OF ALGERIA—CHARACTERISTICS OF THE KABYLES CONTRASTED
WITH THE ARABS—SUPERSTITIONS—INDUSTRY—MANUFACTURES—MANNERS
—WEDDINGS—WOMEN—ADMINISTRATION—LAWS—AUTHORITIES—THE MARA-
BOUTS—THE ZAOUIAS—THE ANAYA—ILLUSTRATIONS OF SCRIPTURAL AND
CLASSICAL ANTIQUITY.

THE existing Mussulman population of Algeria is much like that of Gaul
when conquered by Cæsar, forming one great community with one
dominant language and religion; but there exists no durable tie, and there
are many divisions. "In Gallia," says Cæsar, "non solum in omnibus
civitatibus . . . sed pene etiam in singulis domibus factiones," &c.*
Cæsar fomented these discords, and conquered Gaul; the Turks did the
same at Algiers, fomenting the natural antipathy of the Kabyles and Arabs.
Divide et impera was their motto, and it succeeded, their instinct having
taught this principle to the Ottoman rulers.

Numerous revolutions have visited North Africa; but the populations
that they have deposited have not, generally speaking, gone far from the
coast, and the older races remain commonly in the Sahara and the Atlas.
An exception is found in the Aouress mountain, which seems to be inhabited by a tribe of Vandal origin. The Biskris and Mozabites, who have a
colony at Algiers, are pronounced by some authorities the same people as
the Gætulians of the ancients, to whom Rome gave the right of citizenship.
They live in the Sahara, and have not meddled with the quarrels of the
people of the Atlas. We shall shortly examine their characteristics more
minutely.

The ancients have not given a very flattering picture of the Kabyles, to
whom we shall first direct our attention. These tribes, belonging to the
Berber race, are the aborigines of Algeria, living chiefly in the Atlas, particularly the Djorjora and the Darha; and they have been thus described by
Procopius: "Inured to hardships, they live in little huts in which it is
scarcely possible to breathe; in winter or summer alike regardless of snow
or sun, or any other necessary evil. They sleep on the bare ground, or

---

* De Bell. Gall. i. 6, c. 11.

occasionally the more lucky among them may put something under them. They are forbidden by law to add additional clothing according to the weather ; but their dress is torn and dirty, and they wear a rough tunic in all weathers. They are without wine, bread, and all the other usual necessaries of life; but either roasting or kneading into flour wheat, corn, or at least barley, they devour it after the fashion of wild beasts." *

The Turks looked upon them as a barbarous and perfidious race, without fear of God and without faith to men, keeping peace only with those who kept them under by terror. Similar was the opinion entertained of them by the ancients. "They have neither any fear of God or respect for man, nor do they pay any regard to their oath. . . . Lastly, they have no peace with any one, save with those who coerce them through fear."†

Let us compare these statements with their actual position.

Having given a description of the topography and population of Kabylia, we proceed to lay before the reader a compendious account of the character, manners, and customs of the Kabyles, and of the productions of their territory.‡

The dominant characteristics of this region have been : 1. Independence of the Turkish or French yoke. 2. The use of the Berber tongue. 3. The stability and relative luxury of the habitations. 4. The cultivation of fruit-trees and the exercise of professional arts.

The Kabyles delight in a sedentary life ; some inhabit huts of mud and turf or rough stones, and others reside in solidly and well-constructed villages. They are a highly industrious people, great cultivators, and make their own agricultural implements, arms, gunpowder, haicks, carpets, leather, &c. Yet this race is very unsociable with strangers ; and while the Arabs correspond to the French families that speak the *langue d'oc,* with southern imaginations, personifying material forms,—the Kabyles have a northern precision of thought and expression, confining themselves to a precise and critical statement of facts.§

Patriarchalism is the dominant principle with the Arabs, communism with the Berbers or Kabyles. They are not acquainted, like the Arabs, with the distinction between the terms Oulâd and Beni, as applied to noble and servile tribes. The only distinction that they make in employing

---

* Marusii duris assueti, in parvis tuguriis ubi vix respirare licet degunt, hyemis ac æstatis temporibus, neque nivibus, neque solibus, neque alio quocumque malo necessario curantes. Dormiunt nudo humo ; si qui beatiores inter eos, aliquid substerniunt. Vestes insuper secundum tempora variare ex lege prohibentur ; sed laceram vestem atque crassam, tunicamque asperam in omne tempus induunt. Pane vinoque et aliis bonis omnibus usui necessario carent, sed triticum, sive selaginem, sive hordeum minime aut coquentes aut in farinam terentes more belluarum passim depascuntur.—*Proc. De Bell. Vand.* i. 2.

† Illis neque Dei metus est ullus, neque hominum reverentia, neque item jusjurandi aut hominum ulla cura. . . . . . Denique cum nullo pacem habent, nisi cum his quorum metu coerceantur.— *De Bell. Vand.* i. 2.

‡ See vol. i. chap. xiii. of Captain Kennedy's *Algeria and Tunis.*

§ La Kabylie proprement dite, by E. Carette, in the Exploration Scientifique.

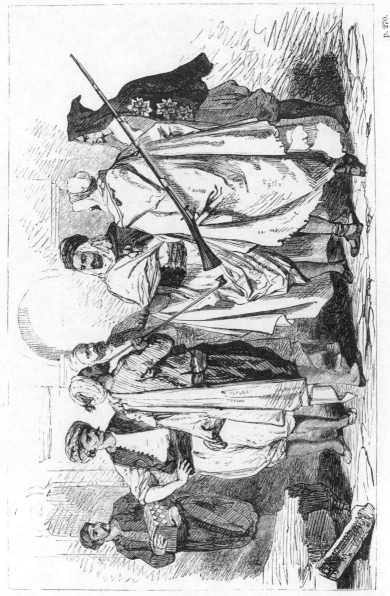

TYPES OF ALGERIAN POPULATION.

p. 270.

these Arabic designations is, that they commonly style the lay tribes Beni, and reserve that of Oulâd for the marabouts. The Berber generic term for tribe is *ait*, which they give without distinction both to nobles and villains, for *ait* has not so much of a family meaning as Oulâd and Beni. It signifies properly the people, the followers, while Beni and Oulâd imply direct descent ; thus familism is not such a dominant influence among the Kabyles as among the Arabs.*

The Kabyles are very frugal in their habits, their principal food consisting of pancakes, called galette, baked upon a plate of clay; milk, honey, butter; figs soaked in oil, of which they consume great quantities ; and the everlasting cous-coussou. †

The moral and physical characteristics of the Arabs and Kabyles are thus contrasted by Colonel Daumas:

" The Arab has black eyes and hair ; many of the Kabyles have blue eyes and red hair: they are also generally *fairer* than the Arabs. The Arab has an oval face and a long neck ; the Kabyle, on the contrary, has a square face, with the head approaching the shoulders. The Arab never shaves ; the Kabyle shaves till he has attained his 20th or 25th year: at that age he is a man, and lets his beard grow ; it is an indication of the judgment that he has acquired, and of his reason which is maturing. The Arab covers his head at all seasons, and clothes his feet whenever he can. The Kabyle, in winter and summer, through sunshine and shade, goes barefooted and bare-headed." ‡

The Kabyles differ in all things from the Arabs. The first live under roofs, the last under tents ; the Kabyle fights in preference on foot, the Arab on horseback. Their languages have no analogy. The Arab flies our contact ; the Kabyles of the tribes that are most hostile to the French do not hesitate to come and seek labour in the towns, and the Amazirghes of the Riff in Morocco have latterly immigrated in considerable numbers into Oran. In short, the Kabyles are the conquered, and the Arabs the conquerors ; hence their hereditary hatred.§ If by chance you meet a Kabyle with his feet covered, it is accidentally, and merely with the skin of a beast just killed. When they cover their feet, which is unusual, they wear a slight sandal of raw hide, whilst a kind of buskin of the same material is often worn up the leg. || Those who border on the plains sometimes wear the *chachia* (Tunis cap).

The Kabyle has for his only clothing the cheloucha, a kind of woollen shirt which falls below the knees, and costs from 7 to 8 fr. (6s. 8d.) ; he protects his legs with footless gaiters, knitted in wool, which they call *bougherous*. When engaged in work, he puts on a large leathern apron cut

---

* La Kabylie proprement dite. General Daumas, La Grande Kabylie. Baron Baude's Algérie, vol. iii. p. 221.

† Dawson Borrer's Campaign, &c.   ‡ La Grande Kabylie, p. 21.

§ Baron Baude, vol. iii. p. 221. || D. Borrer, chap. i.

like that of the French sappers; and he wears the burnouse when his means allow him, keeping it an indefinite period, regardless of spots or rents : he received it from his father, and he bequeathes it to his son.* Some authorities entitle the Kabyle shirt *khandoura*, and describe it as having loose sleeves ; and their burnouse they describe as a white, or black and white, woollen mantle with a large hood.†

The Arab lives under his tent—he is a nomad on a limited territory ; the Kabyle dwells in a house—he is fixed to his spot of ground. His house is built of dry stones or unburnt bricks, which he puts together in a somewhat rude fashion. The roof is thatched, but among the rich it is covered with tiles ; and this sort of cabin is called *tezaka*. It consists of one or two chambers; the father, mother, and children occupying one-half the building to the right of the entrance-door. This family dwelling is called *âounès*. The other part of the house, which they name *âdaïn*, to the left, serves for a stable for the cattle and horses. If one of the sons of the house is married, and requires a ménage of his own, they build him a dwelling above by running up another story.‡

Whoever undertakes a journey, ought to set out on a Monday, Thursday, or Saturday : these days smile on the traveller. Happy the man who begins his journey on a Saturday ; the prophet preferred that day to the other two. They travel, it is true, on Wednesday, Friday, and Sunday ; but then the traveller is never free from anxiety during his whole transit. You must never begin a battle or skirmish on a Tuesday. Thursday is the day on which the bridegroom ought to introduce his bride to the conjugal roof : it is always a good augury ; because the wife awakes on a Friday, which is the Sunday of the Mussulman. No one is to be lamented who dies during the Rhamadan,§ during which the gates of hell are closed, and those of paradise always open. It is a happy presage if you see a jackal when you rise in the morning; and two crows at the moment of setting out are a sign of a prosperous journey. It is a bad sign to see a hare at night ; and a single crow before commencing a journey is a reason for anxiety. The Kabyles, so incredulous on the subject of witchcraft, are less so on the question of demons. They say there are some in all seasons, except during the Rhamadan ; because God compels them to remain in hell during the sacred month. They fear them extremely. A Kabyle will never go out of his house at night, without conjuring them in the name of the all-powerful and merciful God. He will do the same also when he passes near a spot where blood has been shed, because the demons love blood, and are sure to resort to those spots. There exists also, if it be not a prejudice, at least a universal contempt for the she-ass ; and to such an extent

---

* Daumas, chap. i. Dawson Borrer, p. 1 et seqq.
† Dawson Borrer, chap. i.　　　　　　　　‡ Daumas, La Grande Kabylie, p. 22.
§ This word signifies the sacred month of the Mussulmans, during which they fast till sunset.

do they carry it, that amongst certain tribes a Kabyle would not see one enter the house for any thing in the world. They have a legend which would explain this aversion by an act against nature in the time of the ancient Kabyles. The Arab detests work ; he is essentially idle ; during nine months of the year he only thinks of his pleasures. The Kabyle labours immensely, and at all times ; idleness is a disgrace in his eyes. The Arab tills the land a great deal ; he possesses a great number of flocks which he tends ; but he plants no trees. The Kabyle grows less corn, but he gardens a good deal ; he spends his life in planting and grafting ; he has lentils, grey peas, beans, artichokes, turnips, cucumbers; onions, beet-root, red pepper, water-melons, and melons. He also cultivates tobacco ; he has for some time grown potatoes ; he has fruit of all kinds—olives, figs, nuts, oranges, pears, apples, apricots, almonds, and raisins. The chief riches of the country consist in its olives, many of which are grafted, and attain sometimes the dimensions of the walnut-tree. The olives, which are of excellent quality, form a great part of the Kabyle's nourishment ; but an enormous quantity remains to be sold either as fruit or as oil. The latter is exported in goat-skins to Algiers, Bugia, Dellys, to Setif, and to all the markets in the interior. The arable land not being very abundant in proportion to the population, the Kabyles do not neglect a morsel of it. They give two ploughings to the ground, and manure it, but seldom suffer it to lie fallow ; nor do they practise rotation of crops. Generally speaking, their fields are kept pretty clean, and some of them yield as much as twenty-five for one. The wheat is threshed in a barbarous fashion by means of bulls, which work in a circle on the barn-floor ; and being winnowed coarsely with the end of a board, does not pass through the sieve : it is preserved, like that of the Arabs, in *silos* (in Arabic, *metmora*) ; and also in large osier-baskets, which are very wide at the bottom and narrow at the top. The Arab travels sometimes in search of pasturage, but he never goes beyond a certain circle. Among the Kabyles, one of the members of the family goes away for a time to seek his fortune ; thus one sees them every where—at Algiers, at Setif, at Bona, Philippeville, Constantina, and at Tunis. They work as masons, gardeners, reapers ; and they tend flocks. When they have gained a little money, they return to their village, buy a gun and an ox, and then marry.

Baron Baude says we call Kabyles all the inhabitants of the Atlas and of the shore whose establishment preceded that of the Arabs, and who do not speak their language. This definition is in the main correct, though likely to give rise to sundry inaccuracies.

The Kabyles, on their part, do not distinguish the European nations respectively, and think us the same people as the ancient Romans, hence they call us Romni ; and the native Kabyle who serves in the regiment of Zouaves, a mongrel force raised by the French in Algeria, is thought to

S

serve in the Roman troops. The Baron was disposed to think, however, that they had as many shades of difference among them as the different nations of Europe : in some places they present us with dark skins and fine hair, in others with light hair and fine clear complexions. This remark is true, if we give the same extent as the Baron to the term Kabyle ; embracing the blue-eyed tribes of the Aouress and Mount Edough, and the Kabyle Jews.

He further states that their habits and customs change according to the tribes : some, like the Mezayas near Bugia, and the mountaineers of the Chiffa, have no other industry than robbery, and no other law than the sword ; others, again, are superior in honest industry to many European populations. The inhabitants on the slopes of the Djordjora, who are correctly thought to be the descendants of the Vandals, build houses which remind one of European structures, and have no resemblance to Moorish edifices. They work mines, know how to extract the ore of iron and lead ; manufacture gunpowder, steel weapons, and firelocks ; they also make a great part of the haicks, of the coverings, and burnouses, that are used not only all over Algeria, but also in the empire of Morocco and in the regencies of Tunis and Tripoli. They have factories (*des comptoirs*) like those of the Pisans in the middle ages ; and if we take into consideration the simplicity of their implements and the finish of their work, we must confess that their workmen are not less dexterous than ours. They seem to have institutions like those of the ancient Germans of Cæsar and Tacitus. Thus, from the age of twenty to twenty-five years, all the male population are subject to military service ; after the age of twenty-five they make a kind of mobilised reserve in case of war ; and after a certain age they cannot be called out except in cases of great danger.*

Strictly speaking, the Arab is not industrious, although he manufactures saddles, harness, horse-bits, &c.† The Kabyle, on the contrary, is industrious : he builds his house, he is a carpenter, he forges weapons, gun-barrels, and locks, swords called *flissas*, knives, pickaxes, cards for wool, ploughshares, &c. They moreover manufacture gun-stocks, shovels, wooden shoes, and frames for weaving. The burnouses and habayas (woollen garments) are also made by them, together with the haicks for women, and the white chachias (caps). Their earthenware is renowned ; and they make the oil from their olives, which they gather on their own property, besides preparing the mills themselves for pressing them. The following is the most usual form of the Kabyle oil-press : a large basin, formed of one piece of wood, having at each extremity of one of its diameters a vertical post, which works in a horizontal bar ; the latter being pierced in the middle, a wooden vice is let through, terminated by a millstone of a diameter little inferior to that of the basin. The vice presses upon the olives, which, having been previously boiled, are placed under the millstone.

* Baron Baude. † Daumas, chap. i.

The Kabyles also prepare the hives for their bees, and extract the wax ; and in preparing flour for their bread, they only use portable mills at home. They are acquainted with the art of baking tiles, a hundred of which cost from two francs (1s. 8d.) to two francs fifty cents. In certain localities they make cork soles; and they are also familiar with the preparation and use of lime ; but they are very careful of it, only using it to whiten the mosque and the koubbas (the tombs) of the marabouts. They make use of plaister (whitewash) for their houses, this article seeming to be very plentiful in their territory : the quarry of Thisi, among the Beni-Messaoud, at a league and a half from Bugia, furnishes a great quantity of it. They prepare black soap from the olive-oil, and salt-wort of sea-weed or the ashes of the laurel-rose ; they weave baskets in which to carry loads, and prepare table-cloths of the dwarf palm, besides spinning cords from wool and goat's hair. In short, they carry their industrial cleverness to such a pitch, that they manufacture even false coin.*

We shall proceed to enlarge on some of the branches of industry previously mentioned, beginning with the last. From time immemorial the Kabyles who were established at Ayt-el-Arba, a considerable village of the tribe of Beni-Janni, gave themselves up to this guilty practice. Other less noted gangs are still found in the village of Ayt-Ali-ou-Harzoun, 15 leagues south-east of Ayt-el-Arba, 40 leagues (100 miles) distant from Algiers. The spot to which these coiners repair is the summit of a mountain protected by a very narrow and almost inaccessible defile. It is there that, sheltered from all attack, they imitate the copper, gold, and silver coins of all the countries of the world. Their first materials are partly furnished by the neighbouring mines. Copper and silver they have brought to them from all the barbarous parts of the country, even from the Sahara, by men who not only transfer the produce of their country to Ayt-el-Arba, but also come to buy adulterations. They pay them with monies of good alloy, on the footing of 25 per cent. The simple inspection of a piece of counterfeit proves that the procedure employed to obtain it is generally that of fusion. In fact, all the pieces present a diameter slightly inferior to that of the models; a result occasioned by the contraction which they have suffered in casting, after extraction from the mould containing the impression of genuine pieces. The relief of the figures and the letters is generally badly wrought, and the aspect of the metal is faded or coppery. It must be owned, however, and all who have seen them will bear out the assertion, that the greater part of these false pieces effectually deceive you at first sight, and some really require a very minute examination.

The methods of prevention employed under the Turks, in order to oppose the uttering of false coins, were conformable in every thing to the despotic and arbitrary procedures which the authorities at that time

* La Grande Kabylie, p. 27.

sanctioned. The people of Ayt-el-Arba, and those of Ali-ou-Harzoun, never going from their retreat, were obliged to confide to others the care of hawking about their products ; for though the Kabyles protect the manufacturers of false coin, they are quite merciless towards any man who would try to circulate it in the country. It is therefore necessary to send it out of Kabylia ; and the Beni-Janni, the Beni-Menguelat, the Beni-Boudrar, and the Beni-Ouassif, were generally charged with this mission. The estrangement of the other Kabyles from these tribes proceeds, no doubt, from this cause. These people were watched with peculiar jealousy, and could not travel in the interior of the district without the permission of the caid of Sebaou, who never granted it without imposing a duty of two Spanish douros (9s.). If he omitted to show this permission, which they moreover refuse to all who are suspected of trafficking in coin, the first traveller who arrived was obliged to submit to the confiscation of his merchandise, mules, &c.

Three years before the conquest of Algiers by the French, false coin had multiplied so excessively, that the Agha-Yahia, who had a great reputation among the Arabs, furious to find his vigilance of no avail, caused, in one and the same day, men of all the tribes who were known to have devoted themselves to this profession to be arrested in the markets of Algiers, Constantina, Setif, and Bona. They imprisoned in this way a hundred individuals, whom the pasha sentenced to death if they did not deliver up the moulds which they used in their manufacture. The people of Ayt-el-Arba, in order to save their brothers, sent all their instruments, and the prisoners were not set at liberty until a large fine was paid. This check which the false coiners experienced did not give them any distaste for their trade. Ayt-el-Arba lost no portion of its prosperity ; and the number of merchants who came there to supply themselves from all parts,— from Morocco, Tunis, from the Sahara and Tripoli,—did not at all decrease. A Kabyle taken in the act of issuing false coin was put to death without any formal process. It was the only case in which justice was inexorable, and in which the money which redeemed all other crimes was not able to weigh down the scales on their side. Those branches of labour which are more honourable, but not so exciting to the curiosity, are perhaps not so well known.

The manufacture of powder is confined to the tribe of the Reboulas : they make it in great quantities there, and by processes similar to our own. Saltpetre abounds in natural caverns, and it is found incrusting their walls. Being collected like our sweepings of saltpetre, it is first washed, and then obtained by evaporation. Charcoal is procured from the laurel-rose, and possesses the best qualities. Sulphur is imported from foreign countries. The proportions are regulated as with us, and the drying is performed by the sun. This Kabyle powder, which is not quite so strong as ours, is neither smooth nor equally granulated ; but it does not stain the hand, and answers as a good powder for war. The Kabyle cartridges are well rolled, and they are much boasted of in the market. The lowest price of the cartouch is 40 cents (4*d.*), which appears extremely high. The balls are of lead, and very irregular in size. The working of lead-mines is carried on upon a considerable scale in the tribe of the Beni-Boulateb, near Setif. This metal is found also in a mountain near Msila, and in another called Agouf, also amongst the Reboulas : this last is reported to contain silver ore. In all cases they obtain it by simple fusion, and it is exported in pigs or balls. Copper is also found in Kabylia. It is extracted, and employed in making female ornaments. Melted with zinc, it forms a brass which is very useful for powder-horns, the mountings of flissas, handles for poniards, &c. Two very abundant iron mines are renowned in Great Kabylia; one amongst the Berbachas, the other amongst the Beni-Slyman. The vein of ore is smelted in furnaces heated by charcoal, after the Catalan method ; the bellows are made of goat-skin, and plied by men. The tribe of the Flissas prepare steel weapons, bearing their name, with the iron of the Berbaches and the steel brought from the East. The principal manufacturers of fire-arms are the Beni-Abbas. Their gun-locks, which are more celebrated than their gun-barrels, unite elegance with solidity ; they are exported as far as Tunis. Their gun-stock is made of the walnut-wood, and they mount the whole of their steel weapons. In the midst of this vast industry of the men, the women do not remain idle. They spin wool, and weave it into a sort of white stuff, which serves for clothing for both sexes. Their trades are established upon the model of those of Algeria. The flax, gathered in little bunches, then dried in the open air, and lastly pounded and spun by the women, makes a coarse cloth which is employed for many uses. The women co-operate in making the *burnouse*, which in some tribes, for instance the Beni-Abbas and Beni-Ourtilan, becomes an object of exportation, these people having more than they require for their own use.

The Arab occupies himself very little in preserving his arms; it would require some care: "a black dog," he says, "bites as well as a white dog." The Kabyle, on the contrary, considers his gun his chief luxury; he preserves it from rust; and when he takes it out of the case, he holds it with his handkerchief, that it may not be soiled. The Arab, physically idle, is somewhat inert even in the impulses of the heart; but amongst the Kabyles anger and conflicts attain inconceivable proportions. The following is a recent example. A man of the tribe of the Beni-Yala met, at the market of Guenzate, another Kabyle, who owed him a *barra* (seven centimes). He reclaimed his debt. "I will not give thee thy barra," replied the debtor. "And why?" "I do not know." "If thou hast no money, I will wait still." "I have some,—but it's a kind of whim which has taken hold of me not to pay thee." At these words the creditor, quite furious, seized the other by his burnouse, and threw him on the ground. The neighbours joined in the struggle. Two parties were soon formed, and they had recourse to arms. From 1 o'clock till 7 in the evening, it was impossible to separate the combatants; 45 men were killed, and that for about a halfpenny! This quarrel happened in 1843; but the war which was kindled through it is not yet extinguished. The town has since been divided into two hostile quarters, and the houses which stood on the frontier are now deserted.

The Arab is vain : he appears humble and arrogant alternately. The Kabyle remains always wrapped up in his pride. This pride gives importance to the smallest things of life, imposes on all great simplicity of manners, and exacts a scrupulous reciprocity for every deferential act. For instance, the Arab kisses the hand and the head of his superior with forced compliments and salutations, troubling himself little whether or not his politeness is returned. The Kabyle never pays compliments; he kisses the hand and the head of a chief or of an old man; but whatever be the dignity or the age of him who has received this politeness, he is bound immediately to return it. Si-Said-Abbas, a marabout of the Beni-Haffif, was one day in the market, on a Friday, of the Beni-Ourtilan. A Kabyle called Ben-Zeddam approached and kissed his hand; but the marabout, no doubt not thinking about it, did not return the salutation. "By the sins of my wife," said Ben-Zeddam, who placed himself in front of Si-Said with his gun in his hand, "thou shalt instantly return me what I gave thee, or thou art a dead man." And the marabout performed the act. The Arab is a liar; the Kabyle considers lying a disgrace.

The Arabs in war usually proceed (say the French) through surprises and treachery. The Kabyle acquaints his enemy with his intentions; and this is done in the following manner : the token of peace between two tribes consists in the exchange of some article,— it may be a gun, a stick, or a bullet-mould, &c.; this is called the *mezrac* (the lance). All this leads to the conclusion, that before the invention of fire-arms, the deposit of a lance was, in fact, the symbol of a truce and good faith. Should one of the tribes wish to break the truce, the chief simply returns the *mezrac*, and war is declared. The Arabs are satisfied with the *dia*, the price of blood in expiation of a murder committed on one of the members of the family. With the Kabyles, the assassin must die. His flight will not save him; for *vengeance* is a sacred obligation. Into whatever region, however distant, the murderer may fly, thither revenge follows him. A man happens to be assassinated; he leaves a son very young; the mother teaches the child very early the name of his father's murderer. When the son is grown up, she gives him a gun, and says, "Go, revenge thy father!" If the widow has only a daughter, she makes known that she will receive no money for her,* and will give her only to him who kills her husband's murderer. There is a striking analogy between their manners and those

* The Kabyles buy their wives, as we shall show further on.

of the Corsicans ; and it is still more delineated in the following traits. If the really guilty man escapes vengeance, and evades all pursuit, it passes over to the nearest of kin ; whose death, in its turn, requires new reprisals. Hatred thus enters the two families, and becomes hereditary. On both sides, friends and neighbours marry, factions ensue, and actual wars may even result from it. The Arabs practise hospitality; but there is more of policy and ostentation than of heart in it. Amongst the Kabyles, though their hospitality is of a less sumptuous nature, you can nevertheless perceive in its forms the existence of good feeling. A stranger is always well treated, whatever may be his origin. These attentions are still more marked towards refugees, whom nothing in the world could induce them to deliver up. The Turks and the Emir Abd-el-Kader have always been frustrated in any demands or efforts contrary to this noble principle. The following is a generous custom amongst them. When the fruits, such as figs, grapes, &c. begin to ripen, the chiefs publish a decree that no one, during fourteen or fifteen days, under pain of a penalty, shall touch any of the fruit on the trees. At the expiration of the time fixed, the proprietors assemble in the mosque, and swear on the holy books that the command has not been violated. He who cannot take the oath pays the fine. The poor of the tribe are then consulted, they make out a list, and each proprietor by turn feeds them till the fruit-season is passed. The same thing takes place during the bean season, an article much cultivated by the Kabyles. At these periods every stranger may enter the gardens, and eat as much as will satisfy him, without any interruption; but he must not take away any thing with him: for a theft is doubly culpable on these occasions, and might cost him his life. The Arabs cut off the head in combat; the Kabyles, amongst themselves, never do this. The Arabs are accustomed to rob wherever they can, and especially in the day-time. The Kabyles commit robberies chiefly by night, and only amongst their enemies. In this case it is an act worthy of praise; otherwise, quite the contrary. The Arab has preserved some traditions concerning medicine and surgery. The Kabyle has neglected them; consequently we find many chronic diseases amongst them. The Arab does not know how to increase the value of his money; he buries it in the ground, or uses it to increase his flocks. The Kabyle, contrary to the Mussulman law, puts it out at large interest,—for instance, at 50 per cent per month ; or he buys at a cheap rate, and forestalls the harvests of oil, of grain, &c. The Arabs class musicians in the rank of buffoons; and the man amongst them who would dance is dishonoured in all eyes. The Kabyle likes to play on his little flute ; and every one dances, men and women, relations and neighbours : the dance is performed with and without arms.

When a marriage is celebrated among the Arabs, they perform equestrian games before they bring home the bride. With the Kabyles, the relations or friends of the bridegroom shoot at a target. The mark is generally an egg, a peppercorn, or a flat stone. This custom causes a great deal of gaiety, for those who miss the mark are subject to much joking. When a Kabyle wants to marry, he informs one of his friends, who seeks the father of the girl of his choice, and makes known the desire. They fix the marriage-portion which will be paid by the husband ; for he literally *buys* his wife, and a great number of girls is considered to constitute the wealth of the house. These portions amount to upwards of a hundred douros (25*l.*). It sometimes happens that the future husband does not possess the entire sum ; he is then granted a month or two to collect it, and during that time he may visit the house of his future wife. When he has succeeded, he leads her, as his *fiancée,* first through the village, armed with a yatagan, a gun, and a pair of pistols ; after which he takes her under his own roof. This ceremony is performed with great pomp. Each village has its band, composed of two kinds of Turkish clarionets and drums ; and these musicians figure in the nuptial cortége. They sing as they go, and the women and children make the air resound with their joyous cries, " You! you! you!" They fire a number of guns ; and the young people of the village, all or a part of them, according to the wealth of the husband, are invited to a great repast.

Amongst the Arabs, when a male child is born, they rejoice and make compliments, but the fête is held in the family alone ; if the mother has a female child, the women alone rejoice. The birth of a male child amongst the Kabyles causes an assembling of all the neighbours and friends of the surrounding villages. They fire guns and shoot at the target ; and seven days after, the father gives a great feast. Circumcision does not take place till the age of seven or eight years. If a girl is brought into the world, there is no change in the habits of life or appearance of the house, because she does not at all increase the force of the tribe ; since the child, when old enough, will probably marry, and will perhaps leave the country in order to follow a new master.

When one of the family dies amongst the Arabs, the friends and neighbours assist at the burial, and then each one returns to his business. Amongst the Kabyles, the whole village is present at the funeral. No one must work ; and with the exception of the relations of the departed, all unite in giving hospitality to the Kabyles of other villages, who have come to add their tribute of grief. The dead are not placed on a bier: after being carefully washed, they are wrapped in a sort of cloth, and are then committed to the earth. The Kabyle women enjoy much greater liberty than the Arab women ; they are more considered in society. For instance, the Kabyle woman goes to market to get provisions for the house, to sell and to buy. The husband would be ashamed to enter into household details like the Arab. The Arab woman cannot appear in the assemblies of men ; she always holds her handkerchief, or veils herself with the kaik. The Kabyle woman seats herself where she chooses ; she talks, she sings, and her face remains uncovered. Both from infancy wear a small tattooed pattern on the face ; but that of the Kabyle women presents a remarkable peculiarity : it has generally the form of a *cross*. The usual position of it is between the eyes, or upon one of the nostrils. The Kabyles continue this custom, without knowing the origin of it, which appears to have been derived from the early Christian times. A fact worthy of remark strengthens this conjecture : it is, that no taleb or marabout will marry a woman thus tattooed, until he has made the sign disappear through the application of lime and black soap. It is right, however, to observe, that the Koran prohibits all tattooings, branding them with the name of Ketibet-el-chytan (writing of the devil). The Arab woman never eats with her husband, and still less with her hosts. The Kabyle woman takes her meals with the family, even when strangers are present. The Arab woman is never considered free in her actions. The Kabyle woman, if abandoned by her husband, returns to the house of her father or her brother ; and as long as her isolated mode of life lasts, enjoys perfect freedom from moral restraint. A woman who is divorced acts precisely in the same way. This license will explain the pretended custom which is attributed to the Kabyles by several historians, of offering their wives or daughters to distinguished guests. Owing to a certain number of free women being found in each tribe, the Kabyles appear to have been preserved from a kind of debauch contrary to nature, and so frequent amongst the Arabs, but which with them would be punished with death. In certain tribes, and especially amongst the Yguifsal, the women and girls who live by prostitution pay each year, on new year's day, a sort of duty, which does not amount to more than five douros (1*l*. 5*s*.) : this money is thrown into the public treasury. They cease to pay when they marry or give up their condition. But this custom is not general. After what has been said, it will not appear surprising, that the Kabyles attach much less importance to the virginity of the young girls they marry than the Arabs.

The Arab woman who receives no news of her husband during one or two years, or who has nothing at home to live upon, asks for a divorce, and the law directs the cadi to grant it. The Kabyle woman can only be married again on having a certain proof of her husband's death. If her position is unhappy, they give her work, or the tribe gives her assistance. Still, divorces are very usual amongst the Kabyles ; but they are in a great measure at the whim of the husband. "*I leave thee for one hundred douros,*" says the man who wishes to be divorced from his wife ; and the wife retires to her parents with that sum. If she marries again, she is bound to restore the money to her first Benedict ; but if she does not form new ties, she keeps it. This measure is necessary, as girls have no right to inherit property, owing to the chance of their being married to husbands of strange tribes. The more daughters a Kabyle has, the richer he is,* as each of them brings him in a dowry, and he has to give none. The common women amongst the Arabs are generally dirty. The Kabyle women are cleaner, and they are obliged to make two toilets in the day : in the morning they wash; in the evening they adorn themselves with all their ornaments, they apply the henné, &c. This custom, it seems, results from their appearing at the guest's table. It is possible that this attention to their persons has contributed to establish the reputation which the Kabyle women have of surpassing the Arab women in beauty. This renown has always existed ; but it refers principally to the distinction of forms. In short, not only are the Kabyle women more free, more considered, more influential than the Arab women ; but they can even aspire to the honours, the odour, and the power which appertain to sanctity. The koubba of Lella-Gouraya, which stands above Bugia, immortalises the memory of a girl who was celebrated for her science and piety. The legend relates, that after her death, she returned to instruct her faithful disciples, who assembled again round her tomb. In Kabylia there are also other koubbas consecrated to women; and without departing from living examples, we may cite, as enjoying a high reputation of this kind, the daughter of the famous marabout Sidi-Mohamed-ben-Abder-Rahman Kafnaoui,† who receives religious offerings at the tomb of her father, and whom all the Kabyles recognise under the name of Bent-el-Sheikh‡ (the daughter of the Sheikh).

Politically speaking, Kabylia is a sort of wild Switzerland. It is composed of tribes independent of each other, at least in rights, governing themselves, like the Swiss cantons, as distinct states, but whose federation has no permanent character or central government. So many tribes constitute so many unities ; but these unities group together variously, according to the political interests of the day. From this result offensive and defensive leagues, which bear the name of *soff* (rank, line). The tribes thus allied say, We make but one rank, but one single line. Common interests, old or new alliances, relations of neighbourhood, of transit, of commerce,—such are the causes which determine the formation of a soff. The soff obliges the contracting tribes to share in the common good or bad fortune. It is proclaimed in a general assembly of their chiefs. The latter regulate also the plan of military operations, the number and the order of the combatants, their points of reunion ; and finally, they elect a supreme chief. When it is one particular tribe which has summoned the soff, in order to secure itself against danger, or be revenged on an

* Strange contradictions in humanity ! The Rajpoots regard many girls as a curse, and practise extensively female infanticide. See Ward's *View of the History of India ;* Blaquière's *Asiatic Researches ;* Millar's *Inquiry into the Distribution of Ranks,* &c.

† *Sid,* or *Si* by abbreviation, sieur, lord. *Sidi,* my lord. *Abd,* servant. *Rahman,* mercy. *Abd-er-Rahman,* servant of mercy.

‡ *Sheikh,* old, venerable ; and chief.

enemy, it furnishes in general the chief of the expedition. The auxiliaries who come and fight on the territory, and for the cause of an ally, bring with them also their arms and provisions. The succoured tribe does not furnish them with any thing, unless the war is prolonged beyond their expectation ; they then beg their defenders to remain with them, after they have consumed their provisions. Certain tribes pass frequently from one soff to another, whether it be from temper, from a political fluctuation inherent in their situation, or sometimes because they are induced by money. In this last case, they lose much in the public esteem ; they *use* them, whilst despising them. Soffs are formed in consequence of enmities common to many tribes, when these latter war against each other. It resembles the league of the Catholic cantons against the Protestant cantons in Switzerland. There are accidental, momentary soffs ; while others have motives of such stability, that they last for ages ; and in cases of universal peril, great soffs are spontaneously constituted to preserve a common defence. Let the marabouts preach the *djehad* (holy war), let them dread the invasion of the Christians, and all Kabylia, in this emergency, forms only one soff. Many soffs may spring out of this single one, but all animated with the same spirit, if they learn that the enemy is going to pour in by a number of points at once. The tribes menaced in each direction concentrate themselves then into so many particular soffs, who seek as much as possible to unite their operations. But egoism and rivalries continually oppose this. In too-numerous gatherings certain rival families aspire to command. Sometimes they separate, having decided on nothing ; and sometimes those who disagree abandon the common cause. There exist, in fact, amongst the Kabyles (strange disparity in the midst of the most republican manners) some great families of religious or military origin, whose uncontested influence rules many tribes at a time : they are those who furnish chiefs to all the soffs which have some little importance. Every other candidate retires before their members. It is also in the bosom of these families that all governments aspiring to hold sway over the Kabyles are forced to take their instruments : they have accordingly conferred on these the titles of khalifas, of aghas, &c.* This policy was that of the Turkish pashas, and afterwards of Abd-el-Kader ; and it has now become that of France, by the force of circumstances.

We shall not dwell in detail on these preponderating families, though they play a considerable part in the course of Algerian history. That which it concerns us here to prove is, the essentially fickle character of the confederations, the absence of any permanent tie, of all central administration ; and, to conclude, that one must descend into the bosom of the

---

* *Khalifa,* lieutenant. Employed alone, this word signifies lieutenant of the supreme chief, or even of the Prophet. In this last sense, we have translated it by *Calif*. *Agha,* chief ; quite inferior, almost always military.

tribe, properly speaking, to *begin* to discover the appearance of a regular government.

They call *Arch* or *Kuebila* one entire tribe. The fractions, *Ferka*, of the tribe are called moreover *Krarouba, Fekhed, Areg* (carob, thigh, veins). These fractions sometimes, in their turn, are resolved into *Déchera*, villages. According to the Kabyles, the tribe, *arch*, is the body of the man ; *fekhed, âreg*, are its members or veins ; and *déchera*, the fingers which terminate the feet or the hands. The tribe and its fractions find equally their image in the fruit of the carob-tree ; for it is composed of one *cosse*, in which are contained several grains, *krarouba*. Each *déchera* appoints a chief, whom they call *Amin*.* This election depends on universal suffrage ; all Kabylia takes part in this, and the general wish is not in any way limited ; notwithstanding which, they very well know there, as elsewhere, how to influence in favour of rights of birth, to intimidate by influence, to seduce by riches, and to captivate by eloquence. These great assemblies are *Djemmâs* ;† but, in a more special sense, the djemmâ of a tribe is an assembly of all the amins elected, as has just been said, by these divers fractions—deliberating in common upon the national interests, giving judgments, and taking general measures, &c. This same djemmâ proceeds also to the election of a president amongst the members who compose it, who bears the name of Amin-el-Oumena (the amin of amins) ; who becomes also the regular chief of the whole tribe, and to whom the command of the warriors they set on foot belongs on the day of battle. His prerogatives are otherwise very limited, unless an illustrious birth confers others founded on the moral aid of public opinion.

In all cases, however, and were it only for the sake of form and precedent, this president takes the advice of this djemmâ upon the smallest affairs. ‡ In it, properly speaking, resides the government. The duration of power granted to the chiefs is not the same in all territorial districts. Amongst certain tribes they are renewed every six months; with others, every year ; but with all, bad conduct would cause their immediate removal, just as any signal services would cause them to be prolonged. In every case the people must pronounce. The amins are charged with the maintenance of public order, as well as the observance of the laws and customs; and in this connection we shall introduce a series of facts all peculiar to the Kabyles. Alone amongst all Mussulman nations, this singular people possess a code of their own, whose prescriptions are derived neither from the Koran nor the sacred commentaries, but from past customs which have been maintained through ages, even throughout the changes of religion. It is this customary right which the amins consult on all occasions. The old men, the greybeards and the Solomons, have re-

---

* This title answers to that of Caïd amongst the Arabs.
† Djemmâ signifies also mosques, εκκλησια = Angl. *meeting*.
‡ European presidents have rarely shown such modesty.

ceived it traditionally, and they preserve the deposit to transmit it intact to their children. The following are the penal arrangements for the most frequent faults :—

	Boudjous.	£	s.	d.
1. To draw the yatagan (sword) without striking .	8	0	12	0
2. Do. and to strike	16	1	4	0
3. To cock a gun without firing	10	0	15	0
4. Do. and to fire	30	2	5	0
5. To raise a stick without striking . . .	1	0	1	6
6. Do. and strike	3	0	4	6
7. To brandish a sickle without striking . .	2	0	3	0
8. Do. and strike	4	0	6	0
9. To threaten to throw a stone at some one . .	1	0	1	6
10. Do. and to hit the person	6	0	9	0
11. To strike with the fist	0¼	0	0	4½
12. To injure without a motive	4	0	6	0
13. To be convicted of theft	100	7	10	0
14. To enter a house during the master's absence .	100	7	10	0
15. Not to have mounted guard . . .	1	0	1	6
16. Appearing at the washing-place of women .	2	0	3	0

Among the Arabs, you see men and women mixed together at the fountain ; but with the Kabyles, they appoint one place for the men and another for the women. A stranger, should he present himself at the latter, would not be fined for that infraction of the law, for he is supposed to have been ignorant of it. All these fines the amins impose, and levy up to a certain rate, above which they must deposit the amount with the amin of amins, or chief president, who employs it in buying powder ; and on the day of battle this powder is distributed to the most needy of the tribe. The rest of the fine is employed in relieving the poor. Nothing ever remains abandoned to the waste and extravagance of the chiefs, as in the Arab administration. In all circumstances, and whatever authority he may have, an amin is constrained to apply himself most rigorously to the legal text. No arbitrary arrest can be made ; equality before the law forms also the first article of the Kabyle charter. This charter is not written, but it has been observed for two thousand years. We have remarked that a penalty exists for theft; but there is none for the receiver of stolen goods. These authorised receivers, who are called *oukaf*, sell publicly the things stolen ; and it appears that the motive of this injurious legislation is to enable the wronged proprietor to regain his own at a low price. One may imagine that otherwise, considering the small dimensions of each state, all the products of theft would be immediately exported, and all chance of recovering them impossible. We have not spoken of murder : the Kabyle law on this subject is well deserving of the attention of a civilised people. We know that the Koran absolutely prescribes the penalty of retaliation : "a tooth for a tooth, an eye for an eye." Still the Kabyle djemmâ never pronounces a sentence of death. The executioner of capital punishment is not known in this barbarous society. The murderer ceases to belong to his tribe, his house is destroyed, his goods are confiscated, he is an exile for ever : this is the public revenge. But the field still remains open for private vengeance; it is for the parents of the victim to apply the retaliation in all its rigour. The law shuts its eyes on these bloody reprisals : opinion exacts them, and prejudice absolves them. One more remark only remains to be made on the preceding code: there is no bastinading. Contrary to the ideas received amongst the Arabs, this punishment is considered infamous in the eyes of the Kabyles. No amin can dare to order it in the whole extent of his jurisdiction. We may judge by that how dangerous it might be to employ agents not familiar with the customs of the Algerian races. We have remarked, that the office of the amins is limited to the interior police of the tribes; and that their privileges being very restricted, their influence does not suffice to preserve order and public peace in the country. Accordingly they are not required or expected to exceed the limits of their little authority, because for graver matters there exists a vague power raised very much above their petty jurisdiction : this is the power of the marabouts. *Marabout* * comes from the word *mrabeth* (united). The term Marabouts signifies a people united to God. When enmities arise between two tribes, the marabouts alone have the right to interfere, whether to establish peace or to obtain a truce of longer or shorter duration. At the time of the election of chiefs, the marabouts have the right to propose to the people those who appear to them the most worthy. They then recite the *fatah* † over the elected. When one tribe has gained an advantage over another and weaker one, and this last is resolved to perish rather than surrender, the marabouts compel the victorious tribe to declare themselves vanquished. Admirable skill of the human heart, which knows how to apportion to all their due share of vanity! Actions of this kind are not rare; and such is the character of this people, that there is no other method of preventing their weak pride from destroying them. When important circumstances require a gathering of the tribes, the chiefs order it to be made public in the market-places; and with the exception of the sick, of old men, women, and children, no one fails to attend the meeting, however far they may have to go. On the day fixed, the tribes being grouped separately, the marabouts advance to the centre, and explain through the public crier the cause of the meeting, demanding what advice they should follow. Each man has his say, each is respectfully listened to, whatever be the class; and the various opinions having been received, the marabouts unite in a committee, and the public crier makes known to the people their decision. If no voice is raised to make any new remonstrances, they invite the assembly to clap their hands in sign of consent. This being done, all the Kabyles discharge their pieces,

---

* The French have given, by extension, the name of *marabout* to the little monuments which enclose the tombs of the marabouts, and which are called in reality *koubbas*, domes.

† *Fatah*, special prayer to obtain success for any undertaking ; the first chapter of the Koran.

which they call *el mëiz* (the decision). The things they relate of the influence of the marabouts in the Kabyle land are so very surprising, that one hesitates to believe them. The mountaineers, they say, do not fear to butcher their own children, if they receive the order from a marabout. The name of God invoked by a wretched being whom they intend to rob, does not protect him; that of a venerable marabout saves him. The marabouts command the markets; and the authority of the amins falls to the ground before theirs.

Not only are the markets free, exempt from all customs, taxes, and rights; they are also inviolable. With the Arabs, a man who has committed a fault or a crime may be arrested in the open market; in the Kabyle markets the marabouts do not tolerate arrests or acts of revenge, for any reason whatever.

This influence of the marabouts is the more remarkable, as the Kabyle people are much further removed from religious ideas than the Arabs. They know nothing of prayers; they do not properly observe fasts or ablutions; they limit their religion nearly to this : " There is but one God, and Mahomet is his Prophet." It is said, that there are Kabyle tribes where the poor people do not fear to eat the flesh of the boar ; and they almost all drink brandy of the fig, made by the Jews, of whom there is a great number in the country. The precepts of religion are only followed by the marabouts, the chiefs, and the tolbas.

The cause of this passive obedience of the people is found entirely in the industrial spirit, which makes them comprehend of what importance order and peace are to commerce.

The marabouts, moreover, have taken advantage of this general respect to institute one of the most beautiful customs of the world, the *Anaya*, with which the reader will become acquainted further on. The public veneration for the marabouts does not solely display itself in honours, deference, and privileges. These holy men live *on* the people, and *by* the people, as in Christendom ; one might almost say, that all the riches of the nation belonged to them. Their zaouias, or common habitations, of which we shall speak hereafter, are repaired and provided, without their even paying any attention to it, nay without their expressing a desire to that effect. All their wishes are anticipated ; the community interest themselves in all the details of their private life ; they bring them water, wood, food, &c. If they are going to beg in the villages, each one hastens to them, and inquires concerning their wants, offers them horses, and loads them with presents.

The Kabyles pay taxes, which are the *zekkat* and the *achour* prescribed by the Koran, and fixed at a hundredth for the flocks and a tenth for grain. But, contrary to the Arabs, who give these contributions to the Sultan, the Kabyles, organised as republics, bring all to their mosques. They employ it in defraying the expense of schools, in succouring the poor, in feeding travellers, in keeping up worship, in practising hospitality, or in buying powder and arms for the distressed members of the tribe, who are called, like the others, to march on the day of battle. For with the Kabyle people, as soon as it is meditated to revenge an injury or repel an aggression, all must rise up, whether they have arms or not. Those who have no guns take sticks, throw stones, and keep within reach of those engaged, their duty being to remove the dead or the wounded. The women sometimes take part in these bloody dramas, in order to encourage their brothers and husbands : they bring them ammunition ; and if one of the warriors has fled, they put a large mark with charcoal on his burnouse, or woollen shirt, as a symbol of general contempt.

The general recruiting or conscription for the public defence is regulated by a formality which approaches a good deal to the French recruiting system. When a boy has completed his first rhamadan, that is, his fourteenth or fifteenth year, according to his constitution, he presents himself to the djemmâ. He is then declared fit to carry a gun. They inscribe him as one of the defenders of the tribe, whose good and evil luck he is henceforth to share. They read over him the fatah ; and if his father is poor, they buy him a gun from the public funds.

Consequently every man must be considered as a soldier, who serves from the age of fifteen till the age of sixty at least. It is a strange mistake, and too common to be passed over, to estimate Kabyle population according to the number of guns, or reciprocally in the proportion of one warrior to every six persons, as is done in Europe. The combatants in this country must evidently form a third of the entire population ; and calculating on this datum, we shall not depart widely from the truth.

The Kabyles, besides, are accustomed to labour (*souiza*) imposed by the state ; but not like the Arabs, who must do it to increase the goods of the beylik. The Kabyle only labours for the mosque, his marabouts, the common fountain, or the roads, which may be useful to all. He will labour also to dig a grave for one of his compatriots.

These are all the debts due from the Kabyle to the state. We see how he contributes with his person and his purse to public affairs ; but what we seek in vain for is, an administration capable of regulating all these efforts, and of deriving from them the greatest good possible. Another thing wanting is, a competent public authority to enforce them when needful. It seems that *opinion* is the only tribunal before which all delinquencies against the state can be summoned.

Such is the pride of the Kabyle, such is his instinctive inclination towards absolute equality, and perhaps also his supercilious defiance, that he looks upon it as his duty, so to speak, to suppress all depositories of social power. The marabouts, who possess the principal part of it, exercise it with discretion and in a persuasive manner. As to the amins, the smallest abuse of authority on their part leads to a refusal of obedience, expressed in the most energetic terms. *Enta cheikh, ana cheikh,* literally, "Thou chief, I chief." If it were possible to form a correct idea of what the actual life of the Kabyle would be according to the *probable* consequences of a government such as we have sketched, what a fearful picture would be presented to our eyes! No unity in power, no cohesion in the masses; every where intrigue and political rivalries, every where private prerogative braving the general interest; no social hierarchy, no preventive foreseeing authority endowed with the initiative, as in our happy rate-paying parishes; opinion without any consistency, the impunity of the strong, the oppression of the weak, all disorders at their height: this is what would, of course, await them. But happily this primitive society is saved by a phenomenon quite the reverse of that which characterises old nations. Whilst our admirable laws and philosophical constitutions are unaccountably crippled through the irregularities of our morals; here, on the contrary, religious institutions and inviolable customs admirably correct the insufficiency of the political machinery. Thus, this sadly republican people, who carry democracy to the length of individualism, have a terrestrial providence and a sultan. Its providence is the institution of the *zaouias;* and its sultan is a sacred custom which bears the name of *anaya.* We will attempt to describe these institutions clearly.

Every *zaouia* is composed of a mosque; a dome (*koubba*) which covers the tomb of the marabout whose name it bears; of a place where they read the Koran; of a second, reserved for the study of sciences; a third, serving as a primary school for children; of a habitation destined for the pupils and tolbas, who come to perform or perfect their studies; also of another dwelling in which they receive beggars and strangers; and sometimes there is a cemetery at hand, designed for pious persons who may have solicited permission to lie near the marabout. The zaouia is, altogether, a religious university and a gratuitous auberge. Under these two points of view it offers a multitude of distressing analogies with the monastery of the middle ages, with which it is impossible not to be struck in reading the following details.

Every man, rich or poor, known or unknown in the country, who presents himself at the door of any zaouia, is received and provided for during three days. No one can be refused; no example of any refusal of this kind is on record. The people of the zaouia, strangely enough, never take their meals, either morning or evening, without being first assured that their guests have had all their wants satisfied. The principle of hospitality extends even to such childish and unmanly lengths, that if a horse or mule has wandered, and arrives by chance without conductor, they are always received, installed, and fed, till the owners reclaim them.

It is to be regretted that this unconditional reception of unsheltered strangers in the house of God causes the misery of hunger and general

T

destitution to be, properly speaking, unknown to the Kabyles, the life of the poor consisting in a long pilgrimage from zaouia to zaouia.

Considered in the light of colleges, all the zaouias include three degrees of instruction.

The primary school is unhappily open to all children, whether Kabyle or Arab. Some parents send them from great distances, rather than have recourse to the small schools of their tribes. They pay six douros (30s.) beforehand for each child, providing, however, that they are fed, lodged, and clothed at the expense of the establishment, till the time of their leaving school: this is the common rule; but we shall see later, that the rich add to this very considerable presents.

The child is first taught the religious formula of Islam : " There is no other God than God, and Mahomet is his prophet;" afterwards, half a dozen prayers, and some verses of the Koran. The greatest number of the Kabyles learn no more than this; they return to the bosom of their family, to take part in their labours as soon as their physical strength permits.

Those who prolong their education learn to read and write, to recite the text of the Koran, &c. After six or seven years, this secondary education allows them to enter again the tribes as tolbas, and to open small schools for the children of the people.

When a pupil quits the zaouia, his masters meet together, and one of them reads the fatah over him. The young man, in his turn, thanks them, and he usually does it by this form, which is almost prescribed : " O my master, you have instructed me, but you have suffered much evil on account of me. If I have caused you any pain, I ask pardon for it on the day of our separation." We must just mention that the neighbourhood of the zaouias, like Oxford or Cambridge, suffers sometimes from the turbulence naturally consequent on numerous reunions of young people ; such as quarrels, thefts, besides the frequent visits of the Kabyle women whom the law has emancipated, &c. The chiefs of the zaouias spend their lives in settling the disputes which each day brings forth, owing to some new folly of their disciples.

Finally, the transcendental branches of study, especially in some of the most renowned zaouias, attract tolbas from distant regions. They come not only from various parts of Algeria, but from Tunis, Tripoli, Morocco, and even from Egypt. These learned men pay at their entrance four boudjous* and a half for the whole of their stay, which is entirely at their own option. They learn in the zaouias :

1. Reading and writing.

2. The text of the Koran, so as to be able to repeat it completely without fault, and with the proper psalmody or intonation, which tends to preserve the purity of the language.

---

* A piece of money of about the value of 1 fr. 75 cents.

3. The Arab grammar (*Djayroumia*). They do not teach the Berber language : its elements, as a written language, no longer exist, save in a few ancient inscriptions lately discovered among the Tuaricks.

4. The divers branches of theology (*Touhhid-el-tassaououf*).

5. The law ; that is to say, the commentary of the Koran in a legal point of view, by Sidi Khelil, who is in credit with all the rite of Maleki, and in consequence with most of the Arabs of Barbary.

6. The conversations of the Prophet (*Hadite Sidna Mohammed*).

7. The commentaries on the Koran (*Tefessir-el-Koran*) ; that is, the interpretation of the holy text. They reckon seven or eight commentaries which have authority : El Khazin is the most esteemed.

8. Arithmetic (*Hacal-eb-ghrobari*) ; geometry (*Haçab-el-member*) ; astronomy (*Aem-el-faleuk*).

9. Versification (*Alem-el-Aaroud*). Almost all the tolbas are poets.

The different zaouias entertain amongst themselves dissensions and college rivalries : opinion classifies them, the *esprit de corps* mixes up with them, and a Taleb would never leave his own zaouia for another; he would not even be received in it. The most celebrated zaouias are those of Sidi Ben-Ali-Cherif (amongst the Joullen) ; Sidi Moussa Tinebedar (amongst the Beni Ourghlis) ; Sidi Abd-er-Rahman (near de Bordj-el-Boghni) ; Sidi Ahmed-Ben-Driss (amongst the Ayt-Iboura). These reckon a considerable *personnel*, or household.

Sidi Ben-Ali-Cherif, for example, contains permanently two or three hundred pupils and tolbas, with a variable number of passing guests, of whom the mean daily amount may be valued at more than a hundred, and the maximum at four hundred. The zaouias are then, properly speaking, benevolent institutions; they furnish gratuitous hospitality, education for almost nothing; they do it on a large scale, and necessarily at a great expense. In what consists their resources? The zaouias are an object of especial veneration to the people. It is there that the Kabyles resort to oaths, when they have some claims, or any discussions with regard to debts, thefts, &c. The Kabyles upon whom many misfortunes press go to them from afar, in order to ask of God (through the mediation of the holy marabouts) an end to their afflictions. The mother who cannot bring up her children. who sees them about to die young, comes and prays God to preserve them. The woman who is barren is conducted there by her father or husband, hoping for the blessing of offspring.

The mosque of Koukou is the most celebrated for miracles of this last description. They attribute them to the stick of Sidi Ali-Taleub, which the barren woman must agitate in every direction in a hole made in the very centre of the mosque. They also rub the backs of the sick with it, in order to cure them. According to tradition, Sidi Ali-Taleub has only to aim at the cheek of his enemy with this wonderful stick, in order to make him fall down dead. This, if true, would be a powerful case of rapping. The sick also use as remedies the stone of the sacred tomb, which they pound, and then swallow together with many other things. Superstitious beliefs vary at each zaouia. In seasons of drought, they make large processions around all, without distinction, to ask for rain (a striking similarity to the Catholic requests). In short, although each tribe has its mosque, religious persons never fail to go and say their prayers on a Friday in the nearest zaouia.

The zaouia that has once obtained a position receives a portion of the *achour* and of the *zekkat*, otherwise usually appropriated to the mosques. Besides, there are certain tribes of the neighbourhood who, in many cases, have declared themselves its *servants*, and consider it an honour to make it presents (*ziarah*), bringing to it a constant supply of oil, honey, dried raisins, figs, fowls, &c. They also send sheep, goats, sometimes even money to the zaouia. The pilgrims, and above all those who implore a celestial favour, make rich presents. A family whose children are instructed at the zaouia give according to their means. There are also accidental profits; but the zaouias, not trusting the munificence of the voluntary system, have moreover landed property, which the founders have either settled upon them in estates belonging to themselves, or that they have obtained through the extinctions of the *Habous*.* They confide the cultivation of those lands to their own servants, or, according to the Arab custom, to farmers, who deduct a fifth of all the produce.

In case of need, they appeal to the piety of the believers, and the latter furnish them with a general contribution in labour (*touiza*). But the fixed revenues are nothing in comparison to the produce of voluntary offerings.

A zaouia which does not possess an inch of land may be much wealthier than those possessing the largest landed property.†

Each zaouia is placed under the authority of a supreme chief; and this authority passes hereditarily from male to male, in the family of the founder. When this family becomes extinct, all the tolbas of the zaouia assemble, when one of them is elected chief for one year only. If this person justifies the choice of which he has been the object, if he maintains his reputation for sanctity in the establishment, he retains his power, and becomes the stem of a new family of chiefs. On the other hand, should he prove unworthy, they renew the election every year, till it falls on a man really deserving of the situation.

It is the permanent chief of the zaouia who administers the smallest details, through his tolbas and servants; but when the chief is only an

---

* The *habous* is the donation of a fixed property to a religious institution, which is bound to yield a usufructuary maintenance to the testamentary heirs till their extinction, when it reverts *in toto* to the institution.

† The voluntary system amongst a religious people would naturally work better than with us.

annual officer, the tribes who serve the zaouia choose themselves the administrators of its property.*

It is well known that there are religious orders existing amongst the Mussulmans, and that they are scattered over Algeria. Amongst the Kabyle zaouias, only a small number belong to the Brothers (Kouan); we shall, however, say a few words concerning them.

The order which is by far the most widely spread is that of Sidi Mohammed Ben Abd-er-Rhaman, bou Koberein. This surname is founded on a marvellous legend, though recent enough. Sidi Mohammed had just died, and had been buried in the Jurjura, when the inhabitants of Algiers, where his virtues were in high repute, went to pray at night by his tomb. By some neglect, they were not watched ; and these people, through a pious fraud, appropriated to themselves the body of the marabout, which they placed near the road to Hamma, a little before you arrive at the Café of the Plantains,† in the spot where now stands the koubba of this marabout.

But this event was soon made known to the Kabyles by public rumours. They felt a terrible indignation at it ; and a long-enduring revenge would no doubt have followed, when they were luckily advised to examine the tomb which they possessed. They opened it, and, marvellous to relate, a second edition of the remains of the marabout was found there also.

The *Derkaouas*, or rebels, are the puritans of Mahometanism, and, like our dangerous liberals, always in revolt, and perpetually struggling against the authority of the Sultans and the social hierarchy.

In Kabylia they are especially found near Zamora, amongst the Beni-Yala. Their chief is an important man, Hadj-Moussa-bou-hamar (master of the ass), whom we have seen lately joining in the struggle against the Emir.

The *Anaya* is the sultan of the Kabyles : no sultan in the world can be compared to him ; he does good, and raises no taxes. A Kabyle will abandon his wife, his children, his house, but *never* his Anaya.

Such are the passionate terms in which the Kabyle expresses his attachment for a custom truly sublime, and which we find amongst no other people.

The *Anaya* bears some analogy to a passport and safe-conduct, with this difference, that the latter derive essentially a legal authority from a constituted power, whilst every Kabyle can give the Anaya ; and with this additional difference, that as much as the moral support of a prejudice may be carried beyond the watchfulness of the police, so much the security of him who possesses the Anaya exceeds that which a citizen may enjoy under the common guardianship of the laws.

Not only does the stranger who travels in Kabylia under the protection of Anaya defy all present violence, but he also braves for a time the

* Castellane, pp. 183-4 ; La Grande Kabylie, p. 67.   † See Part I. chap. vi. p. 106.

veugeance of his enemies, or the penalty due to his former acts. The abuses which might arise from so generous an extension of the principle are limited in practice by the extreme reserve of the Kabyles in making the application of it.

Far from lavishing the Anaya, they limit it to their friends; they accord it once only to the fugitive; they regard it as a counterfeit if it has been sold,—in short, they punish with death the usurped declaration.

In order to avoid this last fraud, and at the same time to prevent all involuntary infraction, the Anaya manifests itself generally by an ostensible sign. The man who confers it delivers as a proof of his support any object that is well known as having belonged to him, such as his gun or his stick; he often sends one of his servants, and he himself will not unfrequently escort his protégé, if he has any particular motives for suspecting that the latter will be annoyed.

The Anaya naturally enjoys a consideration more or less great according to circumstances, and especially it extends its influence according to the quality of the person who gives it. Coming from an inferior Kabyle, it will be respected in his village and in the neighbourhood. On the part of a man in credit amongst the neighbouring tribes, it will be renewed by a friend who will substitute his own; and thus it proceeds from neighbour to neighbour.*

Granted by a marabout, it knows no limits. Whilst the Arab chief cannot extend the benefit of his protection beyond the circle of his government, the safe-conduct of the Kabyle marabout extends even to those spots where his name would be unknown. Whoever is the bearer of it can traverse the whole Kabyle country, whatever be the number of his enemies, or the nature of the complaints existing against his person. He will only have to present himself on his route, successively, to the marabouts of the divers tribes; each one will be anxious to do honour to the Anaya of the preceding patron, and to give his own in return : thus the stranger cannot fail to reach the end of his journey happily, going from marabout to marabout. A Kabyle has nothing more at heart than the inviolability of his Anaya ; not only does he attach to it his own individual point of honour, but that of his parents, his friends, his village, his entire tribe, answer also morally for it. A man who would not find a second to aid him to take vengeance for a personal injury, could raise all his compatriots, if there were a question about his Anaya not being

---

* A very similar institution to the Anaya exists among the Circassians, by whom the protector is called *konak* (*Revelations of Russia,* vol. ii. p. 295). It may have arisen at the time when Arian Christianity shed its light over the desolate shores of the Black Sea ; but it is probably of more ancient date, and must be attributed to the gallant and generous character of these martyrs to European civilisation and orthodox covetousness. (See Bell's, Langworth's, and Spencer's *Travels in Circassia.*)—Another illustration of the Anaya may be traced in the custom of *tayo* at Otaheite, the Friendly Islands, &c., before the missionaries, the whalers, small-pox, and whisky, had done their work. See Cooke's *Third Voyage,* vol. ii. p. 139 ; La Peyrouse; *The Mutiny of the Bounty,* &c. &c.

recognised. Such cases must rarely occur, owing to the force of prejudice; nevertheless, tradition has preserved this memorable example.

The friend of a Zouaoua* once presented himself at his dwelling to ask for the Anaya. In the absence of the master, the woman, rather embarrassed, gave to the fugitive a bitch very well known in the country, and the man started with this token of safety. But the bitch soon returned alone, and covered with blood. The Zouaoua was greatly troubled ; the people of the village assembled, they followed traces of the animal, and discovered the dead body of the traveller. War was declared with the tribe in whose territory the crime was committed; much blood was shed; and the village compromised in this quarrel still bears the characteristic name of *Dacheret-el-Kelba* كلبة ال دشرات 'village of the bitch.' The Anaya attaches itself also to a more general order of ideas. An individual who is either weak or persecuted, or under the stroke of some pressing danger, invokes the protection of the first Kabyle he meets. He does not know him, nor is he known himself,—he has met his protector by chance ; but this is of no consequence, for his prayer will be rarely rejected. The mountaineer, delighted to exercise his patronage, willingly grants this accidental Anaya. Women invested with the same privilege, and naturally compassionate, seldom refuse to make use of it. They cite the case of a woman who saw the murderer of her own husband about to be butchered by her brothers. The wretched man, struck with many blows, and struggling on the ground, managed to catch hold of her foot, crying out, "I claim thy Anaya!" Whereupon the widow threw her veil over him, and the avengers let him go. It is known throughout Bugia, that in the month of November 1833 a Tunis brig was wrecked in going out of the roads, and that all the shipwrecked persons were put to death as friends of the French, with the exception of two Bugiotes, more compromised than the others, but who had the presence of mind to put themselves under the protection of the women. These scattered traits, which might be easily multiplied, prove that a great influence is given to sentiments of fraternity and of mercy among these people. Their existence in the midst of Mussulman society, invariably so severe in matters of justice, might cause some surprise, did we not remember that amongst a people very much distributed, under very little control, proud, and always in arms, and where consequently internal dissensions must abound, it was necessary that customs should supply the want of spies and police, in order to give security of transit to industry, commerce, and cheating. The Anaya produces this effect. It suppresses also many revenges, by favouring the escape of those who have excited them. In fact, it extends to all the Kabyles an immense net-work of reciprocal benefits.

It must be admitted that these people are very far removed from that

* A tribe of Kabyles inhabiting the Jorjora.

inexorable fatalism, that rigorous abuse of force, and that complete sacrifice of individualism, which have followed the march of the Koran every where throughout the globe. How is it, then, that here we meet with tendencies so much more humane, charitable impulses, sudden movements of compassion? Some respectable authorities consider them, with emotion, as a feeble gleam of the great Christian light which formerly illuminated Northern Africa, before the Church was developed into Catholic and evangelical perfection.

We have now given a broad sketch of Kabyle society.

We shall be much deceived if the picture speaks only to the eyes; it will clearly develop to the mind the great *mixture* of races and creeds which has been working for ages upon this obscure part of the African coast. A single impression results from this whole delineation, which it is easy to sum up. The natives whom the French have found in possession of the Algerian soil constitute really two nations. These nations every where live in contact, and every where an insurmountable abyss separates them; they agree only on one point: the Kabyle detests the Arab, the Arab detests the Kabyle. An antipathy so enduring can only be attributed to a traditional resentment, perpetuated from age to age, between conquering and conquered races. Corroborated by the indelible existence of two distinct languages, this conjecture becomes a certainty.

Physically the Arab and the Kabyle are so dissimilar, as to prove their diversity of stock. Besides, the Kabyle is not homogeneous ; he presents, according to the spots that he inhabits, different types, some of which betray the lineage of the barbarians of the north.

In their morals, also, there are varieties. Contrary to the universal results of the Mahometan faith, in Kabylia we find the holy law of labour obeyed, woman nearly reinstated in her rights, and a number of customs which, unlike those of modern Christendom, breathe equality, fraternity, and Christian piety. Some of these advantages may possibly result from the influence of the ancient Christian Church on the Kabyles, before pluralists and cant had defaced its fair form, and disgusted all honest and honourable men with the mask of sanctity allied to rottenness and atheism. Yet the greater part of their beautiful customs we would attribute to the palæological socialism of the primitive races on this planet, when men held converse with their God, when they entertained angels, and before the love of gold and glory had drawn a veil between heaven and earth.

The following customs among this interesting people, gasping for breath in the *accolade fraternelle* of France, have appeared to us worthy of record as memorials of Christian and classical antiquity.

The institution of *zaouia* has been minutely described in a former place; but we have reserved the account of one of its affiliated societies for the present occasion, on account of its remarkable approximation to Christian monasticism. A certain class of religionists among the Kabyles are called *derouiches* (detached), men detached from the world, and form a very remarkable sect, having striking points of affinity with the ascetic hermits of the Thebaid. In the country of the Beni-Raten, a distinguished marabout, Sheik-el-Mahdy, affects to lead his followers to a state of holiness by the following process :—Each candidate is rigorously shut up in a little cavern or cell, in which he can scarcely turn or stand upright. His food is gradually diminished during forty days, till at length it does not exceed one fig; some even bring themselves to take nothing but a carob in the twenty-four hours. In proportion as they gradually lose their relation with material life, the disciples acquire a second sight; they are visited with dreams from on high; and at last the mystical relation is established between them and the marabout, when their dreams coincide, and when they are visited by similar visions. When this crisis has arrived, the Sheik-el-Mahdy gives a burnouse, a haikh, or some other object as a sign of investiture, to the accomplished adept, and sends him forth into the world to make proselytes. There exist accordingly affiliated lodges, so to speak, of the great master lodge among the Beni-Ourghlis, the Beni-Abbas, the Beni-Yala, &c., amounting, perhaps, to about fifty. Their praxis is always based on the severest asceticism; and all pleasures, such as women, tobacco, &c. are scrupulously proscribed. The state of prayer and contemplation is perpetual among them.*

The philosophic inquirer into the phenomena of human nature might be inclined to attribute this institution to the spontaneous disposition that exists in certain individuals, in all ages and countries, for the mystico-ascetic life. He would probably remind us of the Hindoo yogi, and the bonzes of Buddha, and give them all a common origin in the instincts of the human heart. A local examination of the Kabyle institution gives a different version to the story;† and the same facts here again, as in so many cases, are adduced in support of the most opposite theories of the closet. Thus some authorities would persuade us that there is no sufficient evidence to establish the tradition current among them, that the institution is derived from Ali-Ben-Ali-Thaleb, the celebrated son-in-law of the Prophet; adding that it is quite certain that it was imported from Egypt by Sidi-Abd-er-Rahman, a disciple of Sidi-Salem-el-Hafnaoui; and reminding the reader that Christianity has left in Egypt the deepest traces of the mystical ecstasies and the prodigious abstinence of its cellular hermits.

It is the opinion of General Daumas, who has long been conversant with the subject, that the deeper you dive into the mysteries of the Kabyle life and society, the more traces you find of their having once been a Christian people. One of the strongest apparent evidences of this statement is found in their usages and customs, which have the force of laws. All other Mussulmans over the whole globe look to the Koran as the com-

* La Grande Kabylie, p. 69. † Ibid.

plete and universal code, embracing the whole life of man, and regulating the smallest details of public and private life. The Kabyles, on the contrary, observe particular statutes derived from their ancestors, and which they attribute to a pre-Saracen period. On many important points, such as the repression of thefts, murders, &c., these statutes do not agree with those of the Koran; they seem to approximate more to our penal notions; but a circumstance that appears to give conclusive evidence of their Christian origin is, that the name these statutes bear is *Kanouns.**

We have previously adverted to the prevalence of the sign of the cross, which is tattooed on the faces of the women in many parts of Kabylia. No less than three of the most eminent French authorities† have attested this fact. These fleshly inscriptions are an incarnate proof of the Christian past of many of the Kabyles, particularly such as are probably of Vandal origin. They are found especially among the tribes of the Gouraya, are probably a result of the Vandal invasion, and consist in the mark or sign of the cross on their forehead, cheeks, and the palms of their hands. It appears that all the natives who were found to be Christians were freed from the burden of certain taxes by their Arian conquerors; and it was arranged that they should profess their faith by marking the cross on their persons, which practice was thus universalised. These crosses do not exceed $\frac{15}{1000}$ of a metre (·58 inches) in size. The tattooing is of a beautiful blue colour, and is in much better taste than the patches worn by our grandmothers; its effect is far from displeasing on the faces of their women, who are remarkable for grace and simplicity.

Our final inference from these facts is, then, that the Kabyles universally have preserved strong traces of their primitive convictions and customs, which in certain cases and among certain tribes are clearly attributable to a Christian origin.

All travellers who have visited the hills and valleys of Numidia bear witness to the identity existing between the habits of its present population and those recorded by the pens of the classical authors. It is natural to suppose that, before the aboriginal Numidians and Libyans were driven to seek refuge in the fastnesses of the Atlas by the Arab irruption, they roamed over the plains at their feet, where the genius of the country would force upon them the same mode of life that is now led by their conquerors. And the two peoples being moreover families of the same Semitic variety, there would necessarily be but a slight difference between the habits of the pastoral Libyans of old and the modern Bedouins. Hence the fol-

---

* It will be evident to the reader that the resemblance of this word to the Greek χανων, rule, canon of the church, must be more than accidental. The expression has, however, long been used in Turkey. See Von Hammer's *Geschichte des Osmanischen Reichs*, band iii. p. 481, *Kanuni Raja*.

† Marshal de Castellane, General Daumas, p. 40 and Baron Baude; also Captain Kennedy, vol. i. p. 276.

lowing description given by Virgil of their mode of life admirably illustrates the habits of the wandering Arabs :

> Quid tibi pastores Libyæ, quid pascua versu ?
> Prosequar, et raris habitata mapalia tectis
> Sæpe diem noctemque et totum ex ordine mensem.
> Pascitur, itque pecus longa in deserta sine ullis
> Hospitiis ; tantum campi jacet. Omnia secum
> Armentarius Afer agit, tectumque, laremque,
> Armaque, amyclæumque canem cressamque pharetram.*

Flocks still constitute the sole riches of the southern tribes on the confines of the desert; hence they still preserve the nomadic habits of their forerunners, and the nature of the soil does not, in fact, admit of any other. Now, as in the days of Virgil, their flocks and shepherds plunge into boundless and shelterless solitudes; days, nights, months are passed in the pasturages ; and no change could be traced if bows, arrows, and quivers were substituted for guns, powder, and balls. Nor is the previous description inapplicable to many tribes of Kabyles in the present day, especially those inhabitants of the vast district of the Aouress and the clans of Little Kabylia or the Dahra, who lead chiefly a pastoral and wandering life, and whose principal riches and industry consist in herds and flocks and in the produce of the dairy.

Having now given an imperfect sketch of the physical, moral, and social characteristics of this interesting people, we pass to the Arabs, who still remain the dominant race in Morocco and Tunis, though they now lie prostrate at the feet of France throughout Algeria.†

---

* Georg. lib. iii.

† On the Kabyles see General Daumas's *Grande Kabylie;* Castellane's *La Kabylie,* p. 395 of his *Souvenirs;* Captain Kennedy, vol. i. chap. xiii. ; and Captain Carette's *Kabylie proprement dite,* 2 vols. of the *Exploration Scientifique.* See also, on the Berbers or Kabyles, the Appendix, p. 144, of Wilde's *Narrative of a Voyage to Madeira, Algiers, &c.* 1844.

Leo Africanus and Marmol deduce the etymology of *Berber* from the Arabic *barbar,* 'hot,' and from *Ber,* a proper name. Dr. Pritchard states that *Barbar* was an Egyptian name for the maritime country on the Red Sea. The Coptic Βερβερ, meaning *hot,* may be the root of the name, which is derived by Gibbon from Βερβωρ, meaning to cast out, *i.e. outcasts.*

CHAPTER XV.

The Arabs.

AGRICULTURISTS AND BEDOUINS—TENTS—FURNITURE—WOMEN—DISTINCTIONS OF ARAB LIFE—PATRIARCHALISM—FEUDALISM—DOUARS—HORSES—FALCONRY—ILLUSTRATIONS—MARKETS—LEGENDS—SCRIPTURAL CUSTOMS—THE ARABS OF CONSTANTINA—ADMINISTRATION OF THE TRIBES—BEDOUINS—OFFICIALS—STATISTICS—BUREAUX ARABES.

THE Arabs are in general agricultural or pastoral in their mode of life. This difference of pursuits begets a difference of characters and of manners. In the former case, their stationary habits reduce them into subjection to a regular form of government, and they present a social state approaching our own. The Arabs of this class are the descendants of those Saracen hosts who, under the first caliphs, took possession of a great part of Africa, and even invaded Spain. The pastoral Arabs are only bound to the soil by a transitory interest; pitching their tents at random, they are not the slaves of any cumbrous law-machinery, and they lead a mode of life foreign alike to that of polished and of savage nations. Hence their interest. The latter class constitute the Bedouins,* or nomadic Arabs, who are the principal inhabitants of the vast plains and deserts that stretch over North-western Africa, and who, though divided into independent, and often hostile tribes, form but one people, as is evidenced by the community of language subsisting among them.

The Arabs of both classes are of the middle height, and remarkably strong. Their physiognomy is expressive; they have a quick and animated look, and brown or olive complexions, but seldom black, like that of the negroes. The type of the women is the same as that of the men, whose manly faces are more oval than those of the Moors, with much more prominent, but less agreeable features. Their step is light and elastic, and their attitude often recalls the nobleness of antique statues. Their hair is generally black.

Extremely adventurous and daring, the Arabs meet their enemies in the field with assurance; they treat the vanquished with harshness, but

---

* The word Bedouin, pronounced *bedaouy*, written بدأوى, comes from *bedou* بدو desert. Gorguos, *Cours d'Arabe vulgaire*, vol. i. p. 183.

without indulging in the cruelties practised by the Berbers. Their habitations are very well built of branches of trees kept together by cement, and occasionally consolidated by unhewn stones, which, however, are made to fit together perfectly; these huts are grouped to the number of ten or twelve, and sometimes even of thirty or forty, forming villages, surrounded by hedges of cactus growing to a great height. In the midst of this group stands the hut of the scheik, or chief of the tribe, and a mosque, which is nothing but a lodge built like the others, only of larger dimensions.

In speaking of these habitations, our remarks must be confined to the first class, or that of the agricultural Arabs; for the Bedouins* live invariably in tents, named *hymas* or *kymas*. A collection of these hymas, which are generally placed in a circle of ten, twelve, or fifteen, forms a *douar.*† These tents are composed of black or brown stuff, are of an oblong form, and supported by stakes, which moreover answer the purpose of suspending clothes, arms, harness, &c. No beds are found in them, the Bedouins rolling themselves up at night in a haikh. The middle of the douar is commonly empty, like a court; and each family possesses in general two tents, one for the family, the other for the cattle.

The simplicity, or rather poverty of the family is remarkable, their household only comprising the following articles: some camels, goats, and fowls, a mare and its harness, a tent, a lance thirteen feet long, a curved sword, a musket, a pipe, a pot, a hand-mill, a coffee-pot, a mat, some clothes, and some gold or silver rings for the women's wrists and ankles. With these the Arab is rich.

The best clew to unravel the mysteries of Arab life and manners will be found in their religion. We shall soon go astray in estimating their character, if we lose sight of this mainspring of all their thoughts and actions. Unlike the anxious, bustling, and prosaic populations of modern Christendom, the Arab still holds to the faith of his sires with a glowing devotion; he sees the arm of the Lord and of his angels in all the accidents of time; and conscious of the measure of man's power and days, reverence and submission become the predominant elements of his nature. This feature of the Arab temperament constitutes what will be regarded as puerile weakness or sublime philosophy, according to the favourite bias of the critic; but all who have observed Mussulmans in general, and the children of Arabia in particular, under the stroke of affliction and in the hour of death, bear witness to the manly resignation and dignified bearing that they display in those seasons of distress and trial. The wisdom of this world and the metaphysics of the nineteenth century having decided that the laws of nature are the supreme arbiters of all things, and that the idea of a special providence is an idle dream, we can only regret that the

---

* For an excellent estimate of the Arab character and of Mohammed, see Sismondi's *Fall of the Roman Empire*, vol. ii. chap. 13. See also *Diary of a Tour, &c.* vol. i. p. 52

† *Douar* comes from the Arabic word دار *dar*, house; diminutive دويرة *douira*, little house.

happy scepticism of our free-thinking Europe cannot imbibe some of the comfort and the faith of the Oriental and the Arab, without falling victims to the deplorable heresy of predestination and other mysticism.

Notwithstanding their faith, always ready to fight, the Arabs go about armed *cap-à-pied*, with a musket in a sling, a yatagan, and pistols. Every man must bear arms; and in some cases women and children do so too. Their mode of fighting resembles that of the Berbers. They ride up to the foe in groups, and drawing near, they break into a gallop describing an eccentric curve. After reaching the farthest point they fire, and ride back to the main body to load and dart off anew. If the affair grows serious, and they must come to close quarters, they put their gun in the left hand, draw the yatagan, and set-to bravely. When celebrating a fête, they are fond of mock fights resembling tournaments, and called fantasias.

Those too poor to have a horse are armed with muskets, blunderbusses, yatagans, and clubs. They are good pedestrians, and stand privations well. It may be said that the common people live in perpetual misery.

The founders and the first-born of Islamism, the Arabs are sincerely religious, though some are negligent in their devotions; they are extremely superstitious,—suspend wooden hands to their children as a talisman against the evil eye, and amulets on themselves and their animals, consisting of bits of parchment with texts from the Koran, &c.

They are laudably respectful to the aged, who, if infirm or blind, are always escorted by one or two lads. The Bedouin cemeteries are rather neglected; but if one of their warriors dies, all his relations congregate on horseback and celebrate his obsequies for eight days around his tomb; and they will encounter almost any risks to carry off the fallen in battle.

The rich Arab women dress sumptuously, wearing chemises of fine linen, drawers, and a kind of silk vest, over which they place a long coloured robe reaching to the knee and having large sleeves. In ceremonies they throw over them a long red or blue cloak, fastened round the shoulders by silver hooks; and they have anklets and bracelets of the same metal. The Bedouin women are commonly ugly, tattooing and painting their faces, breasts, feet, and hands blue; which, added to their dirtiness and sweat, makes them horrible. They look on the patterns tattooed as ornaments, or rather national crests, pricking them into their skin with needles made on purpose. They leave their faces uncovered; and it is only on long journeys that they wrap them over with a piece of linen. The dress of the common Bedouin women consists of a long chemise of white wool, with short sleeves, and a rope for their girdle. Their hair, rolled up on the head, is surrounded with a great red cord in a few coils, imitating a turban; but frequently long tails of hair fall on their shoulders, while other smaller ones hang on their foreheads, with bits of red riband tied to the end of them. Their woollen chemise, hanging loose like an apron, is kept straight by an immense copper pin and ring fore and aft; and massive copper rings adorn their ears.

They commonly look miserable, withered, and old when still young; yet some girls of fifteen display the beauty of regular features and the comeliness of youth.

They have little feeling for each other; and in the case of accidents at Algiers, when five or six Arabs have been buried under ruined houses, &c., not an Arab, even a relation,. has been seen to raise a hand to help them. Fatalism and their want of socialism explain this.

Hospitality is, according to some severe writers, only a name amongst them, and the power of the sheikh cannot protect travellers among them from theft. Algeria is roved over by hungry Bedouins anxious to pounce on unprotected males and females; and it is added, that in dealing with men of different religion, they, like many Christians, do not scruple to lie and cheat.*

In receiving these and other statements of French writers about the Arabs, it is necessary to observe great caution, as it is the interest of the conquerors to represent their victims in the most odious light possible, in order to justify their own injustice and cruelty.†

The reader has already learnt much in the preceding chapter regarding the contrasts between the Kabyles and the Arabs, or the primary and secondary strata of Algerian population. Summing up these differences, Baron Baude has happily expressed them in one sentence : "In short, the Kabyles are the vanquished, the Arabs the victors—hence their hereditary hatred." The story of the Cumri and Sassnach is wide as the poles, and Ireland's complaint is an eternal truth.

The Arabs appeared in the seventh century, when they finished the conquest of the Roman establishments in Africa, so well begun by the Vandals, and upset the dominion of the latter. The superiority of their cavalry made them masters of the plains ; whilst the mountainous regions, where the attack was less easy, and the defence more feasible, remained in the possession of the oldest inhabitants. The limits of the Arabian establishment have thus been cut out according to the irregular relief of the territory ; and this tide of humanity has spread and been broken, almost like a fluid that has only reached a certain level.

The Arabs, or Saracens, were organised like all other peoples that have installed themselves by force of arms in a new country. They consisted of chiefs and soldiers, practising the command and obedience of feudal institutions ; which were, and still remain, identical in the barbarous form of government and society once prevalent in Europe, and in that now dominant in north-west Africa. These institutions were preserved, from the necessity felt by the masters of maintaining a firm footing among a conquered people ; and the Turks subsequently made use of them also.

---

* Berbrugger, part iii. ; Captain Kennedy, vol. i. p. 144, vol. ii. pp. 203-214.

† See the Chapter on History, and *Diary of a Tour in Barbary*, vol. i. p. 277.

In short, feudalism is the organic law of transitive races in the early stage of development. Thus, each tribe forms a little state, subject to an inflexible hierarchy. Power is hereditary among them, and military service is looked upon as a tribute due, in the same way as the fruits of the land and taxes. Differing from the Kabyle tribes, whose government is fluctuating and various, the Arab form of society seems to flow from one single principle.

Nevertheless, the Arabs of Barbary are divided into two sects, those of the east and west : the former profess the rite of Hanefi ; those of the west the rite of Maleki. The Hanefis acknowledge the Sultan of Turkey as their spiritual head, and the latter bow to the Emperor of Morocco ; and in countries where creeds are the great causes of separation between populations, this difference has important consequences.*

The Arab tribes may be divided into three classes: those inhabiting the Tell ; those holding the plateaux in the more elevated districts; and, thirdly, the Djeridi of the oases. The first, who are agriculturists, inhabit that part of Northern Africa called the Tell, bounded by the Mediterranean to the north, and often by the mountains of the Lesser Atlas to the south, though, as we have previously seen, the district called Tell stretches farther inland in the east than in the west of Algeria. This country is in general very fertile, with good crops. The second class, belonging to the pastoral society, live in the plateaux between the Tell and the oases, which, though not so rich in grain, afford very good spots for pastures : they also roam over the vast plains of the Sahara. The third class inhabit the ksours, and carry on the barter and carrier trade of the interior. A simpler and shorter division is that into Tellians and Saharians, previously noticed.†

The influence of blood-relationship, aristocratic government, and the love of roaming, are common to all these classes and subdivisions. The Tellians, being agriculturists, are less addicted to roving than the Saharians, who, being shepherds and carriers, are always on the move for fresh pastures or for speculation. Many tribes are powerful and numerous; but war, disease, &c. have much reduced others. Heads of families are treated with great respect by their offspring, who, settling around the patriarch with their slaves, form a douar or circle of tents, خَيْمَة kyma, of which he is sheikh شَيْخ,‡ having an independent authority over its domestic economy. Several douars uniting for safety form a *farka,* and the sheikhs form together a *djemaa* (or council) to watch over the common interests of the farka ; one amongst them, on account either of his superior nobility, age, intellect, or energy, being generally appointed the head or president of the assembly.

---

* Baron Baude. † Capt. Kennedy, vol. i. pp. 203-212.
‡ Meaning elder, senior, signor, seigneur, lord.

Nobility of blood is much respected among the Arabs, who are a nomadic phase of the pastoral society. All descendants of Fathma, the daughter
of Mohammed, and of Sidi-Ben-ebn-Thaleb, his brother, are regarded as
noble, and are called *sherif* or *sidi* (meaning lord or master). Amongst
other privileges, they can only be judged by their peers. The descendants
of the tribe of the Koraiche prophet, and those of the first invaders of
Western Africa, are also noble, but of the military class. The marabouts*
are the lords spiritual, whose influence is far greater than that of all others :
they are commonly men of austere lives, devoted to the study of the Mahometan law; and they are reported not unfrequently to have the gift of miracles and prophecy. Surrounded with a halo of holiness in life, after death
their koubbahs or tombs become places of prayer and pilgrimage, besides
being sanctuaries for criminals. Upon the woody slopes of the Atlas and
the large plains of Algeria the white domes of these sacred sepulchres often
attract the eye of the wanderer. The marabouts frequently unite and form
a douar, or even a farka, near some chapel erected to the memory of a
deceased member of the fraternity. There they instruct youth in the law,
&c., forming a zaouia, such as we have described in the last chapter.

Before we give an outline of the present administration of the tribes,
we shall introduce another description of a douar, or village of tents, from
the pen of Baron Baude :

"The arrangement of all the douars is similar, consisting of about
20 huts or tents, according to the season, one of which is devoted to each
family. The tent is made of a black and very thick woollen tissue, which
swells with the damp and keeps out the rain, requiring much labour in its
manufacture. The weather being very fine during two-thirds of the year,
they only require a roof of branches, supported on pickets of wood, for
their huts, brushwood being piled up on the weather side. These huts,
placed at about 10 metres apart, form a circle, with the cattle in the centre,
and contain numerous savage dogs as guardians. The douars are moved
when the neighbouring pastures are exhausted, seldom remaining in one
place above three months together. The great quantity of dung accumulated by their cattle forms the only manure they employ."†

Baron Baude describes an Arab entertainment in the douar of the
Merdes, near Bona, in the following terms : "After sunset, the Mussulmans take a meal ; and the preparations for ours were being completed
as we came back from our walk. Mats were laid on the ground, and all
the guests crouched round, excepting the host. The meal was served up in
pots made of old wood, and consisted of hard eggs, honey pancakes, boiled

---

* Borrer, ch. 16, derives the word from the Arabic *rbth* (Daumas says mrabeth, *liē*),
'to devote oneself;' the participle of the verb *mrbth*. M. Gorguos describes the word as
m'rûbot مرابط participle of the 3d form of the verb *r'bot* ربط 'to tie.' *Cours d'Arabe
vulgaire*, vol. i. p. 237.   † Baude, i. p. 174.

U

A–K

fowls, and cous-coussou. The pancakes were to be dipped into a copious sauce of an ochre-colour; and Mahmoud (the sheikh) began to stir up this mess with his greasy paws, which induced the Baron's party to decline taking any. The fowls were awfully peppered; but they found the cous-coussou excellent.* After the meal, some water and soap were brought, all washing their hands and beards in it, and some of the company rinsing their mouths with the same foul water.

A night in a douar† is distressing to Europeans, the fleas and mus-quitoes allowing their victims no rest. This veritable plague is so delete-rious, by depriving you of your rest, that it greatly debilitates the French troops, colonists, and visitors, rendering them unfit for work and ill, many having died in consequence. The Arab women anoint themselves with oil to keep off these enemies. Hard is the life and sad the slavery of Arab women, like all females in the pastoral phase of society. They go often to the wells, carry heavy loads of wood, have to grind the corn unceasingly by day, whilst at night they often have no rest, being obliged to spin wool, and weave the cloth for their tents. The Arabs are very jealous of the effects of civilisation in emancipating their women ; but Baron Baude informs us that the ties of family are not felt among them. This remark evidently re-quires qualification, familism being the pivot of pastoral society; but he pro-bably alludes to the spiritual ties and tender regard for women appertaining to the Christian and Germanic phase of civilisation being unknown to them.

Madame Prus gives the following description of an Arab encampment near Bona in 1850 : "The tents were very low, and of an extremely thick tissue of camel's hair, by which they were enabled to resist the scorching rays of an African sun, and also the torrents of rain which inundate the whole country during the wet season. Not a single tree, not a plant, re-lieved the eye from the monotony of this arid spot. One of the extremi-ties of the plain was bounded by the Bourzizi mountains, whose rocky summits appeared in sad harmony with the still and deathlike face of surrounding nature, though the valleys which are occasionally found among the table-lands offer a scant pasturage for cattle.

"The tent of Abdallah was divided into two rather large apartments. One of these was occupied by his family, wives, chickens, cats, and dogs. The other, into which I was introduced, was his hall of reception, which was also used as his dormitory when he wished to sleep alone, and became

---

* Baude pronounces cous-coussou better than English puddings, and a good addition to European cookery-books. It forms the bread, soup, bouilli, and dessert of the Arab, and is made of wheat bruised by the women in hand-mills, and then thrown into a great vessel shaped like a kettle-drum, a little oil being mixed with it, till it forms lumps of the size of millet grain; after which it is boiled over steam, and mixed with milk, broth, butter, &c. (vol. i. p. 176.) For Arab life, see Capt. Kennedy, vol. i. pp. 138, 205-212 ; and *Diary of a Tour*, vol. ii. pp. 184-9.

† A douar is called a *smalah* when it is the residence and contains the household of a noted chief. Kennedy, vol. ii. p. 99.

his private sitting-room, in which he smoked from morning till night, cross-legged, according to the fashion of the Arabs."*

You see their horses to the best advantage in the province of Bona. The equipment is well known. They have a saddle without crupper, with the high pummel and cantle of the Mamelukes, bridles with blinkers and chain bits, besides very short stirrups placed farther back than ours. In riding the leg is accordingly much bent; and they use iron spikes, 15 centimetres (5·85 inches) in length, instead of a spur. The seat is very fatiguing to those unaccustomed to it, nor would it answer for trotting—a pace never used by the Arabs, who stand in their stirrups at a gallop, leaning slightly on the top of the cantle. Common people ride bare-legged; but the sheikhs wear red morocco boots, in shape something like those of our knights of old: they have a very good appearance.† The Arabs are like children, and abuse their horses; whence they indulge in the fantasia, a kind of mock fight to welcome a guest, goading their steeds at full speed with bleeding flanks. Fine horses can be obtained from Tunis for from 500 to 800 fr. (20l. to 36l.), with large limbs like English horses; and a good breed of this sort is found at La Calle, and is in request for the 3d regiment of the Chasseurs d'Afrique, at 242 fr. (9l. 14s. 2d.) each. Castellane, in his trip through the Sahara, writes: "In the course of our ride we admired, as ever, the boldness of the riders, and the beauty of their steeds. We were especially struck with one mare of Mohamed, the friend of our friend Rhaled, so light that it might, according to the Arab phrase, have galloped on a woman's bosom. As we were praising its beauty, Rhaled said to us: ' She had a sister, who alone could contend with her. They were the envy of all, and the pride of their master, when Mohamed was led up a prisoner by the horsemen of the Emir. He managed to escape; but scarcely had he reached his douar, ere the chaous‡ of the Sultan were signalled as coming. Mohamed immediately vaulted on his good mare; and when the horsemen arrived at the tent, they found the chief flown. It was impossible to overtake him; yet one of them, as the only chance, leapt from his horse and ran to the other mare that was still tied by a rope; but Mohamed's boy shot her dead with a pistol. This mare alone could overtake her sister, and the child had saved his father.' "§

A mare is looked upon as the best property, as a fortune.|| She is

---

* P. 91.

† Baron Baude, vol. i. p. 203; Dawson Borrer, p. 18. Lamping informs us that the Arab chiefs consider the skin of the tiger(?) and panther as one of their principal ornaments. The head of the animal is generally fastened to the saddle-bow (the head and teeth are essential), and the skin waves to and fro with every motion of the horse; so that at a distance one might almost imagine that some wild beast had just taken a deadly spring upon the rider.—*The French in Algiers*, p. 9.

‡ Constables and executioners.   § P. 256.

|| The Ulemas relate that when God wished to create the mare, He spake to the wind: " I will cause thee to bring forth a creature that shall bear all My worshippers, that shall be loved by My slaves, and that will cause the despair of all who will not follow My laws." And He created the mare, saying: " I have made thee without an equal; the goods of this

preferred to the horse, entire or gelding; for she does not neigh, she is docile, and supplies milk on an emergency. All who can afford it possess one of these animals, and they pass hours in looking at them. They are, in fact, the Arab's companion and friend, sharing his adventures, perils, and wars, and bearing him from danger with the speed of the wind. Accordingly, wonderful is the care their owners bestow on their grooming and toilet, combing their manes and tails coquettishly; and after washing their legs, smoking their pipes in ecstasy while they gaze at and admire them. When idle the grass of the pastures is their food, but under toil they are indulged with a light feed of barley.

The Arab race of horses is admitted to bear off the palm, the Arabs boasting that they proceed direct from the studs of Solomon. But the Bible contradicts this Bedouin boast, for we read that Solomon "had a thousand and four hundred chariots, and twelve thousand horsemen . . . and Solomon had horses brought out of Egypt."* Hence the great king had no studs, Solomon being satisfied with improving his proverbs.

However this may be, the beauty of the Arab breed is proverbial, which they attribute to the care they take to prevent misalliances, and to their genealogical trees. The birth of a foal always takes place before six witnesses, who all sign an act of nativity in a proper form. It is an article of belief with the Arabs, that they would be unlucky in this life and punished in the next, if they practised any deception in all relating to the pedigree and parentage of the horse.

English jockeying might profitably study the ethics of the Arab turf.

Mediæval manners are still dominant amongst the Arabs, and our chivalric ancestors would feel quite at home in the Sahara. Falconry is quite in fashion among the chiefs, and is thus described by Castellane: "Farther on, two hares, frightened by the sound of our horses, darted from their cover, and the falcons were let loose again. As long as the hare can run, it escapes its enemy; but when it begins to waver, that it may seek a refuge, the bird darts on its back and begins eating its brain and its eyes. Falcons are like men; some are good, and others bad. It was good fun to hear the Arabs banter, jeer, and abuse the latter; and it was amusing to see the pride of the proprietor of the best bird. It is during the summer that preparations are made for the winter hunts. The bird when first learning to fly is caught by the fowler; and even before it is tamed, it is taught to run after its prey: it is initiated at first into easy hunts, and taught to wait for its master's order, to recognise his voice and his signal, to dart at the skin of a hare thrown into the air, and to answer different cries, which the voracious bird obeys with an unparalleled

world shall be placed between thy eyes; every where I will make thee happy and preferred above all the beasts of the field, for tenderness shall every where be in the heart of thy master; good alike for the chase and retreat, thou shalt fly, though wingless; and I will only place on thy back the men who know Me, who will offer Me prayers and thanksgivings; in short, men who shall be My worshippers."—*Castellane*, p. 255.

* 1 Kings x. 26, 28.

ardour. In this manner the falcon of the Arab becomes once more the bird of the middle ages, surrounded with attentions, with glory, and even with honour.

" The chiefs had their right hand armed with a glove called *smegue*. This glove has no fingers. The Arab exquisites have them made of tiger and panther skin. On this the falcon perches; not unfrequently two or three find room,—one on the shoulder, and another on the strings of camel's hair surrounding the cowl of the haikh."*

Here follow a few illustrations from the pens of eye-witnesses.

The following night-scene described by Baron Baude, on his trip from Bona to La Calle, gives a good idea of the poetry of Arab life : " The moon in all her splendour silvered the surface of the lake, and its light was mixed under our black tent with the glare of the fire of dry rushes. Above the heads of the crouching Arabs stretched out those of the Numidian horses, the faithful companions, who seemed to take part in their recreations. The animated faces of our hosts, as attentive to the story being told as to the roast mutton, seemed lighted up with the departed elegance of their race ; and we, the descendants of those barbarians who learnt civilisation from them, were almost envious of their present condition."

A good insight into Arab life is given in Marshal Castellane's description of a market held every Thursday at the little magazine-post of Khamis, among the Beni-Ouraghs, built, like all such French posts in Algeria, on a line parallel with the sea near Mostaganem. Markets in Africa are not only places of sale, but bazaars of news ; and the whole population, Arab and Kabyle, frequents them. On market-days, breaking their repose and silence, multitudes of Kabyles and Arabs were seen trooping in from all sides, from the mountains and the valleys, from every path, some driving sheep, others horned cattle, many carrying loads of corn, beans, wool, or manufactured stuffs, but all armed, and many with their muskets only, or that knotty stick, one blow of which can break the hardest heads. Jews, with dirty turbans, drive in their half-starved mules, displaying their goods at the spot pointed out by the caid and police, and erecting a little tent of bad cotton to guard them from pillage. The first hours were usually devoted to business. The butchers skinned the sheep they had killed, uttering *bis-millahs*,† and suspended the flesh to little fir-trees, whose branches served as skewers ; cattle-dealers were standing near their beasts, awaiting purchasers ; the corn and bean merchants were shouting and quarrelling about a halfpenny ; but the noisiest of all was the Jew. As every where else, he was here the agent and the jobber, over-reaching, selling, and stealing. In Algeria the Jew supplies cotton, pepper, cloves, sugar, and coffee ; antimony for the women's eyes, *henna*

---

* Page 255.

† *In the name of God :* an invocation always employed in slaughtering any thing in Mussulman countries.

for their nails ; a gunsmith, he mends their arms; he repairs rings and makes jewels, and silver ornaments for the chiefs. Nothing comes amiss to him ; he crawls through all trades. You may see him every where, hurried, agitated, thrusting out his dirty hand, greedy, quarrelling, indefatigable, asking for justice from the caid, whose decision is law. A volley and tempest of shouts, eloquence, and special pleading, is at once silenced by his verdict.

The first hours and business over, the hum of men increases. The groups thicken, the state of affairs is canvassed,—sometimes general politics, at others disputes between tribes.

Envoys of the Emir gliding in among these groups used to fan the flame of rebellion at the market of Khamis; and the religious fraternities of the *zaouias* exchanged the messages confided to their fanaticism.*

The religious colouring and phase of the Arab character is finely developed in the following description of Castellane : "In the winter of 1841-42, whilst General de Lamoricière, on the side of Mascara, was striking rude blows at the power of Abd-el-Kader, the authority of the khalifat of the Emir, Bou-Hamedi, was seriously shaken in the west of the province. Mouley-Chirq-Ben-Ali, of the tribe of the Hachem, had been the instigator of this movement. His influence was great, for he had long commanded the country as lieutenant of Mustapha-Ben-Tami, an ancient khalifat of the Emir.

Dismissed from office by Bou-Hamedi, when the latter replaced Mustapha-Ben-Tami, Mouley-Ben-Ali had sworn to avenge himself ; and he kept his word in the following manner. Ben-Ali was patient in his vengeance ; he knew how to await the hour and the moment. His first care was to go the round of the tribes, and by his words to prepare their minds for a change ; and then as soon as the moment seemed favourable, feeling that his own authority was not sufficiently strong for him to raise the standard himself, he cast his eyes on a man whose religious prestige might increase his power. Si-Mohamed-Ben-Abdallah, of the great tribe of the Ouled-Sidi-Chirq, was chosen by him. The religious influence of this tribe of marabouts extends from the oasis to which they have retired to the sea-shore. Having been established many years in the country of Tlemsen, Mohamed-Ben-Abdallah enjoyed a great reputation there. His piety was proverbial ; and the people of the douars related, that every Friday he went barefooted to the tomb of Si-Bou-Medin, passed the night there in prayer, and that the words of God came from his mouth when he quitted the holy place, because the Spirit from on high had visited him. This belief soon became general, and all were prepared to recognise him as chief. The old chief Mustapha-Ben-Ismail, informed of the agitation which reigned on the side of Tlemsen, knowing that Bou-Hamedi began to conceive serious alarms, and had not been able to succeed in getting possession of Mohamed-Ben-Abdallah,

* Castellane, p. 181.

thought that the marabout might serve as a powerful lever wherewith to attack the Emir. Upon the report of Mustapha, General de Lamoricière authorised our old ally to put himself in relation with Mohamed-Ben-Abdallah.

Assistance and protection were guaranteed to him, and their first interview was arranged ; but on the 3d of December, at the very moment when it was about to take place, Bou-Hamedi stopped the advance of Mohamed-Ben-Abdallah.

Three weeks later, recovering from this check, Mohamed demanded another interview ; and Colonel Tempoure, supporting the *goum** of Mustapha with a little column of infantry, set out in the midst of tremendous weather. On the 28th, accompanied only by some officers and people of Mustapha, he marched forth to meet the new chief.

The horsemen extended in long files over the precipices of a lofty mountain ; at their feet the valley of the Tafna spread itself out, with its rich cultivation. On the horizon appeared the white walls of Tlemsen, the town of the sultans. Suddenly, at the winding of the mountain, they perceived the hills and undulations covered with the people of the tribes. The standards on both sides halted, the horsemen remained immovable, and the chiefs advanced between these living hedges. Mustapha first put his foot to the ground; he thus rendered homage, in the presence of all, to the religious character of Mohamed-Ben-Abdallah : but the latter dismounting, pressed him in his arms, without permitting any other mark of deference. Those who assisted at the interview have since related that General Mustapha, after having bowed before the French chief, Colonel Tempoure, pronounced these words : " This is the day of my life which has afforded me the greatest happiness, for I perceive that my endeavours have succeeded in establishing esteem and friendship between the French and so venerated a character. Thanks to the omnipotent God, this day is the commencement of the union which ought to be sealed between the two races, under the protection of the Great Sultan of France. As for myself, my few remaining days cannot be better employed than in labouring for the peace of the country and the elevation of the house of Mohamed,— of thy house, O Mohamed, already so illustrious amongst us."

Mustapha then, with that dignity which never deserted him, pointed to a clump of dwarf-palms ; and all seating themselves in a circle, the conference commenced. It was short ; and the conditions were soon settled. The last addresses having been delivered, Colonel Tempoure presented the Arab chief with the donations brought to do him honour. All then rose. The chiefs mounted their horses, and kept united round Mohamed; whilst standing up in his stirrups, the marabout pronounced the prayer which was intended to call down the Divine blessing on their enterprises. His eye was burning, his features pale and worn by fasting and vigils, his voice

* A contingent of irregulars furnished to government by the tribes.

deep and hollow. It was an imposing and majestic spectacle. " O gracious and merciful God," exclaimed Mohamed, "we entreat Thee to give peace to our unhappy country, laid waste through a cruel war." And the voices of the two thousand horsemen repeated at the end of each line : " O merciful and gracious God, we entreat Thee to give peace to our unhappy country, laid waste through a cruel war." " Have pity," continued the chief, raising his eyes towards heaven, "have pity on this population reduced to misery ! Grant that abundance and happiness may again be restored to us! Give us the victory over the enemies of our country ; and may the holy religion revealed by thy Prophet be always triumphant!" And with one voice all the warriors responded : " Give us the victory over the enemies of our country, and may the holy religion revealed by thy Prophet be always triumphant !" The murmur of these prayers reached even the horsemen of Bou-Hamedi, announcing to them the greatness of their danger. The hour was in fact approaching when Tlemsen was for ever to become French.*

Fairy and religious legends are dear to the Arab heart, amongst which the following combines a Semitic mysticism with a European rationalism, and is not unworthy of the wisdom of Solomon.

Sidi-Mohamed-ou-Allal, be it known, was a man of God celebrated for the pious legends which he loved to relate. The following is one which the traveller never fails to hear, who stops for the first time near the venerated marabout the last abode of the holy man :

One day Sidna-Aissa (our Lord Jesus Christ) met Chaytan (Satan), who was driving four asses before him, heavily laden, and said to him : " Chaytan, why, thou art become a merchant, then ?" " Yes, Lord; and I've so much business on hand that I cannot do justice to it."

" What business do you carry on, then ?" " Lord, an excellent business; just see. One of these asses—and I choose them amongst the strongest in Syria—is laden with injustices; who will buy them of me?— the sultans. The second ass is laden with envies; who will buy them of me ?—the learned. The third is charged with thefts; who will purchase them ?—the merchants. The fourth carries, together with perfidies and wiles, an assortment of seductions, which are related to all the vices; who will buy them ?—the women."

" Wicked one, may God curse thee !" replied Sidna-Aissa. " What is that to me, if I gain ?" replied Chaytan.

The next day Sidna-Aissa, who was saying his prayers at the same spot, was disturbed by the swearing of a donkey-driver, whose four asses, overwhelmed by their load, refused to go on; and he recognised Chaytan by their load. " Thank God! thou hast sold nothing," he said, addressing him. " Lord, an hour after you left me, all my panniers were empty; but, as usual, I had difficulties about the payment."

"The Sultan caused me to be paid through his khalifa,* who wanted to cheat me about the sum. The sages said they were poor. The merchants and I called each other thieves. The women alone paid me handsomely, without bargaining."

"And yet I see thy panniers still full," objected Sidna-Aissa. "They are full of money; and I am carrying it to the kadi (to justice)," replied Chaytan, driving on his asses.

"O my brothers," added Sidi-Mohamed-ou-Allal, "the free man, if he is grasping, is a slave; the slave is free, if he lives on little. Choose tents to repose in, and for your last dwelling the cemeteries. Nourish yourselves with the produce of the earth, satisfy your thirst with the running water, and you will leave the world in peace."†

Many prophetical legends are current amongst the Arabs relating to the rise and fall of the French power in Algeria. A holy marabout, Si-Akredar, many years before the conquest, had announced it in these widely-diffused couplets.

"Their arrival is certain in the first of the 70th; for by the power of God I am informed of the matter. The hosts of the Christians shall come from all sides; the mountains and the towns shall shrink from us. They will come from all quarters, horsemen and foot; they will cross over the sea.

"They will descend on the strand with a host like a raging fire, a flying flash.

"The hosts of the Christians shall come from the side of their country; verily it will be a mighty kingdom that shall send them forth. Verily the whole country of France shall come. Thou shalt have no rest; and the cause shall not be victorious. They will arrive like a torrent in a dark night; like a cloud of sand driven before the wind.

"They will enter by the eastern wall. The churches of the Christians will be raised; the thing is certain; and then thou shalt see them spread their doctrine.

"If you wish to find protection, go to the land of Kairouan; if the Christian hosts advance, and their coming is certain. And the Christian expedition will smite Algiers, and they will spread themselves abroad there. They will rule over the Arabs by the all-powerful command of God; the daughters of the land will be in their power.

"After them will appear the powerful of the golden mountain; he will reign many years, according as God shall will and ordain. Every where, all inhabited places shall be in anguish, from the east unto the west. Verily, if thou livest, thou shalt see all this."‡

The departure of the French has been similarly announced by Si-Aissa-el-Lagahouati, another venerated marabout, in these terms:

* Lieutenant.
† Le Grand Désert of General Daumas, pp. 50-51. Castellane's Souvenirs, pp. 276-7.
‡ Castellane, pp. 128-9.

"Publish, O crier, make known what I saw yesterday in a dream.

"The coming evils shall surpass all imaginable evils; the eye of man shall never have seen the like. The man shall abandon his child. A bey shall come to us subject to the Christians. His heart shall be hard; he will rise against my master, a man of noble origin, whose heart is gentle, who is handsome and prudent, and whose commands are just.

"Make known and say: 'Quiet yourselves; he who has come hath dispersed them; they have fled behind the salt lake; they have mounted to the summit of the Kahars; the Christians have left Oran.'"[*]

Though such oracular language might seem to savour largely of fanaticism, and be thought to fall under the head of remarkable coincidences, the march of science warrants us in throwing out the conjecture that it may not improbably belong to the category of natural prevision; and that a more searching analysis of the anthropological development of all religions will classify it with the sibylline and ecstatic phenomena of all ages and of all degrees.

We shall present the reader with one more specimen of Arab legends, and then proceed to other matters: "In past ages the kings of Tlemsen had dealings with the fallen angels. These sovereigns were called Beni-Meriin (interpreted, the language of thunder); and by mysterious combinations of figures, or by throwing sand on a black table, they predicted the future, chastising those who had offended them by the aid of the devil, their ally. Now it came to pass that one of the Beni-Meriin was struck by the appearance of a girl whom he once met drawing water from the Tafna. Proud of his power, he thought that a nod would give him a new slave; but the girl, plighted to a warrior of the tribe, was deaf to the sultan's bribes. Furious at this refusal, the king swore to revel in the tears of her who refused him her smiles. Accordingly one evening, when the girl glided from the douar to meet her lover under some palms, the sultan called to his aid the evil spirit. At his command the demon seized the two young people, dragged them into the earth, and at once the country changed aspect. It used to be called the valley of flowers; but these were replaced by the dark olive. The palm alone, where the lovers met and vanished, still stands; and a wonderful spring has gushed forth from the spot, consisting of the tears of the unhappy pair, who shed them incessantly in their underground prison, where they are still kept by the devil."[†]

Both the stationary and the Bedouin or wandering Arabs still retain many customs described in sacred and profane history, and are in almost every respect the same people as we find mentioned in the earliest records. The Bedouins in particular still preserve the simple and primi-

---

[*] Castellane, p. 129. Captain Kennedy gives a full version of Si-Aissa's prophecy, which was uttered 130 years ago. See vol. i. ch. 11, p. 235, containing General Marrey's account of his expedition to Lagahouati in 1844.   [†] Castellane, p. 381.

tive habits of their ancestors, and a strong attachment to a pastoral life, so well adapted to the vast plains which constitute their proper home, to the heat of their climate, and to their serene and beautiful nights. They speak the Arabic language, and affect to have preserved the purest pronunciation of this their parent tongue. Of all the nations on the face of the earth, these are probably the greatest conservatives, having scarcely deviated in any particular from the mode of life chalked out by their sires. Save in religion, they are precisely the identical people with the Arabs at the time of the composition of the book of Job, and afford an admirable illustration and verification of the descriptions contained in that book. A traveller arriving amongst them is delighted to find the same dresses, manners, and customs as are portrayed in Oriental romance and historical paintings representing scriptural subjects. Their present habits are also found to be strictly conformable to the statements of Strabo, Leo Africanus, and Pomponius Mela, of whom the latter especially has left such clear and accurate accounts, that you would almost take him to have been a modern traveller in the plains of Northern Africa. We shall presently revert to this part of the subject. A visit to the tents of this interesting people transports you at a bound among the holy fathers of our race, and makes you an eye-witness of the sayings and doings of the ancient of days in the Fore-world.*

Like the daughters of Judah, their women go forth every evening to the distant well,† to draw water for household purposes and for their camels. They are, however, somewhat less gracious than Dinah; and if a stranger approach, they immediately let fall their veil, and cover their face, as Rebecca did on the appearance of Isaac: and should they meet a foreigner on the road, they go aside and sit down, turning their backs to the road.

The Bedouins grind their corn in their tents, employing some millstones with wooden handles; and their women are commonly engaged in this work: which proves the accuracy of Moses' expression, who speaks of the women labouring at the mill; and explains our Saviour's words when he says, that "two women shall be grinding together; the one shall be taken, and the other left."‡

They clothe themselves with a woollen garment five or six feet wide and three yards long, called a haik. It is a kind of blanket or quilted counterpane, of a white colour and fine material, and forms a light and becoming, but a very inconvenient article of apparel; for it gets loose and falls off every moment, needing constant attention to gather it up and fix it again. To this end a girdle or band is required; and probably

---

* Pananti's Avventure, vol. ii. pp. 108-123 : gli Arabi Beduini.

† According to St. Marie, the Arab women lead very hard lives, and have to go forth to the well and draw water many times in the course of the day. Pananti, ubi supra.

‡ Luke xvii. 35 ; St. Marie ; Pananti ; Blofeld. Diary of a Tour, vol. ii. p. 186.

from this has been derived the scriptural expression, " to have your loins girt,"* in order to have strength and activity.

The following patriarchal scene from the pen of Count St. Marie gives a pleasing picture of the poetry of Arab life, and its illustration of scriptural subjects : " We soon arrived at the gates of Medeah," he writes, " where, near a little fountain, there was a caravan which had stopped there at sunset. Camels were feeding on the grass and shrubs on the road-sides. A group of Arabs were at prayers, with their faces turned to the east, in the direction of Mecca; and some children were imitating the pious example of their parents. No one stirred on our approach. A burnouse spread on the ground served as a carpet, on which, barefooted, they alternately lay down prostrate and stood erect. There was something very imposing in the calm dignity of manner with which those men invoked the Deity. When the prayer was ended, they turned to the eldest man of their party and embraced him. Then resuming their slippers, and driving their çamels before them, they penetrated into the sinuosities of a ravine, and were soon out of sight.†

The Arabs are rich in proverbs, many of which present a family likeness to those contained in the sacred oracles of the Hebrews. In all eastern proverbs there is great depth of thought, and they express opinions which are the result of long experience and reflection. We extract the following from the travels of Count St. Marie, as a specimen of this style of wisdom. " If your friend is made of honey, do not eat him up." " If you travel through the country of the blind, be blind yourself." " When you are the anvil, have patience; when you are the hammer, strike straight and well." " He who cannot take a hint, will not comprehend a long explanation." " The mother of the murdered man may sleep ; but the mother of the murderer cannot." " I like the head of a dog better than the tail of a lion." " Take counsel of one greater and of one less than yourself, and afterwards form your own opinion."‡

St. Marie relates, in another part of his light and graceful work, how he saw beneath the shade of some stately old plane-trees an aged Moor seated, wrapped in a white burnouse, and tranquilly smoking his pipe. Before him lay a youth reclining on the ground, passing a chaplet through his fingers. This group, with the surrounding objects and brilliant colouring, was like a Bible scene. Nothing was wanting: the large mountain dog, the curved shepherd's staff, the camel crouched in the foreground, ruminating with upraised head and gazing at his master, all combined to perfect the picture. Save the pipe, the group might have figured in an altar-piece of the old masters.§

---

* Pananti states, however, in another place, that the haik is also bound round the *head* by the men, by a cord that assumes the shape of a turban; while the women fix it with a *fibula* (vol. ii. pp. 111-112).

† St. Marie.   ‡ Ibid.   § Ibid.

We learn from the same authority, that there is a legendary story relating to the river Oued-Kebir, near Blidah, which calls to mind an incident of the Bible. The true believers affirm that once upon a time, water having become scarce in Blidah, Mohamed, the marabout whom I have already mentioned, struck with his stick one of the mountains of the Atlas range, and a spring gushed forth, which has never dried up. This spring is the source of the Oued-Kebir.*

We find in the pages of Baron Baude's *Algérie* several interesting illustrations of scriptural expressions and descriptions, which afford useful materials for future exegesis. On the Baron's trip from the Camp of Draan to La Calle, to which we have adverted elsewhere, he was entertained one night at the douar of the Merdes' tribe, in the tent of their sheikh, Sidi-Mahmoud. It appears that " mats were spread on the ground, and the guests crouched around them, all excepting the host, who, without touching any dish, stood opposite them, watching over the attendants, and anticipating their least wants." It was thus that Abraham received his guests, according to the sacred narrative (Gen. xviii. 8) : ' And he took butter and milk, and the calf which he had dressed, and set it before them; and he stood by them under the tree, and they did eat.' "† On their return from La Calle to Bona, the Baron further relates how " they arrived at the douar of Abdallah-ben-Hassan, and received the hospitality of the ancient patriarchs, such as we read it described in the same chapter of Genesis, verses 6 and 7 : ' And Abraham hastened into the tent unto Sarah, and said, Make ready quickly three measures of fine meal, knead it, and make cakes upon the hearth. And Abraham ran unto the herd and fetched a calf, tender and good, and gave it unto a young man ; and he hasted to dress it.'

" Abdallah himself had, like Abraham (Gen. xxiv. 35), ' flocks and herds, and silver and gold, and men-servants and maid-servants, and camels and asses;' and his French visitors, on seeing his flocks at sunset come trooping in from all quarters, formed a high idea of his riches.

" On reading these descriptions, the image of the venerable Job, or the race of Arab proprietors of whom he was the type, is brought vividly before the mind as a living reality. Stepping over the Mediterranean, we find there are still men whose substance is 7000 sheep, and 3000 camels, and 500 yoke of oxen, and 500 she-asses, and a very great household; so that this man was the greatest of all the men of the east' (Job i. 8)."‡

" The young girls," says M. Lamping, " are to be found every morning at sunrise outside the gate of the town (Coleah), standing by the fountain, at which they assemble, with stone jars on their shoulders, to fetch water for the daily consumption, This truly eastern scene calls to mind Rebecca at the well, drawing water for her father's flocks."§ " The moment the Arab

---

* St. Marie.   † Baude's Algérie, vol. i. p. 176 et seqq.
‡ Ibid. p. 176 et seqq.   § The Foreign Legion, p. 5.

hears the call of the muezzin,* he throws himself upon the earth, wherever he may chance to be, and touches the ground with his brow ; then, rising again, he stretches his arms towards heaven, with his face turned in the direction of Mecca, his white flowing burnouse and his long beard giving him a venerable and patriarchal air.   Thus surely did Abraham, Isaac, and Jacob worship their God."†

The conservatism of Arab life gives us contemporary illustrations of the days of Joseph, Jonathan, and Joshua.   On meeting they still say, like their fathers, placing their right hand on their breast, *Salaam alykum*‡ (Peace be unto you).   Friends kiss each other's hands, heads, and shoulders. On great occasions the women also salute their husbands by kissing their hands.   Inferiors kiss the feet, knees, or clothes of their superiors, and children the heads of their parents.§

Madame Prus observes, there is nothing more singular or picturesque than the appearance of these Arabs arrayed in the majestic drapery of their burnouse, and reminding one strongly of the old engravings representing the patriarchs of the Bible.   Their aspect and bearing are noble and dignified, and their imposing attitude might offer many a model to our actors.||

The Arabs of the province of Constantina differ considerably from those who inhabit the other parts of Algeria; their language, customs, education, and character form a complete contrast to that observable elsewhere.   It is surprising, at first sight, to find that the different shades which thus characterise fractions of the same people should be in perfect unison with the geographical position which these several fractions occupy.   For instance, in the west the Arab is ignorant, coarse, warlike, and rough in the pronunciation of his local idiom, which has undergone greater change than any other ; whilst the inhabitants of the eastern province bear a diametrically opposite character.   To account for this singular fact, we must bear in mind the mode in which the Mussulman conquest operated.   This movement proceeded from east to west ; and thus, as always happens in enterprises of this nature, the most adventurous and hardy advanced the farthest.   It was in this manner that a bold Arab chief pushed on as far as the ocean, accompanied by a dashing train of cavaliers ; but being farther removed than the rest of the nation from the general and original fountain of this stream of population in Arabia, this band of adventurers became more deeply imbued with the spirit of resistance, and it had to sustain longer and more inveterate

---

* M. Lamping says 'marabout;' which may be correct in the country districts of Algeria, where they have no mosques.   But the usual official who in Mahommedan countries call the faithful to prayers, as an incarnate bell, is styled the muezzin.

† The Foreign Legion, pp. 5-6.   ‡ اَلسَّلاَمُ عَلَيْكُمْ.   § Berbrugger, iii. 15, 16.
|| Residence in Algeria, p. 3.

struggles. It is easy to believe that such a state of hostility, so favourable
for the preservation of warlike tendencies, should influence unfavourably
every thing relating to education, to language, and to manners. These

ARABS OF CONSTANTINA.

wanderers, pushed onwards by the tide of conquest, have therefore much
degenerated in these respects, both because of their greater distance from
their compatriots of Arabia, the source of all their national light, and also
because of their closer proximity to the rough and ferocious Berbers, the
ancient masters of Morhereb.* Besides, as the Roman dominion established
itself first in the eastern part of Northern Africa, and had taken deeper
root there than in the west, on account of the greater duration of its sway;
and as, on the other hand, the Vandals came from the west, and their devastating fury received no check till it reached the walls of Carthage,—it
seems to result from these two conflicting causes, the one constructive and
the other destructive, that the provinces in the east at the time of the
Arabian conquest were the least degenerated from their ancient splendour.
Accordingly, it is in this section of the country that the finest remains are
brought to light in the present day, attesting, by their massiveness and
finish, the power, wealth, and good taste of these masters of the world.

It is therefore probable, that in this favoured region, the sight of these

* المغرب el Morreb, *the setting;* the Arabic name for the empire of Morocco.

imposing vestiges of Roman grandeur, and the influence of a civilisation which survived the attacks of the barbarians, contributed in softening the conquering Arabs who settled there, and in vanquishing them, in their turn, by the ascendency of an undoubted intellectual superiority, as the Vandals had been likewise subjected before them. Whatever truth there may be in these explanations, the facts are self-evident, and it suffices to go from Algiers to Bona to be convinced of this.

The difference of language is the first thing noticed : in a general point of view, this consists in a greater softness of pronunciation.

Amongst the Arabs of the province of Constantina, the courtesy of manners corresponds with the softness of the language ; and they are much more disposed to give a friendly reception to the Christians than their countrymen farther west. The rapid progress of the French rule in the east of Algeria testifies to this. On the other hand, the people of Constantina have not the warlike spirit of the western tribes ; and whoever has seen the two races in action can appreciate this difference. A troop of chasseurs dispersed some hundreds of Arabs in 1836, near the marabout of Sidi-Tamtam ; whereas the Beni-Amer and Garbas were seen, in 1835, to stand discharges of grape-shot with the steadiness of veteran European troops.

The costume at Constantina does not differ materially from that found throughout Algeria ; only the people are generally more attentive to dress, and the white burnouse is commonly worn, whereas black and brown prevails in the west.*

We must now proceed to analyse the political and social organisation of the Arab tribes, 1145 in number, composing a population of about three millions, and inhabiting the surface of Algeria, Tell, and Sahara, which is estimated at 390,900 square kilometres, equivalent to four-fifths of the 86 French departments, or 150,000 square miles. We have seen that Algeria under the Turks comprised four provinces, three of which—Oran, Tittery, and Constantina—were governed by Beys, and the territory of the city of Algiers by the Dey in person, on whom the three Beys were nominally dependent. In 1843, by a decision of the French Minister of War, the possessions in Africa were classed in three divisions, Algiers, Oran, and Constantina : Algiers containing two sub-divisions, Algiers and Tittery ; Oran four sub-divisions, Oran, Mascara, Mostaganem, Tlemsen ; and Constantina containing two, Bona and Setif. The latest division is the following :—1st, the province of Algiers is analysed into six sub-divisions, whose *chefs lieux* are, Algiers, Blidah, Medeah, Aumale, Milianah, and Orleansville. The province of Oran has five sub-divisions,— those of Oran, Mascara, Mostaganem, Sidi-bel-Abbes, and Tlemsen. The division of Constantina admits four sub-divisions,—Constantina, Bona, Setif, and Batna. The Governor-general is the supreme head of the

* Berbrugger, part iii. pp. 17-18.

regency, and under him there are, first, the governors of the two other provinces; secondly, each sub-division is under superior officers; and thirdly, each sub-division is distributed into circles, under French commandants. There are eleven of the latter in each province; and a bureau Arabe has been attached to every military commandant since 1844.*

The circle comprehends generally several kaidats, who, when the state of the country allows it, are placed under the direct orders of the superior commandant, without obeying an agha. The khalifa or the bach-agha is dependent either on the commandant of the division or of the sub-division.

The sheikh is appointed by the commandant of the sub-division, on the recommendation of the kaid. The assembly of the notables of the douar, who assist him, is called djemaa.

The kaids are chosen amongst the most distinguished men of the tribe, and appointed by the commandant of the division, on the recommendation of the commandant of the sub-division. The kaids receive no fixed salary; they derive emoluments from the raising of tribute, &c.

The aghas are named by the minister of war, on the recommendation of the commandant of the sub-division, transmitted through legitimate channels. They form three classes, whose salaries were fixed, in 1847, at 1200, 1800, and 3000 frs. (48l. to 120l.)

The khalifas, bach-aghas, and independent aghas are also named by the minister of war, on the recommendation of the commandant of the division, transmitted through the governor-general. The khalifas have 12,000 frs. (480l.), bach-aghas 5000 frs. (200l.), besides perquisites connected with their office.

The khalifas and bach-aghas, when called upon, with their horsemen, to join the French forces in the war, are dependent on the orders of the French commandants. In their judicial capacity they can inflict fines, revise the judgments of the aghas, kaids, &c. of the territory they govern, and are responsible for the regular payment of dues, &c. The duty of the agha is to act on the orders of the khalifa, or direct orders from the French authority; he arranges the military affairs of the khalifat, sees that punishments ordered by the khalifa are duly executed, and has power to inflict fines to a certain amount; renders accounts to the khalifa of complaints made against the kaids beneath him, and arranges the collection of imposts, &c. The office of the kaid is annual; and on his investiture he has to provide a horse fit for military service, as an acknowledgment of

* Tableau, 1850, Appendix, from pp. 713 to 722; also, Dawson Borrer's Campaign in the Kabylie, c. 16, on the Arab tribes. The douar is the basis of Arab social organisation. A certain number of douars, forming a ferka, obey a sheikh; the assemblage of a number of ferkas, sometimes only one, a tribe, commanded by a kaid. A group of several tribes forms either a grand kaidat or an aghalik, under the orders of a kaid-el-kijad (caid of caids), or of an agha. Several aghaliks may form a circumscription dependent on a bach-agha (chief of aghas), or on a khalifa. Such is the hierarchy of the Arab *powers that be.*

vassalage to the French; he then receives the burnouse and seal of office from the government. This feudal custom is derived from the Turkish times. The kaid has to see that the warriors of his tribe are ready for service, and to command them during the war; is responsible for the execution of all orders issued by his superiors; is charged with the interior police of the tribe, particularly that of the markets, where he is obliged to be present, either in person or by his lieutenant. He has to decide controversies between the douars of the tribe in cases of small importance, and to refer to the agha those of greater weight; he has also the power of arresting criminals, of whom he must render account to the agha, and that officer to the French authorities; he has also the right to inflict corporal punishment and fines to a certain amount, &c. The sheikh has to settle disputes in the ferka, to see the taxes levied, and to exercise police functions, &c., which makes his office like that of the *maires* of French parishes. Each tribe has a kadi besides a kaid. The duties of his office are to give judgment upon civil questions relating to person or property, or such infractions of the law as relate to marriages, divorces, successions, the guardianship of orphans, &c. He takes the oath of vassalage between the hands of the procureur-general of Algiers, or the commandant of the sub-division. This office is held for life, unless his conduct is so bad as to render his deposition necessary. The kadis in the towns, and attached to the bureaux Arabes, receive salaries; those of the tribes none, obtaining perquisites and enjoying immunities there. Below the office of the kadi is that of the oukil-bit-el-mel, an officer nominated in each sub-division by the commandant of the division, to watch over certain fiscal interests, to find out vacant heritages, to collect state duties on successions, &c.*

Hence it appears that the government of the Arab tribes is in the hands of the military authorities, aided by the bureaux Arabes. The duty of the heads of this office is most important, all the affairs of the government with the natives being transacted through them. They must know the language of the country, and the laws, customs, and characters of the Arabs; they must also be shrewd and resolute diplomatists, with sound notions of justice. Their duties are laborious and extensive; and the race with which they have to deal is intriguing, sensitive, and hates the French government. They are the intermediaries between the tribes and the administrative commissions in all that relates to fixing imposts, recovering government property, regulating the pay-

---

* The tributes imposed on the tribes are of two kinds, — the *achour* (tithe) on corn; and the *zekkat*, a tax of religious origin on flocks. In the province of Constantina the latter is replaced by the *hokor*, a kind of land-tax in money. In the achour, each *zouidja* (17 to 25 acres) must pay the state a measure of wheat and a measure of barley: it is often raised in money. The zekkat yields one sheep per cent, one ox in thirty, one camel in forty. In the east province the hokor yields 25 frs. per djebda or zouidja. The Saharian tribes pay a *lezma*, a low tax in money, proportioned to their wealth. In the ziban or zaab of Biscara each palm-tree returns 40 centimes (4*d*.). Tableau, 1850, p. 718.

ment of the native troops supplied by the tribes when called for, settling disputes between natives and colonists, attending to markets, listening to reclamations, and administering justice. Theirs is a delicate and a difficult situation, requiring much caution to fill it with propriety. If it falls to a man worthy of it, he becomes a great prop to the existence of the colony, for through him chiefly can some moral ascendency be gained over the Arabs.*

The administrative commissions instituted in each sub-division and all towns, by ordinances published by Marshal Bugeaud in 1842, have to watch over government property, and to regulate tribute, fines, &c. from the Arabs. The proper execution of the duties of the commission depends much on the skill, &c. of the head of the-bureau Arabe.

What we have said will suffice to give a general notion of the tribes; and we shall add below some particulars relative to the political state of the Arabs.†

* Borrer, ch. 16. Castellane observes: "The Arab bureaux represent the centralisation of all the interests of the country in military hands. The chief of these officers represents the old Turkish chiefs, who were the chiefs of the margzhen" (page 367). Tableau, p. 715.

† The population of the province of Algiers consists of 290 tribes, or 900,000 souls. Of these, 175 tribes are administered directly by the commandants of circles and the bureaux Arabes; 35, indirectly by the latter; 52, more distant in the Sahara, acknowledging a feudal vassalage to the French, but with perfect administrative freedom; 28 unsubdued, all Kabyles. The government of the subject tribes is managed by a divisionary direction of Arab affairs, five bureaux of the first class, five of the second, three khalifas, five bach-aghas, and twenty aghas. Auxiliary troops, horse and foot, 570. There is, moreover, at Algiers, connected with the governor-general, a political bureau of Arab affairs, centralising all the administration of the tribes. The province of Oran has 600,000 natives, or 275 tribes, all subdued, 45 indirectly governed, and 28 left to the French commandant, or the great native chiefs of the Arabs. This province has a divisionary direction of Arab affairs, four bureaux of the first class, and six of the second. The native chiefs are analysed thus :—three khalifas (of whom one only in the Tell), and 22 aghas. 335 natives, horse and foot, are paid by France. The province of Constantina contains 1,300,000 inhabitants, or 580 tribes. Of these, 240 are governed directly, 200 indirectly, 80 mediately; and 60 are unsubdued, being Kabyles. This province has a divisionary direction of Arab affairs, three bureaux of the first class, and six of the second. It reckons three khalifas; and three other chiefs exert the power and receive the pay of khalifas, without the title. (Tableau, 1850, pp. 719-20.) The province returns 260 auxiliary horse and foot. In short, France has subdued, and governs directly, 897 tribes, occupying the most fertile part of Algeria. The tribes indirectly governed by means of delegates amount to 160, and the unsubdued in 1850-1 to 88; the latter inhabiting mountains or deserts, and depending for subsistence on the subdued tribes.

CHAPTER XVI.

𝕸𝖔𝖔𝖗𝖘, 𝕿𝖚𝖗𝖐𝖘, 𝕶𝖔𝖚𝖑𝖔𝖚𝖌𝖑𝖎𝖘, 𝕵𝖊𝖜𝖘, &𝖈.

ETYMOLOGY — MOORISH WOMEN — TOILETTE — WEDDINGS — DIVORCES — TURKS—
THEIR GOVERNMENT — THEIR COSTUME — YOUSOUF — THE KOULOUGLIS — THEIR
CHARACTERISTICS AND LAZINESS — THE JEWS — THEIR SERVILITY AND PERSE-
CUTION — THE CORPORATIONS.

VARIOUS opinions have been advanced in connexion with the origin of the
Moors,* who inhabit exclusively the towns chiefly on or near the sea-
coast, except when driven from them. One authority represents them as
the produce of different migrations; † and another describes them as Arabs
in race, but fallen, without the institutions of those in the country, adding
that they have been formed by emigrations from Spain, and of families
separately detached from the tribes.‡ It is very probable that there may
be some admixture of the old Berber blood in the Moorish veins; but it

---

* The name of Moors is very general, and has been used in Europe in the sense of
African Arabs; but it properly denotes Arabs settled in the towns and stationary, in
distinction from the Bedouin Arabs, and those who practise agriculture in the country,
styled *fellahs* فِلَّاح; in Egypt (from the verb *f'lah* فَلَح, to trace a furrow, to plough. *Cours
d'Arabe vulgaire*, par A. Gorguos, pp. 231-2). Dr. Shaw derives the word Moor from
מַעֲבָר, *a ferry;* or, as Genesius (Lexicon, p. 596) interprets it, *transitus, locus transeundi.*
Hence Mauri would mean persons living near a strait or ferry, such as the Straits of Gibraltar. There are, however, other derivations, such as מַעֲרָב, *the west, occidens* (Ibid. p.
603); evidently the same word as the Arabic المغرب *el morreb*, the setting, *i. e.* Morocco. This derivation appears to us much more probable, though not conclusive, as its
claim may be disputed by the Greek μαυρος, μαυροω, which Passow renders *dunkel, un-
scheinbar*, dark,—(Lexicon, 1830, vol. ii. p. 93, where he derives μαυρος from μαιρω and
μαρμαιρω, properly meaning *to shine;* but the derivation, according to him, has passed
into the opposite meaning, signifying the same as αμαυρος, *i.e.* without light),—a word still
used in the Romaic, as μαυρος, mavros, *black*, appearing in such names as Mavrocordato,
Mavrapanagia: the modern Greeks, as is well known, having in the diphthongs, &c. retained the use and pronunciation of the vau or digamma, though they have lost the letter,
which has been transferred to the upsilon. (Eichoff's Parallèle des Langues de l'Europe et
de l'Inde, &c. p. 47.)

† Blofeld.   ‡ Baron Baude.

has been chiefly supplied by the Saracen invasion, and is now far from the purest in the world. Roughly handled and cowed by the Turks, this class has lost that confidence in itself which flows from the spirit of clanship in all the Bedouins. In vain do you look for the noble bearing, the daring chivalry, the love of science, that distinguished their Andalusian and Arabian ancestors, that cast a radiance over the dark ages, and the great qualities of their ancestors is a rare sagacity, dissimulation, and heralded the triumph of intelligence in the better day about to dawn on the West.* The Servant of the Merciful † would no longer recognise in them the same race as those gallant soldiers and glowing worshippers who raised the Alhambra and fought for the Vega. Al-Rashid and Averroes would turn aside and weep over their degenerate posterity. Verily the crescent hath begun to wane. All that remains to the modern Moors of constancy. Yet these men are the offspring of those who filled Spain with monuments of fabulous beauty, and who lighted up the torch of the sciences when the world was plunged in night.

It appears that since the Turks have held sway at Algiers, many Christian renegades have married Moorish women, in such wise that the people called Moors are a very mixed race. Yet many families can still boast of a pure blood; nor is it difficult, notwithstanding these numerous crossings, to detect the true Moorish type. The men are mostly above the middle height; their walk is noble and grave; they have black hair; their skin a little bronzed, but rather fair than brown; aquiline noses, full faces, a middle-sized mouth, large but sleepy eyes. Their features are less marked, and some say less, others more handsome than those of the Arabs and Berbers. Their body is very rounded, and they have generally much *embonpoint*, by which you may always distinguish the race.‡

The Moors have adopted most of the Turkish tailoring with its sway. Their costume, though not nearly so elegant as that of the women, is quite oriental: consisting of very full drawers (*seroual*), leaving their legs bare; a very ample vest (*greliter*), two waistcoats embroidered with gold or silk, and a turban. As chaussure, they wear slippers of morocco leather, called babouches. Stockings, they have none. Some fold a burnouse, and throw it over their arm as an ornament.

Notions of beauty are wide as the poles, and æsthetical disputes betray bad taste. Though Moorish women are not exactly fat, fair, and forty, they are valued by weight, ignorant to excess, and sleepy though not sleeping beauties. Chinese feet, European stays, Bloomer eccentricities, and Hottentot Venuses betray the variable principles of æsthetics; nor need we be surprised to learn that a camel-load of female flesh is thought

---

* Berbrugger, part iii.

† عبد ال رحمان *Abderrahman*, the celebrated caliph of Cordova.

‡ Berbrugger, part iii.

a beauty by the Moors, and that girls are fattened for marriage like turkeys, occasionally dying under the spoon.

The costume of the Moorish women is handsome and various. We will examine its mysteries and different phases of development.

The negligé.—This costume among the poor is extremely simple ; consisting of a chemise of a transparent stuff, and long breeches kept up by a string drawn round the waist. The rich, and even those of the middle class, have a more dressy and a more complicated negligé. They never go bareheaded; and the head-dress of the young girls is generally a little close-fitting velvet cap, called *quonibat*, which only covers the top of the head, and is tied under the chin by a thin bridle. Sequins are frequently placed and fixed in concentric rings in it, and by their number give an idea of the

MOORISH LADY.

wealth of the parents,—or rather of their pride, for you see persons of narrow finances make use of this luxury. The hair falling down in long plaits, or squeezed into a long riband, almost always of a red colour, and of which the two ends fall below the knee, reminds one of the queues of Frederick the Great's grenadiers. With young women, the cap called quonibat, or the little red chachiyah of Tunis, serves only as a support to a coiffure that we are about to describe, and which is not meant to be seen. A cap (*m'hhennah*), almost always black and red, is placed on the head so as to leave all the anterior and superior part far above the beginning of the hair quite uncovered; it is knotted behind on the niddock, and the united ends fall on the shoulders, enveloping them with long curls of black hair, that float and wave gracefully as the fair, or rather unfair, lady moves her pate. Sometimes the hair, instead of remaining free, is squeezed into ribands in the way already described.

The matrons go out often with a more or less lofty *sarmah*, a kind of tiara of gold or silver, which has some analogy to the cap of the French *cauchoises*. Very old women preserve this costume even indoors.

A corsage or bodice of silk (*frimlah*), very tight, compressing the bosom and bringing it unnaturally upward and forward, slightly diminishes the extreme transparency of the chemise. A large and richly-worked zone (*euzame*) of silk and gold also conceals a part of the body above the trousers, which are placed very low.

Let us now attend to Moorish female full-dress.

Over the head-dress of which we have spoken farther back, the Algerian women, when they wish to adorn themselves, place a second cap (*âsisbah*), which covers a part of the forehead, and is knotted in front at the top of the head. Their little bandeau of brilliants (*z'riref*) resembles that of our European ladies; only instead of its being applied immediately on the forehead, it is tied to the rim of a silk kerchief thrown over the head.

They wear on solemn occasions a kind of open tunic (*rh'lilah*), in which gold and silver are married in capricious arabesques on a ground of red or blue silk. They also gird themselves with a long piece of silk with broad stripes (*foutah feuchetanne*), which is tied in front and falls down to the ground.

Besides the clothes of which we have spoken, jewels rich in material, but in bad taste, complete the costume of a Moorish lady, who is seldom able to put on all her ornaments. Long drops (*menagueche*) laden with diamonds hang from her ears. The young girls wear round their neck a collar of sequins, called *mdibehh;* and the married women put on a similar ornament (*queladah*), but composed of diamonds. *Msais*, or gold rings, encircle their arms; the Bedouin women being satisfied with horn rings, called *mquais*. The legs above the instep have gold and silver rings, called *rdise* when they are massive, and *khalkhal* when hollow. Their hands are also laden with rings having brilliants, and a kind of seal called *khouatim*, or *braïim*. In short, to our notions, a Moorish lady in full-dress must be an incarnation of splendid misery.[*]

---

[*] Nor have we yet done with her troubles; for she is, moreover, the victim of gynecidal cosmetics. With a preparation of gall-nuts (*afsah*) they paint their eyebrows black, in a broad band continued across the forehead; and they darken the inside of the eyelid with *g'hhol* or antimony. This gives liveliness to the eyes, but hardness to the face. The juice of the plant henna, yielding a red dye, is put on the nails, on which a deep coat is

Moorish women at home.—If out of doors they are heavily clad, in-doors the Moorish ladies are what we should call almost indelicately, though very gracefully, attired.

When the Moorish lady sallies forth to the bath or her devotions at the marabouts of Sidi-Abd-el-Rahman-el-Tsaalebi, of Abd-el-Quader, or of Sidi Mohammed, &c., she adds to her undress a long white Mameluke trouser (*serouah el zankal*) if marriageable; but of colour if she is not yet nubile. She throws on her shoulders a floating tunic of clear stuff (*hhaik el telhhif*) which slightly hides the transparency of the shift : this tunic is fixed in its upper part by long gold or silver pins (*bzaïim*). She girds the foutah and knots the êudjar, or handkerchief that is to hide the face. Lastly, she covers her head and most of her body with a long and broad piece of white cotton, of which the upper part is put on the forehead, leaving between it and the êudjar only a little space free for the eyes. This piece of cotton or linen (*foutah enta snanig*), which is called *takhe-lilah* when it is of silk, falls back behind half-way down the leg. The Moorish women, pinching the stuff on both sides of the head, bring back the hand under the chin inside, from which it results that they are exactly enveloped on all sides, and only the lower part of their legs is visible. The whole coquetry of the native ladies is concentrated in the movements that they give to this dress. Those who wish to be seen put apart their hands, which hold up the takhelilah, and raise them by removing them from the head as high as the top of the forehead. This sudden manœuvre uncovers all that part of the face which is not hid by the êudjar, and offers a speci-men from which an amateur can judge of what it is not allowed the lady to show. She also displays the rich belt and elegantly embroidered bodice that shines under the transparent tunic. The foutah enta snaniz and the takhelilah are articles peculiar to the Algerines. Every where else the women, when going out, put on a long piece of coarse wool (hhaik) falling to the ground, and showing only one eye.

A few words on polygamy ere we pass to other matters. The position of woman is the key to a nation's social state; and her slavery among Mussulmans accounts for the low scale of their civilisation. Yet uni-formity on this globe is impossible; nor can the morality of the poles tally with that of the line. We have sacred and patriarchal authority for poly-gamy in certain ages and countries; and though it cannot co-exist with Christian civilisation and Germanic chivalry, it has been proved to be the legitimate offspring of the Semitic phase of human nature. In short, there may be circumstances in time and space that justify the institution in the eyes of reason and religion ; though, notwithstanding the arguments of

rubbed, to increase the blush of that part of the nails which is seen. Their hands and feet are also painted with a disagreeable black tincture; and the marks, or characters (*cha-mah*), inscribed by nature on some persons are much valued. Artificial signs are also sometimes made, called *khanat.* Berbrugger, part iii.

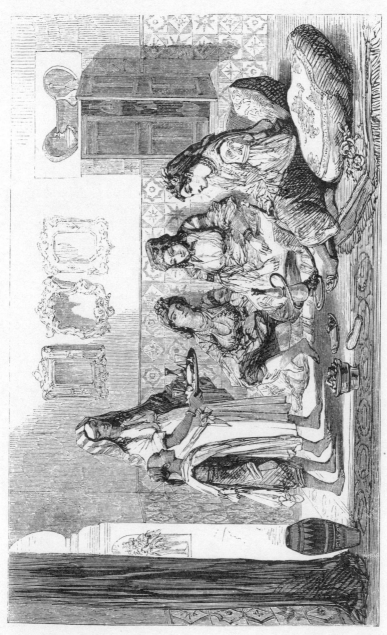

p. 328.

WOMEN OF ALGERIA.

Milton,* it is evidently incompatible with the highest spiritual and social phases of humanity.

Marriages among the Moors, as with most other Mussulmans, are contracted through third parties and gossips, the young people never meeting till the wedding-day. The affair is a regular market, a bargaining, like a London season. The gossip is bribed by a young man to go and examine his ideal mistress, whom he knows only by report; she goes, and gives a coloured report on her return, being bribed by the parents. If the parties are agreeable, and the old folks think the young man has a position, they close.

On the wedding-day she is bathed, painted, daubed with blackened corks and henna, and decked out in her best attire. She is marched through the streets, accompanied by lanterns; and all the women have a grand feast at the bridegroom's house. The men, poor fellows, sup together apart; the wretched bridegroom eating alone, that he may not take too much and misdemean himself. At midnight, when the mosques open, the unhappy pair are left alone, all the guests retiring.

The Moorish women are very fruitful, and marry ridiculously young, being mothers at nine or ten. They nurse their children, and doat upon them when young, but hate them when older,—particularly if they are boys, as they think that they have inherited their father's harsh character.

The Moorish men are mild and lazy, gambling, smoking, and sipping coffee all day. The few that reside in the country live like the Arabs. They do little work; and that little is mostly done by the women, who bruise the corn in hand-mills, before it is sent to the public ovens. The French occupation has ruined many, and injured most Moorish families, by raising the prices of all commodities. The only trades practised by the Moors are those connected with luxury, requiring only dexterity and little strength. Cross-legged, and smoking pipes made of jasmine-wood, seated in little shops like boxes, they embroider, plait silk, or make slippers. This industry cannot support their families; and they are undersold by Europeans. Hence their misery is great; and 8000 francs of alms per month was given to 2000 Moors at Algiers in 1843. The Moorish population of Algiers being 12,000, and many individuals representing families, we may infer that half the Moorish population required charity. Crowds of naked beggar-children infested the streets. Thus African society begins its acquaintance with civilisation where ours ends, —with pauperism, prostitution, and mendicity.†

We shall next attend to the Turks,—that handful of stern Janissary soldiery which held sway for three centuries at Algiers by the terror of

---

* The reader who may be curious to read Milton's clever defence of polygamy is referred to his Treatise on Christian Doctrine, translated by his Grace the Archbishop of Canterbury, 1825, pp. 231-241.

† Berbrugger, part iii.

its arms or name. And now we pass from the Semitic to the Mogul variety, or rather to a mixed race. In 1830, when the French landed at Algiers, the Janissaries did not amount to 3000 men;* but we must remember that recruiting in the Levant had ceased for three years, owing to the blockade of the French fleet, and that many Turks had perished.

From the statistical returns of the 1245th year of the Hegira, answering to A.D. 1829, there were 7876.†

The Moorish population in 1849 amounted to 60,928 souls.

Under the Turkish sway, the Arab and other tribes were made accountable for the crimes of one of their members; nor was this unjust, as the tribe had all the means of keeping up the police of its own district.‡

The Turks, though so small a minority, governed the whole country despotically through the Janissaries; respecting no person, from the Jew to the Dey. The service or duty of the place, the raising of taxes, and piracy, were all carried on by the Turks, and with them; and though a mere handful of men, the other troops of the regency obeyed them.

Their number has greatly diminished since the French conquest, and they are fast disappearing.§

	Turks.
* At Algiers	3976
Province of Oran	1300
Do. of Tittery	250
Do. of Constantina	1700
Movable column to gather taxes	400
Invalids	250
Total	7876

† But there were many superannuated Janissaries besides these. The Odjak (properly Oda), or Company of the Janissaries, only admitted, as is well known, Turks from the Levant, or Christian renegades; no native of the regency being permitted to enter its ranks. Yet they were far from being all natives of Turkey, or true Osmanlis; consisting of a singular compound of Turks, Greeks, Corsicans, Circassians, Albanians, Maltese, &c., and renegades from all Europe: forming an association of piracy abroad, and oppression at home; but acknowledging the sovereignty of the Porte, and speaking its idiom. Yet the Turks had not come at first as masters, but as auxiliaries, like the Anglo-Saxons nearer home. The Janissary force was recruited at Smyrna and Constantinople only, in virtue of a treaty with the Sultan. Berbrugger, part iii.

‡ In 1810, one fine morning, the Turks hung thirty Biskris at Algiers, on account of an insurrection of their compatriots in the oasis of Zaab; and in 1823 the Kabyles near Bugia having made some Janissaries prisoners, the Dey seized all those in Algiers, thereby saving the heads of his Turks. The latter kept with all the tribes kaids, who were overseers, like the *missi dominici* of Charlemagne; and they only used violence to get what was owed them when other means failed, after which a body of Janissaries was sent to enforce compliance. These often disarmed the whole refractory tribe and pillaged it, but respected the women. Baron Baude.

§ Prince Pückler Muskau, who visited Algiers in 1835, and wrote a very amusing account of his adventures in Algiers and Tunis, called *Semilasso in Africa*, saw them still ere their feathers had much drooped; and relates, in glowing colours, the gallant bearing of the *spahis* in their scarlet bournouse, and the dashing chivalry of Yousouf, a Janissary from Tunis, whose adventures are almost fabulous, and may serve to freshen the dryness

We learn that those who remain at Algiers are generally fine men, with fair skins, a stern look, and strong features. They live like the Moors, with whom they may be easily confounded; and you may at times meet them on the walks, in the cafés, bazaars, shops, &c. Their dress is nearly the same as that of the Moors; and they commonly like light-coloured clothes, and a shirt of gauze with full sleeves, large trousers scarcely reaching to the ankles, a silk girdle, and a caftan over it very frequently. The sleeves of the caftan are very broad; the front is decorated with agrafes, and gold and silver embroideries. Old Turks, or those invested with dignities, wear a long pointed beard; the young men only wear mustaches; but all shave the head. The younger Turks wear the fez-cap and no turban; but with the Turks of a certain age, the turban is a long and narrow roll of silk, muslin, or cachmere, wound round the fez; and the arrangement of the folds, as well as the materials of this article of dress, bespeaks the class or profession of the wearer.*

Captain Rozet, who visited Algeria soon after the conquest, has left ample particulars respecting the Janissaries, whose morals do not appear to have been very strict; for he states that, after exhausting all the voluptuous enjoyments natural to man, they were addicted to unnatural practices of classical notoriety; nor is this practice, unfortunately, confined to the Turks.† Children of Turks and Christian women enjoyed the consideration of Turks, and could become Janissaries; but children of Turks and Moorish women were an inferior cast, excluded from the dignities of their fathers, and named Koulouglis,—from the Turkish words *koul*, slave; and *oughli*, sons.

Agriculture was despised by the Turks, whose beautiful gardens were commonly cultivated by Kabyles, or Christian slaves. Most of the Turks engaged or speculated in piracy, and some in trades, selling jewels, essences, perfumes, and valuable stuffs. They also engage extensively in

of barren description. Yousouf is said to be a Christian and an Italian by birth, kidnapped and brought to Tunis in early childhood, when his beauty striking the Bey, he named him one of his pages. An ill-starred intimacy soon sprang up between the handsome Yousouf and the Bey's daughter; and they were surprised in one of their interviews by a slave, whom Yousouf killed, cut up, and concealed, to prevent detection. Yet fearing disclosure and death, he left his Dido and the shores of Carthage by a vessel bound to Algiers, where he was received with open arms by the French; and at the head of the spahis, after many dashing achievements, including the fabulous capture of Bona in 1833 with the Baron d'Annandy, a most gallant French soldier of fortune, he attained the rank of brigadier-general. Yousouf visited Paris some years since, and again lately; and his short but well-knit frame and liquid yet fiery black eyes are said to have done considerable execution among French hearts. In short, there was a Yousouf fever.— *Spahi*, I may add, is the name of the yeomanry or irregular cavalry throughout the Turkish empire. An excellent account of this gallant body of men, who 200 years ago could have ridden over all the Russias in two months, occurs in Sir Paul Rycant's Present State of the Ottoman Empire, 1662, p. 347.

* Berbrugger, ubi supra.

† Rozet, Voyage dans la Régence d'Alger, 1833.

the manufacture of carpets at Algiers, which are not so handsome, but softer and more comfortable than those of Turkey.   Honesty was always the characteristic of the Turk.   The hamal of Constantinople could be trusted any where with your luggage; and the Algerian Turk never cheated in trade.   Friction with Christendom has probably taught them by this time the arts of lying and cheating.*

Most human characters and institutions have two phases, light and shade; nor are the Turks an exception to the rule; and we cannot bid adieu to them without regretting the extinction of their virtues together with their vices.   Like the American Indians, it seems to be their fate to fall before the sickle of European civilisation, or barbarism; and it may not be unprofitable to dwell on their expiring agony, and reap some wisdom from their ashes.   Limb after limb of the mighty Ottoman empire has been lopped off of late, and the axe now lies at the root of the tree.   But though Western somnolence and jealousies may suffer the city and land of the Sultan to expire in the rude embrace of Russia, we will not suffer the majesty and energy of the children of Othman to go unchronicled in these pages.

Though cruel and bigoted in the days of their pride, the Turks had a mighty glowing faith, the secret of their power; while its want is the key to our weakness.   Deficient in science and its appliances, they were emancipated from our scribblomania, and lived in happy ignorance of literary indigestion and repletion.†   Hence they had greater concentration and power of will, more freshness and originality of character.   We have alluded to their honesty; and though piracy would seem little compatible with the love of truth, yet history has shown that the proud Osmanli rarely stooped to lies.   We must not forget their tenderness for the brute creation; which, though apparently incompatible with their ferocity to Christians and foes, is another of the anomalies and twofold phases of national character.‡

Nor can we suffer these few lines on the departed lights of Turkish life to escape us, without expressing a regret that the friction with Indo-Germanic civilisation has not been more profitable to them, and honourable to us.   It is unfortunate that Voltaire and Diderot should be our harbingers in the East, and oracles with the Turkish youth;§ and it savours more of our barbarism than of our Christianity, that we cannot extend the march of mind without abetting the spread of moral and physical poisons; or inoculate the nations with our civilisation, without grasping their terri-

* For ample illustrations of the manly honesty and straightforwardness of the Turkish character, the reader is referred to the pages of Dr. Walsh's Residence in Constantinople, and to Miss Pardoe's City of the Sultan, 1838.

† Montaigne's Essays.

‡ Dr. Walsh's Residence in Constantinople, 2 vols.

§ Macfarlane's Turkey.

tory and enslaving their sons. But Christianity and politics are still wide as the poles; and whilst Birmah and Scinde fall into our toils in the East, we must not complain of the dungeons of Siberia, of the agony of Circassia, or of the northern tempest hanging over Stamboul.

We have observed that alliances frequently took place between the Turks and Moorish women, and ultimately some of the most important families of the regency became allied to them. From these relations sprang the Koulouglis, or sons of slaves, who received this name from their mothers. Filling a kind of transition position between the two races, they occasionally gave much uneasiness to the Turks, and sometimes did good service to both parties. They had enjoyed for 60 years (to 1830) a considerable share in the government, and began to form an imposing military force; and in the last Turkish census, A.D. 1829, we find that the number of Koulouglis fit to bear arms amounted to 8688.*

All the Koulouglis have shown themselves invariably well-disposed to the French from the first, and have accordingly met with much persecution from Abd-el-Kader and the Arab race.

This race possesses many of the qualities of the Janissaries, and is separated from the natives by manners, and by the use of the Turkish language, which they speak almost universally. They answer excellently well as mediators and channels of communication between the French and the Arabs and Kabyles.†

They are generally very handsome men, having regular features, well-shaped eyes, a fair and smooth skin, strongly developed muscles, and a certain embonpoint, proceeding doubtless from their mothers. The marriage of European with African blood can be detected in their appearance; for they have the nonchalence and haughtiness of the Turks, blended with the lymphatic temperament of the Moorish women,—especially the girls, who are also invariably brought up like their mothers. Their costume is the same as that of the Moors and Turks; but they pride themselves on

* At Algiers	.	.	.	.	.	. 2076
Province of Constantina	.	.	.	.	. 1130
Ditto Oran	.	.	.	.	.	. 1402
Ditto Tittery	.	.	.	.	. 1415
On the Oued-Zeitoun	.	.	.	. 2665
Total	.	.	.	.	.	. 8688

† Baron Baude, vol. iii. p. 231. The history of the Koulouglis, though much less interesting than that of the Janissaries, has been intimately connected with the annals of the regency, almost always sharing the good or evil fortune of their sires the Turks. Yet in 1626, under the reign of Maharan, they revolted against the Janissary militia, and devised a conspiracy that almost overthrew the Turkish power at Algiers, and whose discovery led to a horrible massacre. Though far from enjoying any high consideration under the Dey, yet the corps of spahis (Turkish cavalry) was commonly recruited from amongst them.

extreme cleanliness, and even a kind of coquetry in their dress, which is not unbecoming their character, and recalls the Asiatic *tchelebis*.* Almost all rich enough to do nothing, they follow no profession, scarcely taking the trouble to work, and remain for days plunged in apathy, whilst their slaves cultivate their gardens, and receive chastisement if they are not satisfied with their work. The young men study attitudes in walking, to display the beauty of their figure.

The Koulouglis are distinguished above all the races in Algeria for excessive vanity and profound ignorance. In the social machinery, before 1830, they were confounded with the Moors, and had no right to the privileges of their sires ; yet they seldom had cause to fear any persecution from the Janissaries, because of the affinity existing between them. They were only required to take up arms in time of war ; and their pacific character has impeded the just appreciation of their natural valour.

The Koulouglis profess the Mussulman religion, in which they are brought up ; but their faith is characterised by the same indifference that they display in all the acts of life. They are not superstitious, and only attend to the forms of religion to show they believe in God. Exceeding the Turks in apathy, they do not make it a point of conscience to attend the mosques. Whilst on this subject, we must not forget to state that the Turks and Koulouglis, who are all Sunnites, or orthodox Mahometans, observe the rite of Hanefi ; whilst the Arabs and Berbers are Malekites. The Turkish tongue was only used in the odjak of the Janissaries and amongst the Koulouglis, and was employed for all official acts.†

There is yet another tribe of the Semitic race diffused in Algeria, as throughout the world ; a people of riveting interest, and yet generally deficient in the nobler and more gallant characteristics of human nature : I mean the Jews.

The children of Israel are scattered throughout Barbary, and have managed, as usual, to thrive there, notwithstanding greater insults than the Disabilities Act, and harsher persecution than the Ghetto. Those Jews who live scattered among the Kabyle tribes differ from the other Israelites

---

* A Persian word for gentleman or dandy, used only in the East, and occurring in the following Arabic proverb :

حلبي چلبي شامي شومي مصري حرامي

Hhalabi, tchalabi, *châmi, choumi, macri, hharami:* the Aleppian is a fop, the Damascan cunning, the Egyptian a thief. A. Bellemare's *Grammaire Arabe*, p. 98.

† It is probable that in a few years this hybrid race, which seemed born for a gentle, lazy, and voluptuous life, will die out in Algeria, as the recruiting and immigration of Turks, who kept up the stock, has ceased since the conquest in 1830. Berbrugger's Algeria, iii. pp. 8-9. See also Captain Rozet's Voyage, &c.

of Algeria by the period of their establishment, by their manners and language ; and all that they have in common consists chiefly, if not exclusively, in the basis of their Hebrew faith.*

Under the sky of Africa, as in Europe, this wondrous people have preserved their special type : an aquiline nose, a black beard, a magnificent but deceptive eye, a clear but colourless complexion. Their appearance is less scriptural and engaging than the interesting characteristics of the Lithuanian Jews, many of whom present a striking likeness to our ideal of Christ-like and apostolic beauty.†

In Algeria, as in most countries, they can be recognised by their combined look of cheating and humility, the result of the wrongs of ages ; and by their stooping attitudes, their severe features, and the dark rings round their eyes.

As always where they muster strong, they engross almost all the commerce : bankers, brokers, and agents, they are the Rothschilds of Algeria. Nothing can be done without them. They attend to all branches of industry, save agriculture. Active, intriguing, and versatile, they form a great contrast to the apathy of the Moors.

The Jews are forbidden in Barbary to wear gay clothing ; and they continue their partiality for the sable, notwithstanding their emancipation through the French conquest. Their dress consists of several vests, or waistcoats, of grey cloth ; of wide trousers of the same colour, tied round the waist by a blue belt ; and the majority go bare-legged, though a few wear stockings.‡

The Jewish women at Algiers have generally a greater freedom, and are more confidentially treated by their husbands, than the Moorish women : they go out at option, and do their own commissions. They are commonly pretty. Matrons or maids, they go with uncovered faces ; and their coiffure consists of a *sarmah*, or conical head-dress resembling the ancient hennin, and the cap of the French *cauchoises*. The rest of their

* Baron Baude, vol. iii. History has recorded the date and cause of the Israelitish immigration into West Africa, after the destruction of Jerusalem ; but the immemorial establishment of the Scenite Jews, who in the whole extent of Barbary are mixed with the Berber population, would lead us to suppose that it forms the foundation of this immigration from the East and Syria, which Sallust has related in these words : " Afterwards the Phœnicians—some for the sake of lessening the pressure at home, others from motives of ambition and curiosity—built Adrumetum, Hippo, Leptis, and other cities on the sea-shore." (Sallust. Bell. Jugurth. p. 77.) Numerous Jewish migrations occurred during the persecutions of Adrian ; and in the third century these emigrants formed independent tribes in the Hedjaz near Medina, and near Mecca ; and their religion spread in Yemen. If we may believe the Arab historians, most of the African Berbers and Arabs professed the Hebrew faith in the seventh and eighth centuries, and the preaching of Mahometanism made no way amongst them. This would appear to explain the phenomenon of the Jews forming till lately (1843) a fourth of the population of Algiers, and more than four-fifths of that of Oran.

† Marquis de Custine's Voyage en Russie.

‡ Berbrugger, part iii.

costume consists, with the common women, of a full blue cotton gown, without being confined at the waist, with very short sleeves, letting those of the chemise descend below them. The poorer sort put a kind of cap on their head instead of the sarmah, letting the point fall back on the neck. Like most of the men, they generally go bare-legged and bare-footed.

The young girls wear their hair long and plaited in a tail, to which they tie red and blue ribbons. As a coiffure, they wear a small but very elegant cap of green velvet, adorned with a golden tassel, and with a border also of gold, forming the sides of that kind of Greek cap which passes gracefully under their neck, where it is tied. Some sweet faces and regular features are often seen amongst them. Nothing can be more graceful than a pretty Algerian Jewess going to the fountain, and carrying a pitcher on her head. It is not improbable that it was a vision of this nature that inspired the pencil of Horace Vernet, when he designed his admirable Rebecca; in the same way that you find the prototype of his Eliezer, with a parti-coloured white and grey burnouse, in many a Bedouin of the Sahara.

In the kingdom of Fez, the Jews inhabit chiefly the northern provinces, and are even now called Philistines.* Like the Kabyles among whom they live, they take part in war, and are not withered by slavery.†

Under the Turks, the Jews formed a notable part of the population of Algeria; but they suffered grievous burdens and mortifying insults. Hence they gave the French a hearty welcome ; and their condition has been so much improved, that they have turned the tables on their former tyrants, whom they often treat with contumely and harshness. This circumstance, by increasing the hatred of the other native races to them, has led the Israelites to dread greatly the departure of the French, as the Ishmaelites would not fail to revenge themselves bitterly upon them again if they recovered the upper hand. In the present day they have a monopoly of the land-trade and brokerage. Their children frequent the French schools, speak the French tongue, and assume the Frank dress, without losing the spirit of caste. They readily become lawyers' clerks and employés of government ; they are already initiated into French legislation ; and the natives have no other consulting advocates.

The populations of the East in general, and the Jews in particular, have always shown too great a tendency to fence themselves into separate

---

* Berbrugger, part iii. p. 4.

† Kabyle Jews are also found in the regency of Algeria, especially in the Auress mountain, in the province of Constantina. It is probable that they shared the fate of the Libyan, Gætulian, and Numidian populations, when they were conquered and driven back into the mountains by the invasions of the Vandals and Arabs; and this participation in their fortunes gave them probably the right of naturalisation among the true aboriginal natives. See Gräberg d'Hemso, Specchio geografico-statistico dell' Imperio di Marocco, 1836.

races and castes, treating each other as enemies and strangers, though living under the same sceptre.*

Baron Baude found it difficult to obtain a good census of the Jews in 1841 ; and he could not procure any of the Kabyle Jews.†

The whole Jewish-population of the regency amounted in December 1849 to 19,028.‡

We rejoice to think that the sorrows of this mystical race are at an end in Algeria, and that under the enlightened religious code of France,—a model on this point to the nations, though a warning on so many others, —they can once more retrace while they outgrow the steps of their mighty ancestors, by securing that pedestal of all human greatness, self-respect. A fair field is open to them there, and opening elsewhere. It will take time for them to shake off the rust of ages ; but if they put their hands manfully to the plough, and drop the convict's dress and mind, they may yet stand forth once more as "a chosen generation and a peculiar people ;" and should they see the wisdom of disencumbering themselves of their narrow pride and bigotry, a bright future may very probably await this singular people. The luxuriance of their eastern fancy, and the shrewdness of their mother-wit, improved and chastened by an infusion of Germanic chivalry and thought, might lead to massive and brilliant phases of humanity yet unborn. A cross between the seers of Judah and the vikings of Odin would beget a giant progeny.

Other populations, few in number but rich in interest, occur on this soil of Algeria, which having been so long and so often the battle-field and the high-road of nations, must naturally present the relics of sundry warriors and wanderers in its sand-wastes and fastnesses. We have visited the Kabyles ; let us enter the oases, and scale the. Great Atlas ; and among the snows of Auress and the palms of Zaab we shall find pure and independent types of historical and mythical races. Thus, true Germanic sons of the Vandal flood are still preserved in the far south as well as at Bugia.

Most of the great tribes that are not Arabs have representatives at Algiers, where they are formed into corporations of workmen, subject to a rigorous organization and hierarchy analogous to that of our ancient

---

* Baron Baude, vol. iii. Many of the Jews in Algeria have embraced useful trades ; those of tailoring, gold-drawing, and jewellering have the preference, and some are very good masons. The commerce of supplying the tissues of Europe to the tribes is almost entirely in their hands ; and those of Constantina carry on manufacturing industry on a large scale, especially in the preparation of looms. Most of the rich families have houses at Algiers.

† The following were the French returns of Jewish population in 1839 and 1849: (Baron Baude, vol. iii.)

				1839.	1849.
Algiers	.	.	.	6065	7289
Oran	.	.	.	3364	7749
Constantina	.	.	.	698	3990

‡ Tableau, 1850, p. 113.

guilds. These corporations consisted of 5599 souls in 1849.* The three last corporations belong to the Zaab country, the others to the mountain districts of the regency. At all times a great number of Saharians have annually emigrated from the interior to the capital,—like the Auvergnats and the Savoyards in France, and the Tyrolese in Germany,—in order to pocket a few earnings and better their condition at home. These emigrant corporations at Algiers were governed by a chief named Amin, who was charged with the police of the body. The amin was like the agent of a commercial company, the magistrate of a little society, and the father of a family. The three chief corporations are those of the Biskris, the Mozabites, and the Aghrouaths.†

	1838.	1839.	1849.
* Kabyles	2258	2829	1817
Mzitas	185	273	206
Mozabites	702	803	861
Biskris	861	814	756
El-Aghrouaths	91	116	110
	4097	4835	5599

† The Biskris live to the south of the great salt lake of Chott. They have sallow complexions and serious manners; their customs, morals, and character differ essentially from those of the Arabs and of the other tribes. Yet, from their language, a corrupt dialect of the Arabic, it would seem that they are scattered relics of that celebrated people, and that their customs have been affected by admixture with the aborigines. This hypothesis is strengthened when we reflect, that the territory which they inhabit must needs have been crossed by the tides of Arabs who conquered Africa in the seventh century. Their character is complaisant and faithful; they were employed in most houses as confidential servants; they monopolised the baking trade; were the only commissaries at Algiers, and the only employés of the government on public works; and they were, moreover, the commercial agents between Algiers and Gadamez. At present they are porters, and even agricultural labourers, navvies, &c. Blindness is a very common disease in this small nation, possibly proceeding from their propinquity to the glowing sands of the desert. Their religion is Mahometanism. The Mozabites, or the Beni-Mozab, inhabit an oasis in the Sahara of the province of Algiers, at about twenty days' camel's march for a caravan, part of the road passing through the desert, and without any wells. The Mozabites are fair, but their features and type are those of the Arabs. (This statement, which we have borrowed from Berbrugger, is slightly at variance with that of Castellane, who describes the Mozabites as a different people from the Arabs. See p. 110 of the Souvenirs, and p. 31 of the present work.) Their character is gentle and active; and their probity is almost proverbial at Algiers, of whose government they were quite independent. Their privileges and commerce were protected by written covenants, with the sanction of the government; and in civil affairs they only recognised the jurisdiction of their Amin at Algiers. The benefits that they obtained from the Deys were considerable; being privileged agents of the commerce of Algiers with the interior of Africa, and enjoying the monopoly of the baths, butcheries, and mills of the capital. They follow the law of Mohammed, though deviating from it in several particulars; refusing to perform their ceremonies in the usual mosques, and having one of their own outside the town, appropriated to their particular creed. The Biskris are handsomely paid for the services that they perform for the French, and have gained as much as the Moors have lost by their connection with our neighbours. With regard to the native population, it is proper to add, that the Turks and Koulouglis have dwindled down to a handful of men, lost in the body of the Moorish population. For farther particulars consult the Tableau, p. 110.

CHAPTER XVII.

The Negroes.

UTILITY OF SLAVERY — MAHOMETAN AND CHRISTIAN SLAVERY — DEGRADED STATE
OF THE NIGER BASIN—THE SLAVE-TRADE IN AFRICA—THE BLACKS IN MOROCCO
—UNFORTUNATE RESULTS OF THE ATTEMPT TO STOP THE SLAVE-TRADE—THE
DJELEP—NATIVE ARTS AND SCIENCES.

IT is with diffidence that we venture on the disputed question of slavery,
especially as we are led by our honest convictions to differ on certain
points from many venerable names and authorities.

Liberty is such a glorious thing, and the very term slavery is so
odious, that it requires some courage and much solid argument to advance,
and not a little patience to listen to, the advocacy of the comparative advantages of some forms of mitigated slavery. Yet it is only the force of
habit that blinds us to our own; we are so inured to our chains that we
do not feel them; while it is certain that the four historical phases of
humanity up to the present time—savage life, pastoral life, feudal life, and
civilised life—are nothing but comparative modifications of slavery. A
keener sense of justice, a higher relish for the ideal, a better ear for higher
harmonies, would dash the cup of all our present enjoyments with poison,
and banish content from every breast. Our perceptions of a possible
future are blunted, though the chronic throws of revolution bespeak the
occasional awakening of the giant to a sense of his oppression. Revolutions are the symbol and the offspring of the world's social slavery; and
whilst legislation mistakes their cause, and tries to doctor them with political treatment, instead of enlarging the social freedom of the individual,
the streets will grow barricades, and our cities crops of conspirators.

The social movement, like all others, has its scale of degrees, and
slavery admits a serial order. The negro in America is low in the scale
of slavery; in Africa he is often lower. We do not attempt here to suggest the cure for the monstrous evil in the United States. We grant it
at once; but St. Domingo and common sense prove that the evil existing
will be magnified instead of lessened by immediate and entire emancipa-

tion,—an injustice alike to the planter and the slave. Gradual transitions are the law of nature ; and the only safe emancipation is that by degrees.

But slavery in America is not slavery every where ; and the system in Barbary, like that of Circassians and Georgians at Constantinople, works well.* It might seem unnatural for the Circassian parent to sell his lovely daughter to the Turkish merchant; yet pride and ambition prompt him to the act, for she may be the mother of sultans. Nor are we wanting in examples of high-born parents in a place called Britain selling their daughters, that they may wear a coronet.

Mahometanism has ever mitigated slavery ; and the Koran is a check to the master and a comfort to the slave. It recommends manumission as more efficacious for the salvation of souls than much fasting and many pilgrimages; and the marriage of masters and slaves is encouraged by custom.† In short, granting what circumstances have proved, that negro nature cannot bear an immediate transition from the night of the Niger to the comparative daylight of European civilisation, intermediate steps, probations, and preparations are necessary; and of such, Mahometan slavery offers a favourable example.

But since facts are ever more weighty than arguments, we will here produce some evidence, which, with the inductions that may be thence derived, appear to us conclusive.

We are informed, that on the banks of the Niger the average price of a negro is four camels' loads of dates; and it is no uncommon thing there for fifteen or twenty men to be given in exchange for a horse.‡ This is rather a promising beginning, as showing the estimation in which black souls are held at home. But to proceed; we learn from another authority, that the negroes in Algeria possess excellent qualities as slaves, whilst they are equally notorious for their defects when liberated. Immediately that they are emancipated from the yoke, they become thiefs, traitors, liars, bloodthirsty, and subject to the most desperate passions on the slightest opposition. We can already perceive some obvious inductions, which we leave to the reader's skill for the present. Before entering into an examination of the slave's position in Algeria, we will analyse the routes taken by the caravans from Soudan. After having crossed, on leaving Algiers, the Atlas chain and the territories of the nomadic tribes that live to the south of Medeah, you arrive among the Mezzabites, who extend into the Sahara and belong to the Djeridi, or palmers ; and beyond them stretches away the vast tract of treeless country known as the Great Desert, infested by the rovers of the sands, two tribes called Touaths and Tuaricks, the most southern of all the Berber race.§ The latter carry on

---

* See Miss Pardoe's City of the Sultan ; and Revelations of Russia, vol. ii. the chapters on Circassia and Georgia.

† For the Mahometan code of slavery, see Daumas' Grand Désert, p. 419.

‡ Baron Baude, ch. xvii.; Berbrugger, part iii.; Le Grand Désert.

§ Castellane, p. 280; Le Grand Désert; Baron Baude, ch. xvii.

their commerce and depredations as far as the Niger and into the vicinity of Timbuctoo ; and their daring industry in robbing caravans and man-catching constitutes all their riches. Some of them carry on the trade of bagging the negroes who go to seek salt in the salt-lakes of the desert : and more commonly they treat for them by purchase from the petty princes of the basin of the Niger. But, however obtained, man-flesh is always the staple of the goods brought back from these enterprises. Thieving is not an easy trade on camel's backs ; hence the kidnappers and conveyers of stolen men are different tribes, ,and a division of labour is found in the heart of the desert. Those who carry off the slaves do not export them ; they sell them to the Touaths, who carry on trade with the Djeridi and the Sahara. The ordinary wholesale price of the captives, at a rough estimate and without distinction of age or sex, is four camels' loads of dates, or its equivalent in merchandise. Now these sixteen quintals of dates, which amount, they say, in the Sahara to a value of sixteen francs, may, at the moment of exchange, have reached a value of forty francs by the carriage. The caravan performs a journey of seventeen days' march to reach the Mezzabites; and the nearer it comes to the coast, the greater is the danger of being pillaged. At last they arrive at Medeah, whence Algiers used to be supplied (before 1830); and the current prices, on which the merchants gained about 100 per cent, were 200 boudjous for men and 120 for women (360 and 216 fr.).*

Hence what the negro has lost by importation into the regency is a social state in which man is valued at much less than 40 fr. (1l. 15s.) The treatment of the people by the chiefs of Central Africa shews a brutal stupidity. They give twenty men for a horse; their superstitions require bloody sacrifices ; the animadversion of a priest is a condemnation to death; and human sacrifices are thought to honour the funerals of their princes. The prisoner of war, and also the man who does not pay his taxes, become slaves ; it is not thought a crime to sell your relations in a family, because being sold is not thought to be a misfortune ; fathers, elder brothers, and even mothers do not scruple to sell their children, whose life is of so little value; and the latter do not think it a misfortune, to judge from their appearance in the bazaars.†

Moreover, no idea of degradation accompanies slavery among the Mus-

---

* The returns of the bazaars of Egypt, Morocco, Tunis, and Tripoli confirm these data. The slaves brought to Gondar, in Abyssinia, most of them Gallas, are sold to the caravans by the child-stealers for one or two talaris per head (1 talari is worth 5 fr. 4 cent.). Castellane, p. 268 et seqq. ; Le Grand Désert ; Baron Baude, ch. xvii.

† Baron Baude, vol. iii. ch. xvii. on the Negroes. Gen. Daumas, at p. 8 of the preface to Le Grand Désert, inserts letters that reached France from Gorea, dated March 1847, mentioning that a rigorous blockade of the coast had been adopted by the French and British cruisers to stop the slave-trade; the latter undertaking to blockade Gallinas, where a number of negroes were congregated, ready to be shipped off. The proprietors of these slaves finding all means of evading the cruisers impossible, and being unable to feed them,

sulmans. The slave is the companion of his master; he may reach (in Morocco) the high dignities of the empire;* and if a woman, she enters the family by giving birth to a chief. The faithful who manumits his like, says the Koran, saves himself from the punishment of humanity and the torments of hell. Affranchisements by will, after a certain number of years, are very frequent; marriages with slave-women are still more so; and children born of black concubines belong to the condition of the father. Compare this with American prejudices, and infer, reader, where most Christianity is to be found.

Scrupulous Mussulmans think themselves bound to offer liberty to their slaves after nine years' good service, because it is thought that after that time they have paid their value in labour. Hence the law of Mohammed has many ways by which the slave may attain to civil liberty; and thus it seems to bring the slave by violence into the heart of the country, only to incorporate him in it at last by affranchisement. In this manner, slaves brought from the Niger to Barbary have not fallen into slavery, but have changed it. But between the slavery that they have left and that which they have entered is all the difference from Fetichism to Mahometanism, from perpetuity to transition. Thus, to interdict slavery in Barbary would be to throw back the negro into perpetual and hopeless barbarism.†

The origin, history, and language, &c. of the slaves is still in a great measure a mystery; even their country is unknown. In all other countries tradition makes mention of a heavenly origin for man; thus the Indians have their paradise of Brahma, the Jews Eden, the Pagans a golden age. But nothing like this occurs among the negroes. No divine institution has consecrated marriage amongst them; they are just as the ancients left them, even in the shape of their brain. The difference between the whites and blacks seems to be, that we have the principle of our perfectibility in us, whilst they only obey impressions from without.‡ Immersion into white society seems a necessary step to prepare the negro for liberty. But this must be done gradually, or it will lead to mischief; hence the sudden abolition of slavery in Africa would be to desert public duty. There exist plenty of facts to show that the abolition of slavery among the white slave-owners in North Africa would only be to perpetuate and consecrate it among the blacks. The regular army of Morocco consists of 20,000 blacks, regularly recruited by negroes brought by cara-

---

decapitated 2000 of them, fixing their heads on poles raised on the shore. Some French officers, who happened to be on shore, having bitterly complained of this to the chiefs, the latter replied : "What can you expect? If you no longer allow us to make money by our prisoners of war, we shall be obliged to butcher them all." Comments on this fact are unnecessary.

* Gräberg d'Hemso, Specchio geografico-statistico dell' Imperio di Marocco, 1836.

† Baron Baude, chap. xvii.   ‡ Ibid.

vans from Soudan and Timbuctoo. Though purchased as slaves, as soon as they enter the army they become free.

At Algiers they form a corporation under a *kait-laus-fan,* or chief. On the 1st of January 1833 it reckoned 390 persons, and in 1840, 408; whilst the census of 1849 gave a return of 563.* Scarcely one-fourth of them are born in Barbary. In Algeria there were 4177 in 1849.

They become porters, domestics, syndics, or an amin of their own corporation, who is answerable for their delinquencies. It would be impossible to keep up a supply of them without women, experience having shown the physical necessity of their society to prevent the ravages of nostalgia among the male negroes when condemned to the most moderate continence, owing to the amorous idiosyncrasy of their temperaments. Their numbers in Algeria amounted to 1800† in 1840, being principally females. The French guaranteed the observance of their laws and customs to the natives in Algeria at the time of the conquest; yet, by French law, the slave on crossing a French threshold is free. If the slave-trade had not been interrupted, this distinction would have been illusory; many natives would have been under-traders in slaves, and, eluding the law, would have obtained employment for slave-labour from Europeans.

On a broad survey of the case, it appears to us that religion sanctions, and prudence suggests, the adoption of a mild and organised system of negro slavery in North Africa; that it is the only efficient means of inoculating the negro race with a higher civilisation; that it increases the happiness of those exported, and diminishes the fearful slavery of Central African slaughter-houses; and that, if mitigated and checked by the clauses of Christian laws, it could not fail, under God, to be a great instrument in reclaiming the blacks from their brutal degradation, and in preparing them for higher political and social phases. If such results have flowed from Mahometan slavery, what might not be anticipated from a judicious system tempered by a Christian spirit! A superficial survey exclaims against the whole thing as a base injustice; but we are persuaded that a more scrutinising examination will verify our inductions; and that it will be found that justice and expediency both advocate as well as tolerate the practice of a mild system of slavery, as the most charitable and reasonable transitive measure for the negro in his progression from a moral and intellectual death in the basin of the Niger to the intelligence and spiritual life of the Caucasian variety.

In 1845 the state of slavery in Algeria was as follows : The slaves of natives remained in slavery, and when purchased by Europeans enjoyed the full benefit of the French law of enfranchisement ;‡ but the caravans to Medeah to supply Algiers had ceased, though the Arab tribes of the

* Tableau, 1849.

† Gräberg d'Hemso says that the negroes in the empire of Morocco amount to 120,000.

‡ Count St. Marie.

Sahara bordering on the Great Desert still import kafilas of negroes from Kachna Haoussa.* The number annually imported into the regency before 1830 reached from 600 to 800.

At present, owing to the check placed upon the importation of blacks into Algeria, there is much greater outlet for this traffic at Tunis, Tripoli, and in Morocco, into the latter of which countries 12,000 negroes are annually imported. †

Lastly, we are assured by the latter authority, a man of liberal politics and philanthropic heart, personally and experimentally acquainted with the subject, that the negroes are treated as members of the family in Morocco, and are much superior to and better off than the free blacks of Haiti; and he adds, that the negroes regarded slavery among the Mussulmans as a happiness. ‡

The negroes have many singular customs founded on the love of the marvellous and mysterious, universal in man, by which they work on the feelings and open the purses of the Arabs. Magic, witchcraft, mesmerism, and fortune-telling are phases of a general truth and reality, connected with the deepest mysteries of anthropology; and the crawling negro himself, though standing on the lowest platform of humanity, has been taught by instinct the profoundest truths of science and philosophy. Thus extremes ever meet, and Quashy tramples on the stagnant prejudices of Royal Societies and Academies of Science. The most remarkable of their magical processes in Algeria is called *djelep*, whose object it is to make the devil enter the patient's belly, whereby he foresees futurity. Those who wish to obtain prevision consult the chief of the negroes (*el kait-laus-fan*), ask him when the *djelep* will take place, and pay him a consideration to be present,—a favour never accorded to Jews or Christians.

The process can only take place on forty days in the year, which are fixed by the kait-laus-fan. The period begins commonly at the Rhamadan ; and the night before, the intended patients, generally women, resort, in company with an old man and woman, to a house set apart for negro superstitious practices. They are put into a room furnished with cushions and carpets, and concealed by a curtain. The old folks, with the help of some people, make a kind of hell-broth of gum arabic, an essence called sambel, and some pieces of wood called calcari, having previously killed four hens, with whose blood they anoint the joints of the

---

* Baron Baude says the importation of negroes has ceased, vol. iii. ch. xvii. Les Négres. his remark does not extend to the Arabs of Chamblet, &c. in the far south. See Le nd Désert, by General Daumas.

† The diversity in dialects is as great as that amongst the physiognomies of the negroes, some of whom, but especially of the negresses, are very handsome, particularly in figure. In the island of Cuba alone, we are informed by Baron Baude that the languages and types of the black slaves amount to 27. Baude, vol. iii. ch. xvii.; Le Grand Désert, preface.

‡ Baude, ubi supra.

patients; they then perfume them with the hell-broth and dress them variously—long caftans to the heels, belts and caps with shells that rattle together on dancing. That night, or next morning, twenty or thirty musicians arrive with negro instruments, crouch down under the gallery on the ground-floor, and in front of them is placed a carpet for money. The court is swept, but no mats are placed on it, and you must pass over it bare-foot. Those who have asked to be present are brought in one after the other; and one, or at most two of the possessed, accompanied by several negresses dressed like them, are brought into the court where they are perfumed, and then abandoned to themselves.

The musicians then strike up a terrific din, and the possessed begins to dance, at first gently, keeping the measure, and imitated by the negresses. But the movements of the chief dancer soon become quicker, and soon furious; he or she emits screams and displays contortions. This is the moment when the devil enters his body. Those present who wish to share the advantage then step forward, and throw down money, tapers, &c. The music becomes stunning, the dancer is more and more excited, till he falls down in a swoon. Then the old folks advance to perfume him, like the bottle-holders refreshing our prize-fighters, on their knees. The music ceases during this interlude; but presently recovering, the dancer launches forth again, the music recommences, and the same scene is repeated till our Taglioni is utterly worn out, when the devil is thought to have obtained an entrance.

Such is a ballet-dance and a *séance magnétique* among the negroes, which would slightly astonish the *coulisse* of the Opera or the lecture-room of a philosophical institution.

Learning has for many ages been in a great measure neglected by the natives of Algeria, and hardly any education is found among the Bedouins. Most human blessings have hitherto been curses, and most curses blessings, owing to the duality of all human concerns. The Arab has saved his faith at the expense of science, the European has lost it.

The Moorish and Turkish boys, as we have seen, are sent to school, where they learn to know, like Socrates, that they know nothing; but this knowledge, instead of making them Solomons, makes them like our philosophers, big with self-importance; and amidst the miracles and mysteries of creation, conceive themselves giants of learning. After some progress in the Koran, they are initiated with the same care into the ceremonies of their religion; and here at least they are commonly sincere, except after much friction with Europeans.

Their medical skill is small, like that of all allopathists; but they have the modesty not to lay claim to much. Few severe diseases are cured, predestination interfering with precautions, but helping to heal the sores of the soul better than our infidelity. Many trust in *magareah*, or charms; the natural instincts of man impelling them, though blindly, to some great truths, such as the great power of metals over the animal economy. Public baths are used in all diseases by these barbarians, who have condescended to take hints from the Romans in hygiene, and who have invented the water-cure long before Preissnitz. Like our hydropaths, the women are rather too fond of dabbling in water; but as they use it warm, it boils them and soddens them into premature age, instead of turning them into icicles and washing out their vitals.

In rheumatic and pleuritic cases they puncture the parts with red-hot iron, and the operation is repeated if necessary. Decoctions of sandegourah (ground-pine) or globularia fruticosa are administered for fevers; and the common scabious of Algeria, as a salad, frequently removes the ague. A drachm or two of the root of round birthwort or borietum is always given for colic and flatulency, and might be useful to certain gentlemen in Exeter Hall; the root of bookoke or arisarium, dried and powdered, is used for stone and gravel. One drachm of a dark-coloured drop stone, or the same quantity of the powder of the orobanche mauritanica, is reckoned good in diarrhœa. Six or eight grains of alkermes in honey are used in small-pox, and fresh butter is applied to prevent pitting. Inoculation and vaccination are in small repute; the Arabs not having more wit than ourselves in welcoming new blessings, if they upset cherished prejudices and canonical follies. They try to heal all simple and gun-shot wounds by pouring fresh butter almost boiling hot into the part affected; a treatment almost as absurd as the manly system of bleeding and purging employed in England till lately with hysterical females and gentlemen of delicate stamina.

For cases of swelling, bruises, inflammation, &c., the leaves of the prickly pear are roasted a quarter of an hour in the ashes, and applied as hot as possible to the part affected. This application is said to be useful in the Algerian climate; it brings the plague and other tumours to maturity, and it is known to cure gout without any repelling quality. Here is a windfall for the plethoric millionaires and used-up legislators and senators of Christendom. We may shortly anticipate a tour to Algeria as a substitute for the moors, after a gay session or season, for sporting members who live too fast. In bruises, inflammations, &c., some take powder of alhenna, and make it up in warm water into a cataplasm; this tinges the skin, and passes into the blood like iodine, mercury, and other blessings of our pharmacopeia. In fresh wounds the leaves of the virga aurea minor foliis glutinosis, called by them madramam, is found to have a good effect; and the root of toufailet or thapsia, roasted and applied hot to the hips, is of use in sciatica. A few chief medicines or douwas are made use of, in taking which no uniform practice is observed; a counterpart to our allopaths and allopathic treatment, the blind leading the blind, with blue-pill and black-draught as nostrums. Not much more caution is used in their administration than in ours, save that the preparations and infusions are seldom certain death. A handful of dry or green herbs is the usual dose. If taken in a decoction, they pound it in a mortar, and then pour, at a venture, half a pint or a whole, or more, of boiling water upon it. The Moors call these medicines traditional; but the few ingredients in the shops of their *tibeels* give reason to think that their doctors, like ours, are about as ignorant as their victims.*

There is very little knowledge of mathematics among these sons of algebra; astrolabes and other monuments of their sires' genius not being understood or used by the present race. The calendars they possess are the work of former ages; but they are not much used, the hours of prayer being commonly left to the will and option of the muezzin, or crier. The thalebs, or savants, pretend to a great insight into the value of numbers; nor are they singular in this respect, as Neo-Platonists and orthodox prelates have united in attaching special virtues to a trinity.

They have a high opinion of the knowledge of Europeans, asking any thing you like to write as charms.† Nor are they so idle in this superstition; for what wonders can exceed the every-day familiar doings by our fireside? The word was made flesh; and language written or oral is a heavenly telegraph and a standing miracle.

They have several musical instruments, but do not write down their compositions. The music of the Bedouin is rarely more than one strain. Social and domestic harmony is not unknown to the Arabs, though they are innocent and ignorant of imperial spies and republican order; but it must be admitted that their musical harmony is a serious violation of the organic laws of acoustics.‡

* Elephantiasis and blindness are common afflictions in Algeria, as in Palestine. For the native surgery, pathology, and pharmacopeia, see Blofeld; and the Vocabulaire d'Histoire Naturelle, by Dr. Lagger, at the end of Le Grand Désert of Gen. Daumas.

† Blofeld, p. 230.

‡ The *aralebbah* is composed of a bladder and string, and is the commonest instrument; it is very ancient, as well as the *gaspah*, or common reed, open at each end like the German flute, with three or more holes on one side, according to the ability of the person. Its compass does not extend beyond one octave. Strolling dervishes and Bedouins are chiefly conversant with this sort of music. After collecting a crowd of people, they chant over the memorable actions of the Prophet and other worthies. The *taar* is made like a sieve, consisting of a thin rim or hoop of wood, with a skin of parchment stretched over the top of it. This is their double-bass; they touch it skilfully with the fingers or knuckles under the palms. It is the tympanum of the ancients; is used all over the Levant and Barbary; and the shape is the same as that in the hands of the statues representing priests of Cybele and Bacchanals. Blofeld, p. 230; Berbrugger, part iii.

CHAPTER XVIII.

€uropean Population and General Statistics.

EUROPEAN SETTLERS — THE FRENCH COLONISTS — GENERAL CHARACTER OF
EUROPEAN SETTLERS—LATEST TABLES—THE COMPONENT NATIONS—SPANIARDS
—MALTESE — ITALIANS — NATIVE POPULATION.

WE now pass from the Semitic, Mogul, and Ethiopic races of Algeria,
i. e. from the natives, to the Indo-Germanic inhabitants, viz. the
emigrants. The French army requires a special chapter.

From the days of Rome and Carthage to this hour, the Semitic and
Indo-Germanic races have been antipathic, and their agreement on the
soil of Algeria is yet a problem. The various European races living
together under French sway appear to harmonise very well, presenting
specimens of most Christian populations. England is represented by the
half-Arabic Maltese, who are equally industrious by night as by day,
picking pockets at all seasons, but cutting throats after sun-down. The
former propensity and talent they share with the true Briton, the latter
only with the people of Ireland. Of Spaniards there is, and has long
been, a goodly crop in the coast towns; and the Mahonnese·make very good
settlers. The vagabond Swiss are casting sheeps' eyes at the highlands
about Setif; and *la belle* France has made a present of her scum or her
jewels to the marshes of Bona and the thistles of the Mitidja. Let us
examine this motley group, statistical minutiæ being referred to another
page.

The European population in Algeria amounted on December 31, 1850,
to 125,963 persons, thus analysed : — Province of Algiers, 56,784 ; province of Oran, 44,507 ; province of Constantina, 24,672.

In December 1849 the total population was 112,607 :—Province of
Algiers, 57,810; province of Oran, 35,246; province of Constantina,
19,551. The 57,810 of Algiers are nationally analysed as follows:—
30,897 French, 26,913 others. Oran, 35,246=15,959 French, and 19,287
others. Constantina, 19,551=11,149 French, 8402 others.*

* Tableau, 1850, pp. 88-109.

With regard to the French population in Algeria, that country has been a great issue to relieve the plethoric symptoms and bad humours of the mother-country, by draining it of its *mauvais sujets.**

The French do not constitute more than three-fifths of the European population, and are considered to be bad settlers, the only good agriculturists in the colony being foreigners. But, for a considerable period after the conquest (1830), the only French civilians attracted to Algeria were dealers in spirituous liquors, and men of bad lives,—in short, the usual tail of an army.†

According to estimates drawn up in 1842, 1843, and 1845, out of 1000 Europeans there were on the average 415 Frenchmen, 320 Spaniards, 116 Maltese, 103 Italians, and 46 Germans.‡

We shall present the reader with a hopeful description of the European population of Algiers in 1843 in a note below.§

We propose now to lay before our readers a concise tabular view of the development and most recent state of the population throughout Algeria; after which, we shall attend to the French army in Africa. The European civil population of Algeria amounted only to 45,000 persons in 1840, and on the 1st January, 1845, it had increased to 75,867;

---

* Baron Baude, to whom was confided the superintendence of the Paris police, soon after 1830, found that the disturbances (*émeutes*) were occasioned in general by a floating population of from 15 to 20,000 individuals, consisting partly of men out of work, partly of idle vagabonds. The prefecture of police organised a mode of enrolment for Algeria, and thus attracted 4500 of the most energetic of these rascals, who served with great distinction, and did good service to France against the Arabs. Of this more anon. At present we are concerned with the civil population.

† Baron Baude, vol. iii. p. 123.

‡ Dawson Borrer's Campaign in the Kabylie; Baron Baude; St. Marie.

§ French, German, and Swiss agriculturists and artisans have gone to the promised land of Algeria for profitable employment, but have found nothing but beggary, with the immorality that attends it; and depending upon public support, they have become incapable of honest labour. A body of wretched lazzaroni threatens to spring from the families of hardy peasants, who constituted the first emigrants. But the new system has invited a far worse class of colonists than these to Algiers. They are the scum of the seaports of France, Spain, Italy, and Greece: men who have forgotten home, and who speak a jargon of all the languages in Europe; men who have tried all professions, with equal want of reputation and success. Every where and in every thing they have been unfortunate. Each of them has a story to tell of his grievances, and the wrongs he has suffered from his government; and they are all martyrs to liberty. But the fraud is so gross, that when these men meet, they fairly laugh in each other's faces. Such is the higher class of society brought to Algiers. These are the men whom Europe sends to enlighten the poorer colonists, and to be an example to Africa. A third class follows, who will ruin the place, because conduct is as indispensable to success as capital. They are men who have been ruined over and over, by their folly, in all parts of the world. Speculators from England, from the United States, and from France, have flocked to Algiers, contributing nothing to its progress but their evil destiny; and they are most assuredly fated to repeat the failures which were the sole causes of their going there. Their wretched activity is never satisfied, unless when adding to the sum of losses which has always distinguished their career. These are the sorts of inhabitants which France has given to Algiers; and the result is only what might be expected from the acts of such agents. (Blofeld.)

thus doubling itself in less than six years. Of the latter number, 38,646 were Frenchmen, and 37,221 foreigners.*

We have now brought down the development of European population till 1840, after which it began to make rapid strides. Thus, on December 31, 1846, the total general amounted to 107,168 in the towns, and 2232 scattered over the interior.

The increase of the European population during the last three months of 1846 was 3858; but the general increase was in 1845, 20,699 individuals, whereas in 1846 it was only 14,079.†

Baron Baude informs us that the European population of Algiers in 1840 differed from all others. It is chiefly parasitical, few families staying there from choice.‡

Having brought down the movement of Algerian population to 1846,§ we shall proceed to lay before the reader the most recent official tables of the French government, bringing the development down to the year 1850; which will close this branch of our subject.

General European population of Algeria :

Dec. 31, 1846	109,400		
,, 1847	103,893	decrease . . .	5,507
,, 1848	115,101	increase on 1847	10,844
,, '1849	112,607	decrease . . .	2,494
,, 1850	125,963	increase . . .	13,356 ||

---

* St. Marie; Dawson Borrer.   † Borrer, p. 226.

‡ Baron Baude, vol. iii. chap. xiv. on European population, observes that most of the Germans in Algeria belong to the class of migrating journeymen, who pass into France, Italy, &c. under the name of *wandernde Burschenschaft.* You have all temperatures in Algeria. The Vandal race has continued thriving near Bugia; and though the cactus and the palm thrive near Algiers, the Alpine pine would grow on the edge of the snow at the top of Mount Atlas. The climate at this elevation is like that of the south of Germany, and it is there that the Vandal tribes have continued. We are glad to find that the common sense of Baron Baude has told on foreign governments and speculators, and that they are endeavouring to adapt the selection of the emigrants to the localities. People from the Vosges are to be located in hilly districts; and a Swiss company at Geneva is agitating a grand system of emigration to the high cool table-lands round Setif, the healthiest part of the colony.

§ M. Jucheran de St. Denis supposed that the regency at the beginning of the 18th century did not contain more than 2,000,000 of inhabitants, the population of most of the towns having decreased considerably up to 1830. The writer believes that, in stating the number at 800,000, we are very near the mark; but he does not include in this calculation those who dwelt between the Little Atlas and the Sahara, who were never entirely subject to the dey of Algiers, and estimated at 230,000 souls, making a total of 1,030,000, or scarcely 47 inhabitants to every square league, or 5760 acres.

The following comparative tables of European population in Algeria were published by the French government in 1843 : January 31st, 47,150; March 31st, 47,038; May 1st, 47,544; June 31st, 55,122; October 1st, 57,642.

At the latter date the European population was thus composed: French, 24,274; Spaniards, 18,548; Maltese, 6402; Italians, 6332; Germans, 2086: total, 57,642. At the end of June 1842, there were only 40,000 European civilians in Algeria, the military amounting to about 80,000, shortly to be increased to 94,000 or 95,000. Emigration, then greatly on the increase, had more than doubled in the space of three years, as at the end of 1840 there were only 28,736 civilians in the colony. (Blofeld.)

|| Tableau de la Situation, &c. 1849-50, pp. 88 to 109.

The births in the three years from 1847 to 1849 were as follows :

Provinces.	1847.			1848.			1849.		
	French.	Foreign.	Total.	French.	Foreign.	Total.	French.	Foreign.	Total.
Algiers . . .	1380	1141	2521	1307	974	2281	1600	1080	2680
Oran	540	478	1018	659	660	1319	795	832	1627
Constantina . .	500	244	744	455	292	747	595	304	899
Total	...	...	4283	...	...	4347	...	...	5206

This gives in 1847, 4·12 per cent; in 1848, 3·77 per cent; and in 1849, 4·62 per cent of the total births; and in 1847, 4·50 French, and 3·71 foreign per cent; in 1848, 3·77 French, and 4·83 foreign; in 1849, 5·15 French, and 4·05 foreign births per cent; the mean being 4·47 French, and 3·93 foreign per cent.

Passing to marriages, we find 1029 in 1847; 1052 in 1848; and 1097 in 1849; giving, in 1847, 0·99 per cent; in 1848, 0·91 per cent; and in 1849, 0·96 per cent. Of these marriages, in 1847, 553 were French, 175 mixed, and 301 between people of other nations. In 1848, 553 French, 171 mixed, and 328 foreign. In 1849, 619 French, 153 mixed, and 325 foreign.

The deaths from 1847 to 1849 were as follows :

Provinces.	1847.			1848.			1849.		
	French.	Foreign.	Total.	French.	Foreign.	Total.	French.	Foreign.	Total.
Algiers . . .	1683	1283	2966	1416	1097	2513	2112	1806	3,918
Oran	497	722	1219	698	679	1377	1861	1697	3,558
Constantina . .	552	426	978	556	389	945	1916	1101	3,017
Total	...	...	5163	...	...	4835	...	...	10,493

Hence the mean mortality was, in 1847, 5 per cent; in 1848, 4·25 per cent; in 1849, 10·59 per cent. The increase in 1849 was caused by the cholera and the distress of the colonies agricoles.*

* The mortality in some of the chief towns presents the following results:

Towns.	1847.	1848.	1849.
	Per cent.	Per cent.	Per cent.
Algiers	4·87	4·43	5·42
Blidah . . . . · . . .	7·64	5·67	10·59
Oran	5·21	4·49	10·71
Tlemsen . . . . . . ,	4·72	3·29	3·52
Constantina	5·60	4·42	6·10
Bona	4·70	4·68	10·38
Phillippeville	8·20	7·	10·
Bugia	3·83	1·22	3·

Baron Baude states that the greatest vitality among the Europeans is seen in the Spaniards, and the least in the Germans. The Spaniards have also a greater proportion of women than any other European race in Algeria. They long occupied Oran, and have always had a more intimate connection with the coast of Barbary than any other European people.

Algiers alone possessed more Spaniards in 1841 than all the Spanish possessions founded by Cardinal Ximenes in Africa, after an occupation of 150 years. Though few in number at Bona, they form one-third of the population of Algiers and Bugia, two-fifths of that of Mostaganem, and one-half of the European population of Oran.* Most of the fiacre-drivers in Algiers are Spaniards ; and several flourishing villages, peopled entirely by industrious Mahonnese, have been established near the capital.

Turning to the Maltese, the next element of European population, we find that this energetic little island alone has supplied more people to Algeria than all Italy. Great numbers live at Bona, where they act as porters, &c., and drive a thriving trade.†

As there is a surfeit on the island, which shows strong apoplectic symptoms, many Maltese emigrate exoterically; and being Africans by origin and language, they find themselves at home in Algeria. There are 4000 at Tunis, and their relative numbers diminish in proportion as you advance to a distance from their island; they form more than two-fifths of the European population at Bona, at Bugia one-sixth, at Algiers one-twelfth, at Oran one-twenty-fourth.

As regards the Italians in Algiers, they appear formerly to have had important relations with the regency during the bloom of Italian nationality and commerce, fostered by the liberal and republican spirit of

---

The letter of the *Times* Paris Correspondent of Jan. 7, 1853, states, that the deaths among civilians at Bona, from the 1st of July 1852 to the 21st of December, was 683 persons, of whom 544 were Europeans. The mortality from the 1st of January to the 30th of June 1852 is stated to have only been 183 persons.

* Several interesting episodes occur in the Spanish occupation of Oran, displaying the energy, valour, and discipline of the Spaniards in their better days. In April 1622, the Arabs of the Habra assassinated three Spaniards ; and Don Juan de Manrique thereupon starts in the evening with 700 foot and 200 horse, falls on the Arabs at daybreak, and brings back the following night to Oran 319 prisoners and 1200 head of oxen. The distance of Oran from the plain of the Habra is sixty kilometres ; and Don Juan had traversed it twice in the space of thirty hours. The following July the same spirited commander beat in detail a body of 2700 Janissaries, 1400 horses, and a numerous Arab infantry. (Baude.)

† The Maltese population is one of the most fruitful in the world. In 1530, when Malta was given over to the Knights of St. John by Charles V., it contained only 15,000 inhabitants ; and after the desperate siege of 1565 by the Turks, it had only 10,000. In 1590 it contained 27,000 ; in 1625, 40,000 ; the census of 1632 gave 51,750 ; in 1798 it had 90,000, and Gozo 24,000 ; and now it reckons 120,000 Maltese alone ; whilst it contains only 30,000 hectares, *i. e.* about the same surface as the ridge of hills behind Algiers (French *massif d'Alger;* Arab *sahel*). The specific population of all France, by kilometres, is 60 ; for the Departement du Nord, 171 ; for the arrondissements of Sceaux and St. Denis, 357 ; at Malta it is 400. For these facts see Chevalier de Boisgelin's Malte ancienne et moderne, 3 vols. 8vo, Paris, 1809 ; Dapper, Description de l'Afrique, Amstm. 1626 ; Voyage du Duc de Raguse, 4 vols. 8vo, Paris, 1838.

Pisa, Genoa, &c. Nothing can be more interesting than to trace the spirit, intelligence, and industry of Italians under a national and liberal flag, at a time when the rest of Europe, which owes to them its light, was plunged in darkness. It is cheering to be able to produce facts to establish that a manly and industrious spirit is developed in Italy as elsewhere, when its growth is not stunted or blighted by the poison of a hypocritical hierarchy, or trampled on by the foot of the stranger. Dear to us are all such memories of the fine qualities of this historical people, as auguries of a better future yet in store for her.

As regards the present position of the Italian population in Algeria, there is not quite one Italian to three Spaniards in the colony. The Italians form one-fourth of the European population of Mostaganem, one-sixth of that of Bona and Oran, one-fifteenth of that of Algiers; and they are hardly seen at Bugia, whose commerce was in their hands in the middle ages.*

* Native Population.

The sum of the indigenous population in 1845 was 1,983,918; thus analysed:—Algiers, 490,168; Oran, 477,034; Constantina, 1,016,716. It was supposed that the whole native population of Algeria amounted, at a rough estimate, to 3,000,000. See Tableau, &c. for 1846, and Borrer, chap. xv.

Inhabitants of the Three Provinces.

The Negro slaves throughout the Regency were supposed to amount to 10,000; and the number of free blacks was estimated to exceed that amount.

Coming down to the three years ending 1849, we find that, on the 31st Dec. 1847, the native population in the towns amounted to 87,505; and in Dec. 1849, to 84,133. The Moors have decreased, and the Jews have remained stationary. The Negroes have increased from 3348 to 4177 individuals. The Kabyles, Mozabites, Biskris, and other corporations in the towns of the civil territories, amount to 10,742; a diminution from 1846.

Native Population of Towns, nationally and sexually.

Total of Mussulmans in the province of Algiers, 27,773; in the province of Oran, 12,350; and in the province of Constantina, 20,805. Total of Negroes in the province of Algiers, 1714; in that of Oran, 1531; in that of Constantina, 932. Total of Jews in the province of Algiers, 7289; in that of Oran, 7749; in that of Constantina, 3990. Hence the Mussulman population of the towns of Algeria amounted, in Dec. 1849, to 60,928; the Negro population to 4177; and the Jewish population to 19,028 : general total of Algeria, 84,133.

With regard to the tribes, the reader has already been presented with the statement of their general numbers in Algeria, and their particular numbers in the provinces.

According to the last census, the population of Algeria, on the 31st December 1852, amounted to 246,431 individuals (this must mean without the tribes): namely, 124,401 Europeans; and 122,030 natives, inhabiting the territory occupied by the Europeans. The former consisted of 69,980 French; 35,129 Spaniards; 7408 Italians; 5609 Maltese; 3025 Germans; 1323 Swiss; 526 Belgians and Dutch; 483 Irish; 258 Poles; 145 Portuguese; and 515 others : and was composed of 29,451 men, 28,232 women, 40,073 boys, and 26,645 girls. There were 121,226 Catholics, 2561 Protestants, and 614 Israelites : 80,143 resided in towns, and 44,258 in the country. They were divided into 32,826 families, and occupied 16,215 houses.

CHAPTER XIX.

GENERAL SURVEY OF COLONISATION—GOVERNMENT DECREES ON RURAL PROPERTY
—CONCESSIONS IN LAND—DECREE OF THE PRESIDENT, 1851—STATE OF GENE-
RAL COLONISATION IN THE COLONY—PROVINCE OF ALGIERS—CIVIL TERRITORY
—MILITARY TERRITORY—PROVINCE OF ORAN—CIVIL TERRITORY—MILITARY
TERRITORY—PROVINCE OF CONSTANTINA—CIVIL TERRITORY—MILITARY TERRI-
TORY—NEW PROJECTS—PENITENTIAL COLONY AT LAMBESSA—AGRICULTURAL
COLONIES—ST. DENIS AND ROBERTVILLE, ETC.

WE shall now proceed to examine the important subject of concessions of land, and of agricultural and other colonies established by the French government or by private speculation in Algeria.

Though there was a sprinkling of emigrants every year after the conquest, offering a crop of ears and heads to the Hadjutes and other tribes near Algiers, it was not till 1842, after the humiliation of Abd-el-Kader, that the era of serious colonisation began. Before that time there were only one or two villages, with large barracks and hospitals. In 1842 villages sprang up like mushrooms near Algiers, and the high land or massif at the back of the city, forming a barrier between the Mediterranean and Mitidja, was brought into a state of cultivation; whilst other emigrants went to Boufarik and Beni-Mered, stations on the road to Blidah, five or six hours south of Algiers, at the foot of the Little Atlas, across the Mitidja plain.*

Several causes may be perceived for the slowness and dulness of the current of colonisation in Algeria. The port of Algiers is not free,† and much of its trade has been driven to Tunis, Tripoli, Tangiers, &c. through the short-sighted policy of the French government. Another drawback

---

* This plain contains 1,500,000 acres of arable and pasture land, but only a small portion of it is now under cultivation. Vast tracts are still lying waste, sacrificed to the palmetta and squills; whereas, before the French came in 1830, it was cultivated by the Arabs, who grew more corn in it than they wanted. Now there is not sufficient for home consumption, and the price of corn was enormous at Algiers in 1846. Dawson Borrer's Campaign, &c. chap. xiv. p. 222.

† Recent enactments have removed almost all duties on imports. See Appendix.

to colonisation has been the great tardiness of administrative forms necessary for the establishment of emigrants. Though assignments of land (*concessions*) are promised to the colonists, eighteen months will perhaps elapse before you are put in possession of your property.* The calls of the French army on the budget may account for the beggarly sums given to colonisation, just as the heavy items of our sinecures may explain our magnificent tribute to national education. The annual expenses of the colony amount to 100,000,000 fr.; and in 1847 the budget for Algeria placed under the head of colonisation 1,734,000 fr.; but more was doubtless extracted by special *projets de loi.*†

Let us now devote a few words to the machinery of colonisation in Algeria, and then analyse the state of the colonies; and we shall proceed to consider first the decrees of the French government on rural property in Algeria; secondly, the concessions or assignments of land; and thirdly, the government or individual colonies.

The decree of July 21st, 1846, on rural property in Algeria had for its aim the securing of the peopling of Algeria and the fertilising of its land, by placing rural property, which had hitherto been in a vague and disputed state, on a firm and sure foundation. To this end, it ordained the ascertaining, by enactments of the minister-of-war, of the extent of the territories within which the title-deeds of property should be valid; the returns to the government of the names of the Europeans or natives who asserted their claims to properties within those territories; and the verifying of the titles produced, by council of the disputant (at a later date by the councils of prefecture in each province). The decree carefully defined the requisite conditions for the titles to be held valid; and it ordained their application on the spot through a councillor, and their sanction, if necessary, through the council of prefecture. In cases where the property marked out was claimed by many disputants, the council was enjoined to suspend judgment till the civil tribunals could pronounce on the question of right. Lastly, when the title produced did not fulfil the necessary conditions of the decree, the council pronounced it null and void. Yet in this case the government was bound to hand over to the evicted occupant, on his request, a hectare (2·47 acres) of land for every 3 fr. of rent stipulated in the last act of possession, if a certain period had elapsed previously to the

---

* It is a well-known fact, that men of capital coming to Algeria, under the auspices of the minister-of-war, have stayed six years before obtaining the original concession. Others provisionally established on a tract of land have built upon it ; and when at last a definite answer came, the title-deed to it has been refused ; and not being able either to alienate or mortgage it, they have been ruined. Hence many of the poor emigrants have been reduced to a very desolate state ; and the villages on the Sahel and Massif were in 1846, with one or two exceptions, the types of desolation. Perched upon most arid spots, distant from water, victims of the sun and sirocco, they rose among endless tracts of palmetta and prickly bushes. Visions of Utopia terminate in dwarf-palms and disease. Dawson Borrer, chap. xiv. † Borrer, chap. xiv. p. 222.

promulgation of the decree of July 21st, 1846, relative to concessions in land. These measures of the government were much complained of at first in Algeria; but being shortly better understood by the colonists, they ended by submitting, without making any serious difficulties, which have only originated in the ignorance of native proprietors, and the difficulty experienced by the government in discovering the real state of their property.

We shall submit to the reader the following brief outline of the state of European and native properties in the three provinces.

Province of Algiers.—In 1850, the number of properties declared were 592, and marked out 492 ; but as most of the native properties were claimed by a whole djemaa or assembly, and subdivided into small parcels, the real number of properties amounted nearly to 800 in all. Of these, 359 had been finally decided, including 109 connected with natives ; 162* were in suspense ; 176 annulled : total number of decisions by the councils of prefecture, 697. Only five properties had to be surveyed and ascertained in Dec. 1851.†

Province of Oran.—In this province, the civil territory only has been subjected to the decree of 1846. The number of properties declared amounted to 113 ; 54 belonging to Europeans, and 59 to natives. By the rejection of the titles of some proprietors, the number of properties whose limits are ascertained has been reduced to 87, 45 of which belong to Europeans.

The properties whose titles have been verified and confirmed amount to 73 ; whereof 40 are European, and 33 native. There are five in suspense.

Province of Constantina.—The arrondissement (hundred) of Bona, and the banlieues (precincts) of Constantina and La Calle, had been brought under the decree of 1846 in 1849. In December 1850, 425 properties were ascertained in the arrondissement of Bona,—53 European and 372 native,—covering 29,427 hectares (72,683·22 acres) of land. There remained 131—11 European and 120 native—to ascertain. 48 titles had been legally confirmed, 11 European and 37 native ; and 277 protocols,

---

* Of these 162, 65 are native, the greater number of whom are competitors (*coprétendants*) with the state. A special commission is engaged in trying to settle the disputes, and make *partitions à l'aimable.*

† The following table gives at a glance the state of property and of territory subject to the decree in the province of Algiers :—

Plain of the Mitidja, 107,466 hect. 63 a. 40 c. ; Sahel, 29,716 hect. 81 a. 27 c. ; right bank of the Boudaou to the Isser, 31,020 hect. 15 a. ; total, 168,203 hect. 59 a. 67 c. (415,181·41 acres). Property accruing to Europeans, 36,875 hect. 46 a. 86 c. (91,081·25 acres) ; natives, 11,511 hect. 74 a. 57 c. (28,500 acres) ; the State, 94,796 hect. 99 a. 1 c. (234,146·12 acres) : total, 143,184 hect. 20 a. 44 c. (343,664·48 acres). Disputed property between individuals, 7,066 hect. 86 a. 43 c. (17,453·02 acres) ; the state and individuals, 17,952 hect. 52 a. 80 c. (44,341·44 acres) ; total, 25,019 hect. 39 a. 23 c. (61,796·93 acres).

or registers, about delimitations, remained to be settled, 42 European and 235 native : 5 were in suspense, referred to higher tribunals.

14,000 hectares (35,000 acres) of land were voted to the banlieue of Constantina by a decree of the President of the Republic, March 1849. The state of properties had not been finally determined, arranged, or ascertained round La Calle, in November 1849.

The next point we shall analyse is the nature and state of concessions in Algeria.

There are three distinct modes employed there for the alienation of the government property of the colony: 1. Sale by public auction ; 2. Sale by instalments ; 3. Concession.

All town property, cultivated land, or land built upon, is disposed of by the first two methods ; the third only applying to new plots of ground to be built upon in new villages, to uncultivated land, &c.

A decree of 1851 modifies the regulations of 1845 and 1847, according to which, concessions of and under 24 hectares (60 acres) are authorised in civil territories by the prefect, with the advice of the council of prefecture ; and in military territories by the general of division, with the advice of the consulting commission. The concessions of 100 hectares (250 acres) were given by decree of the President of the Republic, with the recommendation of the minister of war and the consultative committee of Algeria. Every concessionary, after a reasonable time, was bound to pay a fluctuating but perpetual rent to the state. These concessions were only given on condition of the grantee being able to fulfil his engagements ; and before his entrance into possession, he had to give 10 fr. per hectare (8s. 4d. per 2·47 acres) caution-money.* Many evils resulted from these regulations in practice, by multiplying formalities, creating delays, &c., especially in the cases of small concessions. To remedy these evils, a project was submitted by the minister to the Council of State in July 1850, intended as a substitute for the old arrangements about concessions. But as it would take time to get this voted by the Assembly, and a reform in the system of concessions was imperative, the government determined to propose to the President a transitory modification of the worst regulations. Hence originated the decree of the Presi-

---

* Every concessionary received, at the moment of being put in possession, a provisional title indicating the conditions imposed, and the delay granted for their accomplishment. During the whole period of this delay the concessionary was not able to consent to any substitution, alienation, or mortgage, without the especial leave of the administration, under penalty of forfeiture. At the expiration of the delay fixed by the provisional title, or before if the concessionary demanded it, a valuation of the labours effected was made. If all the conditions were fulfilled, the provision was converted into a final concession. If the conditions were not, or were only partially fulfilled, the concessionary forfeited, totally or partially, his land ; or, according to circumstances, he could obtain a prolongation of the delay. In the latter case, a new valuation took place, as the prolongation of the delay expired when the concessionary obtained a final title, or was ejected.

dent of the 26th April, 1851, of which the principal articles are inserted below.*

During the four years from 1846 to 1850, we are informed that the colonisation of Algeria by Europeans has slowly but surely progressed ; though accidental circumstances, such as the Revolution of 1848 and the cholera, have retarded and disturbed its advance. The peopling of the old centres of colonisation has continued ; the territory of some of them has been increased, and new creations have taken place ; finally, the concessions that have been made outside the villages, and their occupation, have stamped a new character on the colonisation of Algeria. One circumstance has especially affected it, i. e. the creation of agricultural colonies, which took place in 1848, in consequence of a decree of the National Assembly of September 19th, and to which a credit of fifty millions of francs (2,000,000l.) had been voted.†

A full description of the state of European colonisation in Algeria from 1847 to 1850 falls under two distinct heads. The first relates to the centres of population and agricultural explorations established since 1847, down to the 31st December, 1850, excepting the agricultural colonies. The second gives a statement of the situation of the agricultural colonies in 1848 and 1849.

* By article 2, all concessions under 50 hectares (125 acres) could be authorised by the prefect. By article 3, all future acts of concession in Algeria confer the immediate possession, on condition of accomplishing the required steps. Article 5. If the concessionary does not lay claim to immediate possession after the expiration of three months, at the hands of the local authorities, his title is forfeited. By article 7, the concessionary can mortgage or transmit all or part of his land. By article 8, in the month following the expiration of the term granted for the fulfilment of the conditions, or sooner if wished, the valuation is made by a commission of three members. By article 9, if all the conditions are found fulfilled, the prefect declares the property freed from the conditions exposing the title-deed to be cancelled. If differences arise between the *directeur des domaines* and the prefect, the matter is set at rest by the minister-of-war. If ejected, the property reverts to the state ; but if it has been improved, it will be put up to auction. The prices of the auction, deducting the charges, revert to the concessionary.

If no one bids, the property reverts to the state. Provisional concessions made before 1847 can be exchanged for another title-deed, conformable to the clauses of the decree of 1851, in which the delay for the accomplishment of the conditions imposed will be determined by the original decree. The same regulations are applied to the military territories, the general commanding the division taking the place of the prefect.

† A report of these establishments was laid before the Assembly, at the end of 1849, by a special commission sent to the spot to examine into them. From this report it appears that these establishments have had as results : 1st. to restrict, on economical grounds, the new creation of centres, which would otherwise have been undertaken in greater number on the ordinary credits of colonisation ; 2d. to bring in a check to the demands for concessions of small extent, because the claimants who seek to obtain them are bound to support their claim by pecuniary resources, which has caused many to seek in preference an entrance into the above-mentioned agricultural colonies.

PROVINCE OF ALGIERS.

I. *Colonisation in the Civil Territory.*

1. Centres created by the administration :—

Since 1846, five centres of population have been created in the civil territory of the province of Algiers; *i. e.* Mouzaia, La Chiffa, Arba, the Fort de l'Eau, and Affreville.

Mouzaia. This centre is situated to the west of Blidah, and at the foot of the northern slope of the Atlas. Its population is nearly completed, and its sanitary condition is very satisfactory. Great advances have been made in clearing the land, and the colonists have also planted a great deal. In short, the encouragement that they have received has given birth to favourable results.—*La Chiffa* has the same topographical situation as Mouzaia. The village is in good condition, and the colonists are able to dispense with the assistance of the state.—*Arba.* This place, which was founded in 1849, to the east of Blidah, is a centre intended to continue on this side the colonisation of the area of the Mitidja plain, in connection with Dalmatia, Souma, and Fondouk, which are already created. Though so recently founded, Arba has already made very remarkable advances; it is one of the villages that holds out the most promising hopes.—*The Fort de l'Eau.* This village, situated to the north-west of Fondouk, on the sea-shore, was founded in 1850, and is entirely peopled by Mahonnese. These colonists labour with zeal and diligence, and their produce is already very considerable.—*Affreville.* This centre is situated in the vicinity of Milianah, and its colonists are chiefly devoted to the cultivation of gardens. The colonisation of the surrounding country, which has already commenced, is calculated to secure the prosperity of this village.

The centres created previous to 1847 in the civil territory of Algiers amounted to 32, including seven towns:—Algiers and its banlieue, Cherchell, Ténès, Koleah, Blidah, Medeah, Milianah; and the 25 following villages:—Ain-Benian, Sidi-Ferruch, Cheragas, Ouled-Fayet, Dely-Ibrahim, L'Achour, Drariah, Saoula, Notre-Dame de Fouka, Fouka, Douaouda, Zeradia, Mahelma, Sainte Amelie, Saint Ferdinand, Douéra, Baba-Hassen, Crescia, Boufarik, Beni-Mered, Joinville, Montpensier, Dalmatie, Souma, and the Fondouk.

The administration has been engaged in completing the works at these centres, and in securing their peopling.

We shall here give a rapid survey of their present state in what relates to colonisation.

Algiers.—Fertile market-gardens stretch around this town in a radius of from eight to ten kilometres from the sea; but agriculture, properly speaking, only begins beyond and amongst the Sahel hills.—*Ain-Benian*. The present state of this village is satisfactory enough : it is a half-agricultural, half-maritime station. It had been originally built to receive twenty families of fishermen; but experience having shown that the colonists could not subsist on the mere produce of their fishery, a certain amount of land was conceded to them.—*Sidi-Ferruch* is a maritime village, created, like Ain-Benian, by commercial speculation. Some land has been brought into cultivation and planted in its vicinity.—*Cheragas* is in a flourishing condition. The state of property among the colonists is very good; and those who are the least well off find work among their rich neighbours, which improves their condition greatly. Some colonists, who are natives of the department of the Var in France, have introduced a branch of industry into this colony that has already become somewhat spread, *i. e.* the manufacture of scents. This village has also obtained great importance through the number of isolated farms scattered around it.—*Ouled Fayet*. This village is completely settled, and succeeds well; its prosperity being chiefly due to the presence of several concessionaries in easy circumstances, who give employment to the less-favoured families. —*Dely Ibrahim* is in a satisfactory state, but the development of cultivation is there impeded by the inadequate supply of water, and by the insufficiency of its territory.—*L'Achour* is one of the villages that were earliest created in the Sahel, and its territory is well cultivated.—*Drariah* is one of the oldest colonies, and is also perfectly settled, its success being entirely secured.—*Saoula* is in the same state.—*Fouka* is a centre that was originally peopled by liberated soldiers, on the system of Marshal Bugeaud; at a later date civil colonists have been settled there. Hitherto their situation is rather unsatisfactory; but it will probably improve when the road from Blidah to the sea shall be opened, and the works at the maritime creek of Fouka shall be finished.—*Notre Dame de Fouka* is in a very prosperous condition, and is principally inhabited by a population of fishermen, who do not, however, neglect agriculture. The soil is of a good quality, and it is very productive in cork-trees and oaks. The breeding of swine and the preparation of charcoal constitute the principal elements of its thriving state. — *Douaouda* is a village situated near Koleah, on the road to Algiers; is in a flourishing state, and finds an easy market for its produce. Its chief prosperity is centred in the breeding of cattle; and endeavours are being made at the present moment to extend its territory by the addition of some meadow-lands. The soil is good, although covered with brushwood.—*Zeradia* is not as yet in a very thriving condition, owing to its distance from the great centres of population. The system of irrigation likewise needs studying; but the sanitary state of the place has been improved by the draining of the marshes that exist on its territory.— *Mahelma* is partly peopled by liberated soldiers, and partly by civil colonists. The soil is good, the land is being fast cleared and brought into cultivation, and the future promises well.—*Saint Ferdinand* is a village that has been formed by military convicts. The nearness to Algiers, and the facility it enjoys of getting rid of its produce, are considerable elements of success. Efforts are now being made to increase its territory, so that it may receive other concessionaries.—The same remarks apply to the village of *Sainte Amelie.*—The town of *Koleah* has been entirely repeopled; the state of the colonists is satisfactory; its territory is well cultivated, and attempts are being made to extend it.—*Douera.* In the first instance, this centre was devoted to manufacturing industry, and it suffered greatly by the opening of the road from Algiers to Blidah through the plain. At a later date the colonists have pursued agriculture, and have laid the foundations of a more solid prosperity. The territory is covered with dwarf-palms, but it is well adapted to most kinds of cultivation. The local authorities are preparing to erect near this place two hamlets, by handing over to the purposes of colonisation a considerable stretch of country situated not far from Douera, and previously set apart for the service of the army.—The village of *Baba Hassen* is entirely settled, and on the high-road to prosperity. The dwelling-houses are, generally speaking, substantially erected, with land attached to them, to which additions are made every year. The colonists cultivate, besides corn, olive-trees, vines, and especially tobacco; and they already possess considerable flocks of sheep. — *Crescia* is also in an excellent condition; the colonists are steady and industrious workmen, having opened several quarries, formed several long and difficult roads, and made numerous clearances.—*Boufarik* was formerly an unhealthy locality, but it has recovered its salubrity, owing to the numerous plantations that have been made, and the draining that has been carried out by the exertions of the government. The buildings are substantial and handsome, and it is surrounded with numerous farm-houses.—*Blidah* is an old Moorish city, containing within its circuit several handsome European structures. Colonisation has made rapid strides in the neighbourhood of this town, which has four well-settled villages close at hand.—*Beni-Mered* is a village progressing very prosperously; the surrounding cultivation yields excellent produce, and the colonists are able to support themselves; the clearances and cultivation that they have effected are considerable, and their private dwellings are daily increasing in number. — *Joinville* is an extremely thriving village, whose territory is scarcely extensive enough to meet the wants of its energetic inhabitants; but it is difficult to procure additional land there. The houses are generally well built, and the future prosperity of this colony is quite certain. — *Montpensier* is likewise in a good situation in every respect, such as cultivation, buildings, and plantations. The proximity of Blidah provides it with an easy channel for disposing of its produce.—The colonists of *Dalmatie* work hard and successfully; the village is surrounded with extensive tracts of arable land and plantations; and the implements of labour are in general sufficient to meet the demand. This village possesses two flour-mills on the Oued-Beni-Aza.—The village of *Souma* is in a prosperous state as regards clearances and cultivation, and building is increasing.—The *Fondouk* has advanced slowly, owing to its isolated position in the Mitidja. The approaching continuation of colonisation in the eastern circumference of the plain will rescue this village from its languid state.—*Medeah.* Colonisation begins to extend around this town, to which a territory for settlements has been annexed.—Some capitalists have founded near this town the village of *Mouzaia-les-Mines*, occupied by miners' families engaged in the neighbouring copper-mines. These people are housed in dwellings built with the funds of the company; and there is room at this spot for an agricultural population, which would favour the settlement of 500 workmen united in this village.

Agriculture has somewhat extended around *Milianah;* several farms are being established in its vicinity ; and the district affords numerous and abundant springs, which may be turned to account. Two water-courses, the Oued Boutan and the Oued Anasseur, turn several flour-mills in this district. — Cultivation is developing around *Cherchell* in a territory of 3000 hectares (7500 acres) ; and its prosperity will be secured when it is bound to Algiers and Blidah by a series of villages. This undertaking was begun in 1848, by the creation of several agricultural colonies, of which more anon.—*Tenes* still suffers from the isolation of its position ; the accumulation of troops on this point a few years ago having given birth to an ephemeral activity that departed with its cause. Hence there has resulted some distress. Some plantations of vines, fruit-trees, and mulberries were made in 1849 and 1850, and the growth of cotton has been attempted. The recent annexation of an agricultural territory to Tenes will shortly permit the development of this branch of culture there.*

---

* Besides the state, several private capitalists have forwarded the work of emigration, especially in the Sahel, where there are many scattered farms. The concession of the Trappists at Staoueli deserves a special mention. Its extent is 1020 hectares (2519·40 acres), whereof about 300 hectares (741 acres) were sown or meadow land in 1849; the value of the buildings having risen to 300,000 fr. (12,000*l.*) Several fountains flow in the property; and it contains an orangery of nearly 1100 feet in length, and a *pépinière* (nursery-ground).

That part of the Mitidja, near Arba, called Beni-Moussa, was the first peopled by colonists; and though long afflicted with various plagues and scourges, the perseverance of the colonists has at length overcome the Arabs, mosquitoes, and droughts. Most of these farms are in a very good state now. Free colonisation has extended on the left bank of the Haratch ; and the quarter of the Krachenas, to the east of the Beni-Moussa, contained eighteen European farms at the end of 1849. It has also extended in the plain, in the arrondissement of Blidah. At the end of 1849 these undertakings embraced about 14,000 hectares (34,580 acres), and 380 individuals.

II. *Military Territory.*

1. Administrative centres :—

There are no new creations in this territory; but the old centres now form little, chiefly European, towns.   They are the following :

Orleansville.   This is an important town, founded in 1843, which suffered, like Tenes, from the crisis of the last few years.   Cultivation, however, is beginning to extend around the town, and it is anticipated that the creation of the agricultural colonies in 1846 will give a new impulse to cultivation.—*Aumale, Boghar,* and *Teniet-el-Had* contain but few families of colonists, and the surface to be colonised requires extension there. —*Dellys.*   Down to 1849 the colonists of this place had given more attention to town constructions than to agriculture.   The concessible territory *extra muros* was of small extent ; but lately 197 hectares, given up by the Arabs, have been conceded to colonists and valued.   A centre of colonisation is projected at some distance from Dellys, in the fertile valley of the Oued-Neça.

2. Free colonisation :

Since 1848 free colonisation has extended, in the military territory, chiefly in the subdivision of Blidah, and in those of Orleansville, of Medeah, and of Milianah.

Province of Oran.

I. *Colonisation in the Civil Territory.*

1. Centres created by the administration :

Three centres of population have been founded in this territory since 1847, *i. e.* Valmy, Arcole, and Ain-Turk.

Valmy.   The colonists of this centre are successfully engaged in breeding cattle.   Special branches of culture, such as those of cotton, tobacco, and the mulberry-tree, have also been the object of particular and experienced care.   To sum up all in a word, the state of this centre is good.—*Arcole.*   The same may be said of this village, though it suffered much from drought in 1850.   Public plantations are in the act of being laid out there.—*Ain-Turk.* This centre, situated in the *plaine des Andalous,* has been created within a year.   About thirty families have settled down there; and it is anticipated that the fertility of the territory will recompense the labours of the colonists.

The centres that existed on the civil territory previous to 1847 were *nine, i. e.* four towns,—Oran, Mostaganem, Arzeu, and Mascara; and five villages,—La Senia, Misserghin, Sidi-Chami, Mazagran, and Mers-el-Kebir. The present condition of these villages is as follows : *La Senia.*   Cultivation is very flourishing in this centre, and the colonists were favourably mentioned at the exhibition of 1850 for the fine fruits that they had grown.   They also attend with success to the care of silk-worms; and the favourable state of the public as well as private plantations completes the prosperity of this village.—*Misserghin.*   Some public works have still to be carried out in this village, which has suffered greatly from bad crops within the last few years.—*Sidi-Chami.*   This village has likewise been unlucky during the last year (1849); but this state of things is only transitory.   The colonists are in possession of good implements and a great number of cattle; their future prosperity is secured.—*Mers-el-Kebir.* This is the port of, and an appendage to, Oran.—*Mazagran.*   Near the town of Mostaganem.   The state of this village is satisfactory; the ground is cleared, and most of it tilled.   The success of this centre is secured by the fertility of its territory, and the facility it offers to the colonists for getting rid of their produce.

2. Free colonisation:

The number of farms or private properties founded beyond the centres, in the precincts, or banlieue, of Oran, has more than doubled since the end of 1846. They amounted to 195 in 1850. Some of them proceed from purchases made from the natives, and the others from concessions granted by the state.

This evident increase in agricultural undertakings has resulted chiefly from the creation of villages, where the concessionaries are sure to find work, and which afford a protection to individuals in case of war.   All seems to show that this movement, which has set in within the last few years, will not slacken, but rather increase as population pours in.   The same movement has set-in in the banlieue of Mostaganem.   The number of concessionaries in the *Valley of the Gardens,* which was inconsiderable in 1845-46, amounted in 1850 to 30, and has increased still more since that date.   There has also been a proportional increase of buildings and of cultivation.   A certain number of concessions have been granted likewise in the banlieues of Arzeu and of Mascara, towns which in the course of last year (1849) were attached to the civil territory.

II. *Colonisation in the Military Territory.*

1. Centres created by the administration:

Seven new stations have been created in the military territory since the end of 1846, *i. e.* :

Sidi-bel-Abbes.   As this centre, where some colonists had already settled for several years, appeared to be an important strategical position, a town was founded there by a decree of Jan. 5, 1849.   It is 20 leagues (50 miles) from Oran, and the *chef lieu,* or capital, of a military subdivision, possesses an important agricultural territory, and is being peopled rapidly. — *St. André* and *St. Hippolyte* are two villages near Mascara, founded some time since, and their territory is mostly cultivated; they are both well situated for health and accessibility, and the neighbourhood of Mascara offers a good market for the produce of the land. — Near Tlemsen are four villages : *Bréa* and *Negrier*, each intended to contain fifty families, and peopling rapidly ; *Mansourah* and the *Seysaf*, built on a fine rich soil, peopled partly by old colonists of Tlemsen and veteran soldiers. Though only founded in 1850, these villages are progressing rapidly.

There are twelve other centres in this territory, founded before 1847. 1. St. Denis du Sig placed on the road from Oran to Mascara, and watered by the Sig; both great advantages. In 1850 cultivation had much increased, and many fine plantations had been made here. 2. La Stidia is peopled chiefly by Prussian families, who came in 1846, and were long in a precarious condition; but by the assistance of government they have recovered heart, and are doing tolerably well. 3. St. Leonie, also a Prussian village, whose present condition is much improved. 4. A village of discharged soldiers exists in the subdivision of Mostaganem, to whom concessions in land were granted ; and by help of the state, they are in a satisfactory condition. The territory round Tlemsen is very fertile and well cultivated, all kinds of growth succeeding well there. Most of the colonists are old soldiers of the garrison of Tlemsen, who, on the expiration of their term of service, asked permission to settle there. The civil population of Nemours is still inconsiderable, and it offers little interest in an agricultural point of view; but its commercial position is good, its port being frequented by numerous merchant-ships, which might supply, if necessary, Lalla-Maghnia, Sebdou, and Tlemsen.

Cultivation begins to extend round the military posts of Tiaret, Saida Daya, Sebdou, Ain-Temouchen, and Lalla-Maghnia. These spots offer, moreover, excellent resources in timber, lime, freestone, and clay ; and the only obstacles to their progress are the dearness of provisions, and the distance from the coast.*

* General Lamoricière had vast projects in this territory, which, though made as public as possible, have been chiefly confined to paper. (Castellane's Souvenirs, 1852. Tableau, 1849-50.) 80,000 hectares (200,000 acres) were to be given to 5000 colonists, government helping them in the most essential work ; but only one village, Sainte Barbe, with 2841 hectares, has been founded, and that is a failure. Another decree, in February 1847, created between Oran and Arzeu three new communes, or parishes,—Christina, St. Ferdinand, and St. Isabel. In March 1847 they were given up to French and Spanish capitalists, who undertook to establish 170 families in them ; but they are also likely to prove failures.

In short, colonisation, as a commercial speculation, has failed there, owing to the great calls on capital, and the risks incurred.

A few private undertakings have been accomplished the last few years in the subdivision of Oran ; and the property of M. Dupré de St. Maur, 27 kilometres (17½ miles) from Oran, and containing 940 hectares (2470 acres), to be peopled by twenty families, is pro-

PROVINCE OF CONSTANTINA.

I. *Colonisation in the Civil Territory.*

1. Centres created by the government :

The only new creations in this territory since 1847 are the villages of Bugeaud and D'Uzerville.

Bugeaud is a woodland rather than an agricultural centre, situated at the entrance of the forest of Edough. It is intended to receive some families of wood-cutters, whereof some were already settled and employed by the government in 1850; and the others were shortly expected from the department of the Vosges. As these families are too poor to settle in the village and erect houses at their own cost, the minister-of-war decided that eight workmen's dwellings should be built at the expense of the state, and that a part of the territory should be appropriated to the construction of villas. This twofold measure will rapidly advance the prosperity of the village.—*D' Uzerville.* The creation of this centre is already of some standing; but its progress had been retarded by local causes, and by the necessity of drainage as a preliminary measure. It was in the last three months of 1850 that the colonists began to settle there. The territory is excellent, and the vicinity of Bona must secure its eventual prosperity.

The old centres in the civil territory are nine in number, *i.e.* Constantina, Bona, Philippeville, Valée, Damrémont, Saint-Antoine, Guelma, La Calle, and Bugia.

Constantina. The territory of this centre was long confined to the circuit of the town; but since 1849 it embraces a district of 14,000 hectares (34,580 acres). A portion of land has been conceded, and is on sale. Five groups of habitations have spontaneously arisen in it, corresponding to the principal divisions of the territory; namely, Sidi-Mabrouck, Oued-Yacoub, Cherakat-Bou-azen, Hamma, and Constantina. At the end of 1850 there were 78 undertakings of private enterprise in this district.

Cultivation begins to extend around the town of *Bona.* A certain number of proprietors, instead of confining themselves, as heretofore, to stacking hay and forage, have seriously applied themselves to agriculture. The success of these undertakings must depend entirely on the completion of roads and drainage, which have been hitherto crippled for want of funds.

---

gressing the best. A few other concessions, and fifteen farms, have been granted and founded in this subdivision ; but the whole affair seems throughout slow, though more sure, in this province. The Valley of the Gardens at Mostaganem is the most flourishing spot in the province. Further particulars relating to colonisation in the province of Oran will be found at pp. 214-15 of the *Tableau* for 1849-50.

Philippeville. The territory surrounding this town is almost entirely conceded, is in good order, and fetches a good price; and it is desirable to enlarge it. Independently of the cultivation of the valley of Zeramna, the hilly country situated between Philippeville and the port of Stora, and known as the ravine of Beni-Melek, contains a great number of very productive small farms, where the cultivation of vegetables, fruit-trees, and vines is daily increasing.— *Valée, Damrémont, Saint Antoine.* These three centres, which are situated near and in a radius round Philippeville, have very good prospects. The colonists continue to attend to the breeding of cattle, which is in a thriving state; and the amount of tillage and planting is continually on the increase.

Colonisation had made rapid strides round Guelma in 1849-50. Independently of three agricultural colonies that were created in 1848 in its vicinity, the town has continued to spread, and numerous concessions have been made in its district. The colonists have built dwellings, and founded oil and flour mills; and the cultivation of olive-trees is one of the great sources of wealth in this part of the country, admitting of unlimited extension.*

La Calle. We have already observed that there is no opening for the erection of a village near this place, where the only development that colonisation can take is the working of mines and the care· of the cork-forests, as there is an entire deficiency of arable land in that part of the country.

Two little centres have already arisen spontaneously at spots named *Le Melah*, 12 kilometres ($7\frac{1}{2}$ miles) from La Calle; and at *Oum-Thaboul*, 16 kilometres (10 miles) from that town.†

---

* An agricultural territory has lately been formed round Bugia, in which tillage has begun to make rapid strides, after being kept in check for many years by the hostile Kabyles of the vicinity. European colonists can now find safe and desirable settlements in this district, since the submission of the Kabyle tribes, and the opening of a road to Setif.

† The preceding remarks will give a notion of the results obtained, in matters of cultivation, by private capital, in places under the civil jurisdiction. These results are the most remarkable in the vicinity of Philippeville; but the territory of Bona, almost entirely in the possession of private individuals, and so long in a stationary condition, is beginning at last to be cultivated in real earnest; and Guelma, whose advance was long thwarted by its isolated position, presents a continual progress in the cultivation of its fertile territory. Lastly, within a recent date, numerous concessionaries or grantees have been established in the district of Constantina, and are in a thriving condition. Thus this phase of the colonisation of the province is in a satisfactory state.

Many future plans are, moreover, in agitation for the promotion of colonisation in the civil territory. Thus it is proposed to make Guelma the centre of a great network of colonisation, by the erection of six villages; and it is also projected to build a village near Bugia.

A—M

II. *Colonisation in the Military Territory.*

1. Centres created by the government:

Three villages have been created since 1847 in the military territory, *i.e.* Saint-Charles, Condé, and Penthièvre.

Saint-Charles is situated in the valley of the Safsaf, on the road from Philippeville to Constantina; and building as well as agriculture are progressing there. The territory of this centre has been allotted in such wise as to admit of the establishment of large, middling, and small estates. Situated about half-way from Philippeville to El-Arrouch, traversed by an excellent road, and surrounded by a rich district of wood, arable, and meadow land, the village of Saint-Charles enjoys great advantages.—*Condé* stands a few kilometres from Constantina, on the road to Philippeville; but as yet it has not been much developed; and though a few houses have been built, and some concessions have been made to colonists, cultivation is hitherto not far advanced. This village being, moreover, a halting-place for travellers, the colonists have, up to the present time, devoted themselves chiefly to trading.—*Penthièvre* has been hitherto in a very backward condition, although its position is most favourable, both on the score of farming and traffic. It stands almost midway between Guelma and Bona, which are separated by a distance of 66 kilometres (41·01 miles) ; and it is an indispensable halting and baiting place for man and beast. Water and wood are abundant in the district, which contains much very fertile land. The only thing needful to secure the prosperity of this village is a good road to unite it with Bona and Guelma.

The other centres, existing before 1847, in the military territory, are : El-Arrouch, Setif, Djidjelli, Batna, and Biskara.

The territory of *El-Arrouch* is, at the present time, almost entirely granted and cultivated, and is remarkable for its fertility. The state of the colonists at this spot is satisfactory.

Setif. Colonisation begins to extend in the district of this town, which contained in 1850 twelve private farms, independently of a great number of small dwelling-houses and cottages in the gardens near the town. Hands have been hitherto wanting for agriculture; but it is anticipated that the pacification of Kabylia, and the opening of the road from Setif to Bugia, will attract a numerous population to this centre. At a few kilometres from Setif, a centre of population, called Ain-Sefia, was created in 1846; whose inhabitants possess a number of beautiful gardens and fine plantations, while the surrounding land is extremely rich.*—*Djidjelli*

---

* Four other little villages are in the process of formation around Setif, at spots named Lanasser, Kalfoun, Mezloug, and Fermatou. Neither of them has hitherto been regularly constituted.

having been in a constant state of blockade since its first occupation, the farming in the vicinity is very limited; but the subjection of Kabylia will give breathing-time and a fair field for colonisation. *Batna* has just received the grant of a farming district of 8700 hectares (21,489 acres), consisting of very fertile land, covered with an abundance of wood. A certain number of concessions have been already made, some of which are in a very thriving state. But the future prospects of this spot are as intimately associated with traffic as with tillage; since, owing to its position between the mountains of the Aouress and of the Ouled-Sultan, on the passage from the Tell to the Sahara, it is destined to become some day an important emporium for goods coming from the north and south.

It will be necessary to annex a district for colonisation to *Biskara*, in order to secure the settlement of those colonists who are at present attracted to the spot by the demand for engineer labour, and by the presence of the troops. The soil and climate of Biskara seem well adapted to the culture of tropical plants;· and it is intended to establish there shortly a *jardin d'acclimatation* (nursery-ground for exotics). This town forms, moreover, a good channel for French commerce with the interior of Africa.*

Four centres of population already existed in 1850 in the valley of the Safsaf, *i.e.* St. Charles, El-Arrouch, Gastonville, and Robertville.†

Between El-Arrouch and Condé are already two little agglomerations

---

* In 1846 it was proposed to colonise the valleys of the Safsaf and of the Bon-Merzoug. The first, which was given up to colonisation in 1847, contains about 20,000 hectares (49,400 acres). It has been divided into two parts: 8000 hectares (19,760 acres) on the right bank of the river have been given up to the tribe of the Beni-Mehenna, who were previously dispersed over the valley; whilst 12,000 hectares (29,640 acres) have been reserved for, and distributed amongst, European concessionaries.

It was first attempted to adopt the system of large concessions, many of which were granted, and are still occupied and cultivated; but here, as in the province of Oran, experience has proved that they are not so successful as those on a smaller scale. It has been found necessary to execute a good many evictions; and the greater part of the great lots of land, subdivided into fractions of from 25 to 100 hectares (62½ to 250 acres), are now in process of cultivation. Some of these farms have become notorious for the breeding of cattle in large numbers.

As for the valley of Bon-Merzoug, beyond Constantina, it contained, at the end of 1850, 53 concessions, embracing each from 40 to 100 hectares (100 to 250 acres). Some of these concessionaries have already undertaken very important works.

A system of colonisation is now in process of execution in this district, of which the following are the principal features: first, the creation of two villages as foci of industry, and seats of middling and small properties; whilst nine hamlets, inhabited by petty cultivators, are to irradiate round these villages; isolated farms covering the rest of the valley, containing each from 80 to 100 hectares (200 to 250 acres).

Several private farms in a thriving state are now in existence around Setif. Lastly, colonists are now established in the district of Batna, and their number increases daily.

† A new village and some hamlets were being surveyed at the end of 1850, intended to complete the plan of colonisation in that territory. There was also a plan for colonising a vast territory around Constantina in the form of a polygon, to contain five European towns, each possessing a territory of 1800 or 2000 hectares (4500 or 5000 acres), and 120 families; the towns to stand at the angles of the polygon, on the roads to Setif, Msila, Philippeville, Haractas, and Batna; the interior of the polygon to be filled with farms of all sizes.

A A

of population — Toumiettes and Kantours — the nucleus of future villages.

The road traced out from Bona to Philippeville, by the valley of Fendeck, coasts along the north bank of Lake Fetzara, passes the valley of Oued-Ensa, and abuts at St. Charles, in the valley of the Safsaf. The entrance of the valley of Fendeck being a favourable site for a colony, the village of Jemmapes has been established there.

A little centre of population called Atménia has also been formed on the road from Constantina to Setif. Lastly, a penitential colony has been formed on the site of the Roman town Lambessa, whose name it bears. This colony has been devoted to the political convicts who were condemned and transported after the June insurrection in 1848, and was founded in conformity with the law of the 24th of January 1850. It stands near Batna; the climate is very healthy, water is very abundant, and the local administration comprises above 3000 hectares (7401 acres) of excellent land.*

This establishment, which seems to combine many advantages and some defects, directs our thoughts naturally to that fatality in French governments and economists, which leads them to conceive wonders on paper, and bring forth abortions or monsters in practice. *Parturiunt montes, nascitur mus.* Bravely do they theorise of national workshops, national banks, Icarias, Harmonies, and Utopias; governments talk of progress; presidents and emperors spout much about order and La Belle France; and yet, with a noble colony at their doors, once the granary of Europe, they cannot keep peace at home, and they send a few half-starved skeletons, yclept socialists, from the Faubourg de St. Antoine, to perish miserably in the marshes of Bona. Verily, prevention is better than cure.

Never yet has France fairly faced the subject of colonisation. The Turks kept Algeria with 10,000 men, the French require 100,000. Here

---

* This penitential colony has been founded on the following principles: 1. The creation of a penitential establishment, carrying on external labour in common in the daytime, and separately at night. 2. The addition of a certain plot of land to it, for the purpose of the agricultural department connected with it. The land was intended to be extensive enough to occupy 600 convicts. 3. Other land to be attached to the establishment, on which might ultimately be erected two or three villages, with provisional concessions to the convicts.

The plan for the building, designed by military engineers, was approved by the minister; but the works only began in March 1851. In July 1851 all the foundations of the vast edifice were completed, four buildings intended for stables were occupied, and the barracks were finished to the first story. The main building was in process of erection, to contain 600 cells, in three stories : it was expected to be roofed at the end of 1851. The government offices were only to be begun in 1852 ; but it was hoped that all would be finished in the course of that year.

It was expected that the building would be sufficiently advanced to receive 200 convicts in October 1851.

is a large screw loose; and money that ought to go to establish emigrants is swallowed up by a preventible evil.*

The best cure for the barricade is emigration, and over-pressure at home will ever beget explosions in mercurial Paris. Let the French government export sundry cargoes of blouses to Algeria, and we shall hear no more, I ween, of *la lanterne, coups d'état,* and glorious three days. French Africa is the true issue to relieve the humours of the mother-country ; and a little common sense could preserve order at home, without showers of bullets, hedges of bayonets, and legions of priests, by a spirited, liberal, and earnest encouragement and promotion of emigration among the proletaires of the large cities of France. It is evidently the divine intention, that man should go forth and replenish the earth and subdue it; and should the womb of time bring forth difficulties, and the globe be in danger of a glut, let our brave French theorists muster their Icarias, their Phalansteries, and their *ateliers,* or national workshops, and make the experiment in the field, and not upon paper, of giving, as they promise, a *nouveau monde* (new world).

Nay, we are far from flattering ourselves that our present social system is perfection ; and when China itself is breaking up, that a new world may rise from her ashes, we may safely infer that we are still growing, till we reach our proper and perfect stature. Only we say, instead of breaking windows, heads, and hearts in old Europe, give our visionaries a fair field, and let them work out their problems of economy and fight out their quarrels in a new world, and not in old Europe. If they fail, their folly will have received a sufficient punishment ; if they stand, we shall profit by their wisdom.

To complete our description of the state of colonisation in Algeria, it remains for us to speak of the agricultural colonies (*colonies agricoles*), founded by a law of the National Assembly dated September 19th, 1848, by which a sum of fifty millions of francs (2,000,000*l.*) was voted for their establishment.†

The emigrants, amounting to about 135,000, were disseminated amongst forty-two agricultural centres, which are thus analysed between the three provinces :

The province of *Algiers* comprises twelve colonies, forming successive steps (*échelons*) along the roads, or placed near centres destined to acquire some importance.‡

* A remarkable contrast to the inertia of French colonial administration is presented in the rise and progress of the enterprising Rajah of Sarawak, Sir James Brooke. See Hugh Low's Sarawak, 1848.

† These colonies were to be formed of farmers and of mechanics : the former being made admissible to receive concessions of land of from two to ten hectares (5 to 25 acres), with a dwelling-house, and funds necessary for their establishment ; the latter might enjoy the same advantages, if they showed their intention of settling in the country.

‡ Thus, L'Afroun, the Bou-Roumi, Marengo, and Zurich, are situated on the road from

The twenty-one agricultural colonies founded in the province of *Oran* are grouped into three principal circles, those of Mostaganem, of Arzeu, and of Oran. They gravitate in a small radius round these points ; and the colony that stands furthest from the sea is separated from it by an interval of about twenty-five kilometres (15⅔ miles).*

The agricultural colonies of the province of *Constantina* are nine in number, concentrated in the three circles of Bona, of Guelma, and of Philippeville. They are placed in the vicinity of the roads that lead from Bona to Guelma, and from Philippeville to Constantina, save a few that are at present rather more remotely situated from the great arteries of transit.†

Before any new bodies of emigrants were sent off, the National Assembly wished to learn the state of the colonies of 1848 ; and a commission having been sent by the minister of war to examine them, gave in its report on the 30th June, 1849. From this document it appears that the greatest difficulty found in organising these colonies resulted from the improper choice of the colonists, most of whom, coming from the workshops of Paris, were not fit for farm-labourers; and also from the presence of some idle men, who unsettled the minds of the others.

This state of things has been greatly improved since, chiefly by the dismissal of those unfit, and their substitution by families of agriculturists, and old soldiers inured to Africa. The latest accounts bear witness to the happy change resulting from this arrangement.‡

Independently of the physical wants of the colonists, the administration has not been unmindful of their moral and spiritual necessities. Pro-

---

Blidah to Cherchell ; Castiglione and Tefeschoun on the projected road from Algiers to Cherchell along the shore ; Lodi on the road from Medeah to Milianah ; Damiette and Novi, the first near Medeah, the last near Cherchell ; finally, more to the west, on the road from Tenés to Orleansville, Montenotte ; and near Orleansville, La Ferme and Ponteba.

* Near Mostaganem are Aboukir, Rivoli, Ain-Nouisy, Tounin, Karouba, Ain-Tideles, Souk-el-Mitou. Round Arzeu you find St. Leu, Damesme, Arzeu, Muley-Magoun, Kleber, Mefessour, St. Cloud. In the neighbourhood of Oran have been created, Fleurus, Assi-Ameur, Assi-ben-Ferrah, St. Louis, Assi-ben-Okba, Assi-bou-Nif, and Mangin.

† In the circle of Philippeville you meet with Jemmapes, Gastonville, and Robertville ; near Guelma you find Heliopolis, Guelma, Millesimo, and Petit ; and in the circle of Bona, Mondovi and Barral.

‡ The supplies granted by the state to these colonies are analysed as follows :—

1. A storehouse of one story, containing two rooms, each of from 3 metres 50 centimetres (11·78 feet) to 5 metres (16·40 feet), and a tile roof.

2. To each family have been allotted from 8 to 20 hectares (20 to 25 acres) of land, besides a garden.

3. Ploughs, &c., seed and cattle, have been distributed amongst the colonists.

4. A daily ration has been allowed, since 1848, to all persons of both sexes ; half rations to children. All these supplies were to cease December 31st, 1851. The construction of roads, and the supply of water in these new colonies, have occupied the attention, and swallowed up considerable sums, of the government; but they were indispensable.

visional buildings have been consecrated for divine worship, and provided with ministers and assistants.

Schools for boys and girls were also founded at the end of 1849, several of them confided to the care of nuns and sisters of mercy of Algiers, who also take charge of the local infirmaries.   This arrangement has naturally had the most happy results.

In virtue of a decree of the 20th July, 1850, justices of peace are to be established in the most important of these colonies; St. Cloud enjoyed this privilege in 1850.

These colonies were in such a forward state at the end of that year, that it is supposed they might have dispensed with the supplies from government, but for the deficient crops and the ravages of locusts in 1851.   It was therefore thought probable that it would be desirable to continue the government supplies for some months in 1852.

The law of the 19th of May, 1849, decreed that 6000 additional emigrants should be sent out, for whose reception twelve villages were prepared, and 734 houses built.   These labours were suspended by a new law, of July 20th, 1850, decreeing five millions of francs to continue the colonies of 1848, and people those of 1849.   On March 13th, 1851, the minister of war demanded of the assembly a grant of five millions to finish the public works in the twelve villages begun in 1849.   The government decided to renounce the law of September 1848, and to people them with French or Algerian farmers ; granting them the house and land, but no further supplies.   It was determined to send people of the same department to the same place, and to appropriate the mountainous districts, such as the colonies of Abd-el-Kader-Bou-Medfa, Ain-Benian, and Ain-Sultan, on the lofty plateaux, near Milianah, to French highlanders, while the Bœotians should settle in the plain.   This project received the sanction of the assembly July 10th, 1851, when the sum demanded by the executive was voted.

The last official accounts are, that they were hastening the works in the colony, preparatory to the reception of the colonists, who were being carefully and appropriately selected by the prefects of the departments in France.   It was hoped that they would be installed in October 1851.

The completion of the colonies of 1849 would bring into cultivation a territory of 18,000 hectares (44,460 acres).*

---

* The twelve villages of 1849 are distributed as follows :—*Province of Algiers:* Ameur-el-Ain and La Bourkika, on the left bank of the Chiffa ; Ain-Benian and the marabout of Sidi Abd-el-Kader-Bou-Medfa, on the road from Blidah to Milianah and Ain-Sultan, at 1500 metres (about a mile) from the same road, on one of the outliers to the south-west of the Gontas.   *Province of Oran:* Bled-Touaria, Ain-Sidi-Cheri, Ain-Boudinar, and the Pont-du-Chéli, in the subdivision of Mostaganem; and Bon-Tlelis, distant thirty kilometres (18·6 miles) from Oran, on the road from that town to Tlemsen.   In the *province of Constantina,* Ahmed - ben - Ali and Sidi - Nassar, at some kilometres from the village of Jemmapes.

STATE OF THE AGRICULTURAL COLONIES, DECEMBER 31, 1850.*

PROVINCE OF ALGIERS.

Names.	Population.	Houses.	Land.	
			Hectares.	Acres.
L'Afroun	292	61	603	1489·41
Bouroumi	26	15	134	330 98
Marengo	555	200	1180	2914 60
Zurick	185	90	515	1272·05
Novi	345	110	397	980·59
Castiglione	255	75	154	380·38
Tefeschoun	189	56	118	291·46
Lodi	365	120	1019	2516·93
Damiette	344	120	909	2245·23
Montenotte	323	71	1004	2479·88
Ponteba	188	93	516	1274·52
La Ferme	119	55	413	1020·11
Total	3243	1066	6962	17,196·14

The 3423 colonists are analysed into 1533 men, 1221 women, and 489 children. Of the 6962 hectares of land granted, 3375 (8236·15 acres) have been cleared, and the number of trees planted amount to 173,762.

PROVINCE OF ORAN.

Names.	Population.	Houses.	Land.	
			Hectares.	Acres.
Aboukir	221	86	537	1,332
Rivoli	224	71	412	1,030
Ain-Nouisy	164	76	420	1,260
Tounin	143	51	327	817
Karouba	42	7	100	250
Ain-Tideles	312	107	819	2,047
Souck-el-Mitou . . .	241	94	881	2,202
Saint Leu	134	58	282	705
Damesme	125	41	168	420
Arzeu	135	44	141	70
Muley-Magoun . . .	11	7	23	57
Kleber	274	84	167	83
Mefessour	138	64	211	527
Saint Cloud	789	280	2,689	6,722
Fleurus	207	98	532	1,330
Assi-Ameur	192	72	242	605
Assi-ben-Ferrah . . .	154	46	410	1,025
Saint Louis	345	148	1,122	2,805
Assi-ben-Okba	196	64	612	1,530
Assi-Bounif	143	61	520	1,300
Mangin	118	76	471	235
Total	4308	1635	11,086	26,357

Of these 11,086 hectares, 4205 (10,386·35 acres) have been cultivated, and 88,961 trees have been planted. Of the 4308 colonists, 2001 are men, 1544 women, and 763 children.

---

* It may be acceptable to the reader to be presented with a sketch of some of these French colonies from the pen of recent French visitors ; and we shall begin with St. Denis du Sig, in the province of Oran.

Castellane was there in 1846, at its birth. "The general" (Lamoricière), he says, " wished to ascertain the cause why a village so well placed should not have succeeded better; hence he announced that he would see any of the colonists at five o'clock. As

PROVINCE OF CONSTANTINA.

Names.	Population.	Houses.	Land.	
			Hectares.	Acres.
Jemmapes	512	185	914	2,257·58
Gastonville	350	136	290	716·30
Robertville	420	147	865	2,136·55
Heliopolis	195	100	532	1,314·04
Guelma	265	51	746	1,842·62
Millesimo	208	95	639	1,578·33
Petit	184	72	538	1,328·86
Mondovi.	376	145	667	1,647·49
Barral	315	113	489	1,207·83
Total	2825	1044	5680	14,019·60

Of these 2825 colonists, 1287 are men, 973 women, and 565 children. Of the 5680 hectares of land, 2911 have been cleared, and the number of trees planted by the colonists amount to 48,626.

The grand total for the three provinces is, 10,376 inhabitants; 3745 houses; 23,937 hectares (59,124·3 acres) of land; 10,491 hectares (25,912·77 acres) of land cleared; trees planted, 311,349.

But it is evident that French colonisation in Africa has hitherto been oppressed by the nightmare of bureaucratic interference; and that the French have hitherto shown a much greater aptitude to destroy than to construct. The razzia has heralded the progress of liberty and equality; whilst dreams on paper have been the fruit and ornament of their colonial empire in Algeria.

Hardships are the lot of all new settlers; but it seems surprising that with the example of the Romans, who covered Numidia with flourishing cities, and with the additional light of science, France should be blind enough to her own interests and to the interests of humanity, to offer up her sons, civilians or soldiers, by hecatombs to the demon of fever, for want of judicious outlay and precautions. If some of the millions lavished on triumphal arches, imperial progresses, illuminations, and corruption, had been devoted to great works of national utility in the colony, it would have been more honourable to the crown, and more acceptable to the people of Paris and the enlightened part of the provinces.*

soon as the general had finished his cross-questioning, his mind was made up. Orders were issued to Commandant Chabran to come and bivouac at St. Denis with his battalion; and the soldiers turning carpenters, masons, bricklayers, &c., soon rescued this miserable population from distress. A few months later, the traveller passing through St. Denis could no longer have recognised it, so greatly was it improved." Such is the statement of Castellane. (Castellane, pp. 317, 318.) According to Madame Prus, the colony of Robertvilleh as been entirely destroyed, and the site it occupied is now a vast wilderness; that of Penthièvre can hardly support one quarter of its population; that of the Golden River has filled the hospital of Guelma with its inhabitants; and Mondovi will soon be a desert. Madame Prus, Residence in Algeria, 1850.

* Baron Baude observed, in 1840, that the only workmen who could get good employment in Algeria were the builders and the carpenters : mechanics, who suffer most at home, would be subject to the same fluctuations there; and they are not calculated to change their professional habits, which alone would enable them to get on there. (Vol. iii. p. 123.)

Sanitary and social reform, and colonial empire, are a demand of the age; and the ruler who is indifferent to these matters can never be said to have deserved well of his country, and stands far behind the Romans in public spirit and discernment. Nor are our censures confined to imperial France, the Republican Assembly having lacked the energy and judgment of an experienced senate.*

* On Colonisation, see Diary of a Lady's Tour in Barbary, Madame Prus, and the Tableau for 1850, sect. 6.

CHAPTER XX.

Civil and Religious Government.

THE OOL-AMA—THREE CLASSES OF THEM—SHEIKHS—KHATEBS AND IMAMS—THE
MUFTI—THE SANTONS, AND OTHER ORDERS—THE DEY'S MINISTERS—THE
KAIDS—THE KADIS—FRENCH CIVIL ADMINISTRATION—FRENCH TRIBUNALS—
MUSSULMAN TRIBUNALS AND SCHOOLS.

JUSTICE, like religion, has worn many strange faces on the shores of
North-west Africa. Prætors have succeeded suffetes, and made way
for kaids and kadis; whilst bishops have given up their chairs, once shrines
of Apollo or Venus, to imams and muftis.

We do not pretend here to analyse the various phases of faith and
forms of law historically developed in Algeria. We shall confine ourselves
to the code of the Koran under the Turks and the French, and the present
civil and justiciary administration of the French.

The Koran being a civil and legal as well as a religious code, we
unite the secular and religious in one view in treating of the code and
canon of Islam; and following a chronological order, we shall commence
our analysis by researches into the civil and religious administration under
the Turks; and we shall simplify our statements by presenting a summary enumeration of the names, office, and power of the different authorities.

Beginning with Religion, the basis of every thing in the East and
North Africa—though it is the weathercock in modern European state-
structures,—we find that the califs were pontiffs, judges, and doctors of
the law, having vicars under them named Ool-ama علماء (*learned men*),
divided into three classes : 1st, the Imams, or ministers of religion ; 2d,
Muftis, or doctors of law; and 3d, the Kadis, or judges.*

The Mohammedan divine worship being a more serious and essential
part of life than ours, five prayers are repeated daily in the mosques and
in the country by all faithful believers; and a portion of the Koran, which
is so divided as to be read through monthly, is read daily in the mosques,

* Tableau, 1850, p. 205. Blofeld's Algeria, p. 222-4. Pananti, vol. ii. p. 252.

which are always open, like the venerable fanes of our ancestors and the beautiful temple of Zion, as a refuge and consolation for the afflicted. The Khotba, or profession of faith and prayer, is recited every Friday before the sermon.

The Mesgjeds are always built near cities; and the Khotba cannot be recited in oratories (marabouts), &c. The Ool-ama, consecrated to the service of religion, are divided into three classes. 1st, Sheikhs : the mufti and kadi take this title ; they are the preachers; and in their exhortations on Friday they sometimes even attack the sovereign, in a way that would alarm court chaplains and pet parsons nearer home. 2d, the Khatebs, who preside at the solemn prayer on Fridays, and recite the Khotba. 3d, the Imams, who assist at the daily prayers, excepting on Fridays, and read a portion of the Koran. The chief Imam assists at circumcisions, marriages, and funerals. The Mueddins or Muezzins call the faithful to prayer from the minarets, answering the purpose of incarnate bells; and their sonorous voice sounds solemnly through the tranquil air as it cries, "There is no God but Allah; and Mohammed is his Prophet." The mufti is the head of the code and canon of Islam, at once bishop and judge,— a dangerous union in Christendom.

The expenses of religion are paid out of the proceeds of estates belonging to mosques, in virtue of donations, &c. made by the founders or others from motives of piety. The Imams also receive various fees.

There is no public Mohammedan worship out of the towns, where the people are left to the care of the marabouts, divided into three orders —Santons, Cavalists, and Sunaquites.*

The Santons are under different rules. Some only wear rags; some go about naked, with fanatical gestures; some, of a more composed and rational order, despise these extravagances, only maintaining that good works, fasting, and self-denial refine their minds to the purity of angels; and others maintain that, when they have arrived at a certain degree of perfection, they can no longer sin; which has occasioned many vicious practices, analogous to the excrescences of Evangelic Christianity, when our religion has gone to seed.

The Cavalists are very strict in fasting, never eating animal food. They have a form of prayer for every month, day, and hour. They speak of heavenly visions and conversations with angels, by whom these privileged men, like the saints of other creeds and days, are instructed in the sublimest secrets, and who solve all their questions. Who shall draw the limit between hallucination and reality in these raptures? The negative

---

* Blofeld's Algeria, pp. 222-4 ; Pananti, p. 249. Pananti says, p. 247, that Marabout means men bound by a cord. The sensual Italian has his eye on the Capuchin, and thought of physical binding. The more spiritual Arabs attach to the expression the idea of a moral obligation, a spiritual tie. *Religio* came from *religare*, and is a corresponding term. Hence Marabout answers to the French *religieux*.

prose of European criticism will never root out the instincts of the heart, or extinguish the poetry of Orient faith. A universal conviction must be based on truth. This order, conformably with the science or superstition of all ages, always wear talismans; and they were founded by Beni, an Arabian doctor.

The Sunaquites are misanthropes, go into deserts, and fly the cities of men, living on vegetables. Their doctrine is represented as a compound of Mahometanism, Judaism, Christianity, and Paganism; a coalition apparently formed of the most antagonistic elements, and yet admitting of fusion in the common religiosity of our nature. These vegetarians sacrifice animals, like our modern Platonists,* and do not circumcise before the age of thirty. They say that all religions are originally inspirations from God. Having been guided to the pinnacle of wisdom, and the key of all ecclesiastical history, by the religious instincts of a pure life, they coolly add that they are the most perfect of men, and that they save the world with their prayers; showing the invariable alliance of sublimest convictions and puerile conceits in the heart of man. Some of these men, it is said, deserve the veneration they inspire by their virtues; but a great portion of them, as in other churches, are hypocrites, fanatics, or idiots, vindicating the old story of the noblest institutions ending in tinkling cymbals and whited sepulchres.

The marabouts are not recognised by the Mahometan hierarchy. There are few religious edifices left at Algiers since the French conquest, but enough for the wants of the population. The ministers of their religion have been respected, temples built, and their contents protected (even the entry forbidden to Christians). The French are not now wanting in charity, but in faith; and they might learn to keep one and gain the other, by taking a Christian lesson from a Mahometan priest.†

* The reader may recollect the case of the amiable translator of Plato, Thomas Taylor, who was credibly reported to be a gallicide.

† The reader who wishes to obtain a complete insight into the dey's government is referred to the third volume of Pananti's *Avventure*. His council of state consisted of his creatures and slaves: his ministers were the khaznadji, or treasurer; the michelacci, or secretary for foreign affairs and the marine; the admiral; the kaja, who often acted as vice-dey; the field-aga, or commander-in-chief; the horse coggia, or chief of cavalry; and the aga baston, who inflicted bastinading. There were four hojas, or secretaries of state, registering the decrees of the dey: their advice had much weight with the sovereign. There were also four inferior hojas, who acted as paymasters to the troops, presided over the receipts of customs; some remaining at court, and others accompanying the armies and fleets, &c. There were besides, the doletri, or head of justice, who signed treaties; the mezovard, a warden attending to the police of the city; the checkebeld, who looked after the noble slaves, and punished Moorish women; the pitremelgi, or public registrar of deaths, who looked after the property of deceased persons, and gave certificates of burial; the dragoman, the interpreter of the palace, a man familiar with Turkish and Arabic; and the rais, or captain of the port, noting the arrival of vessels, &c. All these functions being honorary, these worthies, like Austrian and Russian *employés*, were not scrupulous about bribes, &c. See Pananti, vol. iii. pp. 25-27; Revelations of Russia, vol. i. ch. vii. and viii.

We shall next analyse the administration of justice among the natives, 1st, before, 2dly, since 1830,—begging the reader to remember that Mahometan priests are also judges.*

The divan at Algiers was their parliament and chambers, representing, as with us, a small minority of the dominant caste. It was composed of the old agas, of the yiack bashaws, of 300 boulouchi bashaws, and of 200 oldak or odjak bashaws,—in all about 700 persons; and on some momentous occasions all the old manzoul, or retired agas, and the whole Turkish militia, attended : in fact, it became a Norsk *thing*. The oldest aga was president. It sat every Saturday at the Casbah, and on the summons of the dey. The members were unarmed ; and business was transacted in the Turkish tongue. When proceeding to vote, they bawled out together, creating a confusion to match the Frankfort diet. At one period, all decrees required the sanction of the diet ; but latterly, like the French senate of to-day, it became a mockery. The kaja pronounced the will of the dey, which was law to the members. *Coups d'état* were frequent and unscrupulous at Algiers, as at Paris. The beys governed the provinces, were appointed by the dey, and were almost absolute. Every two or three years they had to give an account of themselves, and pay in money to the dey's coffers; but they looked after themselves when at home, and fleeced their unhappy subjects unmercifully, to heap up the immense wealth which they sometimes accumulated.

The reader will be prepared now to receive some strange disclosures concerning Algerian justice.

Next to the beys came the kaids, or governors of towns, who were in the habit of buying and selling all posts. Thus their political machinery was much like that of imperial France, and of most churches in Christendom. What the beys had spared was devoured by the kaids. Like master, like man : the country was a nest of petty despots and thieves, passing from the dey, through the beys and kaids, to the chiaous, governed by two bashaws. The chiaous were Bow-street runners—Turks for Turks and Arabs for Arabs ; and though unarmed, such was the prestige of their name and power, that all bent the neck to their mandate, be it death or a bastinado.

There is honour among thieves; and the government of Algeria had

---

* The danger that might seem to result from this powerful combination is much neutralised in the cities, because it is not the man but the office that is respected in the Mahometan church, where man has never taken the place of God, as with us. Moreover, the pay of the priests is so small, that they need not dread the oppression of a golden hierarchy like that of our Christian Brahmins. The mollahs, or parish priests, in Affghanistan are so poor, that, like our curates, they have to keep school to support their families; and the imams, unlike our pluralists, are men of narrow finances. (See Elphinstone's *Travels in Cabul*, 1809.) As regards the tribes, the marabouts, as we have seen, are paramount ; but the institution is democratic, open to all, and humanising in its effects.

its bright side, in escaping the spirit of caste and the exclusive selfishness of oligarchies ; but it combined the usual anarchy and despotism of democracies. Yet this applies only to the Turks ; for the latter were to the Arabs like the Norman nobles to the feudal serfs of the middle ages. Nevertheless, their administration was firm, and they secured that panacea of modern civilisation, order. They might smother resistance in tears, and silence it in blood ; but as long as this result was obtained, it was no doubt very desirable, like the *coup d'état* of 1851.

We have said that the Koran is the law of Mussulmans; and the sanction of the mufti, called *tefta,* is necessary for the validity of a law.

The kadi studied, at Stamboul or Cairo, the Arabic pandects, &c., and administered justice, or rather injustice, once or twice per day; but he was open to bribery, and therefore the dey and his officers commonly settled affairs of importance. All causes were quickly, we may say summarily, decided.*

There was a kadi for the Turks and a kadi for the Moors; and they both had clerks called paips, who acted as judges in the villages. All matters of property were referred to them. The kadi traded in justice; for there is a law for the rich and another for the poor in Algeria as well as in England.

In civil processes the decisions were venal, summary, and unjust; an irritable kadi before dinner being apt to administer a hundred blows indiscriminately to both parties,—like some of our magistrates, who, when somewhat touchy, will lock a man up in a hurry, without benefit of clergy or bail. Yet even at Algiers Christendom might learn some useful lessons in jurisprudence. The dey sat all day to hear all complaints and petitions. All causes were carried on in public; there were no ruinous delays, no chancery, no quibbles. The law was clear, and there were no solicitors; and the litigious spirit was kept down by the fear of the bastinado ; just as blackguards were taught good manners a few years back by fear of twelve paces and hair triggers.

Criminal jurisprudence in Algeria had two good qualities : it was prompt and sure. For murder, death. Robbers were mounted on asses and had one hand docked. Christian and Jewish familiarities with Mussulman women were punished with death, if the guilty pair were taken *in flagranti delicto;* otherwise, the man was well thrashed. The guilty woman, seated on an ass, facing the tail, was paraded through the town, then put into a sack, and drowned or smothered in mud. Conspirators were strangled : feigned bankrupts, if Christians, strangled; if Moors, impaled; if Jews, burned. If a debtor was willing to pay, he had to pay double. A debtor refusing to pay was shut up, and his goods and chattels sold, the

---

* See Blofeld (pp. 222-4), who appears to have copied Pananti's observations, and to have very unhandsomely abstained from acknowledging his authority. Compare Pananti, vol. iii. p. 44.

surplus, if any, being given to him; after 100 days he was flogged and released, but if he still owed his creditor, the latter could stop and strip him till repaid. Tribes and districts were answerable for the crimes within their precincts.

Justice in Barbary was unaccompanied by clemency. They avoided the inconvenience of sending monsters pleasure-trips to the Pacific, and of giving regicides comfortable berths in Bedlam; but theirs was a reign of terror. " Crucify him! crucify him!" was the chorus of all the kadis and muftis.

Their punishments were as severe as those of China, but their basti-nades do not seem to have been so vigorous as the flogging of royal and noble colonels nearer home; and they never equalled the gallantry of Aus-trian butchers and British mechanics, in stripping and lashing the fair sex.*

We shall add an outline of the French civil administration of Algeria.

The colony has been variously organised and analysed since 1830; but we shall here confine ourselves to a compendious statement of its present organisation. Since 1848 many of the colonial departments are attached to the government at home. Thus the service of the *domaines,* or go-vernment property, and of registering, as well as the levying of taxes, are effected direct through the minister of finance and his agents. The ministry of war in France is invested with the following functions: the interior, public works, agriculture, commerce and finances, save the cus-tom-houses. The other French ministers exercising a direct authority on Algeria are those of justice, of public instruction and worship, of ma-rine, and of finance through the customs.

A new order of things in the interior organisation of Algeria was pro-mulgated on the 9th and 16th of December, 1848. Algeria is divided into three provinces, subdivided into three departments and three mili-tary territories. The governor-general has the supreme command of the

---

* Pananti's *Avventure*. The best and fullest account of native jurisprudence will be found in the *Exploration scientifique*. Interesting particulars relating to their administra-tion under the Turks occur in the *Nachrichten und Bemerkungen über den Algierischen Staat*, by Rebinder. Altona, 1798-1801, 3 vols.

In a former place we have suggested the expediency of a mild slavery as an elevating transition from the basin of the Niger; and it is now our intention to attack a cruel slavery as a degrading transition for the nations of Christendom. The negro, by slavery in Algiers, became a man ; the Christian a beast. An elevation for the former was a degra-dation for the latter; but the absolute condition of the negroes in the Regency was para-dise to that of Christians, and I might add Jews.

The Christian slaves were of two classes: first, those of the deylik; secondly, those of private houses. They were frequently immersed in debauchery, but were more respected at Algiers than the free Christians.

The redemption of slaves was effected in three ways: first, by the public redemption at the charge of the state of which the slaves were subjects; secondly, by the media of such religious societies as made collections for that purpose ; and by the orders of private per-sons. There were various other duties to pay, such as ten per cent on the ransom to the custom-house.

army and the high administration, especially all relating to colonisation. He governs the military territories through the generals commanding there; and he has with him a secretary-general, centralising the administrative correspondence, and a council of government. The three departments are governed by prefects, with sub-prefects in each arrondissement, civil commissioners and *maires* in the communes. There is a council of prefecture with each prefect. The prefects correspond with the governor-general and with the ministers at home. The military territories are administered, under the generals-in-chief, by subordinate generals commanding subdivisions and circles. There are still consultative commissions at the chief towns of these subdivisions. The judicial functions in this territory are filled either by a juge de paix or by the officer in command; and under them, the functions of magistrates and police-officers either by the commandants or *maires*. The settled natives are administered by the prefects, who name the village sheikhs (16th December, 1848). The wandering Arabs are subject to military administrations.

As regards the municipal institutions. In August 1848 Algeria was subdivided into communes, each having elective municipal councils. Electors are all French and naturalised foreigners, or foreigners and natives holding concessions and above twenty-one. Members of the council are generally all French, but strangers and natives are admitted under certain restrictions. The ballot is used, and answers admirably of course, as at the election of the emperor. At Algiers the council reckons twenty-four members; elsewhere, fifteen, twelve, and nine. *Maires* are named for three years by the governor-general or the executive; and the municipal councils can be suspended, but cannot be dissolved, by the governor.

On the 16th December, 1848, the nomination of the *maires* was given over to the prefects, and their appointment has been suffered even in the military territories.

On the 31st December, 1849, there were in Algeria six great parishes with municipal councils. Several other places have been erected into communes, and have *maires;* and in other localities in the civil territory the civil commissaries take their place. The commandants de place usually fill their place in the military territories; and five new civil commissariats were appointed, on the 4th of November, 1850, in the towns of the interior, till then subject to military rule.*

At present General Randon, minister of war under the Republic, still remains governor-general of the colony.

* By recent accounts, it appears that the prætorian government of Louis Napoleon is about to make some sweeping alterations in the administration; and a brood of hungry Bonapartes are to be let loose, like a plague of locusts, on the promised land. It is reported that Prince Napoleon Jerome is named viceroy, with General Pelissier as commander-in-chief; but the sucking Cæsar has delayed his departure, preferring to sport his plumes among the Elysian Fields, rather than encounter the fever of Bona and the sirocco.

In 1848 the service of justice, which had hitherto remained in the hands of the minister of war, was attributed to the minister of justice by decrees of the 30th May and 20th August, which conferred on the latter what had previously been given to the minister of war, in matters of justice relating to the French and European population of the civil territories. Native jurisprudence remains subject to the minister of war. Two civil commissariats were established the same year, one at Tenes, the other at Bugia. They filled the same functions then as juges de paix, who were wanting. Lastly, in 1849 the administration of justice received a considerable extension. By a presidential decree of July 9th, a tribunal de première instance was established at Constantina. Four juges de paix were also created by the same decree, each attached to a tribunal; their functions extend over a radius of 2000 metres round the chef lieu. They are the juges de paix of Medeah, belonging to the tribunal of Blidah; of Tenes, belonging to the tribunal of Algiers; of Guelma, belonging to the tribunal of Constantina; of Tlemsen, belonging to the tribunal of Oran.

In 1850 Tenes and Guelma were brought into the civil territory.

The following table gives a view of the state of the French tribunals and officers of justice on the 31st December, 1849 :

Court of Appeal.	Tribunals of the First Instance.	Justices of Peace.	Civil Commissaries exercising judicial functions.
Algiers	Algiers	Algiers, north canton . Algiers, south canton . Douera . . . . . . Tenes	Cherchell.
	Blidah	Blidah. Koleah. Medeah.	
	Constantina .	Constantina.	
	Bona . . .	Bona . . . . . . Guelma	La Calle.
	Philippeville.	Philippeville	Bugia.
	Oran . . .	Oran. Mostaganem. Tlemsen.	

One commercial tribunal existed at Algiers before 1847; another was established at Oran on the 1st of July of that year. On the 24th November, 1847, the elective principle of French legislation was applied to the formation of the Algerian tribunals. As regards penal jurisprudence, no modification has been effected in it of late years. According to the decree of September 26, 1842, the French tribunals take cognisance of all crimes committed by men of all nations and religions, in all cases falling under French law. But the Mussulmans remain subject to the jurisdiction of their kadis in cases that constitute crimes according to their law and not to French law. The court of appeal of Algiers judges directly the crimes committed in the civil territory of the province of Algiers; the tribunals of Bona, Philippeville, Oran, and Constantina, give a first judgment in crimes committed within their jurisdiction, but appeals are made from their decision to the court at Algiers. Hence these courts have the same cognisance in criminal matters as the courts of assizes in France. In correctional matters, all the tribunals of the first instance are cognisant of the crimes committed in their *ressorts*. The tribunal of Algiers alone has a special correctional chamber; and the court of appeal of Algiers decides on appeals from the judgment of the correctional tribunals. Lastly, the juges de paix, or, in default of them, the civil commissaries, decide concerning the infractions of police regulations in their canton or district.

The tribunals of simple police gave 7607 judgments in 1849, of which 1106 were acquittals. The correctional police tribunals gave judgment in 1062 cases in 1849.

The court of appeal of Algiers, and the tribunals of Bona, Philippeville, Constantina, and Oran, condemned 180 criminal affairs in 1849.* In 1842 the governor-general had the nomination of the kadis and muftis, who are all paid by the state, according to the admirable code of the French administration, on this point worthy rivals of the Romans. It was only slowly, and by dint of great efforts, that the kadis were brought in 1846 to give an exact account of the administration of justice in their districts.

A new organisation was given to the Mussulman tribunals in the civil territory in 1848. A decree of the governor-general, dated July 29, 1848, regulated the composition of the midjeles, or superior tribunal, and of the tribunals of the kadis of the two sects Maleki and Hanefi.†

The cases that most commonly appear before the kadi for settlement are those for payment of money, supplies or labour due, demands for divorce; whilst there are few law-suits about disputed landed property.

The commonest cases in penal matters are drunkenness, which are much the most numerous; breaking the fasts, blasphemy, and improper behaviour in religious edifices.

---

* Of the 180 affairs, 57 crimes were against persons, and 123 against property. The number of persons condemned was 273; 83 prosecuted for crimes against persons, and 190 against property; 245 men and 28 women, or 11 women to 100 men.

Of the 273, 43 were French, 78 foreigners, and 152 natives; and in matters of religion, 120 Christians, 117 Mussulmans, and 36 Israelites. They present 15 minors of 16 years, 51 minors under 21, above 16, and 207 of and over age. See Tableau de la Situation, &c., 1850, pp. 164-5 and following tables.

† Another decree of the same date appointed *oukils*, or agents, and pleaders, who defended the natives gratis on trial. A third decree settled the charges of suits in the native courts of law, which had led to many abuses. All cases are ordered to be registered now; but this cannot always be effected, and the present accounts of judicial operations among the natives are still very uncertain. The kadi of Blidah has calculated the number of unrecorded decisions in 1846 at from 400 to 500.

The kadis still inflict the bastinade, which is admitted by the customs of the country; and might be advantageously administered in England, in our cases of numerous cowardly assaults on females. But in the towns, the fatal example of European indulgence has led them to substitute, for thrashing, imprisonment, which was formerly very rare.*

The number of individuals in the towns imprisoned in 1849 was 124 ; bastinadoed, 15 ; fined, 7. The number of judgments in litigious cases (*en matière civile*), 2333.†

We have long heard that the schoolmaster is abroad ; and it seems that he has stalked over the Mediterranean, and is prowling about the Sahara. Young Arabs are trying to square the circle, and young Turks are studying perpetual motion.

Primary schools are called *mecids* (مسيد, m'syd), and the masters are named *maâllem* (معلم, professors). These gentlemen have a long stick for the refractory; and the pupils have each a little board (*louhhah*), on which the subject of the day's lesson is written.‡

After acquiring preliminary notions, the children learn to read and write the Koran, according to prescribed rules of a technical poem called *Nahdm-el-Kharraz.*

· To know the Koran by heart is the aim of these primary schools; but this is seldom attained now. Poverty prevents many parents from affording the meagre stipend to the maâllem, and makes them take away their children soon, if sent at all.

If boys know a part of the Koran, without understanding it, they celebrate a fête called *khothmah*, or seal. There were once maâlemat for girls too; but scarcely any of them remain, and most native women are in absolute ignorance.

Secondary schools, *medreçah* (مدرسة m'dres), whose master is named *mondeviice* or *chikh* (شيخ chyk, *elder*), the disciple being called *thaleb.* These institutions were numerous, and had many manuscripts; but the French conquest has almost destroyed them. A few remain, and are preserved in the library of Algiers. Unfortunately many bivouac fires have been lighted with manuscripts taken at the rhazyahs.

---

* These remarks only apply to the Mussulman tribunals; as the rabbinical tribunals, which were also kept up by the decree of 1842, were suppressed on the 9th of November, 1845, only leaving to the rabbis purely religious and administrative functions.

† Tableau, 1850, p. 187.

‡ The first thing taught is the points, called *noqath;* when masters of this alphabet, they are taught to write the letters, *terkibou-el-hhorouf;* and then they learn the *hharekat,* or vowel-points. Instead of " Our Father, which art in heaven," the master's text is : *El hhamdou lillah, reubl-el-âlamin* الحمد لله رب العلمين, Praise to God, Lord of the world.

Constantina had many medreçahs, especially that of El-Salahhyah, founded by Salahh Bay, a philanthropic man.*

The French government have offered to the Arab youth the entrance into its own college;† but few take advantage of its liberality.

* It was, besides, the little mosque of Sidi-Kettani, in the street or çouq (سوق souq, *street*) called Djema. It had many Mss., now in the library of Algiers. The studies in the medreçah were grammar, rhetoric, logic, metaphysics, theology, and law; as also the calendar, in order to know the times of day for the five lawful prayers, in a little book of astronomy called *Keçalah-ebn-Soffar;* arithmetic; and versification in the *El-khaz-radjigah,* a book on rules of poetry, was also studied by the tolbas. Some studied the grammar of Ebn-cina (Avicenna), a kind of *corpus medicum,* and the little treatise of Dawoud-el-Autaki. The works on grammar used were: *El Djemimiyah* by Ben Dawoud-el-Sanhadji el Adjeroumi, and the *Elfiyah* of Ebn-Malek, besides a poem on verbs called *Lamyât-el-Afâl.* A poem called *Nadhan* gave some knowledge on religion: this catechism is by Ebn-Atsir.

Divines also study the last great work, El-Bedhawi's commentary on the Koran, and the *Sakhabb* of El-Boukhani.

In law the student uses the *Riçalah,* or treatise of Abou-Zid; the *Maoutha* of Ebn-Malek, and the *Mokthassar* of Sidi-Khelil.

The Zaouias (زاوية zaouya, *monastery*) are the highest schools; that of Guerronmah, among the Beni-Djad, east of Algiers, on the road to Constantina. is very famous. One of the hotels attached to the zaouias was Zaouia-mta-el-Kechahh, Rue des Consuls, No. 35 (built A.D. 1223); another is now in No. 24 Street or Couq el-Djema.

† Berbrugger, part i. There is now an Arab college at Paris.

CHAPTER XXI.

The French Army.

ROMAN RAZZIAS — STRENGTH — NATIVE TROOPS — ZOUAVES — SPAHIS — FRENCH — CHASSEURS D'AFRIQUE—SANITARY STATISTICS, ETC. — THE AFRICAN CHIEFS — CHANGARNIER — CAVAIGNAC — CANROBERT.

IT is with pleasure that we turn our backs on the statistical tables and the beaten highways of Algeria, to fraternise with the gallant French soldier in the romantic bivouac, or under the friendly shelter of the marquee. Notwithstanding that war is ever a scourge, and desolation has too often marked the track of its columns, the French army has ever upheld its high reputation for prowess in the valleys of Atlas; though it has not always united the humanising spirit of civilised warfare with the innate gallantry of its race. We are fully aware of the fact, that long service in Africa, as elsewhere, has hardened the men into soldiers of fortune, whose regiment is their country, and who do not scruple to trample on liberty at home or elsewhere, in mechanical obedience to their commanding officer. Yet intelligence has ever formed an important ingredient in the French army; and we do not believe that it could long be handled as an engine of barbarous despotism without wounding the engineer. Recent events at Rome and elsewhere might seem to contradict this assertion; yet we have confidence in the ultimate return of the French army to a sense of duty, self-respect, and patriotism. But though the march of mind may ultimately rescue the French arms from national and individual disgrace, we are far from countenancing an idle trust in their present moderation and forbearance. Ambition is evidently now their vital principle, glory their vital air; and they would march to hell to-morrow under a Bonaparte, to wreathe their bayonets in infernal laurels. Justice requires us to admit their gallantry; prudence bids us stand prepared; and now to facts.

The numbers, nature, and composition of the French army in Africa have fluctuated greatly since 1830. At the conquest it amounted to 30,000 French soldiers; in 1848 to 87,704; and in 1850 to 70,771 men.*

---

* Tableau, p. 14.

In order to give the reader a historical view of the nature of African warfare, and the character of the troops, it requires that we should go back to the Romans, having already described Spanish *razzias*.

Cæsar changed the manœuvres of his veteran legions in African warfare.* He says, " Cæsar instructed his troops to fight an enemy of this nature," *i. e.* the Libyans, " not like the commander of veteran troops, and like the victor in the most important actions, but as the trainer educates his gladiatorial tyros."†

The present precarious position of Turkey, exposed to the insidious thrusts of Russia and the secret venom of Austria, appears to justify the following digression on the military spirit and institutions that formerly caused and upheld her greatness. The spirit must chiefly be sought in the firm principle of religious conviction, which will always secure the triumph of its possessors; and in the decay of this spirit must we principally trace the abject state of Turkey, and most other empires that have lost their faith. Many of the institutions that contributed to the early glory of the Ottoman name have vanished, though some remain. The fierce Janissary and the dashing Mameluke are no more; but the gallant Spahis still remain in a few remote districts to parry the thrusts of the Czar, to form a bulwark against the strides of Cossack despotism, and to shield the hallowed martyrs of Hungary and Poland. Yet it must be admitted that Mahmoud and Mohamed Ali struck down some of the main pillars of Islam, the keystones of Osmanli empire. The heroic beys of

---

* De Bell. Afr. 71.

† It appears that the natives were in the times of Sallust, Cæsar, Livy, Strabo, and Procopius, as now, capable of being bent to European discipline. In the army of Hannibal the officers were Carthaginians, but the majority of the troops were Kabyles or Libyans ; and in his great expedition to Italy, he left in Spain a guard of 15,000 Africans. Thus Polybius informs us that "he left with Asdrubal 450 Libyphœnician and African horse (a mixed Punic and African race), 300 Lorgitas, 1800 Numidians, Massylians, Massassilians, and Mauritanians from the shores of the ocean, besides 11,000 African foot.

After the fall of Carthage, the Romans encouraged the enlistment of these same men ; and before he took Gaulish horse into Africa, Cæsar had taken Numidians into Gaul. " Cæsar sent Numidian and Cretan archers and Balearic slingers as a succour to the citizens ;" and again, " Cæsar led over the bridge the whole cavalry, and the light-armed Numidian slingers and archers. (*De Bell. Gall.* i. ii. c. 7, 10.) The war ending with the death of Cato showed how profitable their organisation had been to the party in power. In the battle before Ruspina, Labienus and the two Pacidius's caused the Numidian auxiliaries, consisting of more than 9000 horse and 36,000 infantry, to sustain the chief shock. (*De Bell. Afr.* 13-15, 18, 19.)

At a later period, the Gætulians and Numidians favouring Cæsar made the scales turn on his side, by passing over in bands from the camp of Scipio. Cato made levies in the province of Utica, and Considius besieged Achilla with eight cohorts of natives. Scipio occupied Uzita with considerable Numidian forces, and Juba had adopted the Roman organisation for his forces. (*Ibid.* 32, 36, 42, 43, 59.) Thus the army of Pompey, as previously that of Hannibal, was composed chiefly of Africans, Romans alone being the officers. The zouaves and spahis of modern Algeria, officered by Frenchmen, are much the same thing.

Kahira, with the noble Circassian blood coursing in their veins, were splendid specimens of that chivalrous race which, in defiance of the bribes, the snares, and the blows of the Czar, has resisted the advance of Russia into Turkey and Persia for the last fifty years, converting the Caucasus into a theatre of imperishable glory, and dying its snows and torrents with the blood of heroic generations.

It cannot be doubted that the Sultan and the Pasha bought the advantages of civilisation at a high price. The Turk has been drilled, dressed in frock-coat and fez, taught to drink brandy, and to steal or commit adultery; but it is a question if these advantages are a compensation for the sincerity, simplicity, reverence, and honesty that they have lost. True it is that the Deys had a sly affection for their neighbours' goods; but Christendom had set them a good example long before; and the cool partition of the Ottoman Empire by the great Christian powers in the present century,—awarding Greece to Bavaria (a signal blessing to that classic land); Algiers to France, with her liberty, fraternity, and equality ; and the Danubian Principalities to Russia, with its knout, secret police, and Siberia,—must be admitted as creditable specimens of the self-denial, equity, and honesty of Christendom. The Turkish force at Algiers consisted of Janissaries, forming the infantry, with a few Spahi troopers. The Spahis are described by Sir Paul Rycaut as the gentry of the Ottoman empire, 12,000 in number. The principal officers of the Janissaries, Odas or Chambers, of which there were 162, were—1. the Odabashee, lieutenant; 2. the Wekilharg, commissary ; 3. the Bairacktar, ensign ; 4. the Tchorbagi, or captain. The general was called the Janizar-Agasi ; the lieutenant-general the Kirhaia Begh.

Previously to laying before our readers the nature of the French and native troops in French pay, we shall insert a few curious particulars relating to the militia of the Dey—a title signifying patron, or uncle.

In former times the number of Turks in Algiers did not often exceed 5000 men ; in 1826 it was less than 4000, most of whom were superannuated. At the same time the whole military force of the Algerian government consisted of about 15,000 men in all—Turks, Koulouglis, Arabs, &c. The former were infantry, the latter cavalry ; and the Koulouglis were seldom called out except in cases of emergency. They were excluded from the honour of being deys, aghas, or holding any other official rank. The Arabs, Kabyles, &c. were of little value, though kept in pay, being hereditary enemies of the Turks. *Divide et impera* was the maxim of the Turkish government; and it succeeded there, as in Paris in December 1851. The Arab cavalry could never stand the Turkish infantry. Recruits arrived every five or six years from the Levant, consisting of shepherds, outlaws, &c., all of whom could become dey; for never did the sun see so much democracy and equality, united with so little fraternity, as in every Turkish commonwealth. The officers were, first, the agha or general, 30 yia-bashees or colonels, 800 balluck-bashees or captains, and above half that number of oda-bashees or lieutenants.[*]

Before we proceed to analyse the most remarkable corps of the French army in Africa, we shall present the reader with the following picture of its *personnel,* &c.

The total expenditure of the French army amounted in 1845-6 (a heavy year) to 74,465,527 fr. (2,978,621*l.* 1*s.* 8*d.*) for 60,000 men and 13,896 horses.

The largest army that France has ever had in Algeria was that voted to Marshal Bugeaud in 1846, amounting to 100,000 French and 25,000 natives. The first article of the budget of 1846 placed at the disposal of the minister of war 24,000,000 fr. (960,000*l.*), for the maintenance of 34,000 additional men and 3800 additional horses, over and above the effective force of the army, as fixed by the third article of the law of 1845.[†]

The cavalry in Africa are all well mounted, and the Chasseurs d'Afrique especially, their horses being of the country, and far better suited to the climate and fatigue than the French and Sardinian horses. The Arab horse partakes of the abstemious habits of his master. A little green meat or chopped straw, or even a few leaves of the wild artichoke, will sustain his courage for a great length of time, and a ration of barley is a luxury to him ; whereas the European horse sighs for the leeks and onions of his native land, his city of delights. He must have three rations a day ; and cannot stand the sirocco and want of water. Oats are un-

---

[*] Blofeld, p. 212. Walsh's Residence at Constantinople, 2 vols. 8vo. Miss Pardoe's City of the Sultan, 3 vols. 8vo, 1838. Eton's Survey of the Turkish Empire, 1798. Sir Paul Rycaut's Present State of the Turkish Empire, 1682. Auldjo's Visit to Constantinople, 1836.

[†] In 1838, the foreign troops in the pay of the French in Algeria cost 7,284,147 francs (291,365*l.* 18*s.* 4*d.*). The Arab soldier in French pay receives 1 franc (10*d.*) per day, and maintains himself and his horse. In 1837, the Gendarmes Maurus amounted to 146 men : infantry 96 ; cavalry 50. They are employed as guides ; and they cost, in 1837, 113,000 francs (4,520*l.*). Tableau, 1839.

With regard to the supplies for the army in the years 1846-1849, we find that the total amount of corn consumed by the army in 1846 was 45,326 kilog. 06 q. 2 gr. (1662 bushels). In 1848 it was 41,714 kilogr. 92 q. 2 gr. (1529·51 bushels).

The amount of meat consumed in 1846 amounted to 148,975 quintals 20 kilogr. (32,774,540 lbs.).

In 1848 it was 115,721 quint. 60 kilogr. 4 gr. (25,458,720 lbs.).

These estimates, however, include the whole of the carcass or raw material (*poids brut*). The amount of butcher's meat in 1848 was 67,019 quint. 74 kilogr. 2 gr. (14,744,328 lbs.).

The wood consumed by the army in 1838 was valued at 520,280 fr. (20,811*l.*).

The negligence of the French government is apparent even in the management of woods. In 1845 the fire-wood for the army cost 374,000 fr., and was brought almost entirely from other countries ; yet the lentisk, the carob, olive, cork-tree, &c. flourish well in Algeria if cultivated, besides all the trees of the south of Europe. St. Marie.

known in Algeria, being too heating for the horses, and probably for the men ; barley is used instead.*

The native troops in French pay in Algeria are divided into various classes, according to their organisation and their arms. Some are irregulars, and some regulars ; some infantry, and others cavalry. Like the Romans before them, and the British in India, the French have found it expedient to support their power in the conquered territory on native bayonets or lances ; but they have hung fire in extending the system, and begotten a hybrid force by sprinkling native regiments with French soldiers.†

"The most historical, and I might add fabulous, of the native troops are the Zouaves, a body of men renowned alike for extravagant daring and for disorderly behaviour and rascality. They were organised by M. de Lamoricière soon after the conquest; and their uniform is much the same as the Turkish costume, with green turbans. On all occasions they have been invariably triumphant ; and it is related, that when the Duke of Orleans wished to reward a private with the cross of the Legion of Honour, Cavaignac, then colonel of the corps, said : ' If your royal highness wishes to recompense acts of bravery in this corps, you must provide decorations for every man in the regiment.' "‡

According to Castellane, the regiment was partly formed out of bodies of French troops called *volontaires Parisiens* and *bataillons de la Charte;* and he adds, that these fire-eaters were led up the breach at Constantina, in 1837, by Lamoricière, amidst a tempest of bullets, through springing mines, and a chaos of ruins. § There is scarcely a valley or a hill but has borne witness to their gallantry, and re-echoed the report of their muskets ; but we shall only give one more specimen of the stuff they were made of. At the siege of Zaatcha, in 1848, Colonel Canrobert addressed them thus:

* The average price of the *remonte* for cavalry in the French army in Africa was said to be about 425 fr. (17*l*.) per head in 1846 ; whereas in 1830 the usual price given was 280 fr. (11*l*. 4*s*. 2*d*.). This increased value of horses may be ascribed to the devastation of the country by war, and to the rapid decrease of the animals, as well as the retirement of those who breed them, though this effect has been somewhat neutralised by vast importations from France and Tunis.

† We shall first pass in review and dismiss the native irregulars. The Arabs, besides paying taxes, have either to furnish a military contingent or means of transport to the French. Immediately that the tribes were subdued, the victors required them to supply irregular horsemen, called *makhzen* or *goum*, to attack the refractory. Ill-armed and undisciplined, without any military organisation, they were often useless, and sometimes dangerous in a body ; but they supplied excellent spies and scouts. After victory they were invaluable in hunting down the foe. The goums would never fight without the support of the regular cavalry, through fear of the fanaticism and vengeance of the patriotic party. As to the requisitions for beasts of transport, they were very imperfectly obeyed till the final conquest of Algeria ; but after the fall of Abd-el-Kader, the goums fought with more willingness for the French, the transport service was regularly attended to, and now all is organised. The goums have latterly done good service, especially in the Sahara, and appear to answer admirably as policemen among the tribes. (Tableau, 1850, p. 723.)

‡ St. Marie.   § Souvenirs, p. 76.

"Whatever happens, we must scale these walls; and if the retreat is sounded, remember, zouaves, it is not for you."[*]

Mr. Borrer informs us that the regiment of zouaves was formed by General Clauzel in 1830, and that it was originally composed entirely of natives of Algeria; but that it consists now partly of French and partly of natives, but chiefly of the former. It is divided into three battalions, each composed of nine companies; and the oriental costume of these troops is picturesque and convenient, consisting of leathern buskins and loose oriental trousers. Their arms consist of the musket, bayonet, and a short sword much resembling the ancient Roman *gladius*. These troops scale rocks with the agility of mountain goats, combining the utmost endurance with great hardiness and strength; for they are all picked men, and generally of rather short stature, broad-shouldered, deep-breasted, and bull-necked; much more serviceable men for such fighting than our six-foot grenadiers. They are, however, superlatively cruel, bloodthirsty, and eager for plunder, if we may believe Borrer; neither do they give or obtain quarter.[†] So much for the zouaves, who remind one of the ragged rascals (88th Connaught Rangers) of Picton.[‡]

The next native corps we shall describe is the Spahis, a body of light cavalry, whose name is taken from that applied to all Turkish cavalry. The spahis existed under the deys; and now form a body of regular troops, divided into four regiments, under Yussuf, who was made a general under restrictions, to prevent his competing with other generals for promotion in the French army. We have already described the history and appearance of this remarkable man. The Romans had their spahis in Africa; and the Turks never employed more than 16,000 Ottoman troops, governing Algeria in a great measure through natives, analogous to our sepoys.

[*] Souvenirs. On the *zouaves*, see Captain Kennedy, vol. i. p. 49, 50.

[†] Campaign in the Kabylie. The same authority states, that it was in cheering on the zouaves during the murderous struggle on the breach at Constantina, that Lamoricière, then their colonel, so narrowly escaped a hideous death from the explosion of numerous magazines, the fire from which, falling upon the bags of powder borne by the engineer soldiers, grievously wounded him, blew half his men into eternity, and rendered a portion of the venerable Cirta an infernal chaos of ruins, flames, and dying wretches vainly struggling to draw their mangled bodies from the devouring fire. St. Marie bears witness to the rascality of the zouaves. He says they are *mauvais sujets*, and tells many amusing anecdotes of the clever *plants* that they have performed on honest cits in Algeria, bearing a family likeness to the nugget-frauds and sly doings of old hands at the Australian diggings. As an evidence of their desperate valour, he relates that a stand of colours having been made for the regiment by the Queen of France, Marie Amelie, was pierced by fifteen bullets in the first engagement, where it was baptised in powder, and made four lieutenants on the field of battle, three ensigns having been killed there. Mr. Borrer describes their unsteady behaviour at Bugia thus: "The zouaves especially conducted themselves in a most outrageous manner; all discipline was forgotten by them, and they may be said almost to have sacked the town (then friendly). Not only did they violently attack and clean out the wine and liquor shops *vi et armis*, and commit other gross outrages, but they sacrilegiously broke into the French chapel, and robbed it of the sacred plate." Borrer, p. 96.

[‡] See Memoirs of Sir T. Picton.

The following is the history of the spahis : In 1830 some squadrons of Arabs and French were mixed, and after 16 months' service, separated into three regiments of *chasseurs d'Afrique*. As this plan was defective, the spahis were separated in 1834 from the chasseurs, and concentrated into a native corps of three squadrons, with a sprinkling of Frenchmen dressed like natives. This arrangement has been found to succeed best. In 1836 the corps of spahis consisted of six squadrons. French may be admitted in the proportion of one-fourth ; and as a general rule, the chief of each corps ought to be French, as well as the captain of every squadron, but in exceptional cases he may be a Turk or native. The pay of the *sous-officiers* and spahis is from 3 fr. 50 cents. to 4 fr. 90 cents. (from 3s to 4s.), all included. Contingents of Arab allies, &c. can be temporarily incorporated in the spahis ; and the knowledge of the two languages, French and Arabic, gives advancement.*

Castellane observes : " Two elements unite in the African cavalry to insure success—the French and the Arab, the chasseur and the spahi. Those tall soldiers in blue jackets, notwithstanding their valour, could not alone have achieved their bold strokes. The Indian was found neces- sary to drive the Indians out of the American forests ; and the Arab was wanted to contend with the Arab in Africa. The arm that strikes well home requires the quick eye and the cunning thought. Such was the ori- gin of the spahis. Good pay attracted the Arabs, whose discipline was less severe than the French ; and their only uniform was made to consist in a red burnouse, dropped in a moment. An Arab still, though in French pay, the spahi could do much, unsuspected, as courier, spy, &c. French officers, and a sprinkling of European privates, completed the corps ; and it has often made itself useful. It has been called often, with a smile, the ' refuge of sinners ;' and many free-and-easy characters, who would have kicked at French discipline, are in its ranks,—adventurous spirits, fabulous men, whose history is like an old legend."†

The other regulars consist of the *tirailleurs indigènes* or Turkish bat- talion, who are admirable skirmishers. From M. Borrer we learn that the French and native regular troops in French pay were analysed as follows in 1846:—

	1848.		1850.	
	Men.	Horses.	Men.	Horses.
French	87,704	18,742	70,771	13,189
Native	6,653	3,767	6,437	3,422
Total . . .	94,357	22,509	77,208	16,611

* Berbrugger, page 48. The dress is uniform for the officers and men in the Algiers squadrons, the harness being Arab. Officers in the spahis are obliged to stay in the corps at least two years. Their dress is a dark-blue full trouser ; a djebadoli, or red vest, under which are the cedriga (blue waistcoats) ; and for their head-dress they have a red cap (*fez*), called in Arabic *chachiah*, and a red turban. Berbrugger ; Capt. Kennedy, v. i. p. 99.
† Souvenirs, p. 106.

FRENCH INFANTRY ON THE MARCH.

p. 394.

The French army in Africa consists of infantry, line and light,* and light cavalry, besides artillery, engineers, &c. The infantry of the line is, like all French foot-soldiers, active, nimble, enduring, and dashing. Many regiments have long served there with distinction, particularly the 67th and others.

The light infantry is pre-eminently adapted for African warfare ; and the 2d light, and chasseurs de Vincennes, have achieved many deeds of prowess in the defiles of old Atlas. The former corps was at Constantina in 1836, where we have seen its gallantry under Changarnier. We find it again at the Col de Moussaia in 1840.

The chasseurs de Vincennes are a very remarkable body of men, who were first instituted by the Duke of Orleans, and known as the tirailleurs de Vincennes. After the duke's death, they received the name of chasseurs d'Orleans ; and since 1848 they have been known as the chasseurs de Vincennes. Armed with Minié rifles, they have been highly efficient instruments in destroying the liberties of Algeria and of France.† They form at present ten battalions, and more are to be added.

The foreign legion was established in 1831, and consisted in 1841 of six battalions,—four German, one Polish, and one Italian. A great number of Belgians were also found in its ranks ; and in 1841 it was commanded by a German officer (Meyer).‡

* The line used first to christen the light infantry "zephyrs ;" but the term "zephyr" has latterly devolved on all the French infantry in Africa, who well deserve the name. Berbrugger.

† The average range of their new muskets is 1100 yards. *Faggot of French Sticks* by Sir F. Head, vol. ii. p. 260. The engineers are a special corps, commanded in 1845 by Colonel Lemercier, and form a very intelligent body of men, who have rivalled in energy and courage the finest regiments in the French service. The mountain gun-service has been admirably organised, so that their light field-pieces and carriages can be carried swiftly on the backs of mules to the most impracticable districts, where, being mounted in a moment, they carry terror into the heart of the tribes. A new species of force has also been lately organised in the Sahara, consisting of French soldiers mounted on dromedaries, who are already said to have done good service. St. Marie and Le Pays *Journal de l'Empire.*

‡ The French in Africa, part 1, the Foreign Legion. "The foreign legion presents a singular appearance. It contains specimens of all countries and races. Some who have dropped nobody knows from where, after leading a life of adventures and wandering, like the Wandering Jew, come and seek rest by running wild in Africa. A large number of them, well born and well bred, but wild dogs, and the reprobates of all Europe, having saved courage from the waste of their folly, come and ask for protection against themselves under the French flag and a feigned name. Accordingly, when any family has sought in vain for some lost member, and when all the police of the world is nonplussed, there still remains one hope: write to the colonel of the foreign legion ; he will almost always give you what you seek. I saw, while I was at Khamis, the son of an aulic counsellor, the nephew of a cardinal, and the son of a Frankfort banker, reclaimed almost simultaneously. Chinese is the only language that is not spoken in this modern Babel. Italians, Prussians, Portuguese, Russians, and Spaniards, are all represented in it. An iron hand is necessary to bend into the same shape such heterogeneous elements ; but discipline suffers no indulgence. Woe to the disobedient ! Court martials are pitiless, and justice is rapid."—*Souvenirs*, p. 167. Captain Kennedy gives the foreign legion two regiments and 5000 men, vol. i. p. 287.

We read in the work, *Studies on the French Army,* by Captain Guy de Vernon, of the 8th chasseurs à cheval, the following article upon the African chasseurs :

" The chasseurs d'Afrique have been, ever since their creation, the vanguard of the military movements and operations which have signalised our conquest. . . . Placed as vedettes of our warlike colony, they are posted at the extremity of the thousand arms, armed with fire and steel, of that gigantic occupation, which, covering a surface equal to the half of the territory of the continent of France, will repose for a long time yet under the shadow of our sabres and bayonets. Chasseurs and spahis ! their numbers are written in all epochs, and in the most glorious pages of this modern Iliad. . . . The chasseurs of Africa are real light cavalry in the full extent of the term, and the highest acceptation which it admits. These regiments recruit in a most exceptional manner, by means of young soldiers of two years' service, of sundry other experienced materials, and even of military convicts whose punishments may be remitted without danger. They have thus the advantages of previous training and renewal, which would be sought for in vain elsewhere ; and that degree of force which, in all organised bodies, is the principle and the cause of extreme agility. Mounted on horses of native races, supple, skilful, nervous, bold, indefatigable, and from 4 feet 5 inches to 4 feet 7 inches in height, they will go from 15 to 20 leagues, always at a trot or gallop, without resting and without unbridling. Their equipment is light, since they have diminished the weight and number of the pieces of the harness, and have even suppressed useless effects—the heavy *schabra* of sheepskin and the heavy portmanteau. A sabre in the belt, a gun slung upon the shoulder, and a pistol in his *fonte découverte,* such are the arms of the chasseur d'Afrique ; armed as a pilgrim, he has his gun for his stick and cudgel.

Their discipline and instruction are according to the services expected from them, and they are allowed to enjoy certain salutary irregularities. At Constantina and at Oran there are no vain parades, no military spectacles to amuse the curiosity of the public, but exercises and labours suited to adapt the soldiers for a state of war. . . . If the African chasseurs are skilful scouts, dexterous marksmen, and bold and intelligent partisans ; if they perform wonders in plundering ; if they know how to march and fight scattered,—they know also, as well, how to unite in a body, march in order, and charge in line upon two compact ranks.

See them set out for distant *razzias :* nothing impedes their columns. A little corn for the horse, a little rice for the chasseur : these are the only provisions for the route. No obstacle stops them ; they go on, on : nor neglect one of the skilful precautions of the lion on chase ; and as soon as they perceive the enemy, they spring forward ; for to them, an enemy seen is an enemy gained, overthrown, destroyed. The habit of conquest has rendered them invincible. The Arabs, who have felt in a hundred rencontres the terrible qualities of our chasseurs, their strength and agility, their address, their bravery in action, and still more their generosity after victory, have named them the lions of the desert."*

From official documents,† we find that the number of sick in the military hospitals of Algeria in 1849 was as follows : Entered, 105,469 ; dead, 9745 ; discharged, 91,697.‡

Notwithstanding the important studies and publications of the French medical officers on hygiene, the mortality amongst the French troops in the unhealthy districts, such as Bona, seems proportionally as great as ever.§

The fate of Colonel Coombes, of the foreign legion, the day of the storming of Constantina, is a fine specimen of gallantry and discipline. Mortally wounded by two bullets, one of which had passed through his body, he fought at the breach until assured of success ; then, marching tranquilly up to the general-in-chief, he rendered his account of the progress of affairs ; and exclaiming, *"Heureux ceux qui ne sont pas blessés*

---

* Mr. Dawson Borrer (p. 18) informs us that the chasseurs d'Afrique are the *élite* of the French cavalry in Algeria, consisting of picked men, well mounted. Their arms are the carbine, sword, and pistol ; their uniform and accoutrements being neat, plain, and useful. No portion of the army has distinguished itself more than these bold riders, who are thus described by Castellane : "Bronzed faces with long mustaches, tall men proudly seated on little horses, this regiment was worthy of that cavalry whose name alone appals the enemy. 'Sassours ! sassours !' cry the Arabs, as soon as they catch sight of them, without daring to stand ; and this prestige they owe to their impetuous courage and their firmness. The features of these soldiers, waving their swords as we passed, recalled those manly iron squadrons painted by Horace Vernet at Versailles, and the men whom, at the Oued-Foddha, Changarnier launched against the Arabs, saying, 'That is my artillery' " (p. 346). See also Capt. Kennedy, vol. i. pp. 31, 32.

The average loss of the army up to 1840 was 1 soldier in 12·8, and 1 officer in 54·4. This estimate does not include those fallen in battle. But in battle the losses have been proportionally greater among the officers than the men. The mortality among the Janissaries was not so great as it has been among the French, and each of them was reckoned as equal to 20 Arabs in action. The Spanish occupation of Oran was also less destructive to their troops than Algiers has been to the French ; but the Spanish army consisted chiefly of brave adventurous gentlemen, and not of poor conscripts. Baude, iii. 244.

† Tableau, 1850, p. 70.

‡ The proportion of deaths to the number of sick was, in 1846, five 6-10ths ; 1849, nine 2-10ths. The mean stay in hospital was in 1846, 20 days ; 1849, 17 days. The mean daily movement was in 1846, 6841 ; 1849, 5110.

§ The French correspondent of the *Times*, of January 7th, 1853, whose letter is dated January 5th and 6th, says that "A letter from Bona, in Algeria, of the 23d ultimo, states that the epidemic of that town rages with fatal violence. It has extended out of the town to Edough, situated on a high mountain. There were 730 soldiers and 7 officers confined in the military hospital, and the civil hospital was so crowded that the governor was compelled to refuse further admittance. Four of the attendant physicians were attacked with the malady." It was estimated in 1846 that 21,000 French soldiers perish annually in maintaining the interior of Algeria ; and Baron Baude informs us (p. 296, vol. iii.) that from 1831 to 1839, 22,495 men died in hospitals, whereas only 1412 fell in battle. Some authorities assert the annual loss of the French army in Africa to be 36,500 men and its loss during 15 years' occupation to have been 347,500 men (up to 1845). See St. Marie.

mortellement, ils jouiront du triomphe," he fell dead at the feet of the Duc
de Nemours. On the hill Condiat-Aly, whence the French batteries
played upon the Bab-el-Djedid (new gate), on the west side of the city,
stands a pyramid erected to the memory of General Damrémont and
other gallant officers who fell there. The former was killed by a cannon-
ball, just before the final assault up the breach, and close to the above-
mentioned gate.*

Let us now pass in review the Algerian generals.

Describing Bugeaud's campaigns, Castellane says : " Blows like these
can only be struck by an army that has more than reliance in its commander. It must have respect and love for him. Such were, in fact, the
sentiments that Marshal Bugeaud had succeeded in inspiring in his soldiers.
Who amongst us has been able to forget his noble countenance and his
noble heart ? In their familiar way of speaking, the soldiers had christened
him *Père Bugeaud*. And they were right; for his solicitude for their wel-
fare equalled his affection for them. Easy and communicative, he felt
happy among his troops, as in the bosom of his family; his language, full
of good humour, went at once to the heart of the soldiers. They all felt
indebted to him for losing sight occasionally of his high rank; and the
respect they bore him was only increased by this condescension. It
was in times of danger that ' Richard was himself again.' In those seasons all eyes were turned towards him, being certain to find a direction
and precise orders ; or, if the danger became imminent for all, common
safety."†

The name of Bugeaud is associated with many of the most important
successes of the French arms in Africa. He beat Abd-el-Kader on the
Sikkak, near Tlemsen, in 1836; he overthrew the army of Morocco at
Isly in 1844; and he subdued the greater part of Kabylia Proper in 1846;
showing the greatest decision and the most determined courage throughout. Marshal Bugeaud, who was created Duke of Isly after his victory,
had served under Napoleon at Saragossa (1809), as we have previously
seen, and presided over Algeria as governor-general from 1841 to 1846.
He died at Paris, of cholera, in 1849.‡

" I still remember," says Castellane, " that on our way from Milianah
to Algiers, the Arab chiefs came to salute General Changarnier on his

---

* St. Marie.

† Castellane, p. 247. St. Marie admits that Marshal Bugeaud had great military ability,
and that he was a man of perfect integrity ; but he adds that he was altogether a soldier,
and jealous of his authority. The minister of war himself did not know always how to
deal with him. He was heard to say, " L'Afrique c'est moi." He used to be the terror
of the Arabs ; and he received Colonel Pelissier with great cordiality, after the latter had
burnt 1200 victims in the caves of Dahra in 1845. The staff-officers who surrounded him
imitated his rough manners; and one of his aide-de-camps is reported to have thrown a
plate, in a café at Algiers, at a dilatory waiter, who thereupon threw back an omelet on
the officer. Other ungentlemanly tricks are recorded of the French officers. St. Marie.

‡ Borrer.

passage; and among them I met a kaid of the Hadjonteo, an old acquaintance of mine. We spoke of the numerous razzias and mighty strokes that had subdued his warlike tribe. 'His name amongst us,' said the kaid, speaking of the general, 'means the leveller of pride, the subduer of enemies; and he has justified his name.' Pointing to the long line of mountains bordering the Mitidja, he added : 'When the storm comes, the lightning darts in a second over all these mountains, and sounds their cavities. Such was his look in searching us. When once he had seen us, the ball does not reach its quarry quicker than his blow smote us.' And the old Arab chief was right. The distinguishing characteristic of General Changarnier in war is a sure and rapid judgment, and an indomitable energy; he knows how to command. In face of danger, his courage rises; then, if you draw near him, his vigour becomes infectious, and you no longer doubt the event. He first showed himself at Constantina, and since then he has not once failed in supporting his glorious reputation. If ever you visit the bivouac of one of the old African bands, and enter the soldier's little tent, listen to the numerous excursions they have made with him, and you will hear what they will say about him."

Perhaps the most brilliant of Changarnier's achievements was the forcing of the Pass of the Oued-Fodha, near Milianah, in Sept. 1842. Never did a French column run such risks. With a thousand men he was enveloped in frightful ravines, while whole populations of hostile Arabs and Kabyles rushed upon him. But Cavaignac was there; the zouaves, the chasseurs d'Orleans, and the chasseurs d'Afrique were there; and Changarnier commanded. "Calm and impassible." says Castellane, "General Changarnier was at the rear-guard, enveloped in his little caban of white wool,* a target for all the bullets; and giving his orders with a coolness and distinctness that gave confidence to the troops, and doubled their ardour. Not a moment's wavering was seen in that daring eye; his heart seemed to swell with the danger. The column advanced, the mountains re-echoing to the tempest of battle. He led a charmed life amidst the showers of bullets, that seemed to increase his coolness. Seldom have soldiers shown more courage; but the chief knew how to command, and his men to obey.†

Lamping informs us (1841) that "General Changarnier, who commanded us, is known by the whole army as a brave soldier, who exacts the very utmost from others as well as from himself, and who accordingly most commonly succeeds in his enterprises. He is more feared than loved by the men, who say, 'C'est un homme dur ce Changarnier.' He appears to be a few years above 50, powerfully built, but with a face somewhat weather-beaten by the storms of life. He has been fighting in Africa ever since the first occupation."‡

---

* Light white coats worn in Africa winter and summer.   † Souvenirs, p. 74.
‡ The Foreign Legion, part i. of the French in Africa.

Changarnier has since shown the same stern inflexible will and decision at the barricades of the Faubourg St. Antoine and in exile. Whatever his political opinions or errors may be, he has valiantly supported the reputation of the French arms.

Equally brave, more dashing, and less cool, the gallant Lamoricière is the perfection of a hussar officer. His very conversation has all the *entrain*, the lightning speed of a charge of cavalry. As governor of the province of Oran, he was remarkable for the incessant activity, promptitude, and rapidity of his intelligence ; and his bold spirit loved to indulge in brilliant paradoxes, in discussing and studying the questions of colonial empire and emigration.* Appointed governor in 1841, in that and the following year he sapped the power of Abd-el-Kader in the west, his stronghold, by his indefatigable razzias ; while Changarnier, the *montagnard*, as Bugeaud christened him, subdued the province of Algiers.

The name of Lamoricière appears early on the lists of fame after 1830. He engaged in daring exploits at the capture of Bugia in 1833, when he was an officer of zouaves.† We have seen him eating fire on the breach at Constantina ; and he appears again, ever foremost, in the fray at the Col de Moussaia under the eyes of the Duke of Orleans.‡

Cavaignac is a man of very different stamp; austere, silent, embittered, full of the glowing but concealed fire of disappointed ambition. A republican to the back-bone, he won his most glorious laurels under a king; and he committed his greatest fault as dictator of a republic. " Absolute in command, energetic in action, slow in deciding because slow of comprehension, but concealing the laborious working of his mind under a solemn silence, and only speaking when decided, General Cavaignac was esteemed by all, loved by some, feared by many. Those, however, who had had any relations with him, were unanimous in allowing that if you appealed to his heart, the haughty dignity with which he loved to surround himself disappeared, to make room for a quite paternal kindness ; but these moments of forgetfulness were rare. The silence in which he lived, the separation from others in which he liked to move, elevated his imagination; and the smothered fire of his eye disclosed a man who thought his life a sacrifice, even when the rank and dignities of the state were thrust upon him; for we must do him the justice to say that those dignities were received, but that he had far too much pride to seek them. Thus Cavaignac, by raising before his mind an ideal to imitate and worship, and by preferring his own esteem to that of others, ended by giving a false development to his naturally frank and kindly disposition. In his military career, Cavaignac has given many proofs of his cool obstinacy. He obtained his rank of commandant at Tlemsen in 1836, at the time of the expedition of Marshal Clauzel, when he held the place for six months, de-

---

* Souvenirs, p. 302. † La Grande Kabylie, p. 83. ‡ Berbrugger.

prived of all succour and news. This was one of the bright actions of his soldier's life; and it is but fair to add, that though he sadly mistook his mission as a politician, he was never found wanting in war, in the day of danger and strife. In 1840 he held out the whole winter at Medeah, with two battalions of zouaves, and was relieved in April, after five months' imprisonment, by General Bugeaud: his firmness, his noble example, the paternal encouragement of his advice, had been their great support.* We have seen his name in the gorges of the Oued Foddha, when, amidst a hail-storm of shot, Changarnier handed him some splendid wild grapes that he had just picked, with the words, "Here, my dear colonel, take this; you must want refreshment after such glorious fatigues."

We have still one more of the exiles, Bedeau, to consider, ere we pass to the Imperial generals. We have heard Borrer's opinion of Bedeau, whom he met in 1846, when governor of Constantina. (See Chap. XII.)

In 1841-2, General Bedeau was made commandant of the subdivision of Tlemsen. "Established in Tlemsen," says Castellane, "General Bedeau showed that regular and methodical spirit which makes such a useful instrument, as soon as the precise nature of his duties, and limit of his authority, have been accurately determined. To prove this, it is only necessary to observe, that Tlemsen soon rose again from its ruins, that barracks were built as it were by magic, and that the whole country received a wise and systematic organisation. General Bedeau was obliged to fight several times; but as there was no hesitation in his mind, his success was never doubtful."† We have only to add, that since Dec. 2d, 1851, General Bedeau, finding the air of France too close for him, has sought for a freer atmosphere amongst people who have a more vigorous constitution.

It was a lucky thing for the Chasseurs d'Orleans to have as their chief commandant Canrobert (April 1845). The quickness of his *coup-d'œil*, the precision of his orders, his energetic enthusiasm, the reliance that he had long inspired in all, rescued them from danger.

Commandant Canrobert was particularly distinguished for his presence of mind in critical circumstances. The following anecdote is a good specimen of his coolness. In 1848, being colonel of the zouaves, he was on his march from Aumale to Zaatcha to take part in the siege. The cholera had infected and was decimating his column ; and they advanced with difficulty, so laden were the mules with the dying soldiers. At the most trying moment he was informed that the nomadic tribes of the south were preparing to attack him. An engagement was, above all things, to be avoided, for they would have no means of transport for the wounded. The colonel immediately made his arrangements for fighting, and then marched forth alone to meet the nomades with his interpreter, and ad-

* Castellane, pp. 71, 103. † Souvenirs, p. 373.

c c

dressed them thus : " Know, good people, that I carry the plague about me ; and if you do not suffer me and mine to pass, I shall throw it amongst you." The Arabs, who had traced the column for many days by the newly-made graves it left behind it, were seized with terror, did not dare to attack, and let them pass.*

General (now Marshal) St. Arnaud appears first on the stage as a colonel, in which capacity he assisted at the judicious razzia which so happily reduced the Darha to order. We shall shortly allude to this brilliant affair (1845), which was far from ending in smoke. He was naturally promoted for his prowess and chivalry in smothering old men, women, and children in a cave ; and he made a wholesome example of all rebels who troubled the reign of order in the subdivision of Mostaganem in Great Kabylia, and in the Elysian Fields.

Appointed governor of Paris by the Prince President, he was quite at home in the night razzia which swept away the liberties, honour, and ornaments of France. He is one of the pillars of the imperial throne, and holds the portfolio of war. He was greatly opposed to the marriage of the emperor, and also to the liberation of Abd-el-Kader ; and when offered the command of another expedition against his old friends the Kabyles, he refused to go unless he obtained unlimited discretion, or in other words, license to extinguish the last spark of liberty in smoke. Being refused, disappointment nearly broke his gentle heart; but he has since recovered and returned to the war-office, whence he proposes shortly to make a trip to our Horse Guards, which are sadly out of order.

General Pelissier is another of the African chiefs who has attained fame as well as infamy in Algeria, but who, unlike his peers, has not danced a hornpipe on the barricades of Paris. Not satisfied with being a fire-eater himself, he seems to have wished to diet the refractory Arabs on smoke, suffocating 800 men, women, and children in a cave in the Darha in 1845. Nor did the affair end in smoke, as it materially broke down the spirit of the Arabs, and built up his notoriety as a man of decision and cruelty.

We find him a general, and the commandant of the subdivision of Mostaganem, in November 1846.

In the autumn of 1852 we find General Pelissier besieging Laghouat, in the Sahara of the province of Algiers, which he stormed (Dec. 2) with some loss. The flags taken at Laghouat by General Pelissier were deposited on the 30th December, 1852, at the Invalides.†

Marshal Count de Castellane commands at present at Lyons, where he

* Castellane, p. 133.

† Accounts received from Algeria in January 1853 announced the death of General Bouscaren, who was wounded at the taking of Laghouat, and expired on the 19th December, about half an hour after the operation of amputation had been performed. M. de Perceval, his aide-de-camp, caused the heart of the general to be embalmed, in order to be sent to his sister in France. (The *Times* Paris Correspondent, Jan. 1, 1853.)

gave Abd-el-Kader a grand review the other day, on the Emir's passage to Broussa *via* Marseilles.

Thus the boys of African warfare reach the highest dignities; while the grey-headed warriors live in poverty, exile, and disgrace, because they were true to their country and themselves, and kidnapped in the dead of night, while wrapped in slumber, by the cut-throats of a tyrant.

As for Lieutenant-general Canrobert, he had an infamous share in the *coup-d'état*, and enjoys the unenviable notoriety of having his name coupled with one of the basest crimes recorded in history. Yet he has of late slightly redeemed his blasted character by refusing the portfolio of war, when thrust upon him during St. Arnaud's illness, unless a free amnesty were granted to his old comrades Changarnier, Lamoricière, &c.

We have been loth to condemn such brave spirits; but truth and justice pronounce the verdict against them. Nor do we take up the gloves in favour of the selfish bourgeoisie and re-actionary Assembly, which sealed the fate of France by trampling on liberty at home and abroad, by strangling the infant republic at Rome, by staining the streets of Paris with the blood of its gallant sons, and by putting up a low adventurer and a bastard prince as its president.

The *coup d'état* was the expiation of the reaction ; and when Cavaignac, Changarnier, and Lamoricière extinguished democracy at the barricades, they paved the way for their own downfall.

A cloud hangs over France; but science must eventually demolish the chains of prætorian and jesuitical despotism.*

* On the French Army in Africa, see Captain Kennedy's Algeria and Tunis.

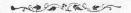

CHAPTER XXII.

The History of Algeria and Barbary.

THE REIGN OF MYTHOS — THE SEMITIC AND INDO-GERMANIC CONFLICT — THE PHŒNICIANS — THE SPIRIT OF CARTHAGE — THE FIRST PUNIC WAR — THE MERCENARIES — THE SECOND PUNIC WAR — HANNIBAL — CANNÆ — SCIPIO— ZAMA — THE FALL OF CARTHAGE—JUGURTHA—METELLUS—MARIUS—JUBA— CHRISTIAN AFRICA — DONATISTS — CIRCUMCELLIONS—TERTULLIAN — CYPRIAN —ST. AUGUSTINE—THE VANDALS—BELISARIUS—THE ARABS—THEIR DYNAS-TIES—THE TWO BARBAROSSAS — CHARLES V. — PIRACY — LORD EXMOUTH— THE FRENCH INVASION — ROVIGO — TREZEL — ABD-EL-KADER — THE CAVE OF KHARTANI—CAPTURE OF ABD-EL-KADER—HIS LIBERATION — ZAATCHA— LAGHOUAT.

PREVIOUS to the colonisation of Carthage, the history of North-west Africa is involved in mystery and deformed by fables. The story of Hercules * leading a mythical host from the far East to the pillars that bear his name, is better calculated to figure in the stanzas of some African Ossian than to bear the stern test of modern criticism. The theory of an aboriginal race is equally unpalatable to scientific ethnography ; and though we may be unwilling to attach much credit to the obscure traditions of the highlanders of the Atlas, the most plausible theory of the original population of this region is that which coincides with the legends of the Kabyles,

---

* This African Hercules must not be confounded with the Greek. There were several Hercules, some say forty. The travelling Hercules was the Tyrian or Phœnician, who is said to have founded many cities on the coast of Mauritania, including Tangiers. The Libyan Hercules is less known, and is probably the same as the Tyrian. President de Broses, however, thinks that the founder of Capsa was a different Hercules from the founder of Torigis, because the ancients call all great adventurers Hercules. Bochart and De Brosses assert that the name was given by the Greeks to the Tyrian Hercules, because the Phœnicians in their tongue called him Harokel, meaning the merchant or traveller. Barbié du Bocage thinks, with French scepticism, that he was nothing more than a merchant or shipowner, pure and simple. Diodorus Siculus makes him travel in Gaul and Italy, after the conquest of Spain ; but Sallust says he died in Spain, where his tomb was highly venerated by the Phœnicians. The mythical expedition attributed to him must have taken place as far back as the entrance of the Israelites into Canaan, for at that time a part of the Phœnicians was forced to emigrate and colonise the north coast of Africa. Barbié du Bocage's Sallust, Geogr. Dict. p. 215.

in attributing an Eastern origin to the earliest occupants of the soil, who, under the name of Libyans and Berbers, were partially or totally subdued by succeeding waves of population.

It would be equally unprofitable and unpalatable to dive deep into the ocean of ancient mythos; and we shall pass on to the clearer light of authentic history, after casting a transient glance at the poetical legend that has attempted to link its earliest history with the heroic age of Greece. Little aid, it has been justly observed, can be derived from the classical authors, who took more delight in gratifying their imaginations than in storing their minds with knowledge. To them Africa appeared much in the same light as India and China did to the writers of the middle ages; and while they crowded it with wonders of magnificence and splendour, they introduced into it all the monstrous and most terrific productions of nature.* Yet while we naturally feel disposed to smile at the tradition recorded by Sallust, of Hercules passing from the Levant with a host of Persians to the Straits of Gibraltar, inverting their barks on the desolate shore in the shape of the later Numidian huts,† we cannot avoid bearing our testimony to the frequent accuracy and value of the descriptions handed down by the father of history.‡ Many of the facts which he has related have been verified by recent discoveries; and races of lion and dog eaters are still found to people its valleys and oases. Thus it was that the glowing or monstrous descriptions of Marco Polo met with ridicule, and obtained him the epithet of *Millione*, till a more searching inquiry, in a more enlightened age, substantiated most of his statements.§

But though we are disposed to justify many of the relations of Herodotus, we cannot attach much credit to the authority of Procopius when he stakes his credit on his having seen, in the time of the war with the Vandals, when he accompanied the great Belisarius into Africa in quality of secretary, near a fountain at Tangier, two columns of white stone, whereon were inscribed, in the Phœnician tongue, the following words : " We fly from the robber Joshua, the son of Nun." ‖ Though there be no moral or physical impossibility in the existence of such an inscription, or its having been seen by Procopius, it would be worse than idle to attach any value to it. That Africa was very early peopled by emigrants from Asia, belonging to the Semitic variety, can scarcely admit of any doubt. Ethnography, and the natural tendency of an established and populous district to overflow into its more vacant contiguous districts, are sufficient and powerful arguments in favour of this view; and though we are not prone to attach much historical value to vague traditions, it may slightly tend to corroborate this

---

* Dr. Russel's Barbary States, chap. i.
† Sall. Bell. Jug. c. xviii. "Iique alveos navium inversos pro tuguriis habuere.'
‡ Herodotus, Melpomene.
§ The Travels of Marco Polo abridged.
‖ Procopius de Bello Vand. lib. ii.

view, that the Moors narrate that their origin may be traced to Sabæa, a
district of Arabia, whence their ancestors, under their king Ifricki, were
expelled by a superior force, and reduced to the necessity of seeking a new
home in the remote regions of the West.* They would probably drive the
older inhabitants from the more fertile districts to the tracts bordering on
the desert, or to the mountains, where they would seek a natural refuge in
caverns. Even at the present day there are found in southern Numidia
the remains of towns and castles which present an air of very great antiquity. Arabian populations have generally preferred a more erratic mode
of life ; hence the earlier inhabitants expelled by this Sabæan invasion
must have belonged to a different race, though still probably members of
the Semitic variety. Experience and analogy warrant us, therefore, in
arriving at the conclusion, that from the earliest periods waves of Asiatic
invaders have immigrated successively into the plains of North-western
Africa, belonging chiefly, in all probability, to the Semitic variety ; but the
time when these early arrivals occurred, the tribes that composed them,
and the places whence they came, must remain involved in uncertainty.†

---

* Morgan's Complete History of Algiers, p. 9, in Dr. Russel's Barbary States, p. 27.
Sallust, in one of his fragments preserved by Priscian, informs us that the Moors, a vain
and lying nation, like all those of Africa, maintain that beyond Ethiopia there are anti-
podic peoples, just and beneficent, whose manners and usages resemble those of the Persians. It is certainly possible that the Moors may have had some knowledge of central
and southern Africa, but this matter must remain involved in mystery.

The Gætuli, who inhabited the whole of Africa with the Libyans in the most ancient
times, were driven south by the invasion of the Phœnician Hercules, and their territory
corresponded in some measure to the modern Sahara. They were very numerous and
barbarous at the time of Jugurtha, who, when he lost Numidia, retired amongst them, and
instructed them in military discipline ; but they were ultimately subjugated by the Romans.
Barbié du Bocage is of opinion that their race has been preserved in the present Berbers ;
but it is evident that the latter people are the descendants of *all* the aborigines of north-
west Africa previous to the Arab invasion, including the Gætuli and the subsequent infusions of Phœnician, Vandal, and other blood.

President de Brosses derives the name of Gætuli from the Phœnician *geth*, meaning
cattle ; but Barbié du Bocage is of opinion that the Berber tongue had no analogy with
the Phœnician. Though this may be true as regards the fundamental roots of the two
tongues, the names of numerous places shew that the Berber had been a good deal in-
fluenced by Phœnician or Punic. Dict. Geogr. p. 204.

† Sallust describes the invading army under Hercules as consisting of Medes, Persians,
and Armenians ; that the Persians formed the Numidian nation, and the Medes and Armenians the Moors. But Abbé Mignot and President de Brosses think that Sallust was probably mistaken, owing to the errors of the translator of the Punic works of Hiempsal.
They think that the Amorites or Arameans were confounded by him with the Armenians,
and the Phereseans with the Persians. President de Brosses asserts that the name Libyan
comes from the Phœnician *lleaba*, signifying burning climate ; but the term may be derived
from the Greek. The Moors were called Mauri by the Latins, and Maurusii by the Greeks :
and Sallust relates that the Medes and Armenians remaining nearer the sea united with
the Libyans, forming a nation called Moors, by altering the name Medes ; whereas the
Persians united with the Gætulians, and formed the Numidians.

Strabo says that, according to some authors, Hercules brought the Moors from India
into Africa, which is not very probable. Bochart derives the name Moors from the Phœnician *mauharim*, signifying postremi, the last- the western people. It is certain that the

The first immigration of which we have any certain, though still a distorted knowledge, is that of the Phœnician colonists who founded Carthage. The general voice of history represents this event to have occurred B.C. 900 ; but before we launch forth into Carthaginian history, we must warn the reader that almost every particular relating to that Semitic people has come down to us through the medium of classical, *i. e.* Indo-European writers, who have necessarily given it a foreign and often unfavourable colouring. At the taking of Carthage almost all their records perished; and though Niebuhr may have been wrong to strike out all Roman history before the burning, because of the loss of her records, we may safely follow him at a respectful distance, and conclude that the *whole* of Carthaginian history has been an *ex parte* statement, owing to the medium through which it has been handed down.

With this proviso, we shall pass to the shores of Syria, and examine that remarkable Semitic race whose sons went forth to raise the walls and man the fleets of Carthage.

The Semitic race* does not appear in rainbow colours on the coast of Syria and Carthage. God chose cruel and ignoble and impure vessels for channels and instruments of his purposes. On the narrow beach overlooked by the cedars of Lebanon swarmed a numerous people crowded into the islands and close maritime cities.† Their religion was coloured

---

Phœnicians had colonies all along the coast of Mauritania ; but President de Brosses says the name comes from the African (Berber) word *more,* meaning merchant. Barbié du Bocage's Sallust. Dict. Geogr. pp. 252, 239, 238, 180.

* The two pillars of history are undoubtedly the Semitic and Indo-Germanic varieties. Each, in a different walk, has crowned us with honour and glory, and made us little lower than the angels. On the one side, the heroic genius, and that of art and legislation ; on the other, the spirit of industry, navigation, and commerce. In Europe, the prose and philosophy of a critical, negative, and analytical spirit ; in Aramæa and Syria, the halo and aureole of an atmosphere of poetry, the realisation of the ideal, the union of earth and heaven. Greece and Germany have given us the revelation of the head, Syria and Arabia of the heart. (Sebold.) Providence has worked through both to a great end ; and though antagonistic, the two varieties have acted and reacted on each other. Greece got her alphabet from Tyre, which she ruined. Rome conquered Carthage ; but her spirit fell from that hour, emasculated by southern luxury. The struggle between the Semitic and Indo-Germanic races has rolled through the ages and re-echoed to our times. The Arabs, bursting like a hurricane from the desert, swept over Africa, grasped Sicily, Corsica, Spain, Magna Græcia ; and penetrating as far as Tour, held Europe in suspense, till a hammar (Charles Martel) turned the scale. It was a mere accident, or rather Providence, that prevented Europe now bowing to the crescent, our being circumcised, and muftis sitting at Canterbury and York. See Sismondi's Hist. of the Fall of the Roman Empire.

† On the rock of Aradus, to cite only one example, the houses had more stories than even at Rome. This impure race, flying before the sword of Sesostris or the exterminating knife of the Jews, had found themselves driven to the sea, and had taken it for their country. Like our noble ancestors the ancient Britons (Milton's *History of England,*—Prose Works, iv. 68), the immoderate licentiousness of modern Malabar can alone recall the abominations of these Phœnicians. There generations multiplied without certain family, each ignorant who was his father, like the happy population of La Belle France, multiplying

with the licentiousness and cruelty of the age and people; nor can we wonder at the extravagances of the faith of Moloch, when we find even the Jews, with divine daguerreotypes, worshipping the calf, and turning the temple into a broker's shop. The Phœnicians had the sins and virtues of their time and race; and though Moloch loved human victims, and the Tyrians many wives, we find that Christian hierarchies have had the same appetite, and that modern as well as ancient Solomons have been prone to display the latter weakness.

Let us now proceed to examine the principal events of their history in a chronological order.*

We shall present the reader, first, with a compendious history of Carthage,† and thereupon make a few reflections.

When the Romans, conquerors of Tarentum and masters of Magna Græcia, arrived on the shores of the strait, they found themselves front to front with the Carthaginian armies. Thereupon several treaties were concluded between the two republics.‡

Three powers, Carthage, Syracuse, and the Mamertini, shared Sicily. Rome, called on by one faction of the latter, hesitated not to protect at Messina those whom she had just punished at Rhegium. The consul Appius passed legions into Sicily; and Hiero, tyrant of Syracuse, was

---

promiscuously, like the insects and reptiles which after rain-storms crawl about in myriads on their burning shores. Michelet's History of the Roman Republic, p. 138, Bogue's European Library. Ezekiel xxvi. 27.

* Herder's Ideen zur Philosophie der Geschichte, vol. ii. pp. 65-6, 1841. Michelet's History of the Roman Republic, ch. iii. p. 140. Montesquieu's Grandeur et Décadence des Romains, ch. iv. When the Phœnician colonists first landed in Africa, the whole of N.W. Africa, from the Gulf of Leptis, near Barca in Tripoli, to the Mulucha river in Morocco, was called Numidia. One encroachment after another was made on this broad territory: first by Carthage, whose territory was bought or gained from the Numidians; secondly, by Bocchus, who was given one-third of Numidia after his surrender of Jugurtha, his share extending as far as the river Ampsagas, between Igilgilis and Cullu, in long. 32° east from Ferro. Its government was confided, under Cæsar, to our historian Sallust, who, like many other literary stars, is reported to have done a little sly business there in pillage and piracy on his own account. Ancient Numidia corresponded originally, in most respects, with modern Algeria, the regency of Tunis, and part of Tripoli. Barbié du Bocage's Sallust. Dict. Geogr. p. 250.

† Carthage was styled in the Phœnician tongue *Carthadt* or *Cartha Hadath*, which means 'new town,' to distinguish it from Utica, properly Ytica in Phœnician, meaning the 'old town,' and which was actually much older than the city of Dido. The Greeks corrupted the name into Carchedon (Χαρχηδων), the Latins into Carthage, which was founded sixty-five years before Rome, or 819 B.C., if we may trust tradition like the Puseyites. There was, however, it appears, before this time, a collection of habitations on the same spot, called Cadmeia and Caccabe, and attributed to Cadmus, whose origin may have dated from a period anterior to the Trojan war. Barbié du Bocage's Sallust. Dict. Geogr. p. 190.

‡ The first was of the age of Lucius Junius Brutus and Marcus Horatius, created consuls after the expulsion of the kings, twenty-eight years before Xerxes invaded Greece. The Fair Promontory, north of Carthage, was generally made the limit of Roman navigation in these treaties. Michelet's Roman Republic, c. iii. p. 148.

conquered by the Romans *before he had time to see them.* He became the most faithful ally of Rome. In eighteen months the Romans seized sixty-seven places, including Agrigentum with its 600,000 inhabitants; for the democratic forms of Sicily had converted that lovely island into a garden, teeming with a happy population, whose capital, Syracuse, 400 years B.C., was larger and richer than imperial Paris. But the Romans wanted a fleet; and copying a wrecked Carthage galley, in sixty days they put to sea with 160 ships, and beat the Carthaginians. Duillius, the victorious consul, was caressed with life-long torch-light processions and serenades, enough to satiate even a German professor at his jubilee, and a young lady in her teens.*

Rome next cast envious looks at Africa, and soon invaded it under Regulus. A huge boa was the first foe they met, speedily followed by others, who, though as wise as serpents, were not as harmless as doves. But the Carthaginians had no peace-society to teach them to kiss the foot that tramples on them; and happily their Manchester men were not able to talk down common sense and outvote militia-bills.

Two victories gave 200 cities to the Romans. Carthage, at the eleventh hour, was delivered by Xantippus, a Lacedæmonian mercenary, who beat the Romans and took Regulus prisoner. Mutual reverses in Sicily and at sea disposed both parties to peace. Regulus was sent to Rome; dissuaded the senate from coming to terms, though to his own cost; and returning to Carthage, died like a hero.†

For eight years the Romans were conquered in Sicily, successively losing four fleets. At length Hamilcar, Hannibal's father, threw himself on Mount Eryx, a steep huge mass‡ between Drepanum and Lilybæum, and stood firm against the Romans for three years, like Wellington at Torres Vedras.

A naval victory of the Romans decided the Carthaginians to sue for peace.§ The merchants of Carthage, like the Dutch and British, weighing the war by its profits, determined that they were great losers by it, and ceded Sicily to the Romans, agreeing to pay 3000 talents (720,000*l.*) within ten years.||

Though exhausted, the two republics, in the interval between the two Punic wars (241-219 B.C.), grasped right and left, like the Yankees, the

---

* Sardinia and Corsica, where Carthaginian monopoly had forbidden the cultivation of land, soon bowed to Rome. Michelet, ubi supra, c. iii. p. 152. Grandeur et Décadence, c. iv. p. 33.

† The Romans, whose testimony, the only one we possess, is not very trustworthy, record the savage vengeance inflicted by Punic spite on the gallant Regulus, which, if true, does not exceed the equivocal charity of Britain to her mighty foe at St. Helena. Dr. Russel, c. i. pp. 38-9. Michelet, c. iii. p. 152.

‡ Polybius, in Michelet, p. 154.

§ They had lost 500, the Romans 700 galleys in the war.

|| Michelet, c. iii. p. 154.

Directory, and all free and independent commonwealths. Hamilcar subdued Africa to the Straits, and part of Spain; while Rome conquered the Gauls and Ligurians.*

Then came the Mercenaries' war. Like other trading countries, including Britain, the Carthaginians, themselves no military people, subsidised foreign horse and foot to fight their battles and save their country. They paid dearly for this piece of folly; and their Punic faith in withholding payment met with its due reward. The mercenaries were masters of Carthage, which trembled at its peril.† The fate of Carthage seemed sealed; but the war-party happily triumphing over the peace-society, Hamilcar was made general, and cut off provisions from the mercenaries, hemmed them in, and forced them to cannibalism. The war ended in a blood-bath.‡ But Hamilcar was a troublesome customer, and seemed an embryo tyrant. Goaded and worried on all sides, the senate gave him no rest, accusing him of infamous morals,—a strange charge to make in immaculate Carthage. But Hamilcar's soul was too lofty to stoop to empire; and unlike a French autocrat, he was satisfied with saving, without aspiring to ruin his country.§

Boldly pushing his way in the peninsula, he was beaten and slain. Like many honourable members, Hamilcar had his weak side, and bribed suffrages at home. Golden influence obtained the choice of his son-in-law, the handsome Hasdrubal, as his successor, who founded Carthagena in Spain, a town still extant, the Portsmouth of their rotting navy. Hasdrubal being soon after killed by a slave, the army named for its chief Hannibal, twenty-one years of age, the prince of condottieri, the child of camps, the greatest captain of the Fore-world, a man of one idea—vengeance.‖ We cannot profess to detail his achievements, much less analyse his strategy, in these pages. The implacable hatred he swore to Rome on the paternal knee, if a myth, is too descriptive of the man's pith and marrow to be dropped before the lancet of a negative criticism. We

---

* Dr. Russel, c. i. p. 37. Michelet, pp. 154-5. Herder's Ideen, vol. ii. p. 156, a noble censure on the grasping, demoniacal spirit of Roman conquest.

† Spaniards, Gauls, Ligurians, Baleares, Greeks, Italians, Africans, made a confusion worse than Babel; and all was uproar and confusion when Hanno, sent by the republic, tried to obtain a remission of part of the debt. The men marched on Tunis 20,000 in number. Carthage tried to soothe them through Gisco, promising every thing; but the mercenaries became overbearing; they were joined by the African provincials to the number of 700,000. Utica, Hippo, and Zarytis massacred the Carthaginian garrisons; the same was done in Sardinia and Corsica; and Hanno was crucified. Michelet, c. iv. pp. 158-164, an eloquent passage. Grandeur et Décadence, c. iv. p. 29. Dr. Russel, p. 34. Polybius, lib. i. c. 6.

‡ In that sanguinary world of the successors of Alexander, in that age of iron, the war of the mercenaries still horrified all nations, Greeks and barbarians; and it was called the inexpiable war. Michelet, c. iv. pp. 163-4.

§ He went forth to subdue distant nations to the Punic flag, and in one year he traversed all the coasts of Africa and passed into Spain. The Punic courser held sway to the ocean. Michelet, c. iv. p. 165.   ‖ Michelet, c. iv. p. 166.

accept it. Saguntum,* his first exploit, his Montenotte, dazzled Carthage and stunned Rome; it stamped him one of Plutarch's men. He brought 150,000 men to the siege of Saguntum, only 80,000 to invade Italy. A splendid monument this to the valour of the sires of the defenders of Saragossa, a gallant people, once free as the ocean, the soul of prowess and chivalry, laughed into commonplace by their greatest ornament and pride, Don Quixote, and emasculated by a bastard breed of nobles and Bourbons. Saguntum taken, the gauntlet was thrown, the sword was drawn, the scabbard cast away.† Meanwhile Hannibal had marched for Italy. His army, like Wellington's, was a mosaic; like the bastard emperor's, it was gained and gorged with wealth. Like Austria, he kept up his empire by playing off nationalities.‡

Like the march on Moscow, Alexander's invasion of India, and Cæsar's of Britain, Hannibal's passage of the Alps is one of those fabulous feats that resound through ages. Weighing the army he had in hand, the barbarous population through whom he fought his way, the mighty piles he scaled, this march of Hannibal's is probably the greatest triumph of military genius ever achieved.§ Like the Crusades and the revolutionary wars, who can measure the push thus given to humanity by the opening up of the highways of nations? The only great tourists, till lately, have been conquerors.‖

We shall not discuss the pass that he took.¶ It suffices to know that he dared and overcame the icy horror of those regions; that with his swarthy soldiers scaling the pine-clad steeps, he dispersed the mountaineers who sought to oppose him; and spurning the glaciers, plunged into the smiling plains of Italy, five months after his departure from Carthagena, with 26,000 men—8000 Spanish infantry, 12,000 Africans, and 6000 horse, mostly Numidians.**

---

* On the siege of Saguntum, see Livy, xxx. 21. Michelet's Roman Republic, p. 168.

† Quintus Fabius called on the senate to ask them their intentions, and demand an apology. Raising the flap of his toga, he said, " I bring war or peace—choose." The Carthaginians replied, divided between fear and hatred, " Choose yourself." He let fall the toga and replied, " I give you war." " We accept it, and we shall know how to maintain it." Michelet, p. 169. Polyb. iii. Livy, xvi. 18.

‡ He drew Moors and Numidians from Africa, and sent over 15,000 Spaniards to guard the metropolis ; 16,000 men were left in Spain.

§ Michelet, pp. 172-4. Dr. Russel, p. 41.

‖ The march from Carthagena to Italy reckoned 9000 stadia. He studied the route well beforehand, and paved the way with gold, yet had he to cut his way through hostile tribes from the outset. The passage of the Rhine with his elephants and swarthy Moors beats that of the Berezina by Napoleon's European train.

¶ Mount Cenis, the Little St. Bernard, and St. Genevieve have disputed the honour. See the subject discussed in Gillies' Visit to the Waldenses; Eustace's Italy, &c. ; Appendix to Michelet's Rome, xxviii. He thinks Mount Cenis the pass taken by Hannibal.

** The entire number of men capable of bearing arms among the Romans and their allies amounted to 700,000 foot and 70,000 horse ; and the armies disposed on the several frontiers consisted of 150,000 foot and 7000 horse. Michelet, p. 175.

It is probable that the confidence and daring of Hannibal, in thus bearding the Roman eagle in its nest, have never been matched. Yet we must recollect that Hannibal appeared as the liberator of Italy from the Romans.

Hannibal soon trapped and beat Sempronius on the Trebbia, the elephants doing good service.*

Flaminius awaited him at Arretium ; his army plied with prodigies, and himself with persecutions from the senate, being a liberal in politics. The armies met at Lake Thrasymenus, and after a fierce battle, the Romans were cut to pieces ; not without much loss to the Africans, whose beloved horses also suffered severely from hardships, as well as their riders.†

Terror prevailed at Rome, and gave the reins to the aristocratic party, which put forward the cautious Fabius, a man whose coolness foiled the dashing genius of Hannibal, who was nearly entrapped and ruined by him, being saved by the stratagem of burning fagots tied to oxen's horns. Roman impatience, however, could not long suffer this procrastination ; and M. Terentius Varro the plebeian, against the counsel of his rival, Paulus Æmilius the patrician, dared a pitched battle with the Punic chief.‡

The Romans, blinded by dust and wind, met the Carthagenians at Cannæ, and were crushed, as at Thrasymenus, between the two wings, besides being taken in rear by Numidians.§

Rome seemed lost; but she stood firm, as centuries of political and ecclesiastical despotism had not then tamed her high spirit, as they have done since. She scorned to sue for peace. Hannibal wintered at Capua, and Carthage was lost; his troops, laden with spoil, would have found a Capua everywhere. Rome was saved by the sacrifice of her sons at Cannæ, as Russia at Moscow.||

Hannibal was beaten at Nola by the gallant Marcellus ; and Hanno, at Beneventum, lost 16,000 men. Yet the mighty Hannibal wrested Tarentum and a great part of Sicily from the Romans. The year 213 was a period of repose to both parties ; but in 214 Rome levied 335,000 men, to finish the war. A tremendous struggle ensued; but the Romans recovered

---

* This victory gave him 90,000 Gaulish auxiliaries ; and he was forced to pass the winter in Cisalpine Gaul, exposed to constant risk of assassination. In March 217 he marched to Arretium, and soon after lost an eye, through exposure and fatigue. Michelet, pp. 177-8. Dr. Russel, p. 41.

† Michelet, p. 179. Dr. Russel, p. 41. Polyb. iii. Livy, xxii.

‡ Hannibal's situation was at this time critical ; at the end of two years he had not a town or castle in Italy, and only corn for ten days left.

§ Paulus fell with 50,000 men, 2 questors, 21 tribunes, nearly 100 senators, and numberless knights. Hannibal lost 4000 Gauls, and 1500 Spaniards and Africans. Grandeur et Décadence, c. iv. pp. 34-5. Michelet, p. 182. Polyb. iii. Livy, xxii.

|| The Scipios, like Wellington, were in Spain ; victory followed their path. Carthage was jealous of Hannibal's success, and would give him no aid ; and his army, weakened and corrupted, was successfully encountered by the Romans.

Capua and Syracuse, though Hannibal made a dash up to the very walls of Rome.*

Sicily was recovered by Rome, but the two Scipios were beaten and killed in Spain. Young Scipio, the son of Publius, was still alive ; and the Roman people, scenting a hero, named him its saviour at twenty-four. A man of gentle temper and lion's heart, he was an advance on the old Roman angular character, though accused of irregular morals—not an uncommon failing with great captains.†

It was Scipio's eye that saw where the death-wound to Carthage should be struck. Though opposed by the senile conservatism of Fabius, he carried his point, and invaded Africa with a gallant army, whilst the Carthaginians were disputing about his projects. All Italy had furnished him with troops and supplies at Syracuse. Scipio hoped to secure the friendship of Syphax, the Numidian chief, whom he had gained during a temporary visit to Barbary.‡

The Roman consul feigned to listen to his propositions ; but, through spies, learning the combustible nature of the camps of Syphax and Hasdrubal, he attacked and burnt them in one night, though containing 90,000 men.§ The Roman soldiers were satiated with plunder.||

The Carthaginians, deprived of Syphax, recalled Hannibal, who left Italy, shedding tears of rage. We cannot agree with Michelet in think-

---

* It was Marcellus who stormed Syracuse; and Archimedes, after aiding his countrymen with his genius, paid the penalty of his patriotism with his death. Michelet, pp. 184-91.

† He described himself as inspired by Neptune ; he seems to have possessed the gift of prevision ; and we shall soon see him turn the tables. Carthagena was taken, and the head of Hasdrubal cast into the camp of Hannibal, who retired sullenly amidst the Brutii. Michelet, p. 197.

‡ Since then Syphax had married Sophonisba, the daughter of the Carthaginian general Hasdrubal Gisco. The African races, like the French, had the amiable weakness of frequently changing their opinions ; the idol of to-day was the victim of to-morrow. Sophonisba flattered Syphax with the proud idea of becoming arbitrator between the two most powerful states in the world.

§ Michelet, pp. 195-6. Dr. Russel, p. 42. Livy, lib. xxi. c. i.-liv.

|| Scipio had brought over with him Massinissa, the Numidian king. The latter, who was the best horseman in Africa, and who, up to eighty, could remain a whole day on his horse, always succeeded in eluding his enemy. Once, when closely pressed by Syphax, he hid himself, like David and Mahomet, in a cave. Massinissa, brought back by the enemies of Numidia, enjoyed the cruel pleasure of taking his enemy, of entering his capital, and taking Sophonisba from him. This African Catharine of Medicis, formerly promised to Massinissa, secretly sent to excuse herself from a forced marriage. The young Numidian, with the levity of his age and country, promised to protect her, and the same night took her for his wife. The unfortunate Syphax, not knowing how to avenge himself, secretly intimated to Scipio that she who had drawn him from his alliance to Rome might do the same to Massinissa. Scipio saw the soundness of the suggestion, and claimed Sophonisba as his part of the booty. Massinissa thereupon gave her a poisoned goblet, which she drank off calmly, saying, "I accept the nuptial present;" whilst he fled. For this he was highly lauded and crowned by Scipio,—honours somewhat dearly bought. Michelet, p. 196. Dr. Russel, p. 43 See the tragedy of Sophonisba in Livy, lib. xxx. c. iii.-xii. Alfieri has closely followed his narrative in the tragedy of Sophonisba.

ing him guilty of such atrocities in the latter part of his sojourn, or his presence so odious, as the Roman historians have related.* If all historical documents, save the French, relating to the last war had perished, we should have a strange version of the Peninsular war and of Waterloo, of Enghien and of Andrew Hofer.

A few days after his return, he encamped at Zama, five days' journey west of Carthage. He tried first diplomacy on Scipio; but this failing, he was forced to fight, and suffered the most disastrous defeat of ancient times.†

Scipio seeing the strength of Carthage, did not push it to extremities. He took their navy from them, and brought the territory of Massinissa, their ambitious foe, to their gates. When these terms were proposed in the senate, Hasdrubal Gisco advised their rejection; but Hannibal went up to him, seized him, and threw him down. There was much uproar; and it appears from this, that the Punic senate occasionally emulated an American congress, a French assembly, and a British parliament, in unseemly irritation and disorderly tumult.

Carthage gave up 500 vessels, which were burned in the open sea within sight of the citizens. Though equally distressing to them, it was less dishonourable to the perpetrators than the national burglary committed by Britain on her brother Danes.‡ What distressed the Carthaginians most was, however, paying the first term of the tribute, — as great a national curse, and as foolishly encountered, as the national debt of Britain.§ Hannibal entered Carthage as a master, with a clear field before him; was named suffete; and directed his attention and care to the prosperity of the state, agriculture, commerce, and pacific measures. Thus, after he had obtained supreme power, like Sylla, he was too noble to stoop to titles, and he bestowed his mind and means on repairing the disasters of his country; — a noble example, and a bitter criticism on those modern autocrats who grind their people to powder to fill their coffers, and who make use of the popularity of borrowed plumes, and take advantage of a nation's divisions, to cover their country with bloodshed, their pockets with gold, and their name with infamy.

---

* Michelet, p. 197. Dr. Russel, p. 43.

† In the front rank he placed the foreigners, in the second the Carthaginians; the reserve was composed of the veterans of the army of Italy. The mercenaries were first alarmed and overthrown; the second line fell back on the reserve, who drove them away with their spears; and the veterans themselves were at last broken by the Numidians in the service of Rome, who had already conquered the two wings, and who, turning, took the reserve in rear. This same cavalry, the cause of Hannibal's conquests in Italy, decided his fate and that of Carthage at Zama in 202 B.C. Grandeur et Décadence, p. 31. Michelet, ubi supra, p. 199.

‡ Michelet, p. 199. Krigs tildragelserne i Sjœlland paa Major Blom. Kjœbnhœvn, 1845.

§ For what have England and Europe derived from the overthrow of Napoleon?—outrage in Spain, and a fair field for Austrian gibbets and the knout.

Hannibal still lived, but Cato was also alive; and, showing the figs of Africa to the senate, " *he thought, moreover, that Carthage ought to be destroyed*."* Nor was it difficult to achieve, for the spirit of faction was rampant in the senate and the streets. The Romans found an excuse in an infraction of the treaty by the Carthaginians, when they drove out the Numidian faction and went to war with Massinissa, who beat them, killing 58,000 men.†

Utica was betrayed to Rome, and 84,000 men thundered against Carthage. Her doom was sealed. The senate decreed that its citizens should reside more than three leagues from the sea, and that their town should be entirely destroyed! The senate had promised to respect the *city*,—that is to say, the citizens,—but not the *town*.

This unworthy equivocation restored to the Carthaginians rage and strength. They called their slaves to liberty; they made 300 swords, 500 lances, 100 bucklers, a day. The women cut off their long hair to make cords for the machines of war. We are reminded of the bullying Duke of Brunswick and the outburst of enthusiasm in France in 1792. There is sublimity in a nation roused to rage by wrongs, like Turkey now.

A desperate defence was made, and the Roman army was thrice nearly exterminated.‡ Scipio, however, carried all before him, entered Carthage; and after a street-fight of six days and nights, Rome's rival sank into a heap of ashes.§

We read of Marius sitting alone amidst the ruins of Carthage; but this city rose again under Augustus, and eventually surpassed its former splendour, as a Roman colony and provincial capital. So great, indeed, was its luxury, corruption, and effeminacy during the latter centuries of the Roman empire, that a monk could not be seen in its streets without ridicule; crowds of men walked about in the garb and character of women; and it is almost with joy that we hail the arrival of the rough stern Northmen of Genseric, trampling under foot this Sodom with their iron heel.||

---

* An analogy has often been traced between Rome and Carthage, France and Britain. There are several points of resemblance, but more of difference. Carthage was an oligarchical commonwealth, a nation of merchants and shopkeepers, but she had not British tars or British faith. Carthage fell; but save by the predominance of peace-societies, or the party-spirit of protectionism, no breach can be made in our bulwarks. Rome vowed the downfall of Carthage,— *la fière Albion* of that day.

† Michelet, p. 238.

‡ Young Scipio Æmilianus, the son of Paulus Æmilius, adopted by the son of the great Scipio, having saved the army once, was made consul. He walled off Carthage from the land, and dammed it off from the sea. But the Carthaginians pierced another channel through the rock, and launched a fleet made of the wreck of their houses.

§ Michelet, pp. 238-9. Dr. Russel, p. 46.

|| Herder truly remarks : "With Carthage fell a state that Rome could never replace. Commerce departed from its coast, and pirates took its place, which they still occupy, (he wrote about 1800). Corn-growing Africa was no longer under the Romans what it had been so long under Carthage; it sank into a granary for the Roman people, a hunting-ground for their amphitheatres, and an emporium for slaves. Still desolate lie the shores

Having briefly run over the principal events of Carthaginian history in a chronological order, we shall, in a note, present a few remarks on the organic laws that held sway in the same.*

and plains of the most beautiful land in the world, which the Romans first stripped of its inland culture. The very letters of the Punic writing are lost to us; for Æmilianus handed it over to the grandson of Massinissa, one enemy of Carthage to another,"—like Poland and Turkey dissected by the northern powers. See Herder's Ideen, vol. ii. p. 157.

* I. It is manifest that the spirit of monopoly [on the Carthaginian spirit of monopoly, see Montesquieu's Esprit des Lois, c. xxi.-ii.] was a chief element of the Carthaginian laws ; as is proved by their commercial treaties with Rome, and from the fact of its having been the custom to drown the crews of such vessels belonging to other nations as were found in the vicinity of those places with which they carried on the most lucrative trade.

The same principle that led to her rise and prosperity occasioned her decline, by severing from her the sympathy of her neighbours, and leaving her alone in her distresses. Such is the invariable result of the spirit of monopoly in every matter connected with man.

II. At the time of the expedition of Hanno and Hamilcar, 480 years before the time of Augustus, her progress in wealth, population, and refinement must have been very considerable. (See Diodorus on the wealth of Carthage, p. 79; Polybius, p. 80. Herodotus on the land-trade of Carthage, p. 80. Heeren's Historical Researches. Heeren's Reflections on the Politics, Intercourse, and Trade of the Ancient Nations of Africa. See Aristotle; Pol. lib. ii. c. xi. on the polity and constitution of Carthage.)

The limit of Carthage was fixed finally, on the side of Cyrene, at the altars of the Phileni, whose legend respecting the self-sacrifice and inhumation of the two brothers is too beautiful to be soiled by criticism. Later this limit was much contracted, and the kings of Numidia, profiting by the disasters of the republic, recovered much territory that had been lost. Under Roman sway, the territory proper of Carthage was styled *Provincia,* including a country called Emporiæ and Byzacium, which had been a great bone of contention between the Numidians and Carthaginians; but Pliny informs us that the last Scipio Africanus separated the Roman province from the kingdom of Massinissa by a ditch running from the mountains to the sea, near the town of Thenæ. The town of Taphnera also stood on this ditch. The province had little depth in the time of Jugurtha; its limits running from the Tusca along the mountains between Hippo and Vacca, crossing the Bagradas twelve leagues S. W. of Carthage, and ending on the sea at Thenæ. Byzacium was afterwards erected into a special province. (Barbié du Bocage, p. 260.) Sixty large ships with 3000 emigrants sailed under Hanno to form a colony on the N.W. coast of Africa in the ocean. The works of Mago alone, one of the suffetes, on all branches of agriculture, amounted to twenty-eight, a few remains of which are found in Pliny; whilst most of their works have been lost through the neglect or rage of Roman barbarism, only matched by that of Ximenes in destroying Arabic Mss. after the capture of Granada. These treatises were translated into Latin by Solinus. Hence it appears that there was a Carthaginian literature patronised by the great, and which had passed from poetry to prose. How great has been the waste occasioned by human ignorance and fanaticism! and how zealously should we endeavour to preserve the little progress we have made in civilisation from the inroads of Croats and Kossacks, who cannot be regarded as men of taste, save in horseflesh, nor very discriminating in criticism!

III. The government of Carthage was an oligarchy. (Montesquieu; Grandeur et Décadence, p. 25.) That was bad. But it had two advantages to counterbalance this : it governed itself, and it had republican forms. The worst native government is better than the oppression of foreigners; and a republic in any form has more vitality than any other form of government.

IV. An eminent historian and philosopher, comparing Rome and Carthage, remarks that the latter had become rich and corrupt; whilst the former still remained immaculate, and rewarded merit instead of accepting bribes. In this circumstance he justly traces

And now to other matters.

From the ruin of Carthage, and even during the lifetime of the faithful Massinissa, the Romans eyed with suspicion Numidia, the ladder whereby they had entered Africa. Micipsa, the son of Massinissa, was too weak and soft to be feared by the senate, but he was obliged to share his kingdom at his death between his two sons, and his nephew Jugurtha, a bold and crafty Numidian, who aimed at empire, and had the suffrages of his countrymen.* Unlike other imperial nephews, he did not shelter his insignificance under the greatness of his name; and his being the nephew of his uncle was not a sufficient passport for office, or guarantee of honour and honesty. He had worked his own way up, doing wonders at the siege of Numantia in Spain. He was the best horseman in Africa; and his heart was as brave as the lions he slew in multitudes. Barbarian nations have generally had the wisdom or folly of choosing the most worthy of a family for king. Hereditary right did not suffice with them to crown idiots, as with modern czars and kaisers. Like Clovis, Jugurtha was made sovereign by the source of all power, the voice of the people. The Numidians saw that the division of their country was its ruin, and would end in subjection to Rome; and they fought like heroes for their chief and country.†

The people of Rome charged the aristocracy with being bought to acquiesce in Jugurtha's rise, and sent C. Piso into Africa with an army. He took a few towns, but was bribed to retire. The tribune Memmius,

---

a chief cause of the triumph of Rome, and the downfall of her rival ; and, as history is a school for the nations, it would be well to remember, that by the same sin fell Rome herself, and the Orleans dynasty; and that two empires are now festering under the same disease. It is to be hoped that England will avert such calamities, by abolishing the purchase of commissions.

V. Strangely does the wheel of fortune rise and fall, and time plays a curious game of see-saw with the nations.

The early intercourse of Carthage with Gaul is proved by the great number of Gaulish mercenaries, which, during the time of the Sicilian wars, fought in the Carthaginian armies. (Dr. Russel's Barbary States, p. 69. Michelet, p. 147.) Thus Africa once held sway in Spain; and Frenchmen fought her battles, who now bring the blessings of imperial despotism, Christian law chicanery, and philosophic infidelity into the solitudes of the Desert.

VI. The fall of Carthage has been attributed to the neglect of her maritime defence, and to the party spirit in her walls. (Dr. Russel's Barbary States, p. 82.) It were well if Peace-Societies, Protectionists, and Manchester men would attend the school of Pythagoras for a season, and in silence study history, ere they expose our national independence and our constitution, passable with all its faults, by the extravagant theories of disarmament and the selfish contentions of parties. (See some admirable observations and reflections on the strategy and polity of Carthage in Montesquieu's Parallèle de Carthage et de Rome, c. iv. of the Grandeur et Décadence des Romains.)

* Michelet, p. 274. Dr. Russel, p. 48. Sall. Bell. Jugurth. c. vi.

† Jugurtha, assassinating Heimpsal, divided the sway between himself and Adherbal, the surviving brother ; and he soon threw off the mask, attacked and murdered the latter, as well as all the Italian traders with Cirtha. Michelet, p. 275.

backed by the indignant people, summoned Jugurtha to Rome, to justify himself. Relying on the judge's venality, the Numidian went and came.*

Unhappy the land, which, like Spain and Russia, is eaten up with venal *employés!* When justice may be bought, Circassian and Polish triumphs are of no avail ; and Peruvian gold is rubbish, with a death-wound in the heart of the empire.†

Aulus,‡ the consul's lieutenant, had to pass under the yoke.§ This disgrace roused the Senate, which, seizing the reins from the doating and flagitious party of the aristocracy then in power, sent over Cecilius Metellus with a new army. (B.C. 109.) Having re-established discipline, he faced Jugurtha, after taking Vacca; but was nearly beaten by the Numidian's able tactics, which raised the siege of Sicca, and foiled the Roman consul, who sought to bribe assassins to dispatch him. This led Jugurtha to negotiate. He submitted to every thing, giving Metellus 200,000 pounds' weight of silver, all his elephants, &c.||

Jugurtha soon recommenced the war;¶ but Marius snatched victory and consulship from his commander Metellus, who, jealous, insulted him.** Gaining and disciplining his own army, he took Caspa amidst desert solitudes, the inaccessible peak where the Numidian treasures were placed, and beat Bocchus and Jugurtha twice. The former delivered up his son-in-law, rather than perish, to young Sylla, prætor of Mauritania, who, in his first campaign, had the honour of receiving the important captive. His success was the result of his coolness, when Bocchus hesitated a moment if he should not deliver up Sylla to Jugurtha.††

---

* Bribing again, and assassinating competitors, Jugurtha left Rome in safety, exclaiming, " *O venal city, and only awaiting a purchaser!*"—Dr. Russel, p. 50. Michelet, p. 275. Sallust, Bell. Jug. c. xxxv.

† Revelations of Russia, vol. i. See the chapters on the Secret and Common Police, and on the Tribunals.

‡ Albinus, brother of Aulus, who was first sent, did nothing against Jugurtha.

§ Dr. Russel, p. 50. Bell. Jug. c. xxxviii. Michelet, p. 275.

|| Baron Baude is of opinion that the road followed by the French army in 1836 up the valley of the Seybouse, by the Ras-el-Akba and the Ouad Sheriff to the Hammam Meskhoutin, was the same as that of the proprætor Aulus, when he coveted the treasures which Jugurtha had placed in Suthul, and let himself be drawn by the cunning Numidian into the defiles, where being conquered without fighting, the Roman army was forced to pass under the jugum. It is undoubtedly beyond Mjez-Amar, in the gorges of Hammam Meskhoutin, or in those of the Ouad-Sheriff, that the snare was laid into which Aulus fell ; the precipitous rocks that enclose them, and their tortuous character, ought to have warned him of his danger ; but blinded by cupidity, or jealous of the good fortune of Calpurnius, the proprætor, like many modern generals, ran after the money, without caring for the honour of his eagles. Baude, c. ix. p. 2.

¶ Metellus met him by putting to death all adult males, treating thus Vacca and Thala, the repository of the treasures of Jugurtha, who retired to the confines of the Great Desert, disciplined the Gætulians, and called to assist him against the Romans his father-in-law. Bocchus, the king of Mauritania, who was vanquished with him near Cirtha.

** A violent dispute arose at Rome between the partisans of the chiefs; but Marius returned triumphant to Africa. Michelet, p. 276. Bell. Jug. c. 54, 55, 56.

†† Numidia was divided between Bocchus and the two natural grandsons of Massinissa.

The hero or rebel was dragged in triumph through the mob of Rome after Marius, and was starved to death in the prison, shivering with his African blood in the chill climate of an Italian dungeon.* With the virtues of the Italian exiles, he joined the vices of his age and race ; yet the heart swells with rage as we think of the chronic injustice of man in exoteric oppressions. Rome, like Austria, has had her Spielberg; and Kossuths and Mazzinis have groaned for ages under the thumbscrew of bloody idiots or gladiatorial republics.

Carthage, Numidia, and the two Mauritanias were gradually subdued† by the Roman arms, and groaned or flourished under the gentle or oppressive administration of Italian prætors.

The expiring effort of Rome to avoid the disgrace of prætorian despotism was defeated in Africa.‡ The other Scipio, to whom Cato had unwisely yielded the command, had interested in his cause the Mauritanian Juba, by promising him the whole of Africa. This alliance gave him all the Numidians, and with their cavalry the means of starving Cæsar's army; but the latter, by a rapid march, separately attacked the three camps of the Pompeians, and destroyed 50,000 men, without losing 50 of his own soldiers.§

Cato had remained in Utica, a town indisposed to risk the slaves, who were its riches, by arming them to defend it. Cato seeing no hope, sent away the senators who were with him, and resolved to die in conformity with the precepts and practice of the stoical philosophy.‖

* Plutarch, Life of Marius. Eutropius, lib. iv. c. 28. Dr. Russel, p. 51. Michelet, p. 278.

† The Romans at first pursued the usual magnanimous practice of conquerors, in patronising dependant kings of Numidia and Mauritania, in order to swallow them up at the proper season, like Poland and Turkey, for which a modern Cæsar seems to have an inordinate appetite. Hiempsal II., grandson of Massinissa, was the first king whom the Quirites restored to the throne of Numidia, a learned prince, who composed several historical works in the Punic tongue, which Sallust professes to have caused to be translated for his own use, and which he appears to have incorporated neck and crop into his original history, after the fashion of modern historiographers. Hiempsal II.'s grandson, Juba, who reigned over Mauritania, was a prince equally conspicuous for his erudition. The eastern part of Numidia was reduced to a Roman province first under Julius Cæsar, who intrusted its administration to Sallust. Barbié du Bocage, p. 250.

‡ The Pompeians had assembled in Africa under Scipio, father-in-law of Pompey. The Scipios, it was said, would always conquer in Africa. Cæsar accordingly announced that a Scipio should also command his army. He declared that he gave up the command to a Scipio Sallutio, a poor soldier of his, obscure, and altogether despised.

§ Dr. Russel, p. 59. Michelet, p. 361, 362. Plutarch, Life of Cæsar. Dion Cass. xlii. 386. App. de Bell. Civ. l. iv. ch. 108 and following. Montesquieu, Grandeur, &c. p. 96.

‖ This man, whom Cicero and Seneca justly style holy (*sanctus*, Amm. Marcel.), in his life approached nearer to the Christian ideal than most Christians, though his death was not that of a saint. Cato, the last of the Romans, read through Plato's Phædo (on the Immortality of the Soul,) twice the night previous to his suicide, and slept so soundly between his lectures, that he was heard snoring from the next room. When his time had come, he ran his sword coolly into his body ; but being found still alive, and his wounds being bound up, while he was insensible he tore them off, and expired as soon as he came to himself. B.C. 47. Michelet. Dr. Russel, ubi supra.

We shall now attend to the Christian church in Africa.

The vitality of the Christian church in Africa* is attested by its councils, its schisms, and its monuments. Like all other communities of the faithful, they quarrelled fiercely about words and stones; but in the day of persecution they exceeded the courage and endurance of homœopathic students, mesmeric professors, and rappists. Whilst the majority was against them, they were, of course, a band of visionaries or impostors, and their system a gross piece of insanity or fraud, from which no good could be derived; but when Catholic Christianity was established, it fulminated the same charges against all innovation from which it had so grievously suffered itself. Such has ever been the history of human wisdom and charity.

Orthodox pens have recorded gross excesses in the dissenters of North Africa; but we feel doubtful how far we may trust them.† The Anabaptists' and the Suabian peasants' war of the Reformation were probably extravagant in some respects; but the reformers, who provoked and exterminated these men, are not the most trustworthy authorities in recording them.‡ Luther was not so conspicuous for charity as zeal; and if Munzer had succeeded, it is not improbable that he would only be remembered as a harsh, violent, and turbulent monk, who sought to rise, like Ronge, on the ruins of the church.§

---

* Neander is of opinion that Christianity was early introduced into the province of Africa. This church at Carthage becomes known to us first about the last years of the second century, through the presbyter Tertullian; but even then it appears to have been in a very flourishing condition. In his tract to the governor, Scapula, he spoke already of a persecution of Christians in Mauritania. After the middle of the third century, Christianity had made such progress in Mauritania and Numidia, that under Cyprian, bishop of Carthage, a synod was held consisting of 87 bishops. P. 114, vol. i. Clark's edition of Neander's History of the Christian Church.

† We cannot even coincide with the mild judgment passed by Neander on the North African heretics.

‡ D'Aubigné himself, and Luther's Autobiography by Michelet (pp. 180-184), show the human frailties and excesses of the monk of Eisleben, and give us gleams of a brighter light in Munzer's followers and the Anabaptists of Munster and Leyden.

§ We find that there met together, in the Donatist conference of Carthage, A.D. 411, 286 bishops of the Catholics and 279 of the Donatist party. St. Augustine was the chief speaker on the Catholic, and Petilianus on the heretic side. The imperial tribune and notary, Flavius Marcellinus, a friend of Augustine, presided. This took place under the Emperor Honorius. Among the bishops we find the following: Donatus bishop of Casa Nigra in Numidia, the primate of Numidia, Secundus bishop of Tigisis; and a Catholic bishop near Carthage, Felix of Aptungis, Aptugnensis, Aptungitanus, or Autumnitanus, figured largely in the controversy between the two factions of the North-African church. (For these facts see Neander, vol. iii.) We find that in the year 305, the Numidian bishops, nnder the presidency of the above-named Secundus, assembled at Cirta in Numidia, for the purpose of ordaining a new bishop for this city. (Neander, p. 247.) Neander says, "The Donatists were inclined to a separation of church and state, and preached against the ambition and avarice of the Catholic bishops. This inflamed the zeal of the Circumcellions, &c." (vol. iii. p. 260.)

The reader will recollect that the Donatist heresy originated in what has been represented by the Catholics as the abuse of the spirit of martyrdom, or the extension of the

The most eminent lights of the North-African church were Tertullian the Montanist, St. Cyprian, and St. Augustine. The principal feature in

---

spirit of saintship, to too large a number of faithful. A strange charge to proceed from Catholics, reminding one of the beam and the mote in the two brothers' eyes. But the fact is, that the Catholic hierarchy was then a proud aristocracy, and aimed at a despotic autocracy, and the too democratic tendency of the Donatist principle was displeasing to it. Resistance and persecution, as usual, begat opposition and extravagance ; hence arose a singular sect, the Fakirs and Jogis of Christendom, who must have affixed the stigma of madness to Christianity in the minds of all sober pagans.

There existed in North Africa, says Neander, a band of fanatical ascetics, who, despising all labour, wandered about the country among the huts of the peasants (whence they were called by their adversaries, the Circumcellions), and supported themselves by begging ; a very Catholic and orthodox mode of life thus far. They styled themselves the Christian champions, *agonistici*. Under the pagan power, parties of them had often, for no useful purpose, demolished the idols on their estates, and thus run the risk of martyrdom, which they sought. These men were roused by the persecution of the Donatists to all kinds of violence. (Neander, vol. iii. p. 257. In Fritzsche's Ketzer-Lexicon the reader will find a full account of these fiery heretics.) Constantine always treated them with mildness ; and when they demolished a church that he had built for the Catholics of Constantina, he had it rebuilt at his own expense, and demanded no indemnification. But under Constans forcible measures were adopted to convert the Donatists, under the imperial commissioners, Paul and Macarius. (Neander, vol. iii. p. 260) This drove the Circumcellions to further extravagance. They traced all corruption in the church to worldly wealth and power, exaggerating an eternal and apostolic truth. The Circumcellions breathed hatred against all who possessed power, rank, and wealth,—the democrats and socialists of that age. They roved about the country, pretending to be protectors of the oppressed,—a sacred band fighting for the rights of God. Probably, like Munzer's peasants in Germany, they have been much calumniated, for they were not successful. (See the Autobiography of Luther, edited by Michelet. Bohn's edition.) They may have perceived that there was much in the relation of masters and slaves at variance with Christianity ; but the cautious Neander insinuates that, in the way they wished to alter matters, all civil *order* must have been upset. (Neander.) They took the part of all debtors against creditors ; their chiefs, Fasir and Axid, styling themselves leaders of the sons of the Holy One, sent threatening letters to all creditors. Whenever they met a master with his slave, they obliged the former to take the place of the latter ;—this would not suit American stomachs. They compelled venerable heads of families to perform the most menial services,—a gross indignity of course, though Christ washed his disciples' feet. All slaves who complained of their masters were sure to find assistance, &c. Even many of the Donatist bishops, probably pluralists or incumbents of fat livings, applied to the civil power against them. But Donatus, and men of his stamp, encouraged them, and the Catholics, like all powers *that be*, sought to compel them to worship with them. (Neander.) Here we have the Test Act and Nonconformists in the third century, injustice being ever the law and gospel of all civil and religious polity. Many Donatist bishops and clergymen fell victims to this persecution ; but they must have deserved this, for they were the weaker party. Certain it is, says Neander, that many Circumcellions sought only the glory of martyrdom ; finally it came to pass that they threw themselves from precipices, into the fire, and hired others to kill them ; so anxious were these unhappy men to exchange the weariness of earthly proletaries for the rest of heaven. (Neander. Gibbon gives, as usual, a mutilation of the Donatist movement.) In this we see the reaction of over-spiritualism against the excessive materialistic depression of an age of fleshly doubt, analogous to the phantasies or visions of modern supernal philosophy turning the tables on Strauss and Voltaire.

Many eminent Donatist bishops were exiled till the reaction of charity or tolerance took place under Julian in 361. Their situation, however, became worse under the emperors

its history was the Donatist heresy, which, having been chronicled by orthodox bishops alone, is as imperfectly known as Carthage through the medium of her rival's historians. The unsuccessful party is ever wrong; and if Christianity had failed, we should brand it as an infamous imposture. The hero of Hungary is, of course, a visionary and a conspirator, for he has failed; but men begin to think that the imperial perjurer is a Solomon and patriot, for he succeeded.*

The North-African church was one of the earliest offshoots from metropolitan Rome, and soon rivalled its parent in heroism, fanaticism, and factions. Tertullian,† according to Neander, was the first scientific organ of Western Christendom;‡ but Augustine had more of the logical Indo-Germanic critical element; and Tertullian's chief feature was the sway of mystical oriental idealism in his mind, the Semitic element. A great impression was made on his rapturous spirit by Montanism, whose ecstasies and divine dreams were chilled to death when they reached the icy atmosphere of European prose.

The study of Tertullian's writings had manifestly an important influence on the development of Cyprian,§ as a doctrinal writer. Jerome states, after a tradition which was said to have come from a secretary of Cyprian, that the latter was in the habit of reading something daily from the writings of Tertullian, whom he was accustomed to call emphatically the Teacher. And who is original amongst us? Education and tradition are the chief ingredients in the infusion called human character; and Cyprian, on the Ganges or at Siam, would have worshipped cows‖ and carried gold umbrellas.¶ Cyprian's most remarkable work is his *Book of Testimony*, to

---

succeeding Honorius, when they were put down in the conference, like Galileo, Harvey, Columbus, Thomas Grey, Mesmer, and all men who have a very long sight.

When the Vandals, in the fifth century, made themselves masters of the country, the Donatists, as such, had no persecutions to suffer. It was only as Trinitarians that they suffered in that formidable controversy, which had almost established the unity of God as the key-stone of orthodox Christianity. They continued to survive as a distinct party till the sixth century, as may be seen from the letters of the Roman bishop, Gregory the Great. (Sismondi's Fall of the Roman Empire, p. 81.)

* See Schlesinger's War in Hungary, 1850, 2 vols., and Schalcher's Histoire du Coup d'Etat, published in French by John Chapman.

† He wanted the chaste sobriety of mind of Irenæus, and though a foe to speculation, he could not resist the impulses of a profound speculative intellect. He was destitute of regular logical forms of thought, and his genius was chiefly emotional, practical as well as speculative, which remained the principle of the North-African Church till Augustine, in whom Tertullian once more appears under a transfigured form.

‡ Neander's Church History. Hase's Church History. Gieseler's Church History. Dr. Russel's Barbary States, p. 133.

§ For a full account of St. Cyprian, see Neander, vol. i. p. 302-323 ; Clarke's edition. In A.D. 253 or 254, according to his own account, he had administered the episcopal office for six years.

‖ See Major Skinner's Excursions in India.

¶ See Ruschenberger's Voyage round the World.

prove from Scripture that Christ is the Messiah. We commend the book to the advocates of the Jewish disabilities and to Rothschild. Cyprian was arrested, like a Hungarian or Italian, for thinking for himself, and put to death by the Francis Josephs and Haynaus of the Roman Empire. He had persecuted heretics like a man, but he met death like a Christian (A.D. 258).*

St. Augustine (of whom more anon) was born A.D. 356, and was the glory or the misfortune of Christendom. Original sin has been an unlucky legacy, hardly redeemed by the candour of his *Confessions*. He was thirty-six years old, and had been born again for four years, when he was ordained priest at Hippo, in 390, with the acclamations of the people, by Bishop Valerius.† St. Augustine wrote his *Confessions* at Hippo in 397 ; his *City of God* in 413 to 426 ; and the same year he began his *Book of Retractations*. On the invasion of the Vandals, he wrote that Epistle ccxxviii. to Bishop Honoratus, which displays a humility, patience, and courage that would not disgrace the Vatican.‡ He died in 429.

Returning to the secular history of North-western Africa under the Romans, the paucity of remarkable events previous to the Vandal invasion is an evidence of the material prosperity of the *proconsulate*.

Juba king of Mauritania was conquered and taken at Pharsalia, but was restored and protected by the generosity of Julius Cæsar, who, unlike modern Christians, saw the policy of clemency. Caligula, however, put to death Juba's son, took possession of his states, and made a Roman province of them ; and Claudius divided them into two provinces — the Cæsarian and Tingitanian Mauritanias.§

---

* Our space prevents us from dwelling on Lactantius the Christian Cicero, for an account of whom the reader is referred to Neander, vol. i. and Dr. Russel, p. 135. He was intrusted with the education of Crispus, a son of the Emperor Constantine.

† In 394 he founded a community there, from which the most learned and illustrious bishops of Spain issued, including Alipius of Tagaste, Evode of Uzale, Possidius of Calama, Prefecturus and Fortunatus of Cirta, Severus of Mileve, Urban of Sicca, &c. Valerius adopted him as his coadjutor in 395, and died the following year.

‡ It would be dangerous to pronounce the verdict on this orthodox man. Nor would it be easy to say if the influence has been more beneficial or hurtful to Christendom. The elevation and brightness of some of his views are clouded by a considerable infusion of Manichæan views ; yet though he has disturbed the transparency of the waters of life, the purity of his life has partly compensated for the turbulence of his doctrine. A Christian bishop, who lives a mendicant and dies a pauper, is now so refreshing a novelty, that his self-denial covers a multitude of sins. (On St. Augustine and the church of Carthage and Numidia, see a full account in Nachrichten und Bemerkungen über den Algierischen Staat, 3 vols. 1798-1802 ; Dr. Russel, p. 136 ; Neander, vol. iii. ; Gibbon ; Baron Baude, vol. ii. p. 41 ; and Possidius de Vitâ Augustini.)

§ Dion Cassius, lib. lix. ; Seneca de tranquillitate animi ; Plin. lib. v. c. i. ii. ; Suetonius in vita Calig. sect. xxvi. ; Nachrichten und Bemerkungen, vol. ii. ; Gibbon ; and Dr. Russel, pp. 53-6.

Our space prevents us from dwelling on the rebellion of Ternius, the son of Nabal, a Moorish prince, in the third century; who, after considerable successes, being hard pressed by Romanus, and still harder by Theodosius, put an end to himself in the Atlas. The

But Rome was approaching her fall; for liberty had expired with Cato, and a prætorian government had disgraced the majesty of the Roman people, which even two Antonini and a Hadrian were unable to rescue. A fresher, healthier spirit of freedom was to descend from the North with healing on its wings, though by a rough treatment. The race of Odin, Sigurd, and the vikings, drew the sword, throwing away the scabbard, and carved out Europe anew. The Vandals,* forcing their way through France, entered Spain with the Suevi and Alani; and after eighteen years' possession of the coast of Andalusia, hard pressed by the Goths (A.D. 427), passed over into Africa (where they were invited by the Roman Count Boniface) under their king Genseric, and subdued the greatest part of what the Romans then possessed there. A contemporary writer has left us a description of the devastation which they occasioned, of which the sceptical Gibbon seems to doubt the accuracy.†

The Emperor Justinian,‡ after having consolidated his empire in the East, wished to restore it to its ancient splendour by recovering its finest provinces; and the renowned Belisarius was sent with this view, in 534, into Africa. He attacked the Vandals, weakened by divisions and enervated by luxury, conquered their last king (Gelimer), and reduced the whole of that country under the power of the Lower Empire. It remained one hundred years subject to the oppressions of the Greek prefects; but in 647, Othman, third Caliph of the Saracens, sent Hucha,§ his general,

---

second rebel was Gildo, his brother, who announced himself sovereign of Africa; but Stiticho, A.D. 398, crushed him by an overwhelming force among the Atlas; and the usurper, trying to escape, was secured, imprisoned, and committed suicide, like his brother. Again, A.D. 413, Heraclian raised the standard of rebellion, and made an attempt to invade Rome, which failed; and on his return, the Africans had deserted his cause, confiscated his property, 200,000l., and cut off his head. Dr. Russel, p. 95.

* See Dr. Russel, p. 97; Procopius, &c. The best account of the Vandals will be found in Gibbon's Decline and Fall, c. xliii., and also in Sismondi's Fall of the Roman Empire, c. vii. pp. 152 to 156.

† Scarcely any part of this beautiful region escaped their ravages. They found a highly cultivated country, they left it a desert. They rooted up the vines, pulled down the buildings, and demolished the temples. They collected crowds before the cities, and butchered them, that the infected air might cause the besieged to surrender. Procopius says, that when the army of the Greek Empire invaded Africa 100 years after, you might travel for three days without meeting a human being.

‡ Excellent remarks on Justinian's conquest and administration are found in Montesquieu's Grandeur et Décadence des Romains, c. xx. Dr. Russel's Barbary States, p. 106. Gibbon, c. xliv. Cardenne's History of Africa. Sismondi's Fall of the Roman Empire.

§ Dr. Russel calls him Abdallah, and states that he advanced from Egypt by Barca with 40,000 men, and met the Greek prefect Gregory in front of Tripoli. The latter was killed, his daughter taken, and his army beaten; but the loss of Abdallah was so great, that he had to fall back on the Nile. Vie de Mahomet, par Gagnier, t. iii. p. 45. Leo Africanus, p. 585, ed. 1632. Akbah, a brave commander, was sent again in 680, and marched through Mauritania with little opposition, but perished, as well as his successor, Zobeir, in an insurrection of the natives.

Akbar founded Kairouan; and Hassan, A.D. 698, attacked Carthage, but had to retire

with a large army into this province. He wrested it from the hands of the Romans, who, in the open country, unprovided with strong fortresses and armies, were unable to resist the Arabs, animated by ambition and conviction. Christianity and paganism yielded to the Crescent, and the power of the caliphs and the authority of the Koran prevailed in the whole region. The caliphs of Bagdad held it till A.D. 800, when the African Arabs shook off their yoke, and set up an independent sovereignty under the Emir Almoumenim.

Having launched the reader on the tide of Arabian history, we shall give a tabular view of the dynasties in North-west Africa, adding a few remarks on the most eminent sovereigns and revolutions.*

Africa was governed by the caliphs, through their lieutenants and viceroys, from 709 to 800. Alwalid sent Musa, in the first instance, with 10,000 men, who completely subdued Barbary, and in 712 passed into Spain. The followers of Musa ruled over Africa, dependent on the caliphs; and from 720 to 800, they remained subject to the house of Abbas. The last of the followers of Musa who submitted to the caliphs was Ibrahim-Ben-el-Aglab, founder of the Aglabites. Many Christians had fled to Europe in this period. In 750 the lieutenant of the caliphs, called Abdoulrahman, rebelled, without success ; but under Caliph Haroun-el-Raschid, in 800, Aglab made himself independent. Haudenis-Ben-Ab-

---

before a large Greek force; which was, however, ultimately entirely discomfited; when the city of Dido and Utica fell into the hands of the Arabs.

An attempt of the aborigines to resist the invaders, under a princess and prophetess named Catrina, was equally unsuccessful. Leo Afric. p. 575. Africæ Descriptio. Morgan's History of Algiers, p. 162. Nachrichten und Bemerkungen, &c. vol. ii. Dr. Russel, p. 116, &c.

* The history of Africa under the Mussulman comprises three periods—1st, the Arab period, from A.D. 647 (Hegira 27) to A.D. 971 (Hegira 361) embracing two dynasties. The Fatimite Khalif Moez Ziddin then leaves Kairouan to go to Egypt, leaving the government of Africa in the hands of Ben-Zivi-ben-Mnad of the Berber tribe of the Sanhadja. The Arab period lasted 324 years. The Berber period embraces three dynasties, and terminates in Morocco in 1519, in Algiers in 1515, in Tunis in 1570. This period lasted for Morocco 548 years, for Algiers 542, and for Tunis 599. The third period lasts from 1515 to our times. In Morocco it has been Arab throughout; in Tunis modified Turkish; and in Algeria since 1830 French. The modern period in Morocco has lasted 328 years, in Algeria 315, in Tunis 277. The Mussulman rule in Barbary has thus lasted about twelve centuries. The Arab period is divided into two orthodox dynasties, the Ommiades and Abassides; and one schismatic African, the Fatimites. The Berber period embraces, 1st, the dynasty of the Sanhadja ; 2, the dynasty of the Almoravides, or the Lemtouna Sanhadja ; 3d, the dynasty of the Almohades, a combination of the Zenata and the Masmonda. The first dynasty is divided into two branches, a, the Beni-Mnad from A.D. 971 to 1087 (480 Heg.); duration 126 years. b, that of the Beni-Hammad from A.D. 996 (Heg. 386) to 1149 (Heg. 544); duration 153 years.

The third period embraces the prætorian sway of Turkish and French Janissaries.

Under the Arab period, the capital under the Asiatic khalifs was at Kairouan in Tunis; under the African caliphs at Mehadia; under the Sanhadja Berbers, first branch at Achir, on the road from Ben-Sada to Bugia, then at Kairouan. The second branch at Bugia, the Almoravides at Morocco, and also the Almohades, till the division of their empire.

doulrahman, an Arab chief, opposed him ; he was, however, beaten and killed in a battle, and his followers submitted to Aglab, who is represented as a great patron of learning and the arts. About this time the empire of Morocco was also founded by a reputed descendant of the Prophet. Ali-Edris, such was his name, was poisoned ; but the Edrisites held his empire long.*

Then the branch of the Beni-Mrin made Morocco and Fez their capital; that of the Beni-Zeian settled at Tlemcen; and that of the Beni-Hafes at Tunis.

The following tree (extracted from the Exploration Scientifique) representing the Arabian conquests and dynasties, may make it still clearer.

ORIENTAL CALIPHS.

Damascus. Bagdad. Al-Walid. Ommiads. Abassides.

Tarik and Musa conquer Spain, 712.	Fatimites in Egypt deposed in 1211 by Saladin. Kurdic and Mamluk kingdom founded in Egypt, 1250. The Bagdad caliphate melts gradually into the prætorian protectorate of the Seldjuk Turks, till Bagdad itself is stormed, sacked, and ruined by the Moguls, 1258.	Tarik and Musa conquer Africa, which remains dependent till 788.
The Ommiad Abderrahman, 755, founds an independent caliphate in Spain.		In 788 the Edrisites, followers of Ali, found an independent empire at Fez.
The Spanish caliphate split up into ten kingdoms.		Under Haroun-al-Raschid, Kairouan and Tunis become independent under the Aglabites, A.D. 800.
		Edrisites and Aglabites swallowed up by the Fatimite caliphs of Egypt, 908.
		Fatimite empire broken up in 1211. Africa a prey to the Zeirites, Morabeths, and Almohades, 1250 to 1517.

(See Herder's Philosophy of History, vol. ii. p. 372. Schlotzer's Geschichte von Nord Afrika. Cardonne's History of the Arabs in Africa and Spain.)

* The followers of Aglab were his son Abil-Abbas-Abdoulah, 811; his brother Ziadetoullah, 815; his brother Abou-Akkal, 837; his son Aboul-Abbas, 846; his son Ishak, 875; his son Aboul-Abbas-Abdoulah, and his son Ziadetoullah, till 908. The Edrisites also ruled till the year 908. (Rebuhner, vol. ii. Herder's Philosophy of History, vol. ii. Geschichte von nord-westlichen Africa. Exploration Scientifique.)

The new empire of the Aglabites lasted for more than a century, having for its capital Kairouan, and including the ancient kingdoms of Mauritania and Massylia, with the republic of Carthage. Several of these Aglabite caliphs assumed the title of Mohammed, and signalised their reigns by military and naval achievements not unworthy of the apostolic age of Moslem history. Large bodies of their troops served occasionally in the mercenary armies of Bagdad, helping the Abassides to uphold their tottering authority in the East. Crichton's Arabia, vol. ii. p. 48.

The same author also informs us that, in 909 Abu Abdallah, emir or governor of Sicily, defeated the caliph of Kairouan, and drove the Aglabites from the throne, bestowing the vacant caliphate on Obeidallah, one of the posterity of Ali, who assumed the title of Ma-

Many insurrections occurred in their reigns; and separate provinces were formed into separate independent states. Under Ishak, Algiers revolted. Excepting Abou-Akkal, and Aboul-Abbas-Abdoulah, they were princes unworthy of the throne;—Ishak was a monster. The bodyguards of these regents were Negros; and Ishak is said to have had 100,000, according to Cardonne.*

The armies of the caliph of Bagdad, which invaded West Africa at this time to reduce it, were beaten back. Ziadetoullah, the last of the Aglabites, being a weak prince, was deserted by his subjects, and fled; and Obeidoullah usurped his authority. Obeidoullah left it to his son Aboul-Cassem-Mahomet-Ben-Obeidoullah, who took the name of Mahadi, and was the head of the Fatimites, giving out that he was descended from Ali. The Edrisites, in 908, were put down and extirpated by him, and he recovered much of Sicily; in 912 he even attacked Egypt, but could retain only Alexandria. He built the town of Mahadi on the ruins of Aphrodisimus. In 933 he died, having reigned twenty-five years,† leaving the government to his son, who was much inferior to him in capacity.

Zeir, a man of distinguished family, founded under him by degrees a powerful state in Morocco. He built a city called Ashir, by the help of the regent Biemlillah, who gave him an eminent architect. In 935 Abu-Jezid, the king's prime-minister, conspired against him; and he fled to Mahadi, where he died during the siege in 945. His son, who beat the conspirators, was a distinguished orator, according to Arabian authorities, and reigned seven years. His son Moez-Ledmillah succeeded him in 952. He was a man of talent, conquered Egypt in 968, and reigned seventeen years. He gave his African dominions as a fief to Jussuf-Ben-Zeiri, the son of Zeir, when he went to Egypt; and Jussuf was invested with them as a dependency on the caliphs of Egypt. This was an act of unparalleled liberality on the part of Moez, as he had a large family himself. The Zeirites possessed the greatest part of North Africa till the year 1148.‡

Many insurrections occurred under the Fatimite dynasty; they were, however, none of them such monsters as the Aglabites, though the taste for piracy increased amongst the Arabs at this period. The Barbary king-

---

hadi, or Director of the Faithful, built a new city, which he called Mahadia, and claimed the distinction of being the founder of the Fatimite dynasty in Africa, when he soon put an end to the Edrisites.

* Under Ziadetoullah, in 827, Euphemius, a Greek fugitive, led 10,000 Arabs, 700 horse, and 100 ships into Sicily. He himself died; but most of the island was subdued; and Syracuse at last was conquered, together with Crete. Afterwards the Arabs lost ground, and Syracuse. Ishak, in 880, sent a great army to Sicily. Syracuse was retaken, after which they abandoned Sicily, and returned to Africa laden with booty.

† Nachrichten und Bemerkungen, vol. ii.

‡ Yusuf Zeiri reigned from 972-983; Abil-Cassem-Mansour, 983-996; Abou-Menad-Badis, 996-1017; Moaz, 1017-1107; Yahia, 1107-1115; Ali, 1115-1121; Hassan-Ben-Ali, 1121-1148.

dom was a dependency of Egypt under these sovereigns ; but under Moez, the caliph tried to recover Egypt, and a war ensued. Troubles occurred in the reign of Jussuf, who put them down, and conquered Fez and all the possessions of the Spanish caliphs in Africa, except Ceuta. It is said his wives were 1000 in number, and that seventeen children were born to him in one day. Fez rebelled under his son, and the Beni-Hamads got possession of the Fez territory, and maintained it 160 years. The followers of Ali—Mohamedan dissenters, or heretics—were much persecuted during the minority of Moaz-Temin, who had many wars in his reign ; and though a great king, could not preserve order in his kingdom. Under him the Almoravides, led by Abubeker, began to conquer a part of West Africa, and extended their power under his son Jussuf-Tesfin to the Straits of Gibraltar.*

The Sicilians conquered the coast of Africa from Tripoli to Tunis, under the last of the Fatimites, Hassan-Ben-Ali. The different states were now split into separate kingdoms. For 300 years their history is nothing but a chronicle of petty wars between petty states. We shall, therefore, here turn to notice the Almoravides, or Morabites.†

The Almohades arose under the son of Jussuf. Mohammed-ben-Abd-allah, their founder, under the mask of sanctity, obtained much reputation. He was of the race of Mossanides‡ among the Atlas chain ; he had studied theology in the East, and he taught a new system of divinity at Morocco (1129), where he became popular, and began to abuse the Almoravides. He got possession of Telmin by treachery. After this he died, and was succeeded by his friend Abdelmoumein, with the title of prince of the orthodox, who conquered Oran (1163), Fez, and Tremesen, afterwards Morocco ; and ultimately became lord of all the state. Ishak, the last of the Almoravides, was beheaded by his orders. He then took Budschia, and the last of the Beni-Hamads, who had been kings of a territory called Jajah for 160 years.§

This race of orthodox Unitarian priests was ultimately expelled by Abdulac, governor of Fez ; and he, in the thirteenth century, was deprived of his new conquests by the sharifs of Hascen, the descendants of those Arabian princes whom Texefien had expelled.||

* Nachrichten und Bemerkungen, v. ii.

† They were an important race, and the following is their genealogy :—Abubeker-Jusuf-Ben-Tesfin, his son ; Ali-Ben-Jusuf, Tesfin-Ben-Ali, and Ishak-Tesfin ; the last in 1149.

‡ He was a native of Cordova. Hist. of Spain : Lardner's Cab. Cyclop. vol. ii. p. 25.

§ Here follows the list of the Almohade kings :—" Abdelmoumein, 1149-1160. Abu-Jakub, 1160-1184. Jakub-el-Mansur, 1185-1199. Mohammed-Ennasar, 1199-1211. Mohammed-Elmostamir, 1211-1223. Abdelwahid, 1223. Abdoulah-Aladel, 1224. Jahia, 1224-1226. Edris-ben-Jakub, 1226-1231. Abd-Eluahed-Ben-Edris, 1231-1242. Said-Abi-Elhassan, Ali-Ben-Edris, 1242-1248. Omer-Ben-Ibrahim-Ben-Jakub-Mosteda, 1248-1266. Abu-Dabus, 1269." (Nachrichten und Bemerkungen, v. ii.)

|| The sharifs divided his kingdom into a number of small ones—Algeria alone being

Spain, taking advantage of their dissensions, under the influence of Catholic charity, sent a powerful fleet and army against Barbary, under the Count Pedro of Navarre, in 1505; the churchman Ximenes accompanying the expedition, and inspiring the troops in the holy work of slaughter, crucifix in hand. The Christian host conquered Oran, Bugia, and some other places, and so alarmed the Algerines, that they put themselves under the protection of Selim-Eutemi. The Spaniards, however, succeeded in building a castle on the peñon of the Mole in 1509, preventing the corsairs from going in or out; and they made the town of Algiers itself tributary to them. The territory of Algiers was at this time very limited, though it had already made itself independent of Bugia.

When Barbarossa arrived to help the natives against the Christians, Salem-Aben-Toumi, or Eutemi, the Arab sheikh, had been appointed king. Like our noble Saxon ancestors, the Turkish chief put to death the sheikh whom he had come to assist, and Algiers found itself at the mercy of the prætorians, who, as usual, established order by a Reign of Terror.*

Algerine piracy owes its origin in part to the Christian charities of Cardinal Ximenes and the Spanish Catholics, who, not satisfied with ruining Spain by expelling the Moors, strove to Christianise Africa by fire and sword.† Ferdinand the Catholic and the cardinal captured Oran,

---

divided into 4: Tremecen, Tenez, Algiers, and Bugia. These petty monarchies continued for some centuries in comparative peace and amity, till at length the king of Tremecen violated some articles; and Abul-Ferez king of Tenez made him tributary to himself. The latter dying, left three sons, which gave rise to new discords.

Nachrichten und Bemerkungen, v. ii. Leo Africanus informs us that, about 1215, a flourishing kingdom was founded at Timbuctoo, by a Moorish chief. Some of its rulers had the reputation of warlike princes, who maintained a splendid court, encouraged commerce, and extended their frontiers in all directions. Crichton's Arabia, vol. ii. p. 56.

It was during this period, i. e. in the year 1270, that St. Louis, an honour to Christendom and to France, inspired by the pious but mistaken prejudices of his age, led an army into Africa to co-operate in the great work of the Crusades against the infidels. Though a gloomy bigotry disfigured the faith of the Christian host, and its gentle saintly king, yet Europe had not attained at that early period at the happy conclusion that nothing is worthy of belief but what falls under the senses.

St. Louis landed near Carthage, where a few buildings stood, which he began to invest, and captured in 1270, with a considerable host. But Africa even then seems to have been the grave of European adventurers; a violent plague and pestilential wind decimated the Christian army, and its beloved commander breathed his pure soul away on a heap of ashes, near the ruins of Carthage, just as a powerful reinforcement arrived from Sicily, under his brother the king of Sicily, who, piously collecting the earthly remains of the best king France ever saw, brought them back to Europe with the relics of his army. Dr. Russel's Barbary States, pp. 265-266.

* For this part of Algerian history, consult Rebuhner's Nachrichten und Beobachtungen über den Algierischen Staat, Bremen, 1798-1800. v. ii., containing the fullest account of Algerian history that the author has seen. Hist. of Spain : Lardner's Cyclop. vol. ii.

† Crichton remarks :— " This detested nation, whose conquest and expulsion were attended with such atrocities, and such triumphs to the Catholic Church, were by far the most industrious and skilful part of the Spanish population; and their loss was a blow to

the island of the Mole (*Penon*) of Algiers, and Bugia, and drove the Arabs into the arms of the Turks. It is true that the towns on the Barbary coast had occasionally before driven a sly trade in privateering, but it had not become an organised system before the Turks arrived.

Two adventurers, the sons of renegades and corsairs, Baba-Aroudj* and Khair-ed-Din,† courting fortune under the Mussulman flag, spread their adventurous sails to the winds, which bore them to the coast of Algiers, where they soon after became two celebrated pashas, and founded a prætorian sway of Janissaries.‡

We cannot enter here into a minute enumeration of the deys in chronological order, and an analysis of their government has been already given. Charles V., unlike modern emperors, aspired to be the guardian and saviour, rather than the scourge of Christendom. In 1535 he led a powerful armament, comprising the flower of Christian chivalry to Tunis, and after meeting some resistance, found his way into the town, disgracing his triumph by the massacre of the population, who had spared the Christian slaves when safety and the governor advised their extermination. Having liberated the slaves and humbled Tunis, Charles V. returned to Europe, where new trials and triumphs awaited him till 1541, when he resolved to chastise the insolence of Algiers. He led a noble squadron and army against the pirate city in the worst season of the year, against the advice of the Admiral Doria, and landing the armament, made his approaches. But the fleet was shattered by a storm, and the army

---

the greatness and prosperity of that kingdom, from which it has never recovered." Arabia, vol. ii. p. 62.

France is paying a similar penalty for the murder or expulsion of her Huguenots and Socialists.

* Baba-Aroudj slew Eutemi, and governed the Moors in his stead, with the usual brutal ferocity of prætorian soldiery. At length, after the sack of Tlemsen, which he had taken by treachery also, flying from the Spaniards, who, jealous of his progress, were investing the place, he was killed on the banks of the Tafna, though he scattered gold and treasure by the way to retard the pursuit of his foes. His brother Khaireddin, who put himself under the protection of the Sultan, Selim I., was afterwards made Capudan Pasha, and assisted by his Janissaries retook the Penon opposite Algiers, which has ever since acknowledged the sovereignty of the Porte till 1830. (Nachrichten u. Bemerkungen, v. ii. Dr. Russel, p. 267. La Grande Kabylie, p. 9 et seqq.

† These men are reported to have been the sons of a potter in the Isle of Lesbos in the Archipelago. That they sprung from the Greek, and not from the Turkish population of the Levant, is evident from their names. See Russel's Barbary States, p. 267. For a full account of them, see Von Hammer's Geschichte des Osmanischen Reichs, 3ter Band.

Algiers was reduced by Baba-Aroudj in 1511, and the armament commanded by the corsair was supplied by Solyman the Magnificent, Sultan of Turkey. Robertson's Charles V. v. ii.

‡ But these terrible rovers were not always and every where successful. Twice did Baba-Aroudj appear before Bugia (1512-14), and twice was he repulsed, notwithstanding the co-operation of the Kabyles of the interior; but forty years afterwards Salah-Rais, his second successor, gloriously avenged these checks (1555).

discomfited by the foe and the elements, though the gallant Ponce de Balaguer plunged his sword in the Gate of Babazoun. Forced to retire, Charles, in bitterness of soul, embarked with difficulty at Cape Matifou, and returned in sorrow to Europe. Algerine insolence increased and continued till 1830, though the powers of Europe sent many armaments to batter and humble Algiers, most of which were complete failures.*

The only occurrences in the history of Algeria sufficiently important to deserve a special notice during the Turkish rule, after the expedition of Charles V., and before that of Exmouth, are the bombardment of Duquesne and the attempt of O'Reilly. Louis XIV., provoked by the outrages committed by the pirates on the coast of Provence, &c., sent Admiral Duquesne in 1682 with a considerable force to chastise these outlaws, who, unlike the Grand Monarque, carried on robbery on a small scale. The town assailed by bombs,—a recent blessing invented in Christendom,—was soon enveloped in flames; but a change of wind prevented Duquesne from fully accomplishing his purpose, and it was not till the following summer that he emptied his vials of wrath on the devoted regency. Sending showers of bombs into the city, he reduced the inhabitants and authorities to the greatest distress and to terms. These were the surrender of French slaves and hostages; but a revolution in the government and the murder of the dey prolonged the hostilities. Duquesne, enraged at this, reduced the greater part of the city to ashes with his shells; yet the dey was undaunted and unmoved, and after putting to death all the French in his

* A full and eloquent description of the two expeditions of Charles V. to Tunis and Algiers will be found in Robertson's History of Charles V., vol. ii.; also Dr. Russel, p. 325-330. Nachrichten u. Bemerkungen, vol. ii. Von Hammer, p. 239, 3ter Band.

We shall now also present the reader with a table of the marine force of Algiers under the pirate government in a chronological order. The port of Algiers possessed, in 1568, 80 piratical vessels. (Gramaye, Africa Illustrata.) In 1581 it had 35 galleys manned by from 18 to 20 benches of rowers, and 30 brigantines. (Haedo, Topografia y Istoria General de Argel. Valladolid, 1612.) In 1588, independently of some frigates, the number of galleys amounted to 25. (Pierre Dan gives, with his habitual exactness, the numerical state of these. Galleys, 35; benches of rowers, 708. The bench contained, at a mean estimate, 8 men, and the number of combatants was half that of the rowers, so that the service of the galleys alone at Algiers employed 2800 of the first, and 5600 of the second; the latter were Kabyles and Christian slaves. A modification was introduced into their armaments and tactics, in 1634, through the progress of European artillery. In that year there were at Algiers only 9 galleys, with 131 benches of rowers; but it had 70 ships, with from 25 to 50 guns of different calibres. Haedo, c. 11.) In 1659, the Algerine cruisers consisted of 23 vessels of a uniform class, each mounting 50 guns, and with a complement of 400 men. In 1662, the celebrated Dutch admiral De Ruyter blew up 22 Algerine frigates; but notwithstanding this disaster, the naval power of Algiers was maintained on the same menacing footing as previously, until the time of the expeditions of Duquesne and D'Estrees, under Louis XIV. In 1825, the port of Algiers possessed 14 vessels of war of different classes, carrying collectively 336 guns. Now nothing exists in the shape of a native marine, excepting a few small sandals, used as fighting vessels.

Our space forbids our inserting here the substance of a curious document drawn up by the Regency, before the French conquest, on the subject of the tributes levied on European powers. The reader will find it in Baron Baude and St. Marie.

power, he blew their consul to atoms at the mouth of a gun. The admiral, exasperated now beyond measure, reduced the fortifications, shipping, arsenals, and stores to a heap of rubbish.

Such was the first occasion on which a French and Algerian force learnt to appreciate each other's gentleness and forbearance. They have both made considerable progress in these amenities of war at a more recent date.

As regards the expedition of O'Reilly, it was not so creditable to the strategy and science of Christendom as that of Duquesne. In 1775 this general landed with a Spanish host near the town, but owing to the vacillation and misunderstandings of the chiefs and the unsteadiness of the troops, he was compelled to re-embark in haste and with considerable loss, though he commanded a force amply sufficient to have reduced Algiers, if well officered and well disciplined. But Barrosa, Talavera, and the national debt have enabled us to appreciate the skill of Spanish strategy and the conciliatory character of Spanish commanders.*

When Waterloo gave Europe breathing time, it started on surveying the audacity of the Algiers pirates, who, while the Christians were murdering each other, had carried their special branch of industry to great perfection. The coasts of Italy were in perpetual terror, and the pirate state was the scourge and bully of the Mediterranean, invading Elba, Calabria, Malaga, &c. At length they bearded the British lion by insulting our flag, whereupon General Maitland was sent to Tunis and Lord Exmouth to Algiers to ask for satisfaction. Though a few Christian slaves (1816) were liberated on this occasion, the pirates soon resumed their old attitude and practices. When Lord Exmouth went up almost unattended to the dey's levee, the Janissaries were on the point of cutting him to pieces ; and no sooner had his squadron left than the British consul was loaded with chains, and Captain Dashwood and a British surgeon, who attempted to rescue his wife and son, were dragged away amidst insults and blows. Greater atrocities followed at Oran, and the massacre of coral fishers near Bona filled up the measure of iniquity. Britain, whose ear used to be ever open to the cry of oppression, till Austrian gold and insolence, with the Peace Society, emasculated her foreign policy, was roused to magnanimous wrath by these outrages. A powerful armament sailed for Algiers under the gallant Exmouth,† and was joined at Gibraltar by a Dutch squadron under Rear-Admiral Capellen. The Dey Omar, an iron man of stern will, threw away the scabbard and prepared to meet his foe. The batteries of Algiers were mounted with a thousand guns, and 30,000 Arabs were encamped under the walls, when Exmouth drew nigh (August 26, 1816). But Omar did not know British seamen. The fleet sailed

---

* Dr. Russel, pp. 332-334. Nachrichten und Bemerkungen, vol. ii.
† Dr. Russel, p. 340. Salame's Expedition to Algiers.

majestically up within pistol-shot of the mole, crowded with Turks, and at
a given signal discharged one of those crashing broadsides that has never
yet seen a peer. Crumbling walls and crying wretches attested the weight
of British metal; and though the Turks fought well, after a hard conflict
of some hours, the wreck of walls and towers, the explosion of powder-
magazines, and the burning of the Algerine ships, gave the dey a taste of
the British navy. The Turks had held out bravely, and their loss was
great; when the dey, seeing that Algiers was about to be a heap of rubbish,
lowered his tone, proposed negotiation, and came to terms. Algiers was
forced to refund the money extorted from the little Italian states, to
restore without ransom all Christian slaves, and to promise from henceforth to abstain from her iniquitous industry. British cannon had once
again spoken to the purpose in favour of justice and charity. But the
dey's promise was not more binding than a Russian Czar's oath, and
rapine was as dear to the prætorians of Algiers as to those of Paris. Omar
was strangled because he had been unfortunate; and Hussein showed himself
as partial to his neighbours' goods as the free and independent citizens of
the United States.

FRENCH LANDING AT TORRE CHICA (1830).

Our limited space obliges us to compress the account of the French
conquest* into a chronological sketch, with a few notes on the most

* There is reason to believe that Charles X. and his minister, Prince Polignac, were
quite sincere in the assurances given to Lord Aberdeen, that the only object of the

E E

stirring episodes of the war. The landing of the troops took place, almost without opposition, on the 14th June, 1830. On the 19th they were vigorously attacked at Staoueli in the Sahel by the Turks, who were defeated, though mustering 50,000 men. The French loss was 60 killed, 450 wounded. Other skirmishes took place, and on the 29th the trenches were opened in front of Algiers. July 4th, the Emperor's Fort was battered in breach, and partly demolished; and July the 5th, Algiers was delivered into the hands of the French, who captured 15,000 brass cannon, and about 2,028,500*l.** The Turkish troops were disbanded, and the Dey suffered to go to Naples.†

French expedition was the thorough extinction of Algerian piracy, so long the scourge and terror of feeble commercial states ; but it was one of the cruel necessities of Louis Philippe's precarious position—resting, as it did, wellnigh exclusively upon the timid sympathies of the moneyed and middle classes, instead of upon those far more powerful buttresses of continental thrones, the traditions and instincts of a numerous army, and the passions and prejudices of the great bulk of the population,—that he was compelled to temporise with every whim and vanity of the popular mind that happened to be in any way associated with the military glory of France. Compelled by this pressure, the citizen-king's government, after the exhibition of much vacillation and infirmity of purpose, finally repudiated the engagement with Great Britain, and, admittedly against their better judgment, prosecuted the war, sometimes with languid irresolution, at others with remorseless violence, till French Africa, as it is called, nominally comprised an area of 100,000 square miles (ultimately 160,000, as we have seen).

In 1844, after the battle of Isly, an understanding was come to with Great Britain, by which the retention of Algeria by France was acquiesced in, upon the agreed condition that the French dominion should not be extended either east or west ; in other words, that the independence of Morocco and Tunis should be respected.

Our space prevents us from chronicling the particulars of the rupture between France and Algiers in 1827-30. Suffice it to say that, as usual, France had promised the liquidation of a debt contracted under the empire, which she refused to pay after the restoration, though there is generally said to be honour among thieves. Deval, the French consul, being pressed on the subject by Hassan the Dey, at a levee, and giving a somewhat sharp reply, was gratified with a *coup d'éventail*, or box on the ear. The mercurial Frenchman could ill endure this attentat, and leaving Algiers, reported his griefs at home. The French government, on the plea of chastising the piratical insolence, but in reality to support the tottering throne of Charles X. by a *coup d'éclat*, determined, after a three-years' unsuccessful blockade of Algiers by their fleet, to reduce it with an army of 30,000 men under General Bourmont, a Frenchman, and naturally a man of accommodating principles, like Ney, &c., first an imperialist and then a royalist. The fleet was commanded by Admiral Duperré. They left Toulon on the 25th of May, 1830; and cruising off Algiers on the 14th June, determined to land on the peninsula of Sidi-Feredge, or Torre Chica. (Blofeld gives us a minute English version, the Tableau for 1839 the official French version of the invasion.)

* Since the French conquest a great decay has taken place in the Algerian marine, owing to the blockade from 1827 to 1830, and the constant troubles during an occupation of twenty-three years. There are now only about 1000 native sailors left. (Baron Baude, vol. i. p. 61.) Before the conquest the port of Algiers alone reckoned more than three times as many. In 1826, the last good year for Algiers, the port had 14 ships of war of different kinds, carrying together 336 guns.

† On the news of the revolution of 1830, Bourmont, the traitor, proposed to return with the expedition to France ; but Duperré opposed this, and Bourmont fled in a trader to Trieste.

The command was next bestowed on Clauzel, who arrived Sept. 10th, the army amounting at that time to 37,357 men and 3094 horses. Medeah was conquered in November, and Oran occupied Dec. 10th. In Feb. 1831 General Berthezene was appointed commander-in chief; but he was superseded in Nov. 1831 by General Savary, Duc de Rovigo.* Bona was reoccupied (May 1832) by General Yousouf and Captain Armandy.

The French army in Africa consisted in 1833 of 23,545 men and 1800 horses. Abd-el-Kader, who about this time had been appointed Bey of the province of Oran, not without violent opposition from many rival chiefs, is reported by French authorities to have been beaten by them, Oct. 10th, 1833, at Ain-Beida, and at Tamozanat on the 3d of September, after he had raised an insurrection against the French among the tribes. General Trezel was appointed governor Sept. 1833. After a determined attack and defence, Bugia was taken (Sept. 29th, 1833); and Sept. 26th a treaty had been concluded in the west between General Desmichels and Abd-el-Kader. But the French complain, probably without any reason, as usual, that the convention of Sept. 26th was not well observed by Abd-el-Kader; whereupon a remonstrance and fresh hostilities ensued (June 1835), and Trezel was discomfited, and forced to retreat with loss, during this campaign.† It is proper to add, that he had

---

* The Duke of Rovigo (Savary) conferred additional lustre on his name by his brilliant administration in Africa. Trained in a good school, he was aware of the expediency of making an example of the refractory, and he resolved to bring the Arabs to *order* by a master-stroke. He enjoys the enviable distinction of introducing razzias into French strategy; and by teaching the noble youth of France to murder the unprotected and the sleeping in the holy stillness of night, he had the merit of preparing the glorious era when a French razzia strewed the boulevards of Paris with the corpses of women and children, and the thief in the night stole the liberties of his country.

On the night of the 6th of April, 1833, a battalion of the foreign legion and a squadron of chasseurs fell suddenly on the unsuspecting Ben-Ouffias, and the morning's sun rose on the mangled bodies of the entire tribe, surprised and slain *whilst they slept!* Tidings of this Christian massacre flew, as on the wings of fire, through the land, everywhere kindling into flame the yet smouldering passions of the vast majority of the country population, and lighting up the fierce war of despair, which has since cost France so dear alike in men, money, and reputation. So universal was the outbreak, that, in the opinion of the Duke of Rovigo himself, his *great lesson* necessitated immediate and powerful reinforcements. They were granted ; and when the duke's love of order was severely censured by the turbulent and factious deputies, his conduct was defended on the plea of necessity. It is refreshing to find, however, that many French officers entitled to a share of the spoil obtained from the Ben-Ouffias razzia refused to contaminate themselves by its acceptance ; and that Savary arrived death-stricken in Paris, and died there in the June following the example of the Ben-Ouffias. The reader will be reminded by this accident of Lactantius' idle superstitions recorded in his book *De Mortibus Persecutorum.*

† It was in this campaign (1835) that General Trezel was tempted into an ambush by an Arab spy, who led the army into a morass, when it was met by a shower of balls from Abd-el-Kader's troops, and whence it was with difficulty rescued by the rear-guard. Trezel also lost his way on the road back to Arzeu, and became involved in a narrow and dangerous defile, in the midst of which his troops were assailed with an avalanche of stones from the Arabs on the heights, presenting a parallel scene to the war in the Tyrol in 1809. A granite hail-tempest rained on the devoted column, which was entirely broken. *Sauve*

been superseded by Drouet d'Erlon (Sept. 26th, 1834), who was the founder of the Spahis, an auxiliary native force, but who was himself superseded by Trezel again in 1835;* the French government of that day betraying the hesitation and imbecility of the Tory ministry at the time of the Convention of Cintra.

Marshal Clauzel succeeded Trezel in August 1835, the French power having lost ground in Algeria at this time. The Hadjoutes, a turbulent tribe in the Sahel, obeyed Abd-el-Kader, who had concentrated a formidable force in the west of Algeria, which the French government sought to intimidate by an expedition to Mascara, that was projected notwithstanding the absence of the foreign legion, which had been lent to Spain at this period. In Oct. 1835 the island of Harchgoun, opposite the mouth of the Taafna, was occupied; Mascara was taken Dec. 5th, and Tlemsen itself about the same time.

A camp was established on the Taafna in April 1836, and an action took place there on the 25th, when the Tableau states that 3000 French engaged 10,000 natives; and some of the enemies being troops of Morocco, an explanation was required of Muley-Abd-er-Rachman, the emperor, who said that the assistance was given to the Algerines without his knowledge. On July 6th, 1836, Abd-el-Kader suffered a disastrous defeat on the river Sikkak, near Tlemsen, at the hands of Marshal Bugeaud.†

November 1836, the first expedition was formed against Constantina, and on the 9th, as we have seen before in another place. After the failure of Clauzel, General Damrémont was appointed governor, Feb. 12th, 1837; and on the 30th of May the treaty of the Taafna between General Bugeaud and Abd-el-Kader left the French government at liberty to direct all their attention against Constantina, a camp being formed at Medjoy-el-Ahmar in that direction.‡

An army of 10,000 men set out thence on the 1st of October, 1837, for Constantina. On the 6th it arrived before Constantina; and on the 13th the town was taken with a severe loss, including Damrémont. Marshal Vallée succeeded Damrémont as governor.

The fall of Constantina destroyed the last relic of the old Turkish government. The garrison of Constantina consisted first of 2500 men, and

---

qui peut became the order of the day, and the French troops ran for their life, with the loss of 1200 men, caissons, cannon, baggage, &c.; nor would a man have escaped, had not the Arabs been detained by the plunder. This murderous business was Abd-el-Kader's great battle of the Makta; nor is it the first time that the tricolor has been trampled in the dust by freemen fighting for their homes and rights. (Alison's Europe. Chambers's Tracts, 1853. The Tableau gives a palliating version of the affair.)

* Tableau de la Situation, 1839: Précis historique.

† Tableau, 1839: ibid.

‡ Clauzel's pamphlet. Baude, vol. ii. chap. ix. Berbrugger, &c. Tableau, 1839.

then was raised to 5000, the Bataillon de Constantine being formed of native Arabs by Marshal Vallée.

By the 27th January, 1838, 100 tribes had submitted to the French.

A road was cleared in April by General Negrier from Constantine to Stora on the sea. This road, passing by the camps of Smendou and the Arrouch, was 22 leagues in length. The coast of the Bay of Stora, on the site of the ancient Rusicada, became covered with French settlers; and Philippeville was founded Oct. 1838, threatening to supplant Bona.

Abd-el-Kader advancing in December 1837 to the province of Constantina, the French advanced also to observe him; then both retired, without coming to blows. A misunderstanding which arose respecting the second article of the treaty of Taafna was settled in the beginning of 1838.

The province of Constantina was divided into arrondissements by M. Vallée.

1st Jan. 1839. General tranquillity prevailed.

The Semitic spirit and genius have found a fitting representative in Abd-el-Kader,* the man of brave and gentle heart, the warrior-prophet of Islam, a man of the true Saladin stamp. With a rapture of devotion foreign to our icy fogs, and a purity of life unknown to Christian sultans, he has upheld the glory and independence of his race and country with a gallantry and endurance worthy of a Tell, and has yielded to the frowns of fate with the resignation of a Magyar.† His elocution is brilliant and lively, his voice hollow, and his utterance interrupted by ejaculations of *Enshalla* (God willing). He never ventures, like so many Christians, to build on the future without deferring to a Higher Power. Always modest in manner and humble in apparel, he is a living criticism on the Tuileries and George IV. Frenchmen call him an ambitious fanatic; Christendom a visionary adventurer; and Islam a hero-saint. Time will decide which verdict is nearest the truth.

* The most graphic account of Abd-el-Kader is given in Berbrugger's Algérie, part ii. See also Blofeld, p. 389, &c.; Lieut.-Col. Scott's Visit to the Esmailla of Abd-el-Kader; and Les Prisonniers d'Abd-el-Kader, being the second part of Lady Duff Gordon's French in Algiers, 1845.

† Abd-el-Kader is about five feet seven inches in height, and is far from being the Bluebeard that many French authorities have described. He is remarkable for an air of melancholy gentleness, which can be seen even through the severity that he was sometimes forced to assume. The prevailing expression of his countenance is profoundly religious; and he has a certain ascetic appearance that recalls the noble heads of warrior monks handed down to us from the middle ages,—those men who were not too cultivated to feel, or too fashionable to express, the emphasis of a sincere faith. His Arab coat and burnouse give an additional force to this resemblance. His face is long and very pale, his beard black and bushy, his eyes blue and very beautiful; and he has contracted a stoop from the pressure of his cowl.

The name Abd-el-Kader (عبد ال فادر) means 'the servant of the powerful.' He was one of six children, five boys and one girl; and his place of birth, in 1806, was in the

When Abd-el-Kader assumed the royal title of Sultan and the command of a numerous army, the French, with republican charity and fraternal sympathy, sought to infringe the Taafna treaty, and embroil the Arab hero, in order to ruin his rising empire, and found their own on its ashes. The Emir had been recognised by the whole country, from the gates of Ouchda to the river Mijerda; and not being implicated in any foreign

---

vicinity of Mascara. His mother, Leila Zahara, who is still living, and shared his captivity in France, was said to be a beautiful girl. Mahl-ed-Din-Hadj was the name of his father. It is related that on visiting the tomb of a celebrated marabout relative, not far from Bagdad, one Mulei-Abd-el-Kader, who had lived a century, half of which on the pinnacle of a rock, a kind of Stylites, fed by a starling, the departed saint appeared to them, and presented the young hero with a remarkable apple. Abd-el-Kader, on returning home and beginning to eat it, the same halo of azure light encircled his brows that had been witnessed at his birth. Whether this was an odylic or a mythical light, we leave as a problem to the learned to solve. He married Leila Kheira, said to be a beauty according to the Arab standard. Blofeld, p. 389.

We now proceed to analyse the native power hostile to France. Abd-el-Kader, once mighty, is now sadly fallen. In 1840 he was in his zenith. He had at that period 500 infantry and 250 horse paid and clothed by himself; his tent was guarded by a body of 30 negroes; and 100 camels followed the army, carrying the provisions, &c. The principal strength of Abd-el-Kader has always been in cavalry; and he has never had any artillery worth mentioning. The Aga laboured hard to discipline the army, but never succeeded. The Arab tribes always supported Abd-el-Kader with a tribute in kind.

The family of Abd-el-Kader is one of the most ancient of Arab, some say Berber, descent in Algeria. He is the third son of a patriarchal chief, who was a highly-venerated marabout, who made twice a pilgrimage to Mecca; and took his son, then eight years old, with him in his second journey. He also resided in Italy, and young Abd-el-Kader was there with him; and was taught, in the course of his education, Italian, Arabic, reading and writing, and the interpretation of the Koran. At the time that the tribes revolted from the French, they begged the father to place himself at their head; but he referred them to Abd-el-Kader, being too old himself for military duties. (See, for these facts, Col. Scott's Visit to the Esmailla of Abd-el-Kader, &c. 1839.)

Abd-el-Kader being chosen, raised a certain number of cavalry from each tribe, composing the *Union of Defence* formed by the insurrectionary movement. These tribes amounted then to 400 in number. With these levies he commenced his attacks upon the French. Immediately after the Treaty of Taafna, 30th May 1837, between General Bugeaud and Abd-el-Kader, the latter was made Emir. This dignity is identical with prince,—religious and secular chief. Soon after he was made Sultan of the whole territory, from Ouchda (a frontier-town of Morocco) to the river Mijerda. The emir established all things on the European system. Each province had a khalifa, who had a regiment of regular infantry, consisting of about 1200 or 1300 men, and a company of French deserters and renegades, sometimes amounting to 200 men; they had each also about 500 regular cavalry. This system is, of course, now entirely overthrown. Milianah was first chosen as his residence by the Sultan, and with great judgment; as the place is strong, and supplied with water. Being driven out of Milianah by the French in 1840, he retired with his property and treasure to Tekedempta, a rocky fastness of the Great Atlas beyond Mascara. Abd-el-Kader planned to rebuild Tekedempta, which is a ruinous ancient city, and to make it the capital of his dominions. The enclosure of the ancient town is 1800 feet long and 1150 broad. On a hill are seen the ruins of the Casbah. At the foot of this hill flows the Oued-Mina, which is a tributary of the Shellif. The Emir administered justice in a very simple and summary manner. The contending parties were brought to his tent; the complaint was first made, witnesses were then called, the defence was next conducted; and at the end of these proceedings the Sultan decided singly and without appeal.

war, he had established every thing on the European system, save his church and faith, which were as yet innocent of atheism and pluralists.* Marshal Vallée was then governor; and Abd-el-Kader complaining to him of breach of faith, was informed that the Marshal had unlimited authority to act as he thought best. The Duke of Orleans, we are told, also wishing to make *reconnoissance* to Constantina as safely as possible, soiled his noble name and character by causing a Jew at Oran to make a seal of the sultan, which was put on passports for that purpose, and given to the kaids of the country through which the French troops went. The deception being found out, the Arabs attacked them in all directions, not approving of such missionaries of Christian truth and honour. The sultan, still unwilling to commence hostilities, assembled his khalifas, to hold a council as to what was best to be done. One of his khalifas (Ben-allel) was absent, and thought it useless to debate about the matter, and, while the sultan was consulting, attacked the French convoy between Milianah and Algiers, and put an end to all debates by taking off 200 heads. French authorities give a different version of the rupture, but we are familiar with the veracity of French bulletins and *Moniteurs;* and though we believe the government of Louis Philippe to have had more principle than is common in Gaul, his satraps were never over-scrupulous.†

The war was resumed, and many French razzias took place. They once marched a large force from Algiers on Milianah to surprise the sultan's camp. They failed in their chief object, but nearly captured the sultan himself. He was surrounded in the middle of a French square, which thought itself sure of the reward of 100,000 francs (4000*l.*) offered for him ; but uttering his favourite " enshallah" (with the will of God), he gave his white horse the spur, and came over their bayonets unwounded. He lost, however, thirty of his body-guard and friends, but killed six Frenchmen with his own hand.‡

Still, notwithstanding his successes, Abd-el-Kader had been losing

---

* By the 7th article of the Taafna treaty, Abd-el-Kader was to be furnished with whatever arms or ammunition he required ; notwithstanding which, his oukil or agent at Algiers was put in irons and sent to prison in France for having furnished him with the supplies he had a right to demand, and also for having sent over to him some mechanics, which was an infraction of the 10th article.

† These events are differently represented by Blofeld and the Tableau, 1846.

‡ The French suffered severely at this time from sickness and the sword. In December 1841, Marshal Bugeaud (then governor) said to the minister of war, that only 4000 men out of 60,000 were fit for active service. Some time before this, the French, following the frank and manly plan of overreaching Spain in 1807, landed 4000 French soldiers dressed as Turks, with bands of music in the Turkish style, commanded by Omar, son of the former dey of Algiers. They issued proclamations, saying that France had given up Algeria, and that Omar was appointed dey by the Porte; that the war was no longer against infidels, but the faithful ; and inviting the tribes to return to their old masters. Some deluded tribes came in ; and their grain, which they could not remove, was taken by Abd-el-Kader, whilst they, as their reward, found out the deception.

all his former power, as his Arabs, though brave, could not match 80,000 French troops, with artillery and all the other ornaments of civilised warfare. Seven actions were fought at the Col de Mouzaia, where the Arabs were overthrown by the royal dukes, in 1841; and at the Oued Foddha, where Changarnier, with a handful of troops, defeated a whole population in a frightful gorge. It was on this occasion that, having no guns, he launched his Chasseurs d'Afrique against the fort, saying, *"Voilà mon artillerie!"**

Abd-el-Kader had then only two chances,—the support of Muley-Abd-er-Rahman, Emperor of Morocco; or the peace that the latter might conclude with France for him.†

General Bugeaud, who had replaced Marshal Vallée, organised a plan of campaign by movable columns radiating from Algiers, Oran, and Constantina; and having 100,000 excellent soldiers at his disposal, the results as against the Emir were slowly but surely effective. General Negrier at Constantina, Changarnier amongst the Hadjouts about Medeah and Milianah, Cavaignac and Lamoricière in Oran,—carried out the commander-in-chief's instructions with untiring energy and perseverance; and in the spring of 1843 the Duc d'Aumale, in company with General Changarnier, surprised the Emir's camp in the absence of the greatest part of his force, and it was with difficulty that he himself escaped. Not long afterwards he took refuge in Morocco, excited the fanatical passions of the populace of that empire, and thereby forced its ruler, Muley-Abd-er-Rahman, much against his own inclination, into a war with France; a war very speedily terminated by General Bugeaud's victory of Isly, with some slight assistance from the bombardment of Tangier and Mogador by the Prince de Joinville.‡

In 1845 the struggle was maintained amidst the hills by the partisans of Abd-el-Kader; but our limits prevent us from dwelling on its parti-

---

* In 1844, the occupation of Algeria had cost 26,800,000*l.* Government had received—first, the treasure taken at Algiers, 2,189.480*l.*; secondly, the proceeds of the revenue from 1831 to 1844, 1,840,000*l.* The total loss, therefore, had been, up to 1844, 22,670,520*l.* (Castellane, Souvenirs; Berbrugger; Blofeld, p. 389; Tableau, 1846.)

The *Moniteur Algérien* of the 15th February, 1844, said that the last accounts placed the esmaillah of Abd-el-Kader at Kesdir, the west point of the Chott-el-Gharbi (West Chott), eighteen or twenty leagues south of Ouchda, containing barely 1200 tents. The Halleb, a tribe subject to Morocco, supplied them with grain, and the Heumrianes with dates. All the remaining troops of the Emir have assembled in a part of the empire of Morocco called Beni-Jala, consisting of about 320 foot and 120 horse. The tribes faithful to his cause, exclusive of those of the Dahra, were dispersed on the frontier of Morocco, and scarcely reckoned 1000 tents.

† Blofeld, p. 389 et seqq.

‡ Chambers's Tracts. In September 1844, Abd-el-Kader had taken refuge in the inaccessible chain of mountains called Er-Rif, running 200 miles along the whole line of the north coast of Morocco,—a region that was never yet subdued by the sovereigns of that empire. (Blofeld, ubi supra.)

culars, save in one instance, honourable alike to the Christian temper of the French commanders, and the civilised spirit of Christian warfare.*

The most remarkable events of 1847 were—1st, the surrender of Bou-Maza at Orleansville, 13th April; and 2d, Marshal Bugeaud's expedition to Great Kabylia in May, detailed by M. Borrer, who accompanied him, and by Colonel Daumas. This was the *coup-de-théâtre* of Bugeaud, who, after conferring the blessings of capital punishment and infidelity on the Kabyles, embarked for France to aid in strangling liberty at home and throughout the Continent in 1848-9.†

* On the night of the 12th of June, 1845, about three months before Marshal Bugeaud left Algeria, Colonels Pelissier and St. Arnaud, at the head of a considerable force, attempted a razzia upon the tribe of the Beni-Ouled-Riah, numbering, in men, women, and children, about 700 persons. This was in the Dahra. The Arabs escaped the first clutch of their pursuers; and when hard pressed, as they soon were, took refuge in the cave of Khartani, which had some odour of sanctity about it: some holy man or marabout had lived and died there, we believe. The French troops came up quickly to the entrance, and the Arabs were summoned to surrender. They made no reply. Possibly they did not hear the summons; or perhaps the courage of despair had steeled them to await the attack of their foes, however numerous and sure of victory those foes might be, and endeavour to sell their lives as dearly as possible in the holy and vantage ground they had happily reached. Colonels Pelissier and St. Arnaud would certainly not have been justified in sacrificing the lives of the soldiers under their command by attempting to force a passage through windings and intricacies thronged with armed and desperate men; but as there was no other outlet from the cave than that by which the Arabs entered, a few hours' patience must have been rewarded by the unconditional surrender of the imprisoned tribe. Colonels Pelissier and St. Arnaud were desirous of a speedier result; and by their order an immense fire was kindled at the mouth of the cave, and fed sedulously during the summer night with wood, grass, reeds, any thing that would help to keep up the volume of smoke and flame which the wind drove, in roaring, whirling eddies, into the mouth of the cavern. It was too late now for the unfortunate Arabs to offer to surrender; the discharge of a cannon would not have been heard in the roar of that huge blast-furnace, much less smoke-strangled cries of human agony. The fire was kept up throughout the night; and when the day had fully dawned, the then expiring embers were kicked aside, and as soon as a sufficient time had elapsed to render the air of the silent cave breathable, some soldiers were directed to ascertain how matters were within. They were gone but a few minutes; and they came back, we are told, pale, trembling, terrified, hardly daring, it seemed, to confront the light of day. No wonder they trembled and looked pale. They had found all the Arabs dead—men, women, children, all dead!—had beheld them lying just as death had found and left them: the old man grasping his grey beard; the younger one grim, rigid, stern as iron, with fanatic hatred and despair; the dead mother clasping her dead child with the steel gripe of the last struggle, when all gave way but her strong love! (Chambers's Tracts; Tableau, 1846.)

St. Arnaud and Pelissier were rewarded by the French minister; and Marshal Soult observed, that "*what would be a crime against civilisation in Europe might be a justifiable necessity in Africa.*" It is in this manner that Algeria has proved an efficient school for prætorian excesses and imperial incendiarism. The razzia has paved the way for the *coup d'état*, and France has forged her own chains in Africa; but it can neither astonish the chief nor the crowd of the gallant French army, if, with these illustrations before them, their neighbours are not anxious to welcome an irruption, and are alive to the propriety of taking due precautions to avoid a more familar acquaintance with the refined humanity of French strategy. (Compare Southey's History of the Peninsular War, and Alison's History of Europe.)

† The Arabs used to call Marshal Bugeaud the white-headed warrior. Borrer. (See La Grande Kabylie, by General Daumas, chap. xi.)

For fifteen months part of the troops of the division of Oran were watching the deira of Abd-el-Kader. The situation of this noble chief had been very precarious in 1846, at the time of the purchase of the prisoners; but had since improved, through his magical influence over the glowing hearts and minds of the tribes bordering on Morocco. His eloquence, like that of the great Magyar,—fabulous to a hard-cash cotton-spinning age, and a stumbling-block to steam-press and patent reviews, overthrowing the logic and sneers of positivism and mythical theories,— went direct to the manly hearts and faith of the free sons of the desert and puszta, ignorant and innocent of court and cant, and stirred up the mighty embers left within them by the burning enthusiasm of their Saracen sires. The Arabs could not yet be brought to see the blessings of private landed property, of law-suits, of stock-jobbing, of income-taxes, of the reign of order, or of jesuitical infidelity.

However, the march of intellect could not be stayed; the progress of French civilisation was decreed, though based on treachery and bathed in blood; and Abd-el-Kader, like Rome, had to bow to liberty, fraternity, and equality.

A taste of French bayonets at Isly, and the booming of French guns at Mogadore, had brought Morocco to reason, by that best of arguments which has since put the curb in the mouth of the French people. Morocco sided with France, and threatened Abd-el-Kader, who cut one of their corps to pieces, and was in June on the point of coming to blows with Muley-Abd-er-Rahman, the emperor.* But the Emperor of Morocco took vigorous measures to oppose him, nearly exterminating the tribes friendly to him; which drew off many partisans from the Emir, who tried to pacify the emperor, but unsuccessfully. Three columns of Morocco troops pressed on him, whilst Lamoricière with 3000 men and 1000 horses aided them on the side of Oran.†

Abd-el-Kader receiving the emperor's ultimatum, to which he would not submit, determined on a desperate effort for liberty, worthy of his gallant spirit. He resolved to attack the nearest camp to the deira, under the orders of Muley-Ahmed, one of the emperor's sons. He failed, however, in surprising it, as the first camp had retired on the sound, on which the Emir rushed with desperate valour. Many bit the dust; but crushed

---

* A camp of Morocco troops was formed by the emperor's nephew and the Kaid El-Hamar in the Taza district, to operate against the emir, who, anticipating them with 1200 regulars, 400 infantry, and 2000 irregular horse, reached the camp in the Oued-Aslif by a forced march at night. The Moroccans, surprised, ran away, the emperor's nephew escaping to Fas. El-Hamar being captured, had his head cut off by Abd-el-Kader, who tried, according to the French accounts, to gain the tribes by representing himself as named to the sovereignty by the king of France. (Tableau, 1849, pp. 3, 4.)

† The colouring given to these facts in the Tableau is somewhat questionable; as Napoleon I. had introduced a chronic system of lying into French official documents, and the nephew of his uncle is too dutiful to depart from so good an example.

by a force ten times his own, he escaped with difficulty. He did not halt till he reached Aguedin, near the Moulouia. The Duke of Aumale landed at Nemours on the 23d of Dec., and the troops of Morocco being supplied with ammunition by the French, hemmed in Abd-el-Kader again, who was forced to leave the territory of Morocco; but, in passing the Moulouia, the Mencer Kabyles rushed on his deira, which seemed lost. By a gallant effort, in which he lost half his best men, Abd-el-Kader covered its passage, and reached the Oued-Kiss on the French frontier. And now he was lost. He asked to negotiate, offered to surrender; and after 24 hours' discussion he came to Sidi Brahim, the scene of his last exploits against the French, where he was received with military honours, and conducted to the Duke of Aumale at Nemours.*

France has been severely abused for the detention of Abd-el-Kader in Ham, yet perhaps ungraciously by the jailors of Napoleon, though the latter was the terror of Europe, and the former the defender of his faith and home. Never shall we deny the meed of praise when it is due to generous friends or foes, and a gallant marquis and imperial swindler are alike entitled to our laudations for their liberation of the high-minded marabout, especially when we remember that the intimate friends of the conservative soldier and prætorian leader are wont to flog refractory ladies like cattle, and to string up obstinate gentlemen as innocently as if they were dried herrings.†

Thus fell Abd-el-Kader, the modern Jugurtha, who united many of the vices with most of the virtues of his race and faith. Though absolute, his ear was open to justice, his heart to pity. Though a Mussulman, his life was chaste and pure. A patriot, he did not see the sin of drawing the sword in self-defence, notwithstanding the protests of the Peace Society. Though an Arab and a conservative, he did not object to learn discipline and arts from Christians and republicans; and though familiar with European thoughts and manners, he was not ashamed to feel and own a glowing faith like his fathers. In him we behold a fabulous, meteoric character, one of those dreamy beings whom it has not entered into the heart of critics to conceive. The austerity of the Carthusian and the chivalry of Bayard are married in this mysterious being, who looks on with his calm melancholy eyes on the fall of his country and the stagnant faith of Chris-

---

* Depositing his sandals outside the apartment, he awaited the duke's signal to sit down; and then observed in Arabic, " I should have wished to have done this sooner, but I have awaited the hour decreed by God. I ask the aman of the king of the French for my family and for me."

The ex-emir employed the 24th for the settlement of his personal affairs. In the evening he embarked for Oran with the governor-general, and was thence sent on at once to Toulon, with his mother, his wives, his children, and a suite of friends, &c.; in all 97 persons. (Tableau, 1849-50, pp. 3, 4.)

† Schlesinger's History of the War in Hungary, vol. ii. p. 229 and following. Pridham's Kossuth, p. 198.

tendom and Islam, with the unutterable sadness of the man of sorrows, and the noble melancholy of the Magyar.

Saladins and Sidneys are rare in this age of loans, unsteady sceptres, and tottering crowns ; and it is with regret we bid adieu at present to this modern incarnation of chivalry and poetry,—the Star of the South.*

The year 1848 opened well for France, Bou-Maza and Abd-el-Kader having submitted ; but the revolution of February, misunderstood both at home and abroad, occasioned some troubles in the colony on the part of the natives. The Kabyles and the Darha were the theatres of petty insurrections speedily quieted; and whilst Algerian generals were making razzias on the Faubourg St. Antoine because its blouses wished to be free, the burnouses of Algeria were fraternising with the advocates of liberty and equality in the plains of the Tell and the uplands of Sahara.

But in 1849, whilst the friends of order were administering a quietus

---

* As surely as time rolls on and Europe goes ahead, under sovereign people or sovereign purple, it seems the fate of Indo-Germanic progress to grind individual soarings to powder, and to crush original greatness into the undistinguishable conglomerate of a dull uniformity. Oriental life has hitherto held aloof from this levelling process; but we much fear that the infection of a mechanical, unspiritual hemisphere will shortly clip and square Semitic notions, emotions, and sensations, according to the patent verdict of Berlin professors and Paris academies. The aristocracy of nature is being levelled and surveyed by economist theodolites, and all eccentricity will probably give up the ghost under the infliction.

We have given the French official version of the surrender of Abd-el-Kader in the text, which we propose to neutralise by an infusion of British bluntness.

The fallen Emir determined on surrendering himself to General Lamoricière on certain conditions, which were negotiated through the cadi of Tlemsen, who, General Lamoricière states, was of great service to him in this affair. The terms were first agreed upon verbally, afterwards written and subscribed by both parties ; only one condition appeared in it in reality, i. e. that the Emir and his family should be conveyed to Alexandria or to St. Jean d'Acre, places indicated by himself. When the Duc d'Aumale heard the news, he personally assured the Emir that he entirely approved and confirmed the terms, and that they would be religiously respected. The Duc d'Aumale had perfect good faith in this ; but the reply of the French government to his announcement was, that he should be sent over directly to France ! Hence they arrived at Toulon in the Asmodée, December 30th, 1847, and were shut up close prisoners at Amboise chateau, on the left bank of the Loire, between Tours and Blois. Five weary years did he pass there, while monarchy, republic, and empire ebbed and flowed without the castle's walls. At last Louis Napoleon releases him — an act of tardy justice, dictated perhaps by policy, Londonderry, and memory. A captive himself in Amboise, he had shortly before escaped as a day-labourer with a plank on his shoulder. Abd-el-Kader left France just as the news of the capture of Laghouat by General Pelissier reached France. Almost on the spot where Themistocles and Hannibal tasted the bitterness of exile, this noble chief finds comfort in his deep and glowing faith, amidst the fall of dynasties and the wreck of creeds. He is reported to be engaged in writing méditations on the Koran. That Abd-el-Kader signed the vote for Louis Napoleon's election to the empire cannot be laid to his charge. He was a prisoner, and gratitude biassed his judgment. We know not the future that awaits this brave spirit ; but though Turkey should crumble into dust, and Islam vanish in vapour, we feel assured that he will be true to himself, his God, and his country. Glad are we to find that the gallant marabout has offered his sword to the sultan in the day of Turkey's humiliation ; and come what may, we feel sure that if God preserves his life, he will be seen, and his blows resound, amidst the thickest storm of battle in the coming struggle in the East. Later accounts

to Rome, Venice, and Rastadt, the Saharians made a vigorous resistance to the inroads of civilisation. The Kabyles also, especially the Beni-Slimans, at the exhortation of Si-Djoudy, the Zouaoua chief, showed an unaccountable antipathy to registers and taxes, but they were brought to order by the gallant prætorian bands of Bugia.

But the most serious insurrection was near Biskara and Batna, excited by the Cherif Bou-Zian, against whom the French had many complaints, probably as well founded as those of Hapsburg against Hungary. Bou-Zian fled to Zaatcha in the oases, which was unsuccessfully attacked by the French, who were forced to retire, as they pretend, through want of stores and material. This was on July 16th. After the heats were passed, the moment arrived to make an example of Zaatcha,* like Saragossa and Tarragona in the Peninsular war, whose conquest was so creditable to the arms of the uncle. The expeditionary column left Biskara October 6th, and reached the rebellious oasis next day.†

After two secondary actions to preserve their convoys, and two assaults (20th of October and 20th of November), Zaatcha was carried. The houses were defended and carried one after the other, and destroyed; and all the defenders, without exception, including Bou-Zian and the marabout Si-Moussa, fell amidst the ruins of Zaatcha! What impartial pen shall chronicle the panting despair of those dauntless Saharians, fighting to the death for wives and home, and waving palms and liberty! what pen do justice to the gallantry of European discipline, establishing order amidst blazing temples and blood-stained gardens! The sweet South will continue to sigh its perfumes over that oasis; but its gardens are a wilderness, its homes a desolation, for the spirit of freedom has left it. And Canrobert, with his brows crowned with the laurels of Zaatcha, returns to Paris and perpetrates the coup-d'état. These are the men to command the army of England. Bou-Sada also raised barricades in 1849, but was also reduced to order, and spared for a consideration.‡

Nothing worth record transpired in 1850, save a few razzias against the Abaidia, near Tlemsen, and the Beni-Immel Kabyles, on which occasion

---

state that he has abjured all thought of worldly distinction, and lives retired at Broussa, devoted to study. Perhaps, like Solomon, he has found out that all on earth is vanity.

* Tableau, 1849, p. 18. See the siege of Saragossa in Count Toreno's Istoria del Savantamiento, Guerra, y Revolucion de Espana, 1835.

† The Zaouia, or cloister college, at the north of the oasis, was stormed forthwith, and probably razed to the ground; but the labyrinth of palms was so impracticable, and the defence so energetic, that the French were forced to carry on a regular siege on a new plan. Reinforcements were sent from Constantina, Aumale, and Oran; and it took fifty-one days and nights in open trenches, with continual fighting, to reduce the place. They had, moreover, four serious engagements on the 25th of October against Arabs seeking to oppose the cutting down of their beloved palms; and the 30th and 31st of October, and 16th November, against the insurgent nomads of the Sahara.

‡ Tableau, 1849, pp. 10, 11. M. Borrer informs us that the French nation is looked upon as a mere passing cloud in Algeria, and that the Moslem only bides his time.

the French lost General Barral, and killed 200 men (May 21) and a few other insignificant tribes. The great event of the year was the first race at Algiers, at which 1500 Arabs *assisted* with their khalifas, back-aghas, &c., winding up the whole with an immense fantasia (29th September). Thus we may shortly anticipate the introduction of thimble-rigs and odds in the oases.

Since 1850 the only foes the French have had to encounter in Algeria have been the Saharians and the Kabyles, *i. e.* the most intelligent and industrious population of the country. Pelissier and the Zouaves taught the former at Laghouat,* in November 1852, the ingratitude of resisting the blessings of French order and law; and the governor-general was last spring engaged in eradicating republican equality, primitive faith, and a merciful jurisprudence in Kabylia, by reducing their homesteads to ashes and bathing their altars in blood. The French papers add, that bright sunshine was smiling on the gallant host as it left Setif in May 1853, to teach the barbarians the humanity of Christian warfare.†

The conquest and occupation of Algeria is another specimen of Christian justice, another honour added to the lustre of our civilisation ; and will pass down with the oppression of Poland and Circassia, and the annexation of Sind and Texas, to the admiration or execration of a discriminating posterity.‡

* Laghouat, or rather El-Aghrouath, was stormed by the French (December 2, 1852), with severe loss, including General Bouscaren, who died of his wounds, and whose heart was saved and preserved for his sister. Among the latest novelties from Algiers, we hear of corps of French gendarmes mounted on dromedaries, reducing the Sahara to order and fraternity. (Le Pays, Journal de l'Empire.)

† Le Pays, Journal de l'Empire, May 1853.

‡ The most important intelligence relating to Algiers, within the present year, applies to two particulars : first, the project of the formation of a French company analogous to our East India Company, and the framing of a constitution for the administration of Algeria (*Times*) ; secondly, the passing of a law in the legislative council, by a majority of 200 to 1, for the establishment of an electric submarine telegraph from France, through Corsica and Sardinia, to Algeria (*Le Pays*). The undertaking is given to Mr. J. Brett, to whom the exclusive right of management and ownership is granted for a course of years. Objections were raised to the plan in France, on the score of its giving England an insight into the colony ; but the objection is evidently groundless, as she possesses that already. It is talked of extending the telegraph to India ; in which case we may shortly anticipate that the same wire which informs us of the Christianising of Kabylia and the Sahara by fire and sword, will record the generous self-denial of Britain in annexing Birmah or a slice of China. Recent accounts state that the Beni-Mzab tribes are just *annexed*, after being long *protected*.

APPENDIX.

SECTION I.

Antiquities of Algeria.

NATURE and man both change, and act and react, though a day to one is as centuries to the other. Man, though an earthworm on his planet, and a monad to the starry universe, does not depart without leaving a trace behind. He has done much, though he might do much more; and the influence that he has exerted has been the child of combination and conviction. The giant works of Rome were wrought by combination; and the dazzling path of Arab conquest was the work of faith. Mighty men have strode across this little span of earth, and left their godlike foot-prints to the ages; but the monumental voice too often falls on unheeding ears, and Christendom has still much to learn from the past. Natural science has still many truths to tell of Elysian fields and summer climes of yore at the Poles, and of wondrous forms of men and beast long gathered to their fathers, yet pregnant with instruction to us, who might tame leviathan and lead the lion captive. So also archæology, with her hieroglyphs, has much wisdom in her womb, and is at hand to teach our paltry sickly race how to recover some of the lost sap of our constitutions, and the beauty of our climates.

The glories of the Vega, the Alhambra, and the Alcasar, might give a touch of life to palsied Spain and the splendours of Roman Africa, might stimulate a noble rivalry in French legislators and warriors. Each age has a mission, and ours may not improbably be that of stokers, pokers, and nuggets : yet there is somewhat more godlike in man than to wield the hammer of Thor, and belabour this hard reality without; and the soul soars and swells as it dwells in the magnetic sphere of a Cato, or hovers round the memory of Scipio. The simple sublimity and massiveness of ancient thought and art must command the reverence of the ages ; and the cosmopolitan munificence of Roman municipal works put to shame the piecemeal economy and extravagance of Christendom.

Falling back 2000 years up the stream of time, we take ship at Baiæ, or glorious Sicily, and our vigorous rowers soon pull us over to Carthage, and land us amidst golden harvests, waving dates, and gorgeous villas. Magnificent temples, stately porticos, and verdant bowers, are reflected in the deep blue waters along that sunny shore : its valleys are crowded with an industrious population, and its ports with countless masts. We are in the granary of the Roman Empire. Elegant baths and fountains refreshed the cities ; massive drains and aqueducts supplied them with pure water, and removed all insalubrity from their streets. The plains and valleys, drained of all stagnant water, were the parents of plenty, and the homes of health and content ; and the well-wooded hills supplied their nourishing streams, conducted by innumerable channels amongst the verdant pastures and waving cornfields.* In short, Africa and Numidia were once a promised land flowing with milk and honey.†

Barbary, as we have previously seen, comprises what was anciently called the two Mauritanias,‡ Numidia, Libya ; the country of the Massassili, Gætuli, and Garamantes ; and Africa Proper, containing Carthage. Algeria, as the reader is aware, corresponds to Numidia, part of Mauritania, and Gætulia ; and it is with this portion of the Roman settlements in Africa that we are more immediately concerned.§ We shall begin our survey with the province of Algiers, which corresponds to a part of Cæsarean Mauritania and of Gætulia ; and its principal ancient cities were: 1st. Jol or Julia Cæsarea (now Scherchell), which was the capital of Cæsarean Mauritania and of Juba II.; and the ruins of the old town are described as not inferior to those of Carthage. Numerous remains of pillars, capitals, cisterns, &c. give a high idea of its splendour. The water of the river Hasham, according to its present name, was conducted to the city in a noble aqueduct not much inferior to those of Carthage, as is proved by the fragments of its arches in the valleys to the south-east. Other small conduits are still perfect, and supply Scherchell with good water. A tradition reports that the ancient city, as well as its successor, was erected by Andalusian ·Moors. Earthquakes account for the demolition of most of its noble edifices.||

The whole coast from Scherchell to Algiers is strewed with ruins, proving the former salubrity, populousness, and opulence of this region.¶

ˣ See an eloquent passage in Baron Baude's Algérie on the vicissitudes of Numidia. Also Dawson Borrer's reflections on the past of Kalama.
The reader who wishes to obtain minute information relating to the archæology of Algeria is referred to the *Exploration Scientifique*, to Blofeld's *Algeria*, and Dr. Shaw.

† It was, next to Egypt, the most fruitful province of the empire; was styled by the Romans the soul of the Republic, and in the corrupt Latin of the lower empire, *speciositas totius terræ florentis*. The greatest luxury of Roman extravagance was to have a villa on the coast of Africa ; for the land was then a paradise, like Sicily while free. (Pananti, Avventure, ii. p. 4.)

‡ Mauritania was originally separated from Numidia by the river Mulucha, was a fertile country containing some arid spots, and whose mountains were covered with large forests infested by lions, elephants, &c. The Moors resembled the Numidians in pride and treachery, and in dexterity as horsemen, their army consisting chiefly of cavalry. They were governed by kings, who, like those of Numidia, affected to be descended from Hercules. As a reward for his treachery in giving up Jugurtha, the Romans added one-third of Numidia to the Mauritanian kingdom of Bocchus, which, under Bocchus II., was divided into two kingdoms : 1st, that of Bogud, to the west; 2d, the Mauritania of Bocchus. In the civil war of Octavian and Antony, Bogud, siding with Antony, was slain in Greece, and Bocchus occupied the whole of Mauritania, which after his death was given to Juba, of the race of Massinissa. His successor, Ptolemy, was put to death by Caligula because he appeared in the circus at Rome in a purple robe ; and two years later Claudius divided Mauritania into two Roman provinces : 1st, the Cæsarean ; 2d, the Tingitanian Mauritanias. Barbié du Bocage, Sallust, p. 238.

§ This country, once so renowned, contains fewer massive remains than might be supposed, having suffered severely from the fury of wars and revolutions. The relics of ancient Christianity and Paganism have alike fallen a sacrifice to the Gothicism of Vandals and the iconoclast convictions of the Moslems. The principal coins, which are few, are those of Claudius, of the Antonines, of Alexander Severus, and of the two Gordians. The Punic and Numidian coins are scarce and almost illegible, save to German and Runic professors, who are clever at discovering sermons in unhewn stones.
The Roman remains can be generally recognised by the terra-cotta materials, and by the idols and implements of bronze. The Moorish antiquities are known by their neglect and dirt. Blofeld ; Pananti, ii. p. 13.

|| Blofeld, p. 73.

¶ Berbrugger, part i. Dr. Russel, p. 364. Pananti, ii. p. 16, says that the archpriest Borghi, a man of vast learning, proved to him that Julia Cæsarea could not have been situated at Scherchell, as Dr. Shaw states, but at a little village half a day's journey distant. When doctors disagree, the patient has no chance.
The Kabyles of Chenouah used to call it contemptuously, Tekedemt, *i.e. an old thing*, from the Arabic word *kedim*, which they have borrowed and altered by prefixing and affixing, as usual, their favourite *t* or *th*. This designation is common to many old towns, as well as that of Kheurbah (ruins), which has been corrupted into Herba in our maps. Berbrugger, part i. Rozet, Voyage dans la Régence d'Alger, iii. p. 258. Dr. Russel, p. 364.

Farther east, following the coast, was situated Oppidum Novum, a colony of veterans sent into Africa by the Emperor Claudius. Yet nearer to Sidi-Ferruch was Tipasa, whose ruins may be still traced not far from Tefessad; and a little beyond them you come to the celebrated Koubber Romeah, which is most likely the *monumentum commune regii gentis*, being the common place of sepulture of the kings of Mauritania.[*]

Algiers presents but few Roman remains ; and it is still uncertain what name it bore under Latin sway, some thinking it Icosium, and others Jomnium.[†] Mr. Blofeld says that there are Roman ruins on the banks of the Savus (Haratch), south-east of Algiers ; and he thinks this more probably the site of Icosium than Algiers. M. Berbrugger mentions the remains of a Roman via, Rue de la Marine, near the port of the capital, which he thinks must have corresponded in most respects with the old Moorish harbour before 1830. St. Marie informs us that at the quarter of the Gate of Victory, in the old town, there stood on one side of the gate, in 1845, a fountain of white marble, constructed among the ruins of a Roman aqueduct.[‡]

The ruined town on the promontory of Matifou, eastward of the Bay of Algiers, is sometimes known by the name of Rustonium, and Ptolemy calls it 'Ρουστονιον. Baron Baude and most of the French savants make it Rusguniæ, from inscriptions found on the spot.[§] When Charles V. embarked at Rusguniæ in 1541, there were more remains of the ancient city than at present ; and in the twelfth century, according to Edrisi, the Arab town on the site of the Roman city was in a state of decay. " Tamendfos," he writes, " is a fine port, near a small ruinous town. The enclosing walls are almost overthrown, and the population is small : you see nothing but remains of houses, of large buildings, and of idols in stone. It is said to have been once a large town."[||] There are still traces there of an ancient *cothon*, with several heaps of ruins of the same extent as those of Tefessad, and which have also contributed to the fortifications of Algiers. The distance of 15 Roman miles between these ruins and those upon the Haratch is the same as that of the Itinerary between Rusguniæ Colonia and Icosium. Rusguniæ is the same as the Ruthesia of Mela, and the Rusconia of Pliny.[¶]

Oran contains few, if any, vestiges of antiquities ; yet Marmol[**] informs us that it stands on the site of *Unica Colonia*.[††]

---

[*] It stands on a mountainous part of the sea-coast (Sahel), seven miles to the east of Tefessad, consisting of a solid and compact edifice built with the finest freestone, 100 feet high, with a diameter of 90. It is of a circular shape, rising with steps quite up to the top, like the Egyptian pyramids. The elegance and beauty of the shape and materials show it to be older than the Mahometan conquests; and it is thought, not improbably, by some to be the monument which Mela places between Jol and Icosium, and which he attributes to the royal family of the Numidian kings. Berbrugger, part i. Dr. Russel, p. 365. Blofeld. Dr. Shaw.

Not far from the mouth of the Masafran, below Coleah, and near the marabout of Sidi-Fouqa, are the remains of Casæ Calventi. Indeed, the whole of this region may be justly styled a classical soil. Berbrugger, part i.

[†] Blofeld. Berbrugger, part i. Pananti says (ii. p. 16), "Algeri non e, qual comunemente si crede, *Jol* o *Julia Cæsarea*, edificata da Julio Cesare, ma bensi l'antica *Jomnium*."

[‡] The Moorish antiquities of Algiers are rapidly melting away before the march of intellect and the reign of order; but the most ancient are comparatively new. Rozet, in visiting the garden of Mustapha Pasha, in the neighbourhood of Algiers, observed a superb aqueduct carried across a parched valley, and constructed for the purpose of conveying water to the city. The architecture is decidedly Moorish, presenting two tiers of arches, &c. Dr. Russel, p. 366: and Berbrugger, part i.

[§] Baron Baude, i. p. 47.

[||] Edrisi.

[¶] Baron Baude, p 72. Dellys stands on the site of Rusucurrum, which was united to Rusguniæ and Saldæ, or Bugia, by Roman roads that are still traceable. Blofeld. Baude, vol. i. p. 129.

[**] Africa Illustrata.

[††] This name does not occur in the best classical authors; they only speak of Portus Magnus, so named because of its size,—so rare a circumstance on the exposed coast of Africa, which Sallust named *mare importuosum*. This characteristic also attracted the attention of the Arabs, who have recorded it

F F

The province of Oran, as far as the Macta (Amnis Mulucha) corresponds to the ancient kingdom of Bocchus, the grandson of Jugurtha : it was then inhabited by Moors and Massysylians.* After passing the Mulucha began the kingdom of Juba, which, under the Roman Empire, was comprised in Cæsarean Mauritania. If we might draw an inference from the small number of ruins that are found in this province, we should say that it is the portion of Algeria in which the Romans have left the fewest traces. Nor is this surprising : their conquest advanced from east to west, and their stay in the latter region was less protracted; hence they had not the time to form great establishments there. Moreover, the Vandals smote Cæsarean Mauritania first, and their maiden fury vented itself first on this unhappy shore.

The most numerous Roman stations are on the coast. From the Oued-Moulouyah† (Maloua), marking the limits of Tingitanian and Cæsarean Mauritania, to the Moqthâ, or ford of the united Sig and Habrah (called by the French the Macta),—that is, for a space of about 108⅓ leagues according to the Itinerary, but really of about 60 leagues,‡— sixteen Roman establishments, more or less considerable, are found.

The points that appear to be the most accurately determined are, *Siga municipium,* styled by Pliny a royal town of Syphax (*regia Syphacis*).§ It must have stood about a league west from the mouth of the Tafna, probably where are now some ruins called by the natives Tikambrit. The bay where the Tafna falls into the sea, and the island of *Acra* facing it, formed the *Portus Sigensis.* The Tafna is named by Ptolemy Assara, corresponding to the Isser of modern times, which falls into the Tafna some leagues before its mouth, and which may at that remote date have preserved its name down to the sea. In digging the intrenchments (1836) of the camp, they found there a medal of Tiberius.

Opposite Portus Sigensis was, as we have just said, Acra insula, which corresponds to the little island now called Rasgoun or Rachgoun, which is occupied by a small French garrison. The *chef d'escadron,* M. Tatareau, who visited this spot (1832), picked up a gold medal bearing the stamp of a Greek cross; and he attributed this coin to the Lower Empire.‖

in the name *Mers-el-Kebir* مرسى الكبير the literal translation of Portus Magnus. Pliny, describing this locality, says : " Portus Magnus, spatio appellatus, civium Romanorum oppidum." The characteristics of the place at once determine the site of the great port, but what Oran represents is uncertain.

Some persons have thought that Oran stands on the site of the municipium of *Quiza* (*Boniza* according to Ptolemy), or *Quiza xenitana peregrinorum oppidum* (citadel of strangers), whose name, as the reader will perceive, is accompanied by two epithets stating precisely the same thing, one Greek and the other Latin. The Itinerary places Quiza at 40 miles (13¼ leagues) from Portus Magnus or Mers-el-Kebir, whilst Oran is only one league from that place : hence this hypothesis is evidently untenable.

* The Mulucha separated the Moors from Numidia till the time of Marius and Metellus, and served as a limit to the kingdoms of Jugurtha and Bocchus. Some authors spell the name Molochath and Malua : it is now styled Maluia, flowing near the frontier between Algeria and Morocco. The word Molochath, of which Mulucha is a corruption, means in Phœnician the *royal* river. Barbié du Bocage's Sallust, p. 240.

† This stream is generally called *Malva* or *Malvana* in ancient geographies ; but it is probable that Maloua, as Ptolemy writes it, is more correct, because this appellation comes nearer to the name given to it now, and especially because a *v* is inadmissible in the names of places evidently belonging to the language of the natives, since the latter (both Berbers and Arabs) have no *v* in their language.

‡ M. Berbrugger says that the Itinerary of Antoninus, in giving the distances between the towns from the Maloua Flumen to the Mulucha, or Macta, almost always doubles them ; whereas in Eastern Mauritania the distances are made less than the truth. The Itinerary was probably drawn up by different persons using different measures, and the compiler was not careful to rectify the differences. M. Berbrugger also thinks it possible that, instead of understanding here the *passus* generally used in the Itinerary, a double pace of 5 feet, we ought to understand the *gradus,* or single pace of 2½ feet. This would make the distances correct. Berbrugger, part ii.

§ Pliny.

‖ Whilst speaking of Rasgoun, we shall notice the resemblance of the name to *Rasgonia,* a Roman town whose ruins still exist at the eastern point of the Bay of Algiers (Cape Matifou). The syllable *ras*

The *Salsum flumen*, placed by the Itinerary 8 leagues from *Portus Sigensis*, must be the Oued-el-Malahh, called by the Spaniards Rio-Salado.*

Beyond Portus Magnus stood *Quiza*, of which we have already spoken as 13¼ leagues from Mers-el-Kebir, which would place it on the spot where you now find old Arzeu, the Arsenaria of the ancients. M. Berbrugger imagines that instead of 40, which you find in the Itinerary, you ought to read 11 miles, which gives 3⅔ leagues, and brings us to a place called Canastel, where there exist some ancient remains at the present time. A tradition records that St. Augustine was born in this place.

After passing this place, you come to Portus Deorum (the bay of Arzeu), *Arsenaria* (ancient Arzeu), where considerable ruins are still found ; then you reach the Amnis Mulucha (the Macta), which Ptolemy calls *Cartennus flumen;* and a little beyond it the colony of Cartenna, which must have stood near Masagran or Mostaganem.†

We shall now pass to the province of Constantina, formerly part of the kingdom of Numidia.‡

It is somewhat remarkable that Hippo§ was the first spot in Africa visited by the

---

or *rus* (pronounced *rous*) رُوس رَاس signifies clearly enough *cape*, having the same meaning as *caput* both in Kabyle and in Arabic ; and if you wish a proof of its having had the same sense formerly, you need only follow the nomenclature of all the Roman establishments situated along the coast, and it is striking to find the syllable *ras* or *rus* prefixed to the names of all towns standing near capes. Going from west to east, we meet in this manner *Rusadir, Rusibricari, Rusucurru, Rusipisir, Rusgonia, Rusasus, Rusicada,* &c. This proves that the Romans almost always adopted the local designations, which they latinised, and also sometimes mutilated, as the French do at present with the native names.

The ancient Libyan and modern Arabic term *ras* رَاس *head*, is not improbably derived from the Sanskr. *raj* राज to shine, whence *rajah*, prince, Lat. *rex*, Goth. *reiks*, &c. Eichoff, p. 242. Gesenius (Lex. p. 916) identifies رَاس with רֹאשׁ the Syr. ܪܺܫܳܐ and the Æthiop. ርእስ:

* A name adopted by the French, and a literal translation of the Arabic appellation. The identity of meaning (Salsum Flumen, Oued-Malahh, and Rio Salado) gives a great probability to this conjecture.

† Part ii. p. 6, of Berbrugger's Algérie. Captain Despointes says that a very perfect ancient temple is still standing near Arzeu. See Baron Baude, St. Marie, and Blofeld.

In the interior of the province, Tlemsen and its vicinity contain several interesting monuments of antiquity, chiefly Moorish. We have already alluded to the Mechouar and the reservoirs of Tlemsen ; and we have still to mention the beautiful tomb or marabout of Sidi Bou-Medina, near Tlemsen, which has been greatly injured and defaced by the æsthetical charities of the French army. Blofeld.

Not far from Tlemsen you also find the beautiful remains of an ancient mosque (Berbrugger), which must date from an age when learning and art flourished on the sunny shores of Barbary, and cast their lustre and breathed their harmony over the faith of Islam and its magic productions.

‡ Numidia extended originally from the Leptis (in Tripoli, near Cyrene) to the river Mulucha, containing at the time of Jugurtha vast tracts of fertile corn-lands, and also deserts. To the south it bordered on the Gætulians. The Numidians, who are described as a hardy, brave, industrious, and temperate, but treacherous and versatile race, were excellent horsemen, great hunters, practised polygamy, and had despotic kings. Sallust refers their origin to the Persians, who came under the Libyan or Phœnician Hercules, and mixed with the Gætuli. Some derive the word 'nomadic' from the Numidians ; others, the name Numidian from nomad. At the time of the second Punic war they were still a very savage people, divided into two sections : first, the Massæsyli to the west, occupying modern Algeria ; secondly, to the east, the Massyli, in the province of Constantina and part of Tunis. Syphax commanded the Massæsyli, and Massinissa, son of Gala, the Massyli.

§ The reader should be careful not to confound Hippo-Regius with another Hippo situated more to the eastward on the northern coast of Africa, and known by the name of Hippo-Dirutus, Zarritus, or Diarrhytus (Bizerta).

The Hippo with which we are now concerned had been named Ubbo by the Phœnicians ; a word meaning 'pond' or 'bay' in their language. The ancient name of the Seybouse, *Ubus*, is probably referable to the same etymology. The Romans changed the name of Ubbo into Hippo, and added the epithet Regius because it was a favourite residence with the native kings of Numidia. A Latin poet has expressed this circumstance in the following terms :

"Antiquis dilectus regibus Hippo."

Romans, and the last place that they lost in the two invasions of the Vandals and Arabs. Bona has also at all times shewn itself more disposed to submit to the sway of the Franks than any other town on the coast of Barbary. It may, indeed, be laid down as a maxim,— that the resistance to European conquest and rule increases in proportion as you advance west in Barbary, and reaches its maximum in Morocco, whereas its minimum is found in the province of Constantina.

In May 428 the Vandals came over to Africa, on the invitation of Count Boniface; and the description of their invasion from the pen of Bishop Possidius will remove all wonder that the term Vandal has become a synonym for ruthless destruction.*

Genseric soon drew nigh, and laid siege to Hippo about June 430. The Vandals had besieged the town three months, when St. Austin fell dangerously ill. The misfortunes of his people and flock hastened his end, after a sickness of three weeks.†

Save these ruins, some remains scattered at the foot of Mount Edough, the aqueduct of the valley of Kharesas, and the cisterns, no trace of Hippo Regius is left above ground. This classic soil is, however, full of foundations of houses, of remains of ways, of tombs, and of fragments of statues.‡

The most remarkable ruins at Hippo are the cisterns, that cover the top of a hillock. They form a square building 78 feet by 70, which, besides rain-water, was probably a receptacle of the stream brought by the aqueduct from Mount Edough. A narrow corridor bisects the building lengthways, having a staircase at one end descending to the bottom of the reservoir. The section towards the bay is farther divided by another passage. The division of the edifice into stories is only applicable to these corridors, as the reservoirs rise to the whole height of the monument.

One of the longitudinal halves of this building is divided into compartments by seven partition-walls, whose materials have an extraordinary adherence and solidity; since, while the lower part of the walls is destroyed, the upper continues to stand, though it has to support the weight of the vaults and terraces. The other half, which is not sub-divided, has been invaded by fig-trees, wild olives, and bushes of jujubes, which form quite a thicket On this side the terrace has almost entirely fallen down, whereas on the other it is in tolerable preservation, though overgrown with plants and trees. The remains of several small cisterns are scattered round the large one.

These are the only remains of the once opulent and powerful Hippo, the rival of Cirtha. The traveller who paces this classic strand is in danger of missing its few and

---

On the Punic tongue see the Section on Language. The Phœnician *ubbo* appears to have some affinity with the Sansk. *ap*, water, *apnas*, fluidity,—from *ab* or *amb*, to move; and hence also with the Lithuanian *uppe* and the Greek *οπος*. From the Sanskrit अब् *ab*, अंब् *amb*, also come the Latin *aqua*, *imber*, *umbra*, and the Greek *ομβρος*. Eichoff's Vergleichung der Sprachen v. Europen u. Indien. Gr. edit. p. 698. Bochart and De Brosses agree with Berbrugger in giving to the word *ubbo* the meaning of 'bay.' Barbié du Bocage, Dict. Geogr. to his Sallust, p. 214.

* This description will be found in the Chapter on History. Boniface, regretting his folly when too late, and facing the Vandals in the field, was worsted and forced to seek refuge in Hippo, then thought to be very strong. The town was crowded with fugitives, including many bishops.

† He lived 72 years, and had been priest or bishop at Hippo for 40. He made no will, having nothing to bequeath but the episcopal library and his Mss. After 14 months' siege, the Vandals desisted; and when they returned after the second defeat of Boniface, the town was deserted, and they destroyed it by fire. The only monument that escaped was the library of St. Austin, in the basilica of Peace. If we may trust tradition, the remains now seen near the Cattle Run belong to that edifice. The resistance of Hippone lasted 14 months; St. Augustine died August 28th, 429; and the city was taken in December 430. Most of the place was burnt, but the bishop's house and library were spared. Belisarius retook it in 534, and it fell to the Arabs in 697. After that it was removed to Bona, 2000 metres to the north. On Augustine see Possidius de Vita Augustini; Berbrugger; Nachrichten u. Bemerkungen; Dr. Russel, pp. 136, 137; Baude, vol. ii. p. 42.

‡ Baron Baude. Dawson Borrer. St. Marie.

insignificant remains, so thickly are they shaded by the luxuriant African vegetation. With difficulty can he realise the existence of sacerdotal fanes and lordly porticoes on this wild and tangled spot. Yet future researches will doubtless bring to light much antiquarian wealth concealed beneath this pregnant soil.*

The ruins of Kalama are situated about half a league from the Ford of the Seybouse.† To the left of the path leading from the camp to the ruins, you perceive a small square edifice, of which a few stone pillars are still standing. The ground is scattered over with broken shafts of columns, and with marble capitals of rather bad quality.

On the slope of an undulation which is met before reaching the site of the old city, and whose base is laved by the river of Guelma, the Romans had raised a theatre, whose construction must have been rendered extremely easy by the character of the locality where it stands ; for the declivity of the hill presented a natural amphitheatre, which only needed to be faced with masonry in order to form steps, and to be surrounded by an en-

---

* On one side of the ruined aqueduct, on the top of the hill of Boonah, and surrounded with lilacs and honeysuckle, stands a white marble altar, recently constructed, with a bronze statue of St. Augustine, one metre high, *in pontificalibus.* This was erected on the occasion of the transfer of his real or ideal remains from Pavia, with all the pomp and circumstance of the church militant, by the French government, in the year 1842. See the account by M. l'Abbé Sybour, afterwards Archbishop of Paris. Madame Prus, p. 106. St. Marie also gives a correct description of the monument.

"Hippo," observes Baron Baude, ii. p. 41, "was grouped at the foot of two mamelons, one of 80 metres (262·40 feet), the other of 38 metres (124·64 feet) in height, and now called by the Arabs Bounah and Gharf-el-Antram. Its enclosure has not yet been ascertained ; but from the scattered ruins, it cannot have been less than 60 hectares (148·20 acres). 300 metres (984 feet) of old quay still exist, at the distance of 1000 metres (3280 feet) from the present mouth of the Seybouse. It was there that, in the year of Rome 707, was stationed the fleet with which P. Sitius, lieutenant of Cæsar, destroyed that of the fugitive Scipio. The old drains under the town are broken up for the stone : they were built of quarry stones (in a covered channel), and were made as cheap as possible. To the north are the remains of a gate of the town, built in alternate layers of bricks and quarry stones, and its dimensions are very large. The church and convent of St. Austin, seen by sly old Père Dan, have somehow disappeared. The hydraulic establishment is the great monument of Roman construction at Hippone, as in all their colonies ; for, unlike the Christians, they agreed with Mahomet in thinking cleanliness next to godliness. An aqueduct brought the water from Mount Edough, 2600 metres (8528 feet) in length, only a part of it being built on arcades over the valley of Laurels and of Boubgimah." Baron Baude. St. Marie seems greatly disposed to coincide in the Baron's views.

The five arcades nearest Mount Edough are still standing. The basements and the pediments are in reticulated masonry, the interior consists of quarry stones, and the arches of bricks ; the channel is 2 Roman feet (0·589 metres) in breadth, and you can still count all the piles of the aqueduct. Near the town it must have been 20 metres in height. The reservoir, on issuing from which the waters divided, is to the north-west, and halfway up the mamelon. It is divided into two principal compartments, each of which is 17 metres (55·76 feet) wide, and 40 long (131·20 feet) which gives, allowing 1 metre (3·28 feet) in depth, a capacity of 1360 metres (1774 cubic yards). The present depth is 5 metres (16·4 feet) ; but it is probable that by cleaning out the rubbish, you would find that the reservoir contains 10 or 12,000 cubic metres of water (13,059 or 15,671 cubic yards). The partition-walls are of masonry, with an interior facing of brick. The eastern basin is still crossed by two bridges. Baron Baude, iii. p. 39.

The stone employed in all these constructions, except some parts of the aqueduct, is a porous limestone, easy to work, which is found, superposed on granite and marble, in the bay of Caroubiers, and the neighbourhood of Fort Génois. The bricks, tiles, &c. are baked with the greatest care, and the beds of mortar on which they repose are thicker than the bricks themselves. The mortar is harder than stone, is very rich, and contains as many little stones as sand ; but the Romans had architects, and never ventured a National Gallery or a Trafalgar Square. The limestone has been obtained from strata of calcareous saccharoid, which is intercalated in the granite all round Bona, to the Sahel, and to the very mamelon of Gharf-el-Antram. All these works are very simple, without decorations or extravagance. The rusticity of the materials and workmanship, which are a disgrace to a Roman colony, but would be an honour to the British metropolis, show that they were mostly the work of Roman soldiers, who were commonly more usefully engaged than in smoking cigars or playing billiards. Between Bona and Hippo are the remains of the Roman via which formed a part of the great road that followed the coast from Carthage to the Straits of Gibraltar. Another via went towards Cirta, and numerous cross-roads ramified with these. Baude, ii. p. 42.

† See Part I. Chap. XII.

closure so as to embrace a complete arena. It follows from this configuration that the top of the monument is on a level with the summit of the hill; whilst the lower part, or the scene, extends over a slope that is met with halfway up.

The site of the orchestra is overgrown with brushwood, which also covers the part where the stage was situated; and the lentisk and jujube, or Barbary fig, climbing up to the steps, strike their vigorous roots in the intervals of the stones that formed the steps.

Notwithstanding the ravages wrought by time on this monument, it is still easy to discover its principal divisions.

Applying the 'analysis of Roman theatres to that of Kalama, we proceed to remark that in the interior of the enclosure of the *cavea* French visitors have found vaulted corridors, partly fallen in, which must have led to each zone, without its being necessary to pass up the stairs, which would have inconvenienced the noble occupants of the *ima cavea*.

In the upper and middle part of the theatre is a little *cabinet*, surmounted by a cupola, and open on the side facing the stage. In the French army, where it is usual to bestow epithets on all the monuments that they meet, the soldiers christened it the Proconsul's box.

Traces of the use of *vela* or awnings are still perceivable in the theatre of Kalama, and it is easy to perceive the holes in which the poles were fixed that supported the veil. The men commissioned to do this were superannuated sailors, whose profession was thought to have qualified them for the office.*

Halfway from the two extremities of the semicircle, some niches may be seen that probably contained statues. Behind these are some rooms that open on the staircase by which the spectators of the *ima cavea* had their private entrance. The two extremities of which we have just spoken, and that were called by the ancients *cornua* or horns, form the limit between the theatre proper and the stage. Scarcely a vestige remains of the latter part of the structure: some blocks of masonry, half concealed by brushwood, are all that is left. Perhaps the stage in this case only consisted of a modern scaffolding, according to the description of Ausonius:

> " Ædilis olim scenam tabulatam dabat,
> Subito excitatam, nullâ mole saxeâ."

It is quite possible that in this remote province they were reduced to adopt the simplicity of this ancient fashion for want of sufficient funds; or possibly the materials have been carried away to build other edifices.

On quitting the theatre, M. Berbrugger and his companions followed the brink of a ravine, in which flows the river of Guelma, and they arrived near a stone enclosure, flanked by thirteen square towers. At the north-east angle of this kind of fortress arose a great ruinous building; and the remains of ancient Kalama lay stretched at the bottom and on the sides of a ravine, whose slopes were formerly arranged into steps, as it is easy to discover. A wall used to extend towards the mountain; running from the citadel of which we have just spoken to the crest of a mamelon that separates the ravine of the river from that in which the town was built, it served to protect the latter in a quarter where, from the nature of the ground, it would have been easily assailable.†

Without wearying the reader with a dry detail of all the parts of the ancient town, we

* Sir William Gell's Pompeii.
† Madame Prus, in 1850, corroborates this description of M. Berbrugger in 1836. She says, "Guelma, situated on the summit of the mountain Serdj-el-Aouda, must have been a place of considerable extent and importance in the time of the Romans. The thirteen towers still existing in the circumference of the walls, and the divers inscriptions found on the monuments, seem to indicate the sixth century of our era as the period of its construction." It is unpleasant to criticise the opinions of a lady, but the love of truth forces us to point out the inaccuracy of the latter statement, when Madame Prus confounds the original foundation of the city with the later attempts at renovation under the Greek prefects, and after the Vandal devastation. It is likewise our unpleasant duty to point out this lady's error in confounding the wall of the citadel, with its thirteen towers, with the city wall,—page 159.

shall, in a note, lay before him a few general notions of the place, and of its principal ruins and inscriptions.*

We shall now proceed to notice the most remarkable ruins of Kalama. The large fortified enclosure is undoubtedly the largest and the best-preserved ruin, and we shall attend to it first.

A glance shows the date of its erection. The walls are composed of heterogeneous materials, presenting a confused heap of marble and stone, votive and tumular ornaments, often upside down, fragments of bas-reliefs, statues, and even domestic utensils. Such walls can only have been raised in times of confusion and barbarism. The foundation is no doubt more ancient, but nothing above ground can date higher than Belisarius ; for the Vandal Genseric, before the arrival of the lieutenant of Justinian, had dismantled all the African cities, save Carthage, the chief seat of his empire.†

* It is an established fact, that the ruins at Guelma are those of Kalama. Numerous inscriptions containing the expressions KALAMENSES, KURIA KALAMENSIUM, leave no doubt on the matter ; and a passage of St. Augustine settles the question for ever. In his dispute with Petilianus and Crispinus, bishops of Constantina and of Kalama, the saint makes use of the following terms : " Inter Constantinam, ubi tu (Petilianus) es, et Hipponem, ubi ego (Augustinus) sum, Kalama, ubi ille Crispinus est, vicinior quidem nobis, sed tamen interposita." It follows from this that Kalama was between Hippo and Constantina, but nearest to the former : this is the precise situation of the ruins at Guelma.

The Romans built so many towns in Africa, that history could not record them all. It scarcely mentions Kalama, whose remains, however, attest its importance. Paulus Orosius relates that it was under its walls, which enclosed the treasures of the kings of Numidia, that the proprætor Aulus Posthumius, who sought to seize them, was defeated by Jugurtha. Sallust, speaking of the same event, calls the town that the Roman general proposed to besiege Suthul ; hence it has been very naturally inferred that Suthul and Kalama were the same town ; but the circumstantial description given by Sallust of the nature of the ground is in no degree applicable to that of Kalama. It is possible that there may have been two towns of that name in Africa, or Paulus Orosius may have made a mistake.

Baron Baude has indulged in a long dissertation in connexion with the true locality of these cities, and of the various positions that were occupied, illustrated, and disgraced by the contending armies in the Jugurthine war. " Ghelma," he says (i. p. 192), " is on the united but rather steep slope of a hill (côteau), 1500 metres from the river, and its enclosure contains 7 or 8 hectares (20 acres) ; but save the remains of a very large building, the walls are reduced to hardly the height of 2 metres. Outside of the enclosure are the remains of a theatre, a temple, and of some other constructions devoted to the public service. Sallust's description of Suthul is as follows (Jug. Bel. c. xxxvii.): " Quod quamquam et sævitia temporis et opportunitate loci neque capi neque obsideri poterat ; nam circum murum, situm in prærupti montis extremo, planities limosa hiemalibus aquis paludem fecerat." But there can be no marshes, observes Baron Baude, on the côteau of Kalama ; and M. Barbié du Bocage, in his Geographical Dictionary, attached to his excellent edition of Sallust (1813), agrees with Baude that Orosius must have been in error in identifying Kalama with Suthul, whose situation is a complete problem. M. du Bocage is disposed to think that Suthul is the same place as Sufetula, in the Itinerary of Antoninus, still called Sbaitla, a place in a similar situation to that given to Suthul (see Shaw's Travels, v. i. p. 260) ; though President de Brosses interprets the Phœnician name Suthul as meaning the city of eagles, and places it far south of Constantina. See M. du Bocage's Dict. Geogr., annexed to his Sallust, pp. 279-280.

Kalama was more than once a source of trouble to the holy bishop of Hippo. The Christians by whom it was inhabited were principally Donatists ; and a bishop of that sect, Crispinus, filled its episcopal throne. There was moreover a considerable number of pagans there, and they displayed a hatred to the new creed that all the severity of the imperial edicts was sometimes unable to restrain.

† To help our conjectures, we insert the following curious inscription :

VNA . ET . BISSENAS . TVRRES . CRESCEBANT . IN . ORDINE . TOTAS .
MIRABILEM . OPERAM . CITO . CONSTRVCTA . VIDETVR . POSTICIVS .
SVB . TERMAS . BALTEO . CONCLVDITVR . FERRO . NV ... VS . MALORVM .
POTERIT . ERICERE . MAN . PATRICI . SOLOMON . INSTITVTION . NEMO .
EXPVGNARE . VALEVIT . DEFENSIO . MARTIRO . TVeT ... R . POSTICIVS . ILE .
CLEMENS . ET . VINCENTIVS . MARTIR . CVSTOD . INTROITVM . PP. V. 7 .

This barbarous Latin, still more obscured by the mutilation of the characters, is difficult to be rendered by an exact translation. We shall only try to give its general sense, and throw some light on the age of the fortress.

The first line alludes to the thirteen towers. The second expresses admiration at the quickness with which they were built. The third, which is the obscurest of all, seems to imply that later this defensive work was completed by a wall of enclosure under the thermæ, or hot-baths. The fourth,

Hence it was the patrician Solomon who restored the towers and walls of Kalama (A.D. 540). But as this inscription has evidently been displaced from its proper position, the enclosure of Solomon must have been thrown down and rebuilt since his time.

It probably resulted from the great earthquake, of which so many traces remain at Kalama; but in any case, its construction in its present form was posterior to 540.

Another monument remains, which may, very probably, be the church of Kalama, though this is simply a conjecture. It was in a very dilapidated state in the autumn of 1836; and when visited afterwards by M. Berbrugger, a great part of it had been pulled down. Still he thought he could trace the architectural design, consisting of a nave, collateral chapels, and transept. Large stones were employed for the vaults and the facing of the walls, whilst the rest was filled up by small stones mixed with cement, and broken at intervals by a horizontal bed of tiles.

This monument does not present any confusion of design or materials like the castle; hence it is not a reconstruction by barbarians.*

---

which is very clear, can be translated thus: "No one will be able to storm the work made by the patrician Solomon." This man was a general, whom Justinian sent twice into Africa to consolidate the conquest of Belisarius. Procopius, his friend, also informs us that during his second sojourn of four years, he helped to restore the walls of all the towns. The inscription winds up by an invocation to the martyrs Clement and Vincent.

* Baron Baude is disposed to regard the neighbourhood of the stream of Bouinfra, opposite Ghelma, as the spot where Metellus and Jugurtha met in battle (Sall. B. Jug. c. xlvii.-liii.). After he had passed over the same ground three times, his conjecture was confirmed. In order to carry on the severe war (bellum asperrimum) that he prepared for Jugurtha, Metellus had to march into the heart of Numidia, on Suthul and Cirta. He was proceeding from Vacca, the Bedja of the present day, at 15 leagues to the east of La Calle; he had not taken his road by the mountains, because it was by that covered country that the Numidians lost a march on him; he, therefore, must have advanced by the plain of Bona, and the Bouinfra was on his road. Jugurtha, on the other hand, must have wished to defend, in an advantageous position, the towns which were threatened by Metellus; and the hill of Bouinfra was admirably suited for this purpose. From this point, in fact, the Numidian army commanded the plain; it covered Suthul and Cirta; and was so placed, that without leaving the shelter of the mountains, it could watch all the movements of the Romans. Lastly, in case of misfortune, its retreat into the valley of the Seybouse was secured by the Col of Mouelfa. If it were certain that the flumen Muthul is the Seybouse, all difficulties would be removed; but we only know the Latin and Punic names of this river, the Ubus and Rubricatus; and it is not yet proved if the Numidian name Muthul, only mentioned this once in history, is the name of the same river. At all events, the Muthul, which Metellus had left behind him, was a sufficiently large stream to deserve the appellation of flumen, and flowed from the south to the north. But the Mafrag and Seybouse can be the only rivers that answer to this description, and it cannot be the former in this case. Between the Mafrag and Seybouse there does not exist any hill standing out in the plain, and resembling that described by Sallust. If the battle had taken place between the two rivers, the historian would not have failed to mention so characteristic a circumstance. Nor can it be supposed that Jugurtha would have chosen so unfavourable a spot for the Numidian army; and it would, moreover, be impossible to make the manœuvres of Rutilius, the lieutenant of the consul, agree with this hypothesis; for Sallust states, that "Metellus coming, after having passed the Muthul, into a dry and desert plain, caused Rutilius to turn back and go to establish the camp on the river. In this situation the consul had before him a hill covered with myrtles, wild olives, and other shrubs, and which was detached transversely from the mountains, and advanced some distance into the plain." Nor would it be possible to give a more exact description than this of the river, and the hill of Ascours, which terminates like a jetty in the Bouinfra. It was there that the Numidians lay in ambush; and it was, no doubt, from the Djebel-el-Ousth of the Arabs that Metellus discovered them. Coming from the fords of the Seybouse, he was crossing the plain obliquely, and his right flank must have been, as Sallust says, nearest to the enemy. Isolated in the middle of the plain, the Djebel-el-Ousth affords a very strong military position; and it can be easily conceived how, as soon as he was exposed by the movement of Metellus, in coasting along the hill on which the Numidians stood, Jugurtha, in order not to leave at the disposal of the Romans so advantageous a position, caused it to be occupied by 2000 foot-soldiers. The manœuvre by which Bomilcar betook himself there with the mass of the Roman army, and the division commanded by Rutilius, becomes now equally intelligible. A single circumstance seems difficult to explain; it is, that according to the Latin text, the distance from Muthul to the hill that is parallel to it would be about 29,450 metres (32,193 yards), ferme millia passuum xx. (the Roman mile is equivalent to 1472·50

Besides the theatre, the castle, and the enclosure, there was also a pretty antique fountain at Kalama, which has, however, been demolished by the modern Vandals of the place to assist them in their own erections. It contained on two stones the following inscription: M. IVNIVS. RVFINVS. SAB. This monument may have been probably erected by a person bearing that name; and must have contained four basins, presenting at its base a shape similar to a small *x*. M. Berbrugger found some Corrinthian capitals amongst its materials.

Kalama contains many more interesting remains, but our space forbids any farther description of them.*

The ruins of Announah,† near the Ras-el-Akbah, and at no great distance from the road from Guelma to Constantina, present several objects of interest to the archæologist. They were visited by Shaw and Peyssonnel, who give very imperfect descriptions of them,

ROMAN GATE. (ANNOUNAH.)

owing to the cursory nature of their visits, in company with the Turkish tax-gathering forces.

An immense ravine extends from the Seybouse to the Ras-el-Akbah, sending out

metres); while from the Seybouse to the Bouinfra there are only 16,000 metres (17,826 yards). But we may be allowed to remark, that the account of the battle corresponds better with the real state of the place than with this announcement of distances. If there had been an interval of 7 leagues from the field of battle to the camp, how, in the middle of the action, could Bomilcar have feared that Rutilius, informed of the critical position of his general, would have come to his help? How, after a severe struggle, which only finished at the close of day, could Metellus have thought of retiring at night into the camp prepared by Rutilius? It was already a good deal to have gone more than 3 leagues. (*Jug.* c. xlvii.-liii.) Baron Baude, ii. p. 99.

* Berbrugger, part iii.

† On Announah, see Berbrugger, part iii. Baude, vol. ii. c. 9. Dr. Shaw, Peyssonel, St. Marie, &c.

many lateral ravines. One of the latter, passing and widening between two mamelons, leads from the road of Constantina to the ruins of Announah. On the left mamelon lie scattered the remains of a considerable monument, to judge from the number of columns, capitals, &c. that lie strewed around. Leaving these remains to the left, and advancing to the town, you meet, on both sides of the way, a number of tumular stones with inscriptions, the first of which alone was copied by Dr. Shaw.

After issuing from the ravine, you perceive the whole extent of the ruins of Announah. The chief of these consists in a kind of triumphal arch and a number of arcades, whose arch springs directly from the soil, and which appear to be coarse and rude attempts at reconstruction by the hands of barbarians, who have mixed up all styles and materials in a lamentable disorder.   There are, moreover, at Announah, the ruins of a church, which is a still more curious example of this spirit of confusion, being built of blocks of stone and marble of all sizes, while shafts, and capitals of columns, and fragments of sculpture, are fitted into the wall. This specimen of Byzantine architecture speaks volumes on the disastrous effects of the Vandal invasion, which must have destroyed not only monuments, but the very appreciation of art.

The town of Announah stood in a considerable valley, and its circumference can be probably ascertained by the tumular monuments that surround it.   To the left, as you approach it from the road of Constantina, are vast cisterns; and a little farther on, the remains of a Roman road that descends towards the Oued-Cherf, which is probably the road to Hippo Regius, joining the great Roman road whose remains are seen at Hhammam-el-Berdâ (Aquæ Tibilitanæ).*

The reader is already aware that Constantina was once the ancient Numidian city of

---

* The ancient name of Announah is unknown. It may probably be an Arabic corruption of Annona. This town, being in an eccentric situation, is not mentioned in the Itineraries, which only give the names of the stations on the great road. All that the Itineraries record of the road from Hippo to Cirta is as follows :

	Miles.	Leagues.
Hippo ad villam Servilianam	25	9½
„ Aquis Tibilitanis (Hhammam-el-Berdâ) .	15	8¼
„ Cirtam	54	20½
	94	38½

We omit here a fraction of one-fortieth, by taking the Roman mile as 760 fathoms.

A great number of funereal inscriptions were found at Announah by M. Berbrugger; of which 13 give the age of the deceased (two individuals had lived 90 years), and prove the healthiness of Announah. As one specimen will answer for all, we present the reader with the epitaph of a venerable lady named Victoria.

HELVIAE H. F. O.

VICTORIA

V. A. LXXXX.

H. S. E.

Other inscriptions, some of them containing many embarrassing abbreviations, show how considerable was the number of citizens of the tribe Quirina that inhabited this unknown city; e.g.

D   M

POMPEI POMPE VETTIA

S. LELIQVIRIVS LELIQVIR

FIL . QVIR.

ONORASERTIM.

The remaining inscriptions may be seen in M. Berbrugger's valuable work on Algeria, part iii. pp. 24, 5.

Baron Baude describes Announah as lying halfway up the Ras-el-Akba, on a natural terrace bordered with precipices, commanded by vertical rocks, and only accessible on one side. This singular town, of which the ancient name is unknown, seems built in this out-of-the-way situation merely for the sake of the fine view. The ruins are also in a much better state of preservation than those of Kalama. To the north, and under the walls, you find a zone of tombs, consisting of masonry, with simple inscriptions, and many crosses showing the religion of the people. C. ix. p. 2.

Cirtha.* On the right bank of the Ouad-el-Rummel are six arches, being all that remain of an ancient Roman aqueduct. They are built of blocks of calcareous stone, the largest pillar being upwards of 65 feet high.

Among the principal ruins is an old Roman causeway, vestiges of which are found in several places. This road is paved with lozenge-shaped stones of various dimensions; but most of them are about 1 metre in length, 60 centimetres (23·40 inches) in width, and 12 (4·68 inches) in thickness. It is bordered by a little parapet of about 40 centimetres (15·60 inches) from the pavement.

A volume would scarcely be sufficient to describe all the antiquities daily to be found in this remarkable city. But there is one singular edifice that we cannot omit. Within the walls of the Casbah or citadel (belonging to most Algerian cities) is an ancient church of Byzantine architecture. Several of the columns have fallen down, but a portico is still standing, which, however, seems to have been built on the site, and with the materials of a prior edifice.†

Tathubt, bordering on the Ain-yac-coute to the N.E., is about four leagues from Om-Oley (Sinaab), and eight to the S.S.W. of Constantina. This has been formerly a considerable city, but at present it is almost entirely covered with earth and rubbish. Tathubt seems to be the same as the Tadutti of the Itinerary; and lying between Lambæsa and Gemellæ, as the ancients called Tezzoute and Jimmeelah, may lay claim to this situation.

Ten leagues to the south of Tagou-Zainah, and twelve from Medrashem,‡ are the ruins of the ancient Thubana, which may be probably identified by the present name of Tubnah, and Ptolemy's position. These ruins stand almost in the same meridian as Jgilgilis : the city was situated in a fine plain near Bareekah and Boomazooze, but the few existing remains are too much buried in sands for the explorer to be able to estimate its extent.§

* Pananti observes, that " la citta, per le antichita che conserva, piu degna d' esser vista, si e Constantina" (ii. p. 18).   Blofeld (p. 59) and St. Marie (p. 232) both give a summary account of its historical vicissitudes ; and Dr. Shaw had minutely described its principal antiquities more than a century ago.   Bochart and President de Brosses maintain that the name of Cirtha comes from *karth*, which signifies town.   Its long siege and capture by Jugurtha are described in Sallust.   It was then adorned with fine buildings, which it owed to the splendid reign of Massinissa.   The Romans made it generally their winter-quarters, and the capital of Numidia; though Hiempsal II. and Juba, his son, preferred Zama.   Sextius held it when the party of Brutus and Cassius besieged him there.   But he was relieved by the soldiers of Sittius Nucerinus, who sided with Cæsar and Octavian, his nephew, and who gave their commander's name to Cirta, calling it Sittianorum Colonia ; a name which it soon changed for Cirta Julia, till Constantine honoured it with his own.   Barbié du Bocage, Dict. Geogr. to Sall. Jug. p. 195.

† We have already alluded to the ancient palace of the Beys, a quarry of Roman antiquities.   The chief gate of the four is on a neck of land facing the S.W., and about a furlong broad.   This spot is quite covered with walls and ruins down to the river, and along a strip of plain ground parallel with the deep valley, and already described.   Such was ancient Cirta; but the present city is not so large.   In the centre of the city you also find cisterns for the water brought from Mount Physgeah by an aqueduct, a great part of which remains, and is very sumptuous.   The cisterns, which are about 20 in number, make an area of 50 square yards.   The gate before mentioned is built of a beautiful red stone, not inferior to marble, well polished, and shining.   The side-posts are mostly moulded in panels; an altar of pure white marble forms part of the neighbouring wall; and the side of it in view presents a well-shaped simpulum in a bold relief.   The gate to the south-east is in the same style and design, much smaller, and opening on the bridge over this part of the valley.   This was a masterpiece in its kind.   Below the gallery, between the two principal arches, you see in a bold relief, and well executed, the figure of a lady standing over two elephants, and a scallop-shell as canopy.   The elephants face each other; and the lady, whose hair is in curls, raises up her petticoat and looks scornfully on the city.   Pananti, who was *never there*, says that in his time, " V'e un bellissimo arco trionfale che si nomina Cassir Goulah, il castel del Gigante, d'ordine Corintio."   This monument is accurately described, with an engraving, by Dr. Shaw; and has been minutely analysed by St. Marie, Blofeld, Madame Prus, Borrer (p. 352), and all recent travellers.

‡ Two contiguous and ruinous cities 8 leagues south-east of Constantina, with a triumphal arch supported by two Corinthian columns.   Blofeld, p. 64 et seqq.

§ On the plateau of Soumah, near Constantina, north-east, stands a Roman monument, which consists of a thimble raised on a cylindrical base, and surmounted by four broken pilasters, between which

Among the numerous other remains of this province, we have still to notice the building called by the French the tomb of Syphax, and by the Arabs Medrashem, situated near the Ain-yac-conte, or Diamond Fountain, on the road from Constantina to Betna. It was visited in 1846 by Mr. Borrer,* who describes it as having a circular form, the exterior being built of finely-cut stones from 3 to 4 feet square. This exterior coating has been torn away in some places, and you see interior layers of much smaller stones. This monument has a diameter of 40 or 50 feet, and its circular base is ornamented with pilasters, with plain squared capitals supporting a heavy cornice which is perhaps 20 feet from the ground. Above this the roof, made of less massive stones laid in regular gradations, runs up to a point. The east face of the base has been a portico, but is now nearly buried in sands brought by the sirocco winds, as well as by stones fallen from the monument itself. This monument is a great object of superstition with the Arabs, who think that it contains a great treasure. A certain Bey of Constantina is said to have battered it with great guns, but to no avail. They think that it is guarded by Jins, or genii.†

There are the remains of several Roman posts on the route, and one in particular with numerous Roman coffins; and you can trace the vestiges of a Roman road, in some places in a perfect state, between Constantina and Betna.‡

The ruins of Lambæsa are situated about two leagues east of Betna. Borrer§ went there with General Herbillon and 50 dragoons in 1846. This was a fine old city (having 40 gates, from each of which, according to tradition, 40,000 Arab horsemen issued in time of war), lying in a nook at the northern base of the Djebel-Aouress. Its remains are very extensive, the best consisting in a temple of Esculapius, several gateways, and three arches of an amphitheatre. It contains very many finely-chiselled inscriptions, and a furious antiquary might spend a century there with profit. On one stone, in very large characters, is the name of Alexander Severus, during whose reign a council was held at Lambæsa to condemn an unfortunate heretic, Privat, who, like Servetus, was of course in the wrong. Eighty Christian bishops attended this council: and it appears that much blood of martyrs was shed at Lambæsa during the persecutions of the conservative government under Severus, Valerian, Galerius, and others; and much injury was done to this city after the edict of Diocletian at Nicomedia had been promulgated.‖

---

there ought to be a statue, which was in the centre of a rotunda encircled with columns, whose fragments strew the surrounding ground. The details of this construction have not the elegance and correctness of the monuments of Rome.

* Campaign, &c. p. 355 et seq. Blofeld, pp. 64-69. Dr. Shaw.

† Dr. Shaw supposes that the Koubber Romeah, near Algiers, is of the date of the Numidian kings; and this ruin of Medrashem, which is of the same form, may not improbably be of the same period; but it is not probable that it is the tomb of Syphax, who was taken prisoner by Lælius, and died at Alba.

‡ This was no doubt one of their great roads, passing through the city of Diana (now Zanah), and leading from Lambæsa to Sitifis (Setif) and Constantina, and from thence branching to Saldæ (Bugia).

§ Dawson Borrer's Campaign, p. 355 et seq.

‖ Blofeld, p. 64-69. Dr. Shaw's Travels in Barbary, i. 126. Exploration Scientifique, Archéologie.

Our space prevents us from dilating on the numerous and interesting remains in the regency of Tunis, and in the Cyrenaica. The reader who wishes to study the relics of Carthage, Sicca Veneria, Vacca, Sufetula, &c., must consult the works of Dr. Shaw, Dr. Russel, Leo Africanus, Heeren, Russel, the Lady's Diary, Greville Temple, Captain Kennedy, vol. ii., &c. An admirable account of Cyrene will be found in Captain Beechey's excellent survey of north-western Africa; and those who wish to read of the glories of Kairouan, Mehadia, and the African khalifs and emirs, can consult Gibbon (ch. lii.), Abulpharagius, Renaudot, Fabricius, Asseman, D'Herbelot, Casiri, Middeldorpf.

To give the reader a slight idea of the former intelligence and civilisation of the Moors in the days of their majesty, we may just mention that Kairouan was once filled with palaces and schools; that in the library of Cairo, the MSS. of medicine and astronomy amounted to 6500, with two fair globes, one of brass, the other of silver; that the royal library consisted of 100,000 MSS. elegantly transcribed and splendidly bound, which were freely lent to the students in the capital, as well as at Kairouan and

SECTION II.

Language.

WITH reverence we approach the ancient and venerable tongues of Northern Africa, but mostly the Semitic, of yore the speech of angels, and the vehicle of the Almighty Himself when He walked with man and spake unto the fathers. The accents of tenderness and love transcending the heart of man, utterance of a sweetness emanating from higher harmonies, flowed in the soft Syriac stream from Immanuel's lips; and that mysterious writing on the wall, the warning of the despots once again startling the vision of the New World, was traced in the primeval Ninivean characters affiliated with the great Aramæan family; and lastly, the glowing yet sublime language of the Koran must ever command the respect and admiration of Christian charity.

The languages of Algeria fall under four great heads:

1st. The Berber and its dialects.
2d. The Arabic.
3d. The Turkish, now almost extinct.
4th. The Negro idioms, below criticism.

Following a chronological order, we shall begin with the most ancient African tongue, the Berber.

The Berber tongue is subdivided into sundry dialects, including,

1st. The *Zenatia.* This dialect exists among the Kabyle tribes, who, advancing towards the west, extend from Algiers to the Morocco frontier of Algeria.

2dly. The *Chellahya.* This idiom is used by almost all the Kabyles of Morocco.

3dly. The *Chaouiah.* This modification of the Berber belongs to all the Kabyle tribes who are mixed up with the Arabs, who, like them, live under tents and keep numerous flocks. Many Arabic words have naturally insinuated themselves into this dialect, which is greatly diffused in the province of Constantina.

4th. The *Zouaouiah.* This language is spoken from the country lying between Dellys and Hamza as far as Bona, and represents the old national idiom in its greatest purity. A slight difference may always be traced amongst the tribes to the east of Djidjelli, arising from their commerce with the Arabs. Hence these tribes are considered by the *pure Kabyles* as degenerate Kabyles (Kabails-el-Hadera).

The Berber alphabet was long thought to be lost, and at the present time there does not exist a single book written in Berber character. The copies of the Koran, &c., found among the smoking villages of the Beni-Abbess by Dawson Borrer, were all Arabic versions.* The Kabyle tolbas (and they are numerous) maintain that all their Mss. and literary monuments disappeared at the capture of Bugia by the Spaniards (1510). But this assertion cannot stand the test of criticism, though it is easier to refute it than to replace it by another and a sounder theory. At the present day the Berber is only written in Arabic characters; and it is said that the Zaouia of Sidi-Ben-Ali-Cherif, of whom we have spoken before, possesses many Mss. of this description.†

Alexandria; that the city of Morocco at one period contained 700,000 inhabitants; that the Ommiades in Spain formed a library of 600,000 vols.; that Andalusia alone could boast of 70 public libraries; and that Cordova, with the towns of Malaga, Murcia, and Almeria, could boast of having produced 300 authors. Crichton's Arabia, ii. c. 3.

* Campaign in the Kabylie, 1848.
† La Grande Kabylie, pp. 7-9 (1847).

The following are the characteristic differences of the Berber and the Arabic:

Arabic has but one article for all genders and cases,—*el;* the Berbers have the masculine and

Arabic is at once a rich and a poor language. It is *poor*, inasmuch as, being the child of the desert, it has no words to express a great number of ideas that are only imported by

---

feminine. The masculine consists in the letters *a*, *ou*, *i*, represented by the elif $\mid$ and placed as affixes before the word. The feminine article consists in the letter *t*, pronounced like the English *th*, placed as an affix and prefix to the word. We here present a few examples :

	Berber.	Arabic.
Man	Argaz . . .	Er-Radjel.
Woman . . .	Tamettout . .	El-Mra.
Male child . . .	Akchich . . .	Et-Tfel.
Female child . .	Takchicht . . .	Et-Tofla.
Male slave . . .	Akli	El-Khedim.
Female slave . .	Taklit . . .	El-Khâdem.
Young husband . .	Isli	El-Arous.
Young bride . .	Tislit	El-Arouca.
Ox	Afounes . . .	El-Bgueur.
Cow . . .	Tfounest . . .	El-Begra.
Ass	Ar'ioul . . .	El-Hemâr.
She-ass	T'rioult . . .	El-Hemâara.
Camel	Alr'em . . .	El-Djemel.
She-camel . . .	Talr'emt . . .	En-Nâza.
Lamb	Izimer . . .	El-Khrouf.
Sheep	Tizumert . . .	En-N'adja.
Kid	Ir'îd	Ed-Djedi.
Goat	Tara't . . .	El-Ma'za.

The masculine article becomes commonly *i* in the plural, as, *irrgáz-en*, men; and the feminine usually *ti*,—*tifounâs-en*, women. The masculine ending of the plural is *en*, the feminine *ín*. There are, however, many exceptions to this rule; *e.g. akli*, male slave, becomes in the plural *aklán*; *tarát*, plural *tiretten*, goats. There are also many very irregular plurals, such as *vulli*, plural of *tikhsi*, sheep.

Almost all words are hermaphrodite in Berber, and can receive the masculine or feminine gender. They are not, however, used indifferently, but according to natural laws. In all the animal kingdom, save man, civilised or plucked of his feathers, the male commands the female; by his size, beauty, and strength, he is naturally chief and master. The Berber language always reproduces this natural law, the feminine being a diminution of the masculine. Possession or dependence is expressed by an initial prefix to the second word. This is one of the letters *m* and *n*, or the diphthong *ou*. If applied to persons, all three may be used; but in the case of inanimate objects, the second *n* is alone used, and determines the genitive. Example: *Tala-m Bou Hai* (the source of Bou Hai), *Alma-n Bisri* (the meadow of Bisri), *Agmin Aklan* (the country of the negro). ('La Kabylie proprement dite,' in the Exploration Scientifique, vol. i.)

The Berber language, though one of the most ancient in the world, has never yet had a grammarian. This idiom reigns in Algeria over almost the whole of that series of high cliffs which border the Mediterranean from the gulf of Stora to the frontier of Morocco. A few hiatuses in the chain occur about the meridians of Algiers and Oran. In the province of Constantina it is found in the high plateaux that give birth to the Rummel and Seybouse ; and in the plains inhabited by the Harachta, Seynia, Telar'ma, Oulad, Abd-en-Nour, and all that part of the country, it is called, as previously observed, Chaouia. It occupies exclusively the whole ridge of the Aouress.

In the east of the Algerian Sahara, the oases of Ouad-Rir, Temacini, and Ouaregla, are inhabited by a twofold population, some using an idiom called *lar'oua*, which is the Kabyle. It is found also about the centre of the Algerian Sahara, in the oasis of the Beni-Mzâb. In the regency of Tunis it is almost confined to the little island of Djerba, in front of Gabes, about the southern frontier of that state. It occurs again in the little town of Zouâra, where the desert meets the sea, between Tunis and Tripoli, and is there called lar'oua. Going west, it is called *Chelhia* in the desert of Figuîg, and in the high and vast chain of the Miltsin, the Atlas of the ancients. It reappears in the gorges of the Djebel-Nfous, between Tripoli and Egypt, and in the solitudes of the great desert, where it is spoken by the emphatic Touaregs. All the high summits along the coast know no other language; and M. Carette observes that there is really very little difference between the Chaouia and the Chelhia, or all other Berber idioms.

We also learn from the same source that recent explorations of the desert and remotest Berber tribes (Tuaricks), have brought to light inscriptions in the ancient Berber character, which will give us the Berber alphabet, and prove another Rosetta stone to unlock the mysteries of this venerable tongue.

We trust that the French will shortly convert their swords into geological hammers, and their bayonets into antiquarian pickaxes, and that the future fruit of their razzias will be Berber inscriptions rather than barbarous atrocities. (Explorat. Scient.: La Kabylie proprement dite, by M. Carette, vol. i. pp. 49, 27, 76, &c.

civilisation ; and it is *rich*, because it possesses, on the other hand, many expressions to describe the same thought, when this thought was found in the narrow circle of the primitive wants of the Arab people.

When this language became diffused through the world, in consequence of the Mussulman invasions ; and when the Arabs, after the conquest of Syria, Egypt, Barbary, and Spain, became established in these countries, and founded separate empires in them,— it lost somewhat of the uniformity that it possessed at its cradle.

Each Arabic colony was obliged naturally to borrow from foreign and neighbouring tongues new words to express new ideas, more or less numerous according to the intimacy of its relations with more civilised states. They would also be led occasionally to distort some genuine Arabic expressions from their primitive significations, in order to express the new ideas. And since each of these distinct Arab branches led henceforward an isolated and independent existence, and only held mutual intercourse at long intervals, they would find it convenient and less irksome to adopt one or two of the words existing in the Arabic tongue to express the primitive ideas that it admitted, dropping the rest.

Now it was not probable or possible that in this selection exactly the same words should be chosen by these related but distant branches. Their choice was often directed by chance, and particular countries selected particular terms in the great division that took place of the expressions common to this tongue. Thus arose various modern idioms of the Arabic, presenting certain differences among themselves, but all derived from genuine primitive Arabic words.

The differences that may be traced, on the one hand, between the spoken' and the written language, and, on the other hand, between the dialects spoken in Barbary, Egypt, and Syria, result from a more or less accurate observance of the rules of the Arabic grammar ; from the importation of certain words from foreign tongues ; from the more special adoption of particular Arabic words by particular countries to express the same thought ; and, we may add, from idioms peculiar to different regions.

These differences are, however, less considerable than is generally supposed, particularly in what relates to idiomatic peculiarities ; and it must be admitted that these would, in all probability, have been much greater, if the Koran and its language had not been a great bond of union between all the Arab races. Nor can we avoid a feeling of surprise when we behold a tongue that has been handed down through so many ages, and countries, and events, presenting its original form and purity with such slight deviations.*

It is proper to remark,† that the greater part of the variations of the Arabic language‡ may be traced up to a common origin in the learned language, or the idiom of Modhar, which Mahomet employed to write the Koran. It is probable that this ancient language, so very rich in synonyms, of which a great number, however, are mere epithets, only acquired its astonishing richness in expressions by adding to its original fund, which was the dialect of the central tribe of the Qoreichites, words borrowed from the idioms of neighbouring tribes. The Arabs who invaded and settled in Africa brought there the varieties

---

* Grammaire Arabe (Idiome d'Algérie), by M. Alexandre Bellemare, 1850 : Introduction, p. vi.

† Berbrugger's Algérie, &c. part iii. p. 19.

‡ A specimen of the operation of external causes in modifying the Arabic dialects is presented in the idioms of Algeria. Throughout the province of Oran, at Algiers itself, and in Western Barbary, the pronunciation of the Arabic tongue is much harsher and more guttural than in the province of Constantina, and there is every reason to believe that this harshness increases at present in north-west Africa in an inverse ratio to the distance and separation of the tribes from the districts of Tunis and Constantina. It may even be remarked that the idiom of the province of Constantina has attained the maximum of softness of all the Arab dialects ; a circumstance that may be attributed to the softening influence of Roman civilisation in Numidia and Tunis.

Strong aspirations and guttural articulations, so frequent in Arab speech, are uttered with less roughness at Constantina. Some letters have even a different phonic value in the different provinces: thus the word for mountain, pronounced *djebel* at Algiers, is sounded like *jebel* at Constantina, though written the same way in both cases. The variations of idiom sometimes go still further, and entirely different expressions are used in different districts.

that distinguished the mother-tongue of their fatherland, and have made greater or less alterations in it in proportion as their connection with the Berber race has been more or less intimate. Mahomet, by establishing a unity of faith among the Arabs, laid at the same time the foundations for the unity of language, by the adoption of the idiom of Modhar, with which every well-educated Mussulman is partially, if not perfectly, acquainted: but this applies only to the language of religion and science, for in the ordinary intercourse of life every one employs the peculiar dialect of the Mahometan country that he inhabits.*

SECTION III.

Commerce and Agriculture.

ACCORDING to Dr. Shaw, the annual taxes of the regency under the Turks brought in 1,647,000 fr. about a century ago; and Shaler, the United States consul in 1822, estimates the revenue at 2,360,964 fr. (94,438*l*. 11*s*. 8*d*.)

It appears that, since 1830, Algeria had swallowed up in 1846, 100,000,000 fr. (4,000,000*l*.),† of French money; and the whole amount of the tribute squeezed out of the extreme poverty of the Arab tribes in 1846 was, in rough numbers, 5,000,000 fr. (200,000*l*.). But Mr. Borrer shows the extreme impolicy of imposing heavy taxes on the Arab tribes, and of seizing their lands, hereditarily transmitted, without remuneration, in order to found a European settlement on it.‡

The total amount of the revenue derived from the colony during the six years of Marshal Bugeaud's administration amounted to 105,000,000 fr. (4,200,000*l*.).

Since 1835, a portion of the produce of the domaine of the douanes, and of divers contributions, was appropriated to the expenses of the towns and corporations.

Count St. Marie informs us that 5,000,000 fr. (200,000*l*.) are spent every year over and above the ordinary pay the troops would receive if in France; 2,000,000 (80,000*l*.) for the navy; 2,000,000 fr. (80,000*l*.) for persons employed in the different departments of the civil service, viz. the administration of the interior, of finance, of the police, of rivers and forests, and of the clergy; finally, 1,000,000 forming a secret fund for presents

---

* See A. Gorguos' Cours d'Arabe Vulgaire, 2 vols. 1849; Bled de Braine, Clef de la Prononciation des Idiomes de l'Algerie, 1848; Ventura's French and Berber Dictionary; Hodgson's Account of the Berber Language, in the Transactions of the American Philosophical Society, vol. iv. 1834.

† A report addressed to the Emperor of France, and dated August 11th, 1853, states that the law relating to customs of Jan. 11th, 1851, has been a great benefit to Algeria, by uniting more closely the interests of France and its colony. But this law, moreover, contained provisions whose gradual development was destined to procure new advantages to both countries. The application of one of these provisions is urgently demanded at the present time, namely, the establishment of *douanes* on the frontiers of Morocco and Tunis, in order to favour the opening of a land-trade with those countries, hitherto closed. It has also appeared desirable to lower 50 per cent the duties at present levied on certain produce of Morocco and Tunis when brought into Algeria by land. St. Arnaud, minister of war, proceeds in his report to submit to the sanction of the emperor a project of a decree concerning the land-trade of Algeria. This decree, which has become law, contains 13 articles, which, among other enactments, remove the prohibition made in 1843 on the produce of Morocco and Tunis, though it is continued on the produce of a different origin. The produce of Morocco and Tunis must pass to the east through Soukara and Guelma, through Tebessa and Ain-Beida, and through Biskara; to the west, through Lalla Maghnia, Tlemsen, and Nedrouma. Douane offices and bureaux to be established at or near Bona, Guelma, Constantina, Ain-Beida, and Biskara to the east; at Rashgoun, Tlemsen, and Daya to the west. The Saharian frontier will be closed to all produce not the growth of Algeria, or the offspring of Algerian industry. We refer to this important decree for further particulars.

‡ Dawson Borrer, p. 23.

and losses. This makes a grand total of 10,000,000 fr. (400,000*l.*) annually, or 200,000,000 fr. (8,000,000*l.*) in 20 years. Yet this does not represent one-fourth of the real amount, for the 547,500 deaths must be considered that occurred in the army from 1830 to 1845. Each of these soldiers cannot have cost less than 274 fr. (10*l.* 3*s.* 4*d.*) at the hospital, for clothing, transport, &c. The custom-house duties in 1845 brought in about 400,000 fr. (16,000*l.*) per annum. Out of that sum the salaries of the persons employed in the customs' service must be paid. There is no tax on fixed property or on persons; and the contributions from cattle, levied by the troops on the Arab tribes cannot be considered as receipts, for the sale of the cattle produces very little, and the money thus raised is usually distributed among the soldiers, not much to their advantage. Specimens of this are given by St. Marie.*

The intricate web of employés is condemned as a serious evil by St. Marie and Borrer, under Louis Philippe; and this host of locusts is still flourishing under the empire. In the year 1845, 24,000 dispatches were received from Paris by the *administration civile*, and 28,000 were sent to Paris by this branch in Algeria. The number of functionaries is immense, as we have seen; and the pay of the corps in 1846 about 600,000 fr. (24,000*l.*), and since 1830, 5,000,000 fr. (200,000*l.*); whilst the European population over whom they acted only amounted to 100,000 persons. The pay of the native troops in 1845 amounted to 7,000,000 or 8,000,000 fr. (320,000*l.*)†

As foretold by the visitors in 1844 and 1845, there was a financial crisis in Algeria in 1846-7, recorded by Mr. Borrer, when the interest on capital rose to an extravagant pitch. This distress diminished many sources of revenue, save the Arab impôt, whose produce has steadily increased.

The financial legislation of Algeria has undergone great changes, especially since 1839. In September 1st, 1847, the director of finances and the directors of the interior and of public works were suppressed. Directors of civil affairs were appointed in each province, uniting the functions of the suppressed directions; and this movement decentralised the administration of finances, which is now in the hands of the prefects.

The Arab impôts of all kinds, minus one-tenth, have passed from the colonial budget to the budget of the state; and the colonial budget has taken the name of local and municipal budget; and both budgets have been centralised in the hands of the minister of war, who is the only manager (*ordonnateur*) of the local and municipal diet, and the final paymaster of all expenses.‡

The occupation of Algeria by the French appears to have injured French trade to Barbary up to 1838, but since that period matters have gone on improving.

* P. 257.

† Borrer. St. Marie informs us that a certain officer bought a house for 300 francs (12*l.*), which six months after he let for government service at an annual rent of 4000 francs (160*l.*). Usury in 1845 destroyed the trade of Algeria, by banishing all confidence, ruining the unfortunate borrowers. Not a day passes without six bills being stuck up, headed with bankruptcy. If persons in the employ of government could purchase real property ostensibly, there would be more regard to decorum. They would not go boldly to the public notaries, and sign deeds devoid of authenticity, forms for lending money at 20 or 25 per cent interest. The Jews manage better; they make up the rate of usury by bills of exchange; this at least is more modest (p. 265). Other embarrassments tend to depress commerce. For instance, whatever is required for the army, the shipping under the government, has to be accepted by a commission, to which the merchants invariably offer a gratuity to prevent articles of the best quality being rejected as bad. The following fact is an illustration of this abuse. Six vessels laden with corn for the army were in the port. A commissioner went on board to examine the cargoes, which were of the first quality; but the consignee not having paid the required fee, they were rejected. The government, it was understood, would have taken them at 17 francs (14*s.* 2*d.*) per measure. But on change next day, they were purchased all at 30 francs (1*l.* 4*s.*) per measure; and within a fortnight the government was negotiating for that same corn at 32 francs (1*l.* 5*s.* 10*d.*), the new owner having taken care to get it inspected by the right persons. In this case the transaction was good for trade, because the article was in great demand; but it must often be very ruinous.

‡ Tableau, p. 400.

G G

On the 7th December, 1835, the legal interest was made 10 per cent, in the hope of calling in competition, and cheapness in capitals; and from good information Baron Baude learnt that loans gratuitous in appearance varied in interest from 25 to 50 per cent.*

But passing to the middle ages, we are strongly reminded of the glorious union of an enlightened freedom and a humanising commerce in the annals of unhappy Italy. Before foreigners had trodden her spirit in the dust, and the church had crushed the elements of her national greatness, the republics of Italy sent forth their active commercial fleets, manned by hardy mariners of the Columbus and Gioja stamp, who bravely ploughed the Mediterranean, and enriched their native land with the produce of the East and South; while they enlightened Europe with the remnants of the Greek fire smouldering at Byzantium. In those palmy days, Pisa and Siene reckoned above 100,000 happy citizens within their walls, who, thanks to the wholesome agitation of democratic forms, were saved from the stagnation of '*order.*' Industry and science marched in the road of progress; and Italy, in the dark ages, pioneered the road of Europe to the light. But imperial France had not then strangled liberty at its birth : imperial Austria and the Roman Pontiff had not conspired in emasculating the progeny of the Gracchi.

Baron Baude, speaking of Algeria, says this coast once flourished commercially. The greatness of Carthage had no other basis than commerce : at each page of the ancient historians you find traces of the riches of towns which have afterwards fallen into the last state of misery. Such were, in the neighbourhood of Algeria, Bedja (whose markets attracted a crowd of Italian merchants), and Adrumetus, Thapsus, besides Utica, on which Cæsar could impose in passing a contribution of 13,000,000 sesterces (2,665,000 fr. or 106,600*l.*).†

The history of the treaties of commerce with Africa is very interesting. At the end of the tenth century the navigators of Pisa had treaties of commerce with the sultans of Egypt and Damascus : in 1167, being driven from the Levant and Sicily, they sent as their first consul the famous Cocco Griffi to the Emir of Bugia, and to Abdallah Boccoras, sultan of Tunis. From that period dates their establishment on the coast of North-western Africa. The archives of Florence possess the treaty in Italian and Arabic that was concluded on the 14th of the month Hreval, in the year 662 of the Hegira, between the Pisans and the Khalifs. Ultramontane barbarism and bigotry, however, eventually

---

* Baron Baude, iii. p. 7.

The history of the commerce of north-western Africa is a matter of deep interest, and to treat it in detail would trespass too much on our space. We have alluded to the trade of Carthage in the chapter on history; and it will suffice us here to refer the reader to Heeren's valuable work, *Reflections on the Politics, Intercourse, and Trade of the Ancient Nations of Africa;* only remarking that this favoured region was once a garden, the granary of Europe, and the centre of a vast and organised system of esoteric and exoteric commerce. Caravans have for ages ploughed the desert, bringing gold, ivory, and slaves to the north coast; whilst vessels freighted with the luxuries of India, spices of Araby, the fruits of the Levant, and the amber of Persia, have crowded the ports of Carthage and Hippo in the most remote ages.

In ancient times, observes Baron Baude, Carthage carried on commerce with the whole known world; and Dr. Russel asserts that at the time Carthage was most flourishing, she traded northwards directly to Britain, and indirectly to the Baltic; southwards to the Gambia by sea, and by caravans far into the interior of Africa; whilst eastwards she carried on an extensive commerce with all parts of the Mediterranean, and through the mother-city, Tyre, obtained the produce of India. She may have purchased slaves too from the Grecian slave-dealers. Her commercial relations would thus have extended over nearly the whole known world, and would only have been surpassed by those of modern Europe since the discovery of America, and of the passage to the East Indies by the Cape of Good Hope. Dr. Russel's Barbary States, Ed. Cab. Cycl. p. 22. Foreign Quarterly Review, No. 27, p. 225. See Herodotus on the trade of Carthage, p. 80. Heeren's Historical Researches, vol. i. p. 173. Heeren's Reflections on the Politics, Intercourse, and Trade of the Ancient Nations of Africa, p. 53.

† We find that " oppidum Numidarum, nomine Vaga, forum rerum venalium maxime celebratum : ubi et incolere et mercari consueverant Italici gentis multi mortales" (Jug. 47.). For Bedja, see Cæsar. de Bell. Afr. 97. Barbié du Bocage's Sallust, Dict. Geogr. p. 292.

destroyed the prosperity of this trade ; and amidst the severe struggle of the following century's crusade, the Pisan flag almost disappeared from the Mediterranean.

We shall insert below a few tables from the French official documents to show the state of imports and exports and of navigation in the colony from 1831 to 1846.*

As regards the nature of the imports, most alimentary matters have been imported in increased quantities from foreign countries. Materials such as wood, coal, &c., are almost exclusively foreign imports, whilst manufactured goods proceed almost entirely from France.

Algiers alone engrosses three-fifths of the commercial movement. Oran presents the most satisfactory results; the imports at Philippeville are steadily progressing, and those of Bona are about stationary. Mers-el-Kebir is becoming an important emporium.†

It appears that, in 1834, 437 ships entered the port of Gibraltar and 385 left it, whilst 527 anchored in its roads. 1200 vessels are reported to pass each year before Mers-el-Kebir, the best Algerian port.

On the 31st of December, 1839, the Arabs possessed 88 boats of 1123 tons burden, and 493 sailors, besides 405 sailors and 60 ships of 695 tons in the ports not occupied by the French. The number of sandals frequenting the ports of Algeria was, in 1838, 1329; in 1839, 1391.‡

* Receipts of Customs.

Years.	Imports.	Exports.	Accessory receipts.	Navigation.	Total.
	fr. c	fr. c.	fr. c.	fr. c.	fr. c.
1831	281,717 3	11,592 74	9,138 14	22,000 0	324,447 91
1837	764,902 12	6,365 24	1.278 68	220,694 12	990,419 99
1844	1,292,213 71	12,544 99	1.113 46	548,102 43	1,853,974 59
1846	2,417,151 71	8,769 98	14,151 99	868,477 62	3,306,551 30
1848	1,643,035 55	14,186 32	8,321 76	409,291 37	2,074,835 0

The sum-total of the commercial movement in 1845 was 109,851,423 fr. (4,394,056*l.* 19*s.* 1*d.*) The amount of merchandise transported by sea under the French and foreign flags in 1846 was :

In French bottoms   .   .   87,304,195 fr. (3,492,167*l.*), or 72 per cent.
In foreign ditto   .   .   .   33,196,266 fr. (1,327,850*l.*), or 28 per cent.

† France has almost monopolised the river trade, and the imports of cotton, woollen, silk, and linen tissues.

The merchandise derived from French ports amounted, in 1846, to a total of 11,906.753 frs. (476,270*l.*); whereof 563,832 fr. (22,553*l.*) were merchandise from the French colonies, and 11,342,921 fr. (361,720*l.*) were foreign goods. The total value of the exports from Algeria amounted, in 1846, to 9,043,000 fr. (453,716*l.* 17*s.* 7*d.*) The exports of oil, which are some of the most important, fell off in 1846, owing to a bad crop; and that of wool, owing to an epidemic among the native flocks. The branches of export trade that have the most improved are raw hides, leeches, silk in cocoons, medicinal herbs, raw cork from Mount Edough, and leaf tobacco. The exports from Mers-el-Kebir stand first on the list, Bona stood third in 1846, and La Calle fifth.

England exported, in 1846, 249,580 fr. worth more than in 1845, consisting entirely in manufactured tobacco carried from Mers-el-Kebir to Gibraltar. (Tableau, 1849-50, Douanes, p. 454.)

‡ In the 9th century the Kabyles and Arabs established themselves at Malta; in the 10th in Sicily, Sardinia, Corsica, the Balearic Islands, and at length in Spain. The Balearic Islands were conquered and settled by the Carthaginians, who captured Majorca (B.C. 406), whence they were expelled 200 years afterwards during the second Punic war. These islands were subsequently conquered by Gunderic, king of the Vandals (A.D. 427), who passed from Spain to Africa (429), and made themselves masters of Hippone in 435; but whose empire was destroyed by Belisarius (A.D. 534). The Balearic Islands, which were again conquered by the Arabs (A.D. 797), were ultimately attached to the crown of Arragon in 1229 by Don Jayme. Port Mahon was captured by Khaireddin in 1525, and occupied twice by the English, from 1708 to 1756, and from 1798 to 1802; was restored to Spain by the French arms under the Duc de Richelieu and M. de la Galisonniere, July 7th, 1756. Smollett, vol. iii. c. xxv. p. 242. Baron Baude, vol. i. p. 45-6. The sailors of Ivica carry on a brisk trade with Algiers in the

The latest analysis of Algerian navigation presents the following results:

Years.	French.		Algerian.		Foreign.		Total.	
	Ships.	Tons.	Ships.	Tons.	Ships.	Tons.	Ships.	Tons.
1831	123	. .	. .	. .	215	. .	338	. .
1835	341	28,524	495	3,984	1254	103,732	2090	136,240
1837	1129	100,202	1032	13,211	1204	114,664	3365	228,077
1844	2362	216,028	1510	23,340	2281	183,325	6153	422,693
1846	2523	2:7,036	1506	23,587	3078	263,182	7107	533,805
1848	2151	209,992	1846	26,145	2117	145,042	6114	381,179

The most numerous bottoms are the French, and after them the Greek, owing to the importation of grain from the Black Sea. The English and Spanish bottoms have somewhat fallen off since 1848, owing to the slackening of the trade with those countries.*

Around Bona, Djidjelli, Collo, Philippeville, and especially La Calle and Tabarca, have been, are, or will be, the chief settlements for coral fishing. Tabarca belongs to Tunis; but the 5th article of the treaty of August 1830 gives the French the right of fishery as far as Cape Negro, and 7 leagues (17½ miles) beyond.

From Cape Bon to the Zeffanine isles, for 300 leagues (750 miles) there is a fair field and good chance for fishing it.†

---

present day. This little island alone has 20,000 inhabitants, and 60 or 80 xebecs (Mediterranean craft rigged with lateen sails). Ib. vol. iii. p. 91. They were rather populations than armies that went there, and they must have had a considerable naval *matériel*. They also carried on at that time as extensive a coral fishery as the French do at the present time.

* One of the most important branches of maritime commerce on the coast of North-western Africa is the coral fishery. The space set apart for the fishery is situated between Cap Roux and the Cap de Fer. All the French coral-boats come from Corsica; but though they have no duty to pay, they do not seem to gain more than the others. All the Sardinian gondolas start from Rapallo at six leagues from Genoa, and they are either commanded or accompanied by the proprietor. The expense of fitting out is sometimes shared between fifty persons, and amounts to about 800 fr. (32*l.*) per boat. Each shareholder gets a part of the clear produce of the fishery. Two parts revert to the boat, 1½ to the patron or master, and 1 to each of the sailors; a part gives about 200 fr. (8*l.*) profit. The Genoese are the most enterprising fishers, being still sea-dogs, somewhat of the old Columbus school. Most of the Tuscan boats belong to ship-proprietors'(*des armateurs*) of Leghorn; and their crews consist partly of men from Torre del Greco, near Naples, and partly of men from the vintage, after their work is over in the winter. The latter go for 70 cents (sevenpence) per day. If the boat founders, the capital is lost. The money advanced on the armament amounts, on the average, to 4500 fr. (180*l.*) per boat, at a profit of 2⅔ per cent per month. The patron is paid 500 fr. (20*l.*) for the whole fishery; the common sailors 147 fr. (5*l.* 19*s.* 2*d.*)

A few boats attend from Sicily; but Torre del Greco, under Vesuvius, is the seat of the most considerable trade. The proprietors generally go themselves with the boats, and a new Neapolitan boat is worth 800 ducats or 3360 fr. (140*l.*); the construction is perfect; and the material, consisting of oak, is excellent. 800 or 900 fr. (32*l.* to 36*l.*) per boat is the ordinary profit of the summer fishery, which begins on the 1st of April, and ends on the 30th of September. But the expenses last eight months. The amount of coral fished in 1839 amounted to 13,805 kilogrammes (30371 lbs.); the duty on it was 138,074 fr. In 1838, 33,080 kilogrammes (72,776 lbs.) were fished by 245 boats, with a profit of 282,884 fr. (15,315*l.* 10*s.* 10*d.*) The coral is classed into six qualities, the price varying at Leghorn for the last 15 years as follows :—

Per kilogramme (2·20 *lbs. avoirdupois*).

Moutri	from	14 fr. 71 c. (12*s.* 3*d.*) to 19 fr. 4 c. (15*s.* 10¼*d.*)
Sousmontre	,,	11 fr. 40 c (9*s.* 6*d.*) to 15 fr. 20 c. (12*s.* 8*d.*)
Exart	,,	5 fr. c. (4*s.* 6½*d.*) to 6 fr. 72 c. (5*s.* 7*d.*)
Barbaresco	,,	3 fr. 36 c. (2*s.* 9½*d.*)
Tenegliatura	,,	3 fr. 36 c. (2*s.* 9½*d.*) to 2 fr. 72 c. (2*s.* 3*d.*)
Terraille flottante . · . .	,,	1 fr. 63 c. (1*s.* 4¼*d.*)

† The latest statistics of the coral fishery, according to the *Tableau* (p. 540), present us with the following results :—

The coral fishery, as previously, was carried on in 1848 in the sea near Bona and La Calle, the attempts in the western waters not having succeeded. Most of the produce went, as usual, to Tuscany and Naples; and the fishermen consisted chiefly of Tuscans and Neapolitans.

The Leghorn dealers send much coral to Russia; the rest goes to Gallicia, to India, to China, and Japan by London, and to Morocco. They have agents in all those countries,—the enterprising progeny of Marco Polo on all hands vindicating their right to liberty and independence by the light and learning they have shed over northern barbarians. Wonderful are the ways of Providence in girding the earth with relays of " Lombards," who spread civilisation and humanity through trade to the walls of China and Jeddo, by means of a worm's refuse ; whilst at home their unhappy land is trampled under the foot of the stranger, as a reward for their cosmopolitan energies. Yet compensation is the law of eternity; and Italian traders may yet turn the tables on czars and kaisers.

The large round coral is sent to Russia, the pink of the first quality to China, that of inferior quality to Poland, the *barbaresca* and *roba chiara* to India. The Algerian Jews employ per annum about 200,000 fr. (8000*l*.) worth, which they prepare and send into the interior of Africa.*

Beautiful is the provision made by Providence to meet the wants of progressive humanity. Reason and industry unlock the treasures of the universe, and a wise direction of power would strew the earth with affluence. Chemistry, the child of Arabia, is giving us the sovereignty of the mineral kingdom and the gases; nor can we doubt that the animal and vegetable will soon bow to our sway, and that man will be monarch of all that he surveys.

The coast of Algeria is, particularly in its eastern part, one of the most fishy in the Mediterranean. The tunny fishery formed in ancient times the riches of the coasts of Spain, Gaul, Italy, Sicily, and the islands of Greece. The tunny, by a divinely implanted and infallible instinct, follows with avidity the migrations of those kinds on which it preys. The shoals of tunny formerly entered the Mediterranean in March by the Straits of Gibraltar; they then used to follow the coast of Spain to the vicinity of Carthagena, occasioning an immense prosperity to that part of the country, whose population lived on the doomed fish, like the silk-worm on the mulberry-leaf. But since the earthquake of Lisbon in 1755, they have left the coast of Andalusia, to approach that of Africa. The regency of Tunis alone has thought of profiting by this change.

Passing to the internal trade of the colony, we shall still tread on the heel of Italians, who, on the pathless deep and on the trackless sands, had boldly sounded the unknown in search of fame and wealth, at a time when the rest of Europe slumbered and slept a commercial and social death, save the republican Hanse Towns and Flemings, the Albigeois Socialists, and the democratic Arragonese and Catalans.

Before the French occupation, the Regency had its local caravans; thus every time that travellers went from one town to another, they put themselves under the protection of troops, if possible, or they formed associations for mutual protection. This organisation

Years.	Number of Boats.				Proximate value of the fishery.	Customs.
	French.	Neapolitan.	Sardinian.	Tuscan.		
					fr.	fr. c.
1840	1	43	13	38	666,450	102,524 40
1844	3	129	20	47	1,387,000	217,673 20
1848	2	118	15	18	{ 794,600 { £31,784	{ 128,400 0 { £5136
Total of boats in 1848, 154.						

* Baron Baude, vol. i. c. vi.; Blofeld; St. Marie.

led to periodical rendezvous, like the fairs of the middle ages in Europe. All the
caravans from the interior of Africa regulate their movements by the Great Moghrebin
Kafila, which proceeds annually from Fas to Mecca, leaving the Atlas to the north, and
which, since the French advances south, has continually swerved more into the desert.*
From Fez to Gadamez, in the regency of Tunis, this caravan divides itself into several
branches, each of which drops passengers and goods for transverse caravans. In this
manner a chain of relations is formed throughout all parts of North Africa. The Pisans
joined themselves to this chain, and accompanied the caravans more adventurously than
the countrymen of La Peyrouse and Mungo Park. There were three chief points for
the starting or intersection of caravans outward and homeward bound: 1, Constantina;
2, Oran; 3, Medeah. The point of junction of the caravans of Constantina and Algiers,
or, more accurately speaking, Medeah, with the great caravan from Fas to Mecca, was at
Ouerghela, the most southern town of the Regency of Algeria. It is 150 leagues (375
miles) from the coast, and 2° of east longitude from Paris. Leo Africanus, describing
it, says: " It is a very ancient town, built in the desert of Numidia, surrounded with a
wall of raw bricks, filled with beautiful houses, and well peopled with workmen; and its
inhabitants are very rich. Most of the people are black," he adds; "not through the heat
of the climate, but because they have commonly commerce with black slaves, which
occasions their breeding some fine children." Here we have another strong argument
in favour of a mild system of slavery, advocated by philanthropy and physiology as a
valuable means of emancipating the genuine negroes through a noble cross-breed, forming
a bridge for them to civilisation, and a link between white and black. Leo continues,
speaking of the mulatto citizens of Ouerghela: " They are pleasant and liberal, and very

---

* Le Grand Desert. " Since the Christians have appeared in the Sahara, as soon as the Great Caravan reaches their meridian, it strikes much farther south, by Guelea, Ouargla, Souf, and Touzer
whereas it used to pass to the north of the Djebel-amour, and by El-Aghrouat" (p. 121). See also Baron
Baude.

kind to strangers, because they could obtain nothing without their aid; all necessaries, such as grain, salt-meat, suet, cloth, linen, arms, knives, and every material, coming from abroad. They have the same veneration for their lord as if he were a king; he has for his guard about 2000 horse, and his revenue is about 150,000 ducats (25,714*l.* 5*s.*)."* Dapper adds that, independent of the contributions, the Sheikh of Ouerghela gave every year thirty black slaves to the Pasha of Algeria.†

Every summer the flocks that people those regions transmigrate to seek pastures in the north, and escape the burning heat of the sun and sand. The disposition of the soil being, in countries governed by the Koran, an attribute to the sovereignty, as it ought to be in Christendom, the people being sovereign, these migrations only take place with the consent of the beys, and under conditions of tribute. Like Tuggurt, the inhabitants of Zaab require corn from Constantina.

South of Constantina the camel is the most appropriate vehicle for the nature of the country; to the north of that city you find roads, carriages, and beasts of burden of Europe. Thus the means of transport and the produce are quite dissimilar between these two regions. One overflows with corn, the other is without it.‡

The town of Constantina brings an important contingent to the barter trade of Algeria by its industry. Its inhabitants manufacture the clothes and chaussures (or hose) of the people of the Zaab country, and the harness of their horses; and its weavers, tanners, shoemakers, and sadlers are organised into corporations, showing the ineradicable instinct in human nature for association and *ateliers nationaux.* You may reckon in Constantina almost as many spinners as there are women; and it is from their hands that issue those solid and light burnouses, whose elegance none of our European dresses can equal. Strange that the progeny of the meteors of the desert, with a dash of Berber blood, should have achieved the dreams of Louis Blanc, and put Manchester to the blush. The *gandouras* of Constantina are very beautiful stuffs of silk and linen; its carpets are more valuable than those of the Levant; and before the French occupation, 60,000 haicks left the town per annum, without steam-looms: but what power can match industry, or what machine is peer to the hand!§ The activity of these local industries has of course

---

* Description of Africa, vol. iv. A Venetian ducat is equal to 3*s.* 5*d.*

† Folio, Amsterdam, 1686. Ouerghela is like a port on the brink of the desert. Its standing population is about 1000 persons, but the concourse of travellers makes it run up to several thousand tents.

The distance from Constantina to Ouerghela is divided into three parts by the intermediate towns of Biskara and Tuggurt; the latter of which is 45 leagues (112 miles) north-east of Ouerghela, and is the Turaphylum of Ptolemy, the Ticarta of Gramaye, and the Teehort of Leo Africanus. (Baron Baude, vol. iii. p. 43.) At the time of the latter author, the country of Tuggurt paid to the sultan (?) of Tunis a tribute of 50,000 ducats. (8571*l.* 8*s.*) The sheikh's revenue was 130,000 ducats (22,750*l.*). Leo, who gives these details, had lived at Tuggurt in familiarity with the sheikh, and what he relates about the abundance of dates, &c. is still perfectly exact. (Africæ Descriptio.) Under the Turks, a detachment of a dozen Janissaries went every year from Biskara to Tuggurt to receive the tribute. This town is now a considerable market for gold-dust; but the objects of exportation that the merchants of the interior collect there are almost exclusively directed on Tunis.

‡ Baude, v. iii. p. 43. Borrer. Explor. Scient. Etudes sur la Géographie et le Commerce de l'Algérie méridionale, par E. Carette. Dawson Borrer, who estimates the Tell at 16 million hectares, computes that it would admit one person to every hectare, whilst in France the population is only two individuals to every three hectares, owing to the greater richness and fertility of the African soil. Thus, the Tell alone, under favourable circumstances, might easily support 16,000,000 of inhabitants. Nearly all the Saharian tribes pay an annual visit to the Tell. In winter they have water on the southern plateaux and plains, but at the end of spring they come northward and reach the Tell about harvest-time. There they remain in their black tents during the summer heats, and an active commerce is carried on during their stay; but when the summer is at an end, they depart, and go home about the middle of October, when the dates are ripe. (p. 236)

§ Though the hand may be a slower machine than many others, we have overwhelming proofs that its productions are the best, from the illuminated missals of mediæval monks to the exquisitely tempered, carved, and worked blades of modern Circassia and of historical Damascus. Revelations of Russia, vol. ii. pp. 312-13. The silks and satins manufactured at Broussa, in degraded and vile Turkey,

diminished one-half since 1837, the date of the French conquest, the reign of order, and the triumph of civilisation.

In the province of Oran, as in that of Constantina, the corn-lands are all on the northern slopes of the Atlas, and the populations of Angad come there to get their provisions. The caravans of Tafilet arrived formerly with bodies of two or three thousand camels, bringing negroes, wool, feathers, ivory, drugs, gold-dust, and carrying back corn, clothes, and other European merchandise from the Tell of the province of Oran. The tribes to the east of Oran and those of the Sahara of that province are very rich in cattle, of which they formerly furnished great quantities to Spain and Gibraltar.

The country of the Mozabites and of the Biskris produces little corn, but is rich in dates, in land, in cattle. It is a kind of terrestrial archipelago; it is scattered with oases severed from each other by a sea of sand; it advances as a promontory into the desert, and the caravans of Nigritia seek to reach it by its southern part, *i.e.* by Ouerghela, or by Gardeyah.* Biskris, Mozabites, and El-Aghrouaths, emigrating part of their lives, like the Savoyards, and come to Algiers. Not being the slaves of a hypocritical priesthood or of a degrading idolatry, nor infected by the shallow prose of scepticism, they have obtained a reputation in Africa for their fidelity and their industrious habits. Their honesty is also proverbial; and unlike Europe, the Sahara can preserve the moral qualities, even when rewarded by wealth and ornamented by learning. But this godly simplicity of the desert will probably soon disappear before the march of Christian intellect. These Saharians bring back the gains of their labour and economy, and some of the tribes have become very rich.

The beauty, fertility, and cultivation of the territory of Carthage at the time of the Roman conquest, and till the irruption of Genseric, are attested by classical pens; and we have abundant evidence of the care bestowed on agriculture by the Carthaginians, who wrote many ingenious treatises on that subject, now lost.† Not so Numidia. At the time when the Romans first entered into relations with that country, the Kabyles, who are its most ancient inhabitants, were in a worse condition than that in which you see them in the present day, their fertile country abounding in nothing but wild animals. In his long and glorious reign Massinissa succeeded in drawing his subjects from thievish habits; he made soldiers and labourers of them, and renewed "the face of the earth." Nor were his labours extinguished at his death. When subsequently the army of Metella marched against Jugurtha, it found, in places where the French troops would now die of hunger, numerous and clear signs of great agricultural prosperity. The fields were covered with labourers and flocks; the magistrates of the towns and country came to offer to the Romans corn, provisions, and means of transport.‡ In the middle ages these same regions presented a flourishing cultivation. At the very gate of Bugia, among the Kabyles, on the right bank of the Soumah, you may see fields regularly bounded and as well cultivated as many in Europe. The Arab agriculture approaches nearer that of the pastoral state of society. They still live in tents, where the Romans had their most solid constructions. Except near the towns, the tillage of ground was, with a few

---

are also superior to the best in the Western and Christian countries of Europe. See Spencer's Travels in European Turkey, 1851, pp. 22-3.

* Full particulars relating to the tracks followed by the caravans from Soudan to the Sahara will be found in the *Grand Desert*, by General Daumas.

† Diodorus Siculus on the wealth of Carthage, p. 79. Polybius, p. 80. Remains of the Carthaginian treatise on Agriculture are found in Pliny's works. They were translated into Latin by Solinus. Dr. Russel, pp. 7, 8.

‡ Strabo, Geogr. 17. Cum regionem uberem colerent, nisi quod feris abundabat. Ipse intento atque infesto exercitu in Numidiam procedit; ubi contra belli faciem, tuguria plena hominum, pecora cultoresque in agris erant. (Sall. Jug. 46.) . . . Is enim Numidas civiles et agricolas reddidit, et loco latrociniorum eos militiam docuit. (Ibid.) Baude, v. ii. p. 225.

exceptions, a grant from the sovereign to the whole tribe; the only individual property was the flock, the tent and its furniture; the tillage given to the land only gave a right to its harvest, and the next year the same field returned to common pasture. Here we meet another paradox,—Communism in vogue among the aristocratic Arabs. But the French, like Moliere's doctor, who knew not where the heart was, *ont changé tout cela*, and we presume that law-suits will soon flourish as actively in Algeria as in Christendom; indeed, it is not improbable that there may be the blessing of an Arab court of chancery some fine day. Pasturage lasted all the year there, and men lived in independence on their cattle, in happy ignorance of fences, game-laws, spring-guns, and man-traps. They could not build or appropriate any part of land, without usurping the patrimony of all. Thus they could not calculate on above a few months in planting, reaping, &c. This explains partly why they have always lived in tents and not in houses. The Turks levied taxes in a way to continue this usage. They sent armed bodies of troops, whose steps were marked by violence and plunder,—lawyers in another shape, armed with matchlocks instead of writ, but frank and honest in their extortions, instead of fawning on their victims. Wherever the Arabs in Africa have enjoyed calm, they have applied themselves to culture, and their conservatism has preserved its old tastes in this as in so many respects. Spain shows, on all hands, the perfection to which the Arabs carried agriculture. The Vega is still a dream of beauty, and Valencia an earthly paradise, though fallen. That Arab agriculture was equal to their architecture, is moreover proved in the libraries of the Peninsula, especially of the Escurial. Where a Christina wallows in sensuality, an idiot nobility and a slavish people bow to bastard Bourbons, philosophic Moors once thought and taught the inexhaustible riches of industry and of this crust of earth on which man crawls a few hours, while the everlasting heavens tower above and shine on his path of labour, as he toils for the unseen. These noble men anticipated the day when labour should be held the most worthy worship of God, and when universal industry and combination should establish a reign of justice, harmony, and commonwealth on earth. But they had faith and charity unknown to Christendom, which drove them from its bosom; and Spain, the wreck and ruin of her past,* will long regret her insanity.

The treatise of Ebn-el-Awam, composed in the twelfth century in Andalusia, presents alone as complete a statement of agriculture as it is profitable. His remarks apply equally well to the climate and soil of Algeria as to those of Andalusia.

The first branch of agriculture that we shall attend to is that of cereals.

The soil rarely received, under the hands of the Arabs, any other care than a superficial tillage, over which the seed is scattered; a second tillage covers the seed, and the harvest follows. They say that on the vast zone of good land, of which the traveller crosses a part between the Ras-el-Akba and Constantina, they content themselves with the labour of just turning up the soil, without preventing thereby the harvest from yielding sometimes twenty-five for one. Another remarkable feature of this district is the fewness of weeds. In short, the plains and table-lands of Upper Numidia, now the province of Constantina, are naturally among the richest arable land on the globe. You do not find in the zone above noticed those forests of thistles and of great parasitical plants which cover the plain of Bona; and Baron Baude never saw neater corn than that in the silos of Constantina. The natives give long fallows to the land of this district; a system that takes the place of manures. The straw of wheat, which is here interiorly furnished with a nourishing and sweet-tasted gelatine or pith, is gathered as forage, and used in winter as nourishment for the cattle.†

---

* Let the reader consult Hallam's Middle Ages for a proof of the enlightened and constitutional spirit and government of Aragon in former times; for Christian Spain reflected at that time the light shed abroad by Arab science and intelligence. See also the History of Spain in Lardner's Cyclopædia, and Crichton's Arabia, vol. ii.

† From 1742 to 1793 the corn that the African Company exported from the province of Bona or Constantina cost 7 fr. 60 c. the metrical quintal. There is much wheat, maize, and especially millet,

Next in importance to the cereals are the olive-plantations in the vegetable productions of Algeria. They are very numerous and extensive in Africa, especially near Bugia among the Kabyle tribes, near Tlemsen, and at the foot of the Atlas along the Mitidja; but the native oil that is made is of a very inferior quality, owing to its careless preparation.

The olives are also suffered to get rancid, by leaving them many months soaking (*en maceration*). Near the towns, the carpenters make clumsy bruising-machines to press the olives.*

The olive has nothing to fear from cold in Africa; it reaches the size of forest-trees, and, though very productive, requires very little cultivation.

Gramaye counted in the seventeenth century, among families fled from Spain to Algiers, 600 given to the culture of the silk-worm. Peyssonnel found, 100 years later, the numerous plantations that they had made on the coast. Thus, the men who had covered Spain with elegant structures, and converted its valleys into gardens, whilst they shed the light of science over polar darkness, were rewarded, by Christian and Catholic charities, with exile to the coast of Barbary, less barbarous than the heart of Christendom, there to " waste their sweetness on the desert air." Nor were they left there long in peace; the graceful amenities, superior wisdom, and pacific industry of a liberal race being usually trampled in the dust by the iron heel of military despotism. The Turks destroyed or ruined most of the Andalusian proprietors, and now there only remain a small number of magnificent mulberries;† silent monuments of the gentle hands that tended them, a condemnation of the gloomy bigotry of our sires and of their allies the Turks.

The coasts of the Mediterranean carry on a great trade in fruits; some of which, like the citron, have naturally the aptitude for being preserved, which others have not. The figs of Carthage were the envy and the ornament of Roman tables; the Levant gave us cherries, and gives us the Christmas puddings of merry England.

Bona no longer deserves its name of Blad-el-Aneb (the city of jujebs); and there are but few of those fine olives near it that constitute the riches of the environs of Seville.

The numerous gardens surrounding Algiers only contain wild stock (*des sauvageons*), as the Algerians do not practise grafts; and this explains the bad quality of their fruit. They have in this, as in most things, much degenerated from their Andalusian ancestors, whose system of irrigation, in particular, was the most perfect that the world has perhaps ever seen. Anomalous, indeed, is the apparition of this Arabian people, the meteor of the desert, handing over to us chemistry and algebra, the keys of the material universe, and following the plough in the steps of Socrates and the Emperor of China. When will Europe learn the debt it owes to the Semitic variety?

It has been said that the Mitidja would soon yield cotton, coffee, and indigo. This quackery has only made a few dupes; still, cotton is sure eventually to flourish in Algeria. The Moors grew it formerly in Andalusia;‡ and it succeeds now in Sicily, Majorca, Arta, and Malta.

The market-gardens, like the orchards, are very poor. The greatest part of the melons, of the water-melons, and even of the potatoes that are consumed at Algiers and Oran, come from the Balearic Islands and from Spain. The garden of the military convicts

---

in the province of Constantina, and you meet large fields of rice in the plains between Algiers and Oran.

* MM. Roche and Colombon have made great improvements in their olive-plantations in Algiers.

† The mulberry and the olive-tree flourish splendidly on the road from Algiers to Point Pescade. The declivity of the soil adapts this part of the vicinity of Algiers to the cultivation of orchard-trees, for from the Hospital du Dey to Fort Pescade you hardly pass twenty hectares of arable land. A few vines had been planted there, too, in 1840. (Baude, vol. ii. p. 63.)

‡ Ebn-el-Awam, c. xxii. art. 1.

(Marengo) at Algiers, that of the Misserghin near Oran, and the attempts made by the garrison of Bona, show what may be expected from the soldiers turned gardeners.

Among the cultivated trees of Algeria we must reckon, besides olives and mulberries, walnuts, hazel, almond, jujebs, white and black figs, pomegranate, carob, banana (called Adam's fig-tree), palm, sweet and bitter orange, and the melon and other kinds of citrons : the vine, the red mulberry, caper-bush, and almost all the fruit-trees common in England, such as the apple, plum, apricot, and cherry, stock the orchards and gardens of Algeria. The olive, walnut, jujeb, bitter orange, citron, pomegranate, cactus, vine, and worm-wood, are also among the spontaneous productions of the soil. They grow on the mountains, in the valleys and fields. On the northern slope of the Atlas, at an elevation of 2000 feet, oranges grow mixed with the aloe and cactus. On the southern slope figs are found at 1300 feet; and wild pomegranates grow in such profusion near Algiers, that their fruit, when perfectly ripe, is sold at six for a halfpenny. Gardens, fields, and houses near the metropolis, are fenced in with hedges of cactus and aloes. The cactus produces the Barbary fig,* which is eaten by the Arabs during six months in the year. The stems, stript of their numerous thorns, and cut into pieces, are eaten by the poor when vegetables are scarce. The shoots when planted will sometimes take root. Of the leaf of the aloe they make a kind of paper, and the fibres are used as a thread for weaving into cords. Palm-trees are not nearly so common as the other fruit-trees ; they are found on hills, in valleys, and among thickets ; they are propagated mostly by young shoots, taken from roots of full-grown trees. These, if planted with care, will bear fruit in the sixth year, attain to maturity thirty years after transplanting, sometimes continue flourishing for seventy years, bearing yearly clusters of fifteen or twenty dates, each weighing 15 or 20 lbs., altogether between 300 and 400 lbs. of fruit. After this age the tree begins to decline, and falls about the latter end of its second century. It only requires to be well watered once in four or five days, and to have a few withered boughs occasionally lopped off. The dates ripen only in the spring in the Sahara or Blad-el-Djerid ; but they might be brought to perfection in the north of Algeria, if perfectly attended to. The fruit of the dwarf-palm is less esteemed, though the heart of the plant is much in request. The palma - christi (yielding castor-oil), sugar-cane, cotton-tree, cactus without thorns, madder, flax, and alhenna, grow wild. The latter is a beautiful odoriferous plant 10 or 12 feet high, bears small flowers, with a pleasant smell like camphor. The leaves of this plant, dried and powdered, are used by all African women as a cosmetic, being preferable to the bullock's guts and dung with which the Gallas smear and adorn themselves in Abyssinia.† The palma-christi reaches its full height of 16 or 20 feet in one year.

Several English vegetables grow in French Africa, such as carrots, celery, asparagus, parsnips, &c. &c. Numerous odoriferous plants, including myrtles, lavender, Barbary and spurge laurel, &c., are found there in abundance ; and rose-laurels form a purple border on the banks of the rivers. During the winter the hills are covered with tulips, anemones, and ranunculuses. In spring you find large fields full of the star of Bethlehem, asphodel, iris, and yellow lupine. In autumn you meet a large family of squills of all colours.

Apricots are fit to gather in May. The sashee, or male apricot, though better than the female—as appears the invariable but ungallant rule of inferior nature—is a little later,—like little boys, who are generally more stupid than little girls, but shoot up much higher afterwards. The common apricot is apt to generate fevers.

---

* Pananti, Avventure, vol. ii. pp. 26-30. Madame Prus, p. 38. Le Grand Desert, p. 384.

† The Berbers living near the cataracts of the Nile anoint their sleek sable bodies with the palma-christi juice, and sport a luxurious growth of hair, rivalling the bear's-grease crops of the North-American Indians. We commend the study of the comparative merits of these cosmetics and human manures to our modern Calvins and Macassars. See Lord Lindsay's Travels. Catlin's Indians, &c.

Fruit is very cheap in Algeria even now, but prices have probably doubled, with other blessings conferred by the French conquest and occupation, since 1844.

The plantations of oranges near Blidah, &c. formed beautiful groves, till the French came and cut them down or burnt them up, to improve the country and promote the reign of *order*. The oranges of Algeria are as good as those of Portugal, Malta, and Candia. Almonds, pistachio-nuts, and grapes are excellent, and articles of considerable trade. Bona used to be celebrated for its beautiful jujebs. The finest figs in the north of Africa come from Scherschell, whence they are sent to Algiers, Constantina, Tunis, &c. Now the best are gathered in the Middle Atlas.

Tobacco is much cultivated in Algeria, especially near La Calle,* presenting two sorts, the nicotiana tabacum, and the nicotiana rustica; the latter is the most common and the most esteemed. It exhibits a rapid growth and a vigorous vegetation, and might be as good as any if well manufactured; nor would it be long neglected if in the hands of Germans. There used to grow much flax near Scherschell; and alhenna, which grows at the foot of the Atlas, is among the most profitable productions of French Africa. The natives have never made much of indigo, though it is very good there, yielding three crops annually. All the plants used as food for cattle grow in the Mitidja; and we have no doubt that Algeria will some day yield oil-cake miracles, and great exhibitions of prize-cattle, besides aldermen fattened on turtle.

Moorish kitchen-gardens are not so rich as ours, but they yield plenty of melons, cucumbers, pumpkins, onions, calibashes, pepper, and tomatos; and the French grow many peas, lentisks, and beans, of which they consume large quantities.

All vegetables grow to a very large size at Algiers,—like the women, who seem to vegetate themselves: the fennel and carrots are gigantic; parsnip-leaves are nine feet long, cauliflowers a yard in diameter. Grasses grow very profusely, and of very good quality; and you meet thickets of dwarf-palms, mastichs,† and thorn-broom growing to a height of two or three yards, but the cactus much higher. The orange and citron trees too are very fine. The jujeb, olive, and carob tree,‡ reach an eztraordinary size; the stems of the vines are very large, and the bunches of grapes enormous.

Opium, according to an examination by a commission of the French Academy in 1844, is very good, with as much morphine as the best of Smyrna or India.§ This is glad tidings for our allopaths, who may import cheap cargoes of this drug *ad libitum*, and poison their patients à *discrétion*. Algeria grows other narcotics besides opium and French economists,—I mean hashish, or Indian hemp, of which we have already treated.||

* Baron Baude, v. ii. p. 164.

† The mastich is properly the lentisk, *pistachia lentiscus* of Linnæus, *pistachia atlantica* of Desfontaines. It is very common in the Sahel, but does not grow in the Sahara. The flowers are in fleshy membranes; and the fruit is small, globulous, and of a red colour. The resin of this tree, called mestikâ, is greatly used in the East to strengthen the gums and whiten the teeth; but it is almost exclusively cultivated in the island of Scio for this purpose. Olivier (Voy. dans l'Europ. Ottoman, vol. i. p. 292) describes the incisions made in July to extract the resin, which drops down the trunk like tears, and is removed with iron instruments, &c. Grand Desert, p. 408.

‡ The carob is called in Arabic kharoub, and is the *ceratonia siliqua* of botany. It is a rather large tree, growing vigorously on the coast, but seldom in the uplands, because of the cold in winter. Its fruit consists of saccharine particles, sold in great sacks in the markets. Grand Desert, p. 396.

§ Blofeld. The reader will find a scientific classification of Algerian plants in the Algerian Flora at the end of Berbrugger's Algérie, and in the Exploration Scientifique; also in Desfontaines' Flora Atlantica.

|| We must now attend to cattle and pastorals. Horses and oxen are becoming much rarer in Algeria than they were. The fine Barbary horses are very scarce, and only found amongst the most powerful sheiks. The Deys of Algiers had a haras stud at the Rasauta; and the government of Tunis still keeps one, which is celebrated. (Captain Kennedy, Baron Baude, or the Lady's Diary.)

In 1338 there were bought 633 horses, costing 188,064 francs; and 1225 mules, costing 537,990 francs (7622*l.* 11*s.* 8*d.* and 21,616*l.* 12*s.* 6*d.*). The provisions with which the camps are supplied are carried on the backs of mules, imported from Perpignon, which are the only animals that have not suffered by the change of climate and country in Algeria (St. Marie). The victualling department for the army employs for draught large white oxen, from Italy. Some landed proprietors attempted to import cows, &c. from Switzerland, and a premium of 50 francs (2l.) per head was granted them as an encouragement; but the scheme was not successful. The great change of climate and food do not agree especially with ruminative animals.

The Arabs, like clever economists, have but a very imperfect notion of the idea of property in the soil; poor unsophisticated fellows that seem to feel instinctively, with Proudhon, that "*property is robbery;*" but they understand very well the right of property in animals, flocks, and women.

The herds of horned cattle are innumerable in French Africa. Before 1830 meat had a low price in Algeria; an ox was sold for 20 francs (16s. 8d.); and many tribes only obtained money from the leather of their herds. Matters are now much changed. The French consumption of meat has clearly overthrown the balance of the pastoral riches of the country, and when the French are in power the price of oxen is three or four times as high. Here again we have the usual accompaniments of the march of intellect and the advance of civilisation. Excessive competition and inordinate luxury convert plenty into poverty. In 1837 the meat for the army cost 2,873,353 francs; in 1838, 2,185,102 francs (116,334l. 2s. 6d. and 87,404l. 3s. 4d.).

A great part of the wool that commonly went to the markets of Algiers, Bona, and Oran, has taken since the conquest the road of Tunis, or of the ports of Mcrocco. At the capture of Algiers in 1830, there was in the warehouses of the divan an accumulation of 130,000 quintals of wool (28,600,000 lbs.), having a value of 180,000,000 francs, forming 7000 bales. Constantina used to be at one period one of the chief wool-markets of the desert, and fleece weighing about 2 kilogrammes, 4·40 lbs., were only worth, according to their quality, from 50 centimes (5d.) to 1 franc (10d.) per piece: these prices had already doubled in 1841.

Before leaving the subject of agriculture and pastorals, we propose to offer a few brief remarks on the efforts of the government to preserve or improve the breed of horses. The first stud (depôt of stallions) was at Mostagamen in 1342, since which two have been established at Kobat and Alelick, 6 kilometres (3¾ miles) from Bona.

There are three classes of Arab horses in Barbary: 1st, the Tunis; 2d, the Morocco; 3d, the old regency of Algiers breeds. The two first are higher, and stand more fatigue than the latter. "I have known," says St. Marie, "a Morocco horse mounted by a *spahi* (native trooper), travel in 11 hours 50 leagues (125 miles), without a moist hair or the need of the spur. The Algerine horse is shorter and plumper. The ordinary price of a 4 or 6-year old is about 200 francs (8l.). At Oran, stallions brought from Tunis have fetched 2000 francs (80l.); but that is a government price, and one cannot judge from it of the average prices." St. Marie's Visit, 1845.

The Algerian breed of horses being found (1845) to degenerate, the government determined to have recourse to the east. to central Arabia for *nedjis,* and to Syria for *anezis,* which are affiliated by race with the Barbs. They have also attended to the improvement of the breed of mules, by introducing jack-asses from Spain. General Oudinot, of liberticidal notoriety, has devoted much attention to, and spilt ink in this service.

The French hope by these means to improve the breed as they call it, *i. e.* to civilise the hardy Arab horse, and make him like the European, subject to influenza, dyspepsia, and the protæan nervous complaints of our thrice-happy hemisphere. No wonder that the Arabs are rather chary about supplying the studs with mares.

We presume, however, that French grooms of the Oudinot school will eventually confer the blessings of order and civilisation alike on Arab steeds and Roman citizens, and that the regency will shortly be stripped of its animal and vegetable ornaments and treasures.

Borrer, p. 21, describes that at Arla there is held once a week one of the greatest markets of the plain, much frequented by Arabs, who bring horses, cattle, &c. to it. The province of Constantina is more productive of horse-flesh and ass-flesh than the others; and it appears that in the last three months of 1845, 8279 horses, mules, and asses, came to the Constantina markets, whereas to those of Algiers there came only 5023, and to those of Oran 2341.

A matter essentially related to the agriculture of Algeria is the drying up of the marshes and the irrigation of the plains. From the foot of the mountains to the sea, the surface of the basins of the Hamiz, the Haratch, and the Mazafran, is about 140,000 hectares (350,000 acres). The rivers of Algeria only float boats towards their mouth for a very short space, where, mingling their waters with those of the sea, they acquire a little depth. In no country are the gorges of valley, which may be barred at little cost for the establishment of artificial reservoirs, so numerous as in the branches of the Atlas.

The Arabs may be styled the inventors of irrigation, yet the lazy and surly Turks, as every where else, suffered the conduits to go to ruin and the woods to be destroyed. Nature had thrown her pearls before swine, and their successors seem still allied to a nameless quadruped in their neglect of this obvious pillar of agricultural prosperity. The blindness of civilised states in laying the axe at the root of all trees has been well manifested in results, and lashed by cynical pens and wits of the first order. (See Charles Fourier's *Unité Universelle.*) The power of an organised system of planting is an established fact, nor can we over-estimate the influence of man in doctoring the distempers of a restive climate and soil, and in patching up the rickety constitution of our unfortunate planet, the victim of man's neglect and depravity. Large tracts of Persia (see Frazer's *Persia*, also Spencer's *Travels in European Turkey*, and Forbes's *Spain*, where the reader will see how much fine European land lies waste through the crimes of Russia and of Bourbons), once a garden, are now a waste; the sands of Libya have won a broad stretch of the vale of the Nile; and Algeria, the granary of Rome, is in a great part the home of frogs and a nest of locusts. Sublime visionaries have dreamt of reclaiming the desert by irrigation, and practical minds have coolly argued its chances. Nor can this age of wonders dare cough down a yet more wonderful future. The rich lands of Spain lie fallow and deserted, that once waved with Moorish harvests, and the rich pusztas are a desolation whilst Bourbon and Hapsburg idiots reign, and their butchers blight the fair face of glorious nature. Oh, when will the nations learn to combine in works of cosmopolitan utility, instead of filling the air with the shriek of despairing liberty and expiring nationality, and manuring a paradise with the hearts' blood of heroes!

Reverting to Algeria, we find that the natives burn and neglect the woods, yet the lentisk, the carob, the wild olive, the holm oak, the myrtle, the oak tree, though in the form of brushwood, carpet the Algerian soil on all sides. All the forest-trees of the south of Europe would thrive in Algeria; and Baron Baude saw on the Boudjareah pines of Aleppo of the greatest beauty, and near the lake of Tonegue elms and magnificent poplars.

Besides wood useful as fuel, for carpentry, &c., Algeria has a great many forest-trees of valuable timber, and rare in France, such as oaks with sweet acorns, and oak trees (*quercus ballota fructu longissimo*). Desfontaines, of the Academy of Sciences, has given a very detailed description of the former (tome ii. *Voyage dans la Régence d'Alger, &c.*). He saw vast forests of this oak in the mountains near Blidah, Mascara, and Tlemsen. The acorns were sold in the public markets; the Moors eat them raw, or roasted under cinders; they are very nourishing, and have no bitterness; large herds of swine in Spain are fattened on them,—no agreeable analogy for the pig-hating Mussulmans. Baron Baude says, " I have been told that in certain districts of Barbary a very sweet oil was obtained from them, equal to that of the olive. The wood of the ballota oak is hard, compact, and very heavy; and it might be useful in the works of cartwrights, &c. and of carpentry."

We shall now attend to the extensive and valuable cork-wood forests that clothe some parts of Algeria. Bordering the road to Bona, near El Khallah, or La Calle, above the hills of that town, you find about 50,000 acres of beautiful forest, intersected with lakes and prairies, and stocked with cork, elm, ash, and gall-bearing oak-trees. These forests could furnish cork sufficient for the consumption of all Europe. (Blofeld).

Our space will not permit us to dwell on the interesting subject of the forests of Algeria, a great part of which have been surveyed by the *service forestier*, or the engineers. The following tables will show the reader the extent and quality of all the forests of Algeria that have been hitherto explored.

Contents of the explored forests, 1849.

Province of Algiers	. . . . . .	168,645 hectares	421,612 acres.
„ Oran	. . . .	269,764 „	674,410 „
„ Constantina	. . . . .	429,606 „	1,074,015 „
Total	. . .	868,015 „	2,170,037 „

Most of the forests are, however, mixed in the quality of the trees that compose them.

Seven forests have also been ascertained in the Sahara, containing 27,000 hectares, at a proximate estimate (68,000 acres), which, added to the 104,700 hectares (261,750 acres) of the forests of Batna, Aouress, and Tebessa, and to the 868,015 hectares previously noticed in the Tell, give for the whole of Algeria 999.915 hectares (2,498,537 acres). These estimates are given from the *Tableau*, 1849-50, p. 438.

The pépinières are government nursery-grounds, some of which appear to be in a thriving condition, and to have succeeded in introducing many foreign plants into Algeria. In the season 1850-51 they were in a condition to supply the public plantations, &c. with 625,776 shoots of trees, 305,813 herbaceous vegetables, 14,403 kilogr. (316,86·60 lbs.), 992 grams. of different grains. The sum total of these plants represents a value (according to the trade price) of 604,130 fr. 50 cents (24,165*l*. 4*s*. 7*d*.), and according to the tarif of the prices of the administration, 228,152 fr. 16 cents. Attempts are making to introduce the tallow-tree, the china hemp, the bamboo, the camphor tree, the ficus elastica, &c. No care has been spared in advancing the interests of these experiments, according to the Tableau; yet Borrer, in 1846, represents the Guelma Pépinière as particularly remarkable for its flourishing weeds. The cotton plantations are most thriving.

Three railroads are in contemplation in Algeria, 1st, from Philippeville to Constantina; 2d, from Algiers to Milianah, through Blidah; 3d, from Oran to Algiers, with branches to Tlemsen and Arzeu. It is estimated that the latter line could be opened from Oran to St. Denis du Sig within two years.

To complete our picture of Algeria, we must not omit the roads and bridges, the arteries of a country; and here we readily admit that the French have done much in the way of improvement. The annexed table will enable the reader to judge for himself:

Provinces.	Length.	Expense.
	metres.	fr. c.
Algiers	949,441	8,825,895 16
Oran	1,655,450	2,295,262 19
Constantina	466,850	3,208,142 40
Total	3,071,741	14,329,399 75

To fill up our picture of the principal channels of intercourse in Algeria, we shall add the following particulars from the *Exploration Scientifique:* Philippeville is 343 kilometres (213·1 miles) east, 3° north, in a straight line from Algiers, and 354 kilometres (220 miles) by sea; from Constantina 63 kilometres in a straight line, and 77 kilometres by the road. Constantina is in 36° 22' 21'' lat. north, and 4° 16' 36'' long. east, and 664 metres (2177·92 feet) above the sea, according to M. Boblaye ; this refers to the Kasbah; the maps of the *depôt de la guerre* give it 656 metres (2161·68 feet), but this may refer to another point.

Constantina is 320 kilometres (199·4 miles) east, 7° 17' south from Algiers in a straight line, and 431 kilometres (267 miles) by Philippeville, the only road now taken ; 390 kilometres (242 miles) by the Biban or Iron Gates; but the real road is only 365 kilometres (227½ miles), of which 240 (150 miles) by sea to Djidjelli, and 125 (78 miles) thence to Constantina, following first the sea-shore, and thence the Ouad-el-Kebir. The last part of the road will admit of a railway, now in contemplation. Constantina is 118 kilometres (73 miles) west, 290° 21' south from Bona ; the actual distance is 170 kilometres (106¼ miles) by Guelma, and 150 kilometres (93¾ miles) by El Arrouch. (E. Carette's Routes suivis par les Arabes. Explor. Scient.)

The following table represents the bridges built by the French in Algeria :

Provinces.	Number of Bridges.	Length.			Expense.		
		Wood.	Wood or piles.	Masonry.			
		metres. feet.			fr. c.	£	s. d.
Algiers . . .	43	794·45 (2605·7960)	473	91·50	1,126,288 29	(45,050 10	10½)
Oran . . .	18	546·40 (1792·1920)	56	187·00	616,780 41	(25,870 4	6)
Constantina . .	25	321·75 (1055·3400)	78	52·00	272,687 94	(10,907 12	5₁₀⁴)
Total . .	86	1662·60 (5453·3280)	607	330·50	2,045,756 64	(81,830 5	6)

We have reserved for this place our remarks on the mineral treasures of Algeria, which, though yielding metal less attractive than Australia and California, is more likely to hasten the return of a golden age. In the province of Algiers iron, copper, and lead ore has been found in abundance near Milianah. Four mines have been conceded ; 1, those of Mouzaia (iron and copper), employing 400 men, annual yield 17,571 metrical quintals (3,865,620 lbs. avoirdupois), value 271,083 fr. (10,807*l.* 6*s.* 8*d.*); 2, Oued Alalah (copper and iron), employing 120 men, yield 15 or 20 per 100 at a rough value of 350 fr. per ton ; 3, iron and copper mines of Ouad Taffilez, 50,000 fr. (2000*l.*) spent on it in 1850, but no yield as yet ; 4, iron and copper mine at Cape Tenes, in the same state as the last.

The permits to work are five in number, two being in operation, *i. e.* those of the Oued Merdja and of the Oued Kebir. The first employs 20 workmen, with a yield of 70 tons of pyritous copper ore and of carbonated iron; mean richness, 17 per cent; rough value, 22,400 fr. (896*l.*) The Oued Kebir only employed 7 men, with a yield of 50 metrical quintals (10,000 lbs.); value, 1400 fr. (56*l.*)

Constantina. Veins of iron ore have been lately found at the Bou Ksaiba, near Jemmapes, yielding a great quantity of oligist and oxydulated iron ; in Mount Edough, presenting rich brown hematites, and at Ouad el Arong, near La Calle, where are several veins of peroxydated iron. There is a lead mine at the *Nbails,* 40 kilometres (24 miles) from Guelma. Galena has been found at Skikida, near Philippeville. It is inserted in very thin veins, in clayey slates, and yields 50 grammes of silver (31 dwt. 5·150 grains troy) per quintal of slick. Five mines have been conceded in this province: Kefoum Thaboul, Bou Hamra, Karesas, Ain-Morka, and Meboujda.

Kefoum Thaboul is a lead mine, employing 70 to 75 workmen; yield 7 or 8000 metrical quintals (1,768,000 lbs.), independently of the galena, which yields 55 to 60 of lead and 175 grammes of silver (6¼ oz. avoird.) to the quintal ; Bon Hamra had yielded, in 1846 and 1847, 7500 metrical quintals (1,650,000 lbs.); Karesas, 1900 metrical quintals (380,000 lbs.); Ain-Morka, 21 metrical quintals (4620 lbs.).

In 1849 they had stopped working. Medoubja, also an iron mine, had stopped in 1850. Exploration has been permitted in various parts, showing rich veins of magnetic iron at El Mkinem, near Bona ; oligist and oxydulated iron at Filfila; antimonial galena at Oued Cherf; pyritous copper and sulphurated zinc at Ain Barbar ; oxydated and sulphurated antimony at Djebbel Taya ; and oxydated antimony at Ain-Babouch.

Province of Oran.—An important vein of pyritous copper has been discovered among the Oued Ali,

SECTION IV.

Natural History, Geology, &c.

THE *kumrah* is a little useful beast,— a cross between an ass and a cow; its hoofs being like the former, but having a sleeker skin, with a tail and beard like its mother's, but without horns.

These species are, however, greatly interior to the *camel*, an animal of the greatest utility to the Arabs, and to the devout mind a beautiful evidence of design and bounty in the Creator. Its strength, docility, and indefatigable patience are invaluable ; and its ability to go seven or eight days without water make it a treasure in the desert. Two quarts of beans or barley, or a few balls of flour, are enough to support it for a whole day. Pliny's observation, that they disturb the water before drinking, is correct, as also that they are long in drinking. They first put their heads deep in, and then make successive draughts like pigeons. Camels carry 6 or 7 cwt. A day's journey for them consists of from 10 to 15 hours, at $2\frac{1}{2}$ miles per hour. A great many camels are found in the Mitidija and on the sides of the Little Atlas ; and those about Algiers are very fine. The dromedary species, called in Algiers *ashaary*, is much rarer there than in Arabia. It is noted for its swiftness, and the Arabs say it will go in one day as far as one of their horses in eight or ten ; hence those messages which require speed are, in the Sahara and throughout the south, transmitted by the dromedary, the Arab telegraph, which is said to have a finer and rounder shape than the common camel, and also a smaller protuberance on its back. The common camel has its head at liberty when on the turf, or rather sand ; but this species is managed by a bridle, which is usually fastened to a ring fixed in its nostrils. The young dromedaries are born blind, and continue so ten days after birth, whence their name of *ashaary.**

---

8 kilometres (5 miles) south from Arbal. Carbonated hydroxide iron has also been discovered at the mountain of the Lions; lead galena near Sebdou, &c. No concession had been made in this province in 1850 ; but two permits to explore had been given : one, for an anthracite vein of combustible mineral at the foot of the mountain of the Lions ; the second, for oligist-micaceous iron at Djebel Mansour, near Cape Ferrat.

As regards quarries, we find 110 in the province of Algiers in 1849, employing 890 men, yield 196,765 cubic metres of gypsum and stone (1 cubic metre being equal to 35·287,552 cubic feet, or 1·306,946 cubic yards). The province of Constantina had, in 1849, 112 quarries ; workmen, 1006 ; yield 201,806 cubic metres of stone and gypsum. The province of Oran had 105 quarries ; workmen, 812 ; yield 7406 cubic metres of stone and gypsum, besides an enormous quantity of bricks and tiles (3,890,000). Tableau, p. 393 to 396.

* Col. Daumas in his *Grand Desert*, and Castellane copying him, have described the *mahari* of the Touaregs, which appears to be what we call a dromedary. Form much more slender than that of the common camel; elegant ears like the gazelle, the supple of the ostrich, the belly of the hare; head gracefully shaped, black and prominent eyes, long and firm lips well concealing its teeth ; hump small, breast protuberant, tail short, limbs slender below, but muscular from the knee to the trunk; feet not spreading, and its tawny-coloured hairs as fine as those of a jerboa (p. 483).

The mahari supports fatigue better than the camel, and never betrays an ambuscade ; hence they give every care to the education of the young onos. The children play with it in the tent, and it loves them much from gratitude. It is first called *boukuetaa* (father of shearing), after being shorn in the spring ; then it takes the name of *heug*, from the verb *hakeuk*, he has understood or become reasonable, when two years old. He is now broken in and trained, then a ring is put into his right nostril ; a rahhala or common saddle is put on his back, and the master mounts him, sitting cross-legged on his neck. The least movement of the nostril gives much pain, and he obeys at once. If the heug can stop suddenly when going full tilt, or describe a narrow circle round a spear, his education, like that of a modern miss, is complete. Miracles are told of his speed, sobriety, and courage. Castellane, who saw some maharis, corroborates these facts, p. 286. General Marrey in his expedition to El Aghrouath, June 1844, received a gift of 3 mahara. He describes them as a variety of the camel genus; their usual pace is a trot, which they can keep up all day. Herodotus says that the Arabs of the army of Xerxes had camels as quick as horses (book vii. c. 76).